“十四五”时期国家重点出版物出版专项规划项目
先进制造理论研究与工程技术系列
黑龙江省精品图书出版工程
哈尔滨工业大学（深圳）教改经费资助出版

功能性纳米纤维及其应用

邱业君 著

哈爾濱工業大學出版社

内 容 简 介

功能性纳米纤维是一类十分重要的纳米材料，在能源、催化、环保、生物、传感器等多个重要领域展现了巨大的潜力与前景。本书系统地介绍了纳米纤维的特性、制造技术以及多种应用，汇集了国内外学者的最新科研成果，特别是本领域的一些重要的技术突破，较为充分地体现了静电纺丝技术在当今的发展和研究趋势，将为国内外同行提供十分有益的参考。

本书适合作为纳米材料与技术、材料科学与工程、纺织工程等专业本科生选修课程的教材，也适用于材料、能源、催化、环保、生物医学、传感器等领域的研究人员及高等院校的教师和研究生参考使用。

图书在版编目（CIP）数据

功能性纳米纤维及其应用/邱业君著. — 哈尔滨：哈尔滨工业大学出版社，2024.3

（先进制造理论研究与工程技术系列）

ISBN 978-7-5603-9177-9

Ⅰ. ①功… Ⅱ. ①邱… Ⅲ. ①纳米材料-功能材料-纤维素-研究 Ⅳ.①TB383

中国版本图书馆 CIP 数据核字（2020）第 219407 号

策划编辑 王桂芝
责任编辑 刘 威 李青晏
出版发行 哈尔滨工业大学出版社
社 址 哈尔滨市南岗区复华四道街 10 号 邮编 150006
传 真 0451-86414749
网 址 http://hitpress.hit.edu.cn
印 刷 辽宁新华印务有限公司
开 本 720 mm×1 000 mm 1/16 印张 26.5 字数 458 千字
版 次 2024 年 3 月第 1 版 2024 年 3 月第 1 次印刷
书 号 ISBN 978-7-5603-9177-9
定 价 168.00 元

前　　言

在生活中，天然的纳米纤维无处不在，其中，蜘蛛丝就是典型的例子，细的蜘蛛丝的直径可细至100 nm。蜘蛛丝是自然界中产生的结构较好的纳米纤维材料之一。在学术上，纳米纤维可以从两个角度进行理解：一是指纤维直径小于100 nm的纤维；二是在纤维中填充纳米颗粒，对纤维改性，使其具备抗菌、阻燃、电磁屏蔽、促进伤口愈合等功能性。大批量制备纳米纤维材料是科学家梦寐以求的事情。科学家付出巨大的努力，探索出了一些纳米纤维的制备方法，如拉伸法、模板法、相分离法、自组装法和静电纺丝法等。然而，这些制备方法存在设备要求较高、生产工艺复杂以及不利于大批量制备纳米纤维的问题，限制了纳米纤维材料的进一步应用。

近年来，静电纺丝法制备纳米纤维材料成为材料制备领域的研究热点。静电纺丝起源于20世纪30年代，Fomhals在专利中公开了静电纺丝的实验装置，揭示了通过高压静电作用对聚合物流体进行拉伸可以得到纳米级聚合物纤维材料的原理。20世纪90年代，静电纺丝逐渐引起人们的重视，开始蓬勃发展。静电纺丝在材料适用性、形貌多样性和工艺可控性等方面具有极大的优势。静电纺丝所使用的材料可选择性较大，同时，还可以通过改变工艺参数来控制纳米纤维的形貌与性能。此外，还可以在静电纺丝纤维形成的过程中，填充不同的纳米材料，赋予纤维不同的特定功能。采用静电纺丝法制备的纳米纤维材料，在能源、催化、环保、生物医药、传感器以及电磁屏蔽领域发挥了巨大作用。为了推动静电纺丝产业化的进一步发展，非常有必要对近期研究结果进行总结与归纳，为其提供强而有力的理论支撑。

在本书的撰写过程中，一些研究生在收集文献、绘制图表以及整理数据方面做了大量的工作，他们是朱景辉、张高玮、余靓、陈焕辉、曹兴、凌建军、姜明宇、陈文豪、姜宁、韦彦如、刘亚、王心远、戴江浩、刘布等。在此，对他们的贡献表示感谢！

国家自然科学基金（51971080）、广东省自然科学基金（2018A030313182）、深圳市基础研究计划（JCYJ20170811154527927）等为本书的研究工作提供了经费保障。另外，哈尔滨工业大学（深圳）教改经费为本书的出版提供了经费资助。在此一并致谢。

由于作者学识有限，书中难免有疏漏之处，还望读者批评指正。作者将不断完善本书，为促进纳米纤维材料领域的进步贡献绵薄之力。

邱业君

2024 年 1 月

目　录

第1章　纳米纤维的制备方法及分类

对于“纳米纤维”这一概念，可以从“纳米”和“纤维”两个方面进行分析。纳米（nm）是长度单位，1 nm=10^{-9} m。自1990年首届国际纳米科技学术会议召开以来，纳米科技进展迅速。纳米技术研究主要针对的是尺寸在0.1～100 nm范围内的物质，通过特定的方式方法让原子或者分子在纳米级微粒表面以一定规律进行排列组合，进而形成特殊结构，展现出一定的物理、化学特性。而对于纤维的概念，人们通常定义为长径比大于 1 000 的纤细物质。传统纤维主要分为天然纤维、人造纤维和合成纤维，它们具有弹性模量较高、力学稳定性好和机械强度大等优点，在医学、化工、建筑等多个领域应用广泛。随着新材料制备技术的发展，在质量轻薄的基础上，具有一种或者多种特殊功能的纤维逐渐引起了人们的重视，巨大的需求和广阔的应用空间进一步推动了纳米纤维的发展。

从较为广泛的角度来讲，直径在1 000 nm以下的纤维均可被称为纳米纤维。与传统纤维相比较，纳米纤维具有大比表面积和丰富的表面功能性基团，并且展现出纳米材料的诸多特征（如，小尺寸效应、界面效应、量子尺寸效应和宏观量子隧道效应等），使其在能源、催化、环保、医疗、传感器等多个应用领域表现出更为优异的性能。另外，从制备方法上比较，一些传统纤维的制备方法，比如液晶纺丝、胶体纺丝和熔融纺丝等，已经很难满足新型纳米纤维的制备。因此，研究者通过不断深入探索，发展了一些制备纳米纤维的重要方法。

1.1　拉伸法

与传统干法纺丝类似，拉伸法是在显微控制器的辅助下，采用直径在几微米的微型吸液管以牵伸的方式将液滴拉伸成纳米纤维，如图1.1所示。

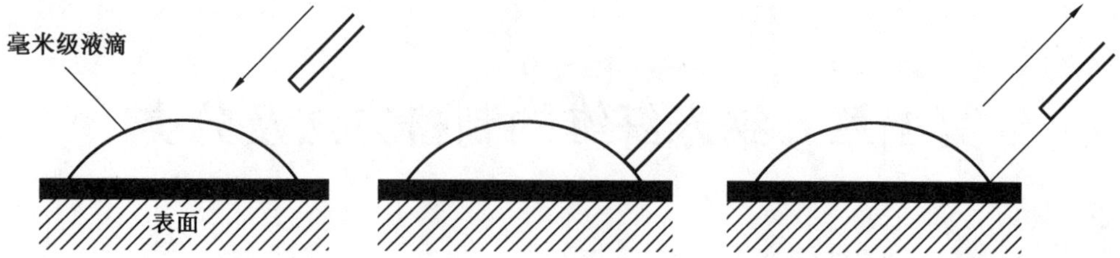

（a）微型吸液管向液滴靠近　（b）微型吸液管与液滴接触　（c）拉出微型吸液管

图 1.1　拉伸法制备纳米纤维

在液滴拉伸过程中，随着液体的不断蒸发，液滴边缘材料的黏度逐渐增大。在蒸发初期，由于毛细管破坏作用，因此拉伸的纤维脆性较大，容易断裂；随着溶剂的不断蒸发，液滴黏度增大到一定程度，纳米纤维结构逐渐形成；但是在蒸发的最后阶段，由于液滴的内聚力破坏作用，因此纤维以黏着方式断裂。鉴于以上原因，虽然拉伸法可制备出较长的单根纳米纤维，但是原材料必须具有一定程度的黏弹性且能够承受较大的形变，所以限制了该制备方法在选材上和应用领域上的广泛性。

1.2　模板法

模板法的概念源自传统的湿法纺丝技术，即通过采用纳米多孔膜（多孔阳极氧化铝、多孔硅和聚碳酸酯膜等）作为模板来获得纳米纤维结构。其优点是原材料来源丰富，包括半导体材料、导电聚合物、原纤维和金属材料等。但是该方法并不适用于制备较长的连续纳米纤维材料，因此，在工业应用中受到了限制。如图 1.2 所示，阳极氧化铝（AAO）模板形成的纳米孔洞大小均一、排列分布有序，研究者常将其作为模板，通过对聚合物溶液施加压力，聚合物溶液经纳米微孔结构挤出，固化后形成了纳米纤维结构。因此，该方法采用纳米多孔膜的孔隙直径决定了所制备纳米纤维的直径。

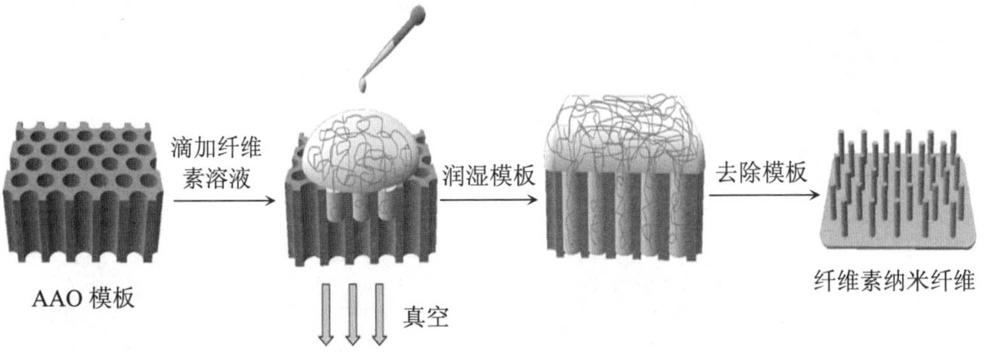

图 1.2　阳极氧化铝模板法制备纤维素纳米纤维

1.3　相分离法

相分离法是通过将物理性质上不相容的两相混合后，对溶剂相萃取，进而留下另一相的方法。该方法的制备过程如图 1.3 所示，主要分为五个步骤进行：

（1）聚合物物料在溶剂中的均匀溶解；

（2）凝胶化；

（3）冷冻分离；

（4）溶剂相的萃取；

（5）移除溶剂得到多孔纳米纤维结构。

在制备过程中，聚合物浓度和凝胶化温度将会在很大程度上影响凝胶化持续的时间。较低和较高的凝胶化温度最终分别会形成纳米尺寸的纤维状结构和类六边形结构。所制备的纳米纤维直径在 50～500 nm 之间，且呈现出多孔结构。同时，所使用的聚合物原材料和溶剂种类、凝胶化温度、持续时间以及热处理过程都会对纳米纤维形貌产生影响。相分离法的原材料价格低廉，且能够制造连续的纳米纤维。但是该方法工艺复杂、耗时长且效率低，目前多适用于实验室级别的制备，同时，所得到的纳米纤维结构稳定性较差、较难保证适宜的孔隙率且原材料选用范围较窄，因此不太适用于纳米纤维的大规模生产。

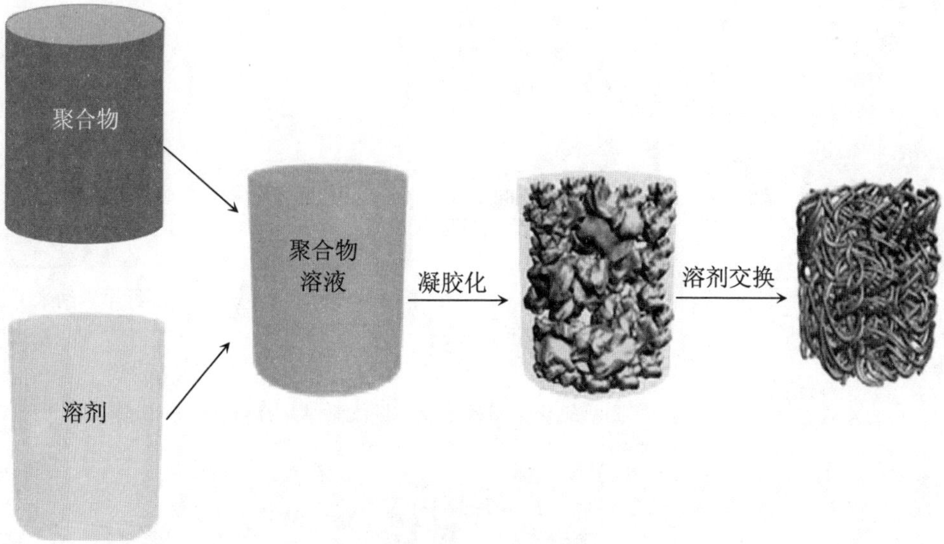

图 1.3　相分离法制备纳米纤维示意图

1.4　自组装法

自组装法是基于分子间作用力，通过小分子间自发组装来制备出预想的纳米纤维结构，其制备过程如图 1.4 所示。

第（Ⅰ）步，首先小分子间会形成带有自由基阳离子的低聚物，之后，该低聚物与溶液中的阴离子发生相互作用，形成静电偶极子；第（Ⅱ）步，溶液中的低聚物自由基阳离子（带有正电荷的纤维“种子”）会由偶极子间的静电作用再进一步被吸附，如果存在足够的单体，低聚物自由基阳离子将会引起聚合反应，最终形成纤维结构；第（Ⅲ）步，由于纤维和“种子”间存在 π—π 键相互作用，使得形成的纳米纤维趋向于表面生长，同时，受阻于纤维间的静电斥力，因此形成的纤维间会保持一定距离，进而形成阶梯状结构，实验数据表明，高温下（120 ℃）纤维间的静电斥力大于常温下静电斥力；纤维在形成阶梯状结构之后，如果仍然能够吸附足够浓度的低聚物自由基阳离子，则聚合反应将被继续引发（第（Ⅳ）步），为了减少整体系统能量（静电能和界面能），纤维将会在“种子”的另一侧紧挨着表面生长，且与已形成的纤维垂直，最终形成网格状结构，该结构有利于使系统整体能量最低。

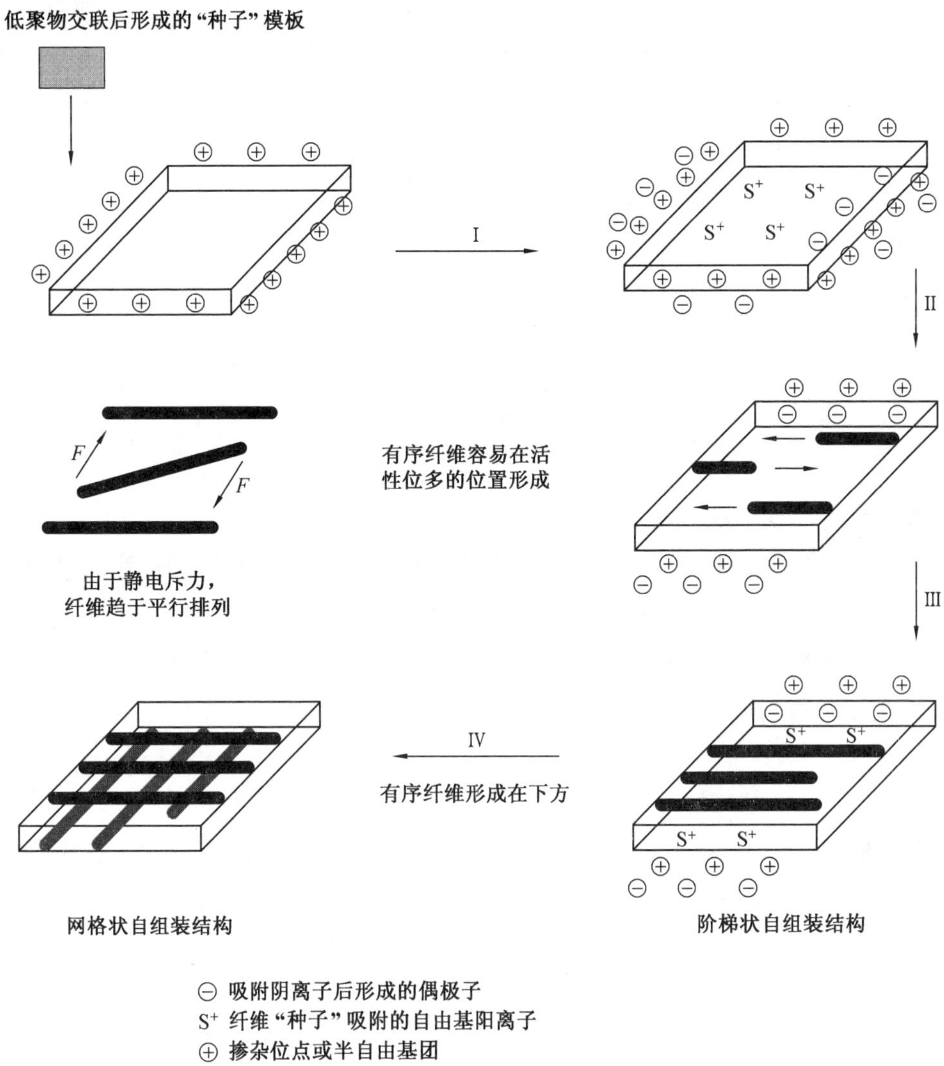

图 1.4　自组装法制备纳米纤维示意图

在该方法中，自组装反应温度和反应状态都能对所得到的纳米纤维形状产生影响。虽然该方法得到的纳米纤维形状丰富，但是其耗时较长，不利于纳米纤维的规模化制备。

除了以上较为普遍的纳米纤维制备方法之外，日本东京大学研究人员使用甲基铝氧烷（MAO）金属催化剂，基于硅石纤维对聚合聚乙烯（PE）链起到集束导向作用，在硅石纤维孔中使 PE 发生聚合反应。该方法得到的 PE 纳米纤维直径为 30～50 nm。同时由于 PE 链为直链结构，因此，所制备的 PE 纳米纤维具有较高的分子量和机械强度，能够被广泛应用在汽车部件、绳索和体育类产品当中。

同时，海岛模型法、分子喷丝板纺丝法、水热合成法等都成功地实现了纳米纤维的制备。但是以上这些制备方法对设备要求较高、生产工艺较为复杂，因此不利于纳米纤维的大规模制备。

1.5 静电纺丝法

静电纺丝，英文为“electrospinning”“electrostatic spinning”，应用过程中通常简称为电纺。自 1934 年 Fomhals 在专利中将该方法的实验装置公开后，首次让人们了解到，静电纺丝是如何通过高压静电作用对聚合物流体进行拉伸，从而得到直径在纳米级的聚合物纤维材料的。图 1.5 所示是一种典型的平行-圆盘静电纺丝装置，形象地展现了静电纺丝过程。

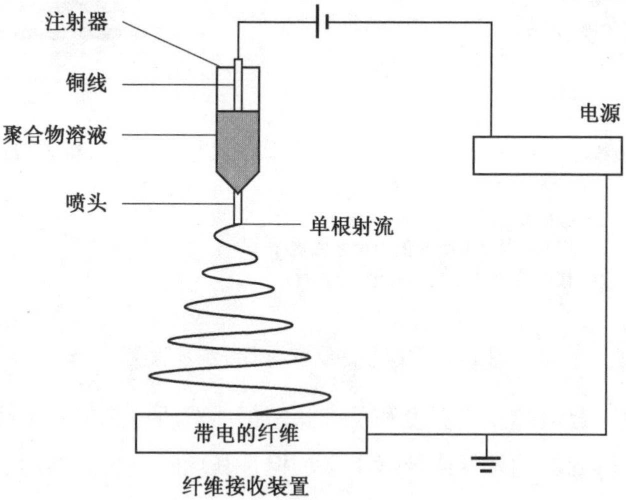

图 1.5 一种典型的平行-圆盘静电纺丝装置示意图

首先，聚合物溶液经泵进入喷口当中，通过施加高压直流电源，喷口处的聚合物溶液获得额外的电荷。其中，当施加的电场较弱时，液滴间的静电斥力会小于液体表面张力，使得端口处的液滴仍保持为球形；随着施加的电场强度增大，液滴间静电斥力增强，球形液滴逐渐被拉伸；最终，当电场强度达到阈值时，液滴间的静电斥力与液体表面张力平衡，从而在端口处形成了“泰勒锥（Taylor cone）”。当电场强度进一步增加，克服液滴的表面张力而形成的射流会被进一步拉伸，如图 1.5 中“螺旋状”结构所示，随着溶剂挥发或熔融体冷却固化，在接收装置上形成了直径在几十纳米至几微米的纤维状材料。

静电纺丝方法可制备出各种各样的聚合物纳米纤维，同时该方法设备简易、操作方便、形貌易于调控，也促进了聚合物复合材料的发展。因此，20 世纪 90 年代后，静电纺丝方法被世界各国进行广泛研究，研究方向也从实验室级别的纳米纤维制备转向静电纺丝工业化设备的制造、多种纺丝工艺和原理剖析以及静电纺丝纳米纤维结构的调控，同时，所得到的具有特殊结构的复合纳米纤维材料展现出更广阔的应用前景。综上所述，表 1.1 比较了在纳米纤维制造领域，静电纺丝法与其他方法的优势和不足。

表 1.1　制备纳米纤维主要方法的比较

方法名称	技术水平	优势	不足	纤维尺度可控性
拉伸法	已产业化	单根纳米纤维长径比高	操作烦琐，原材料受限	较差
模板法	实验室	原材料来源丰富	纳米纤维长度受限	较好
相分离法	实验室	原材料成本低、纳米纤维连续	工艺复杂、耗时、低效	较差
自组装法	实验室	纤维形状丰富	制备时间长	较差
静电纺丝法	已到中试阶段	原材料范围广泛、纳米纤维连续且形貌丰富、孔隙率高达 90%以上，具有超高长径比、操作便捷等	力学性能仍需改善	较好

1.5.1 工艺参数对纤维的影响

1. 纺丝参数的影响

静电纺丝法按原材料自身特点主要分为溶液静电纺丝和熔体静电纺丝。在溶液静电纺丝中，聚合物首先会被溶解在一定溶剂中，配制为具有一定浓度的稀溶液，且该溶液具有良好的导电性，通过纺丝过程中溶剂的挥发最终形成纳米纤维结构；在熔体静电纺丝中，聚合物熔体为浓度较稠的高分子流体，且一般是绝缘性较高的介电材料，在纺丝过程中通过射流和环境热交换固化作用形成纳米纤维结构。虽然两种方式存在一定的差异，但总体上也都要关注以下几个方面。

（1）施加的电场。

1964 年，Taylor 通过研究雷暴的形成，首次提出了液滴在电场作用下呈现出锥形的完整理论，并于 1969 年在研究静电纺丝的过程中成功得出临界电压的概念。如图 1.6 所示，静电纺丝过程中施加的电场对溶液中承载的电荷量有巨大影响。增大所施加的电场，将会促进静电纺丝喷射，同时也会导致大量的液体从喷头处溢出。如果进液速率固定，则会使得喷口处形成的“泰勒锥”变小且更不稳定，最终有可能引起呈现“泰勒锥”的液滴回到进液管中。另外，众多报道论证了较高的电压对所制备的纳米纤维形貌的影响。通常而言，越高的电压会使得溶液被拉伸的程度越大，从而进一步促进了纤维直径变细。低黏度溶液在高电压条件下的电纺会形成二次射流，也会使所制备的纳米纤维直径更细。但是，在这个过程中也存在着例外，有研究表明，当施加一个接近临界电压的电场时，溶液在电纺过程中的飞行时间在增加，从而使得更多的溶液在沉积之前就被拉伸，令所制备的纳米纤维的直径减少。普遍而言，增加纺丝电压能够获得更细的纳米纤维。

溶液静电纺丝采用的是具有一定导电性的牛顿流体，所施加的电压通常为 5～20 kV。而对于熔体静电纺丝，由于熔体自身具有较高的介电常数，因此施加的电压通常为溶液静电纺丝的 2 倍以上，为 20～100 kV，才能使得聚合物熔体形成稳定的射流。

另外，在熔体静电纺丝中，当电场强度超过某一阈值时，将会出现电晕放电或者空气击穿的现象，从而引起纺丝设备报警并使得电压断开。因此在熔体静电纺丝生产过程中，要特别注意这一现象。

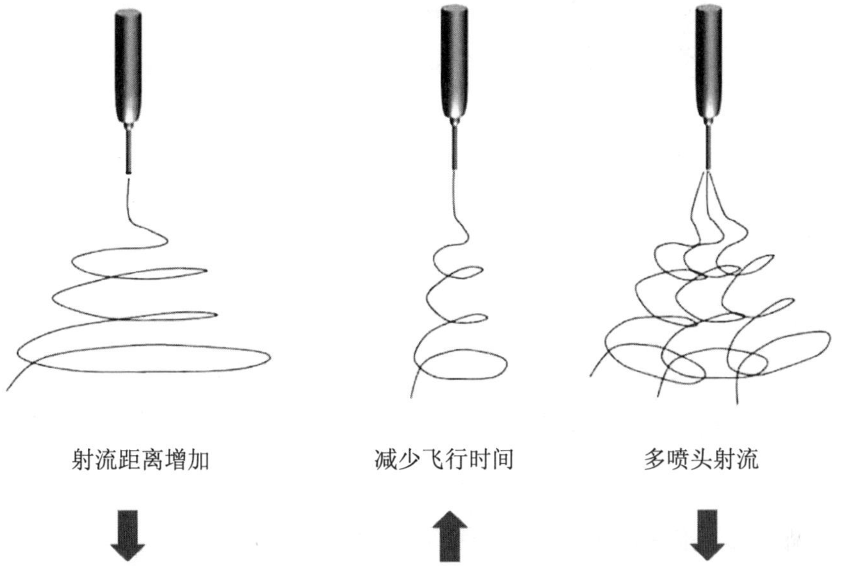

图 1.6　静电纺丝过程中电压与纤维直径关系示意图

此外，静电纺丝中所采用的电压类型（交流电或是直流电）也会对纺丝过程产生影响。通常来讲，大多数使用的电压类型为直流电源，这是因为其能够使聚合物流体携带极性相同的电荷，从而在静电排斥作用下形成纤维状结构。同时直流电源要求接收体是导电的，这样才能将纤维上存在的静电荷传导出去。但是，对于交流电源而言，它能够连续地赋予射流电荷，使射流整体上呈现电中性，使得电荷间的静电斥力减少，提升了射流的稳定性，并能够使得所制备的纤维沉积在绝缘材料表面。另外，交流电源也会使射流的拉伸程度降低，导致纤维直径增大。

（2）喷头与集流体间距。

在静电纺丝中，喷头与集流体间的距离将会直接对电场中射流的飞行时间和拉伸程度产生影响。对于溶液静电纺丝，射流中的溶剂在完全挥发后才会转化形成纳米纤维结构。相同条件下，缩短喷头与集流体间距，将会使施加的电场强度增大，射流的飞行时间减少，从而导致射流中溶剂挥发不完全，纤维间形成部分黏结。Buchkoa 研究发现，当喷头与集流体间距过小时，由于纤维中仍有残余溶剂存在，因此收集到的纤维呈扁平状。当喷头与集流体间距增大到一定程度后，才能获得较细的纳米纤维结构。如图 1.7（a）所示，李欣等研究也表明，虽然随着喷头与集流体间距增大，电场强度会降低，进而不利于纤维细化，但是在一定距离内，由于间距增大会促进纤维牵引，更有利于减少纤维直径。

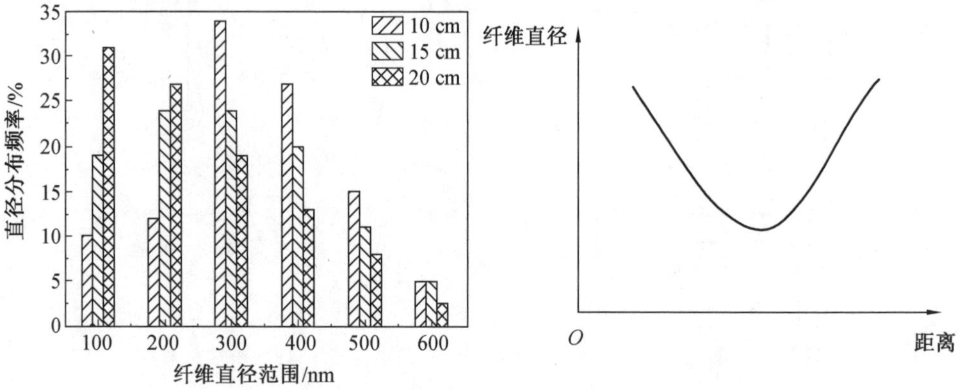

（a）喷头与集流体间距对纤维直径分布的影响　（b）喷头与集流体间距同纤维直径的关系

图 1.7　喷头与集流体间距对纤维直径的影响关系

肖婉红等通过研究聚氧化乙烯（PEO）的静电纺丝过程得出，随着接收距离的增加，纤维直径呈减小趋势，获得的聚合物薄膜面积逐渐增大。接收距离与纤维直径间的关系成因有两点：一是，接收距离增大使得射流的拉伸和扰动更为充分，利于减少纤维直径；二是，施加的电场也随着接收距离的增加而减少，进而降低了射流的加速度，反而使纤维直径变粗。所以，结合聚合物自身性质，确定适当的接收距离是控制纳米纤维直径很重要的因素。

此外，喷头与集流体间距对纤维形貌的影响也与前驱体溶剂的性质紧密相关。若前驱体溶剂沸点较高，则需要相对较大的间距，从而使射流中的溶剂充分挥发、固化，形成纤维状结构；若前驱体溶剂沸点较低，射流中溶剂喷出后即会挥发、固化，使得电场力对纤维的拉伸作用减少，射流飞行时间增长，造成了纤维四处飘散且难以收集的现象。经过对比研究，可得出，在喷头与集流体间距相同的条件下，挥发性低的溶剂会导致所制备的纳米纤维黏结，而挥发性高的溶剂更容易制备出无黏结的纳米纤维。

喷头与集流体间距同纤维直径的关系如图 1.7（b）所示。当喷头与集流体间距较小时，由于射流中溶剂挥发不完全，所制备的纤维间黏结，因此纤维直径较大；

随着喷头与集流体间距增加，射流的飞行时间增长，在电场力作用下被充分拉伸，且溶剂挥发较为充分，所制备的纤维直径逐渐减小；当喷头与集流体间距超过某一阈值后，因为较大间距会降低电场强度，所以射流被拉伸的程度减弱，最终导致所制备的纤维直径增大。

另外，喷头、集流体间距与纤维直径的关系也受到电压范围的影响。当施加的电压较低时，射流的飞行时间对纤维直径起主导作用，因此较大的间距有利于溶剂的充分挥发，从而制备出更细直径的纤维；当施加的电压较高时，对纤维直径起主导作用的是电场强度，若喷头与集流体间距仍较大，反而使得电场强度降低，导致纤维变粗。

综上所述，对纤维直径和形貌产生影响的主要因素分别是前驱体溶剂性质、喷头与集流体间距和施加的电场强度。因此，可根据所需的纤维直径和形貌去调节上述参数。

（3）溶液供给速率。

溶液供给速率也能够在一定程度上对所制备的纳米纤维直径和形貌产生影响，因此，要在纺丝速率和溶液供给速率间寻找平衡。在电压一定时，提升溶液供给速率会增大射流的直径，从而使得所制备的纤维变粗。另外，在端口处形成的“泰勒锥”形状也与溶液供给速率相关。如果溶液供给速率过低，则形成的“泰勒锥”不稳定，影响了射流的稳定性；如果溶液供给速率过高，则形成的“泰勒锥”出现跳动。因此，以上两种情况均会影响到所制备纤维的结构和形貌。

溶液供给速率与纤维形貌之间的关系也受到溶液浓度和黏度的影响。当溶液浓度较大时，其他条件不变，仅提升溶液供给速率，会使得射流更加不稳定，且易造成射流沿轴向固化，在纤维表面或者纤维之间形成珠粒。当溶液浓度或者黏度较小时，静电纺丝演变成了静电雾化，最终得到的也不是纤维而是珠粒，并且珠粒的尺寸随着溶液供给速率的提升而增加。

电压一定的条件下，溶液供给速率的提升使得射流溶液量增大，因而聚合物固化的时间也会随之延长，导致纤维中含有的溶剂量增多，最终有可能会造成所制备的纤维黏结。

因此，为了得到更加均匀的纤维直径和形貌结构，在静电纺丝过程中，务必要根据不同聚合物浓度及黏度，调整好溶液供给速率。

另外，除了以上几点，溶液性质和环境参数都会对纳米纤维在产量和形貌结构方面产生影响，具体如下。

（1）溶液电导率。

选择电导率高的聚合物溶液能够有效增加射流溶液中的电荷含量，进而提升射流的拉伸程度。另外，高电导率的纺丝溶液也能够提升弯曲不稳定性和延长射流的飞行时间。因此，选择高电导率的纺丝溶液有利于提升所制备纤维的质量，例如减少纤维中珠粒的形成、使纤维直径变细等。

（2）溶液黏度。

对于纺丝溶液而言，聚合物分子量和溶液浓度直接决定了纺丝溶液黏度。当溶液浓度较低时，聚合物分子链与链之间的缠结程度较低，采用该条件下的溶液静电纺丝，易在纤维表面形成聚合物颗粒；随着纺丝溶液浓度的增大，聚合物分子链与链之间的缠结程度增加，摩擦增多，在该条件下静电纺丝更容易形成纤维状结构；当溶液黏度继续提升，电纺时射流的电荷载量降低，导致聚合物溶液的拉伸程度不足，进而使所制备的纤维直径增大。

（3）溶剂挥发性。

采用低挥发性溶剂所得到的前驱体溶液，静电纺丝后有可能会得到互相粘连的湿纤维或是无纤维结构存在；采用挥发性高的溶剂所得到的前驱体溶液，静电纺丝过程中会存在着液滴在喷口处固化的现象，导致纺丝过程连续性差。因此，溶剂的挥发性一定要在某一范围内选择。

（4）环境湿度。

静电纺丝过程中，射流的溶剂会与环境介质形成一种双扩散过程，而环境湿度则是介质中的一个主要因素。如果环境中构成湿度的液滴与溶剂相亲，则会抑制射流中溶剂的挥发，从而降低了纤维的固化速度；如果环境中构成湿度的液滴不与溶剂相亲，则促进了溶剂挥发，加快了纤维的固化速度。

（5）环境温度。

环境温度的提升有利于促进射流中分子链运动，进而利于射流的拉伸；同时，较高的环境温度也有助于提升溶剂的挥发速率，使得所制备的纤维直径变细；另外，包括明胶和海藻酸盐在内的生物大分子材料，在室温下的溶液黏度较大，导致纺丝困难，提升环境温度有助于纺丝过程的顺利进行。

2. 接收装置的影响

（1）滚筒或圆盘接收装置。

对简单的平板接收装置而言，越来越多的纤维在接收装置中心位置沉积，会对纤维的均一性产生影响，从而影响纤维的形貌。因此，在采用旋转滚筒接收装置后，会使得沉积的纤维在接收装置表面更加均匀，所得到纤维的膜厚度更均一。

在高速旋转下，滚筒的旋转速度能够用来控制纤维的一致性，纤维会沿着滚筒旋转的方向进行沉积收集。同样地，该现象也出现在旋转圆盘接收装置上。相比于滚筒接收装置，圆盘接收装置有更大的周长，意味着纤维的拉紧程度更多，因此可以在较低的转速下获得更为有序的纤维排布。如图 1.8 所示，E. Zussman 等采用由绕 y 轴旋转的圆盘和铝制块体组成的接收装置收集纳米纤维，该装置除了收集平行排列的纤维之外，还能每隔一段时间绕 z 轴旋转，令收集的纤维排列方向发生转变，进而与之前的纤维一起编织成网络状结构。

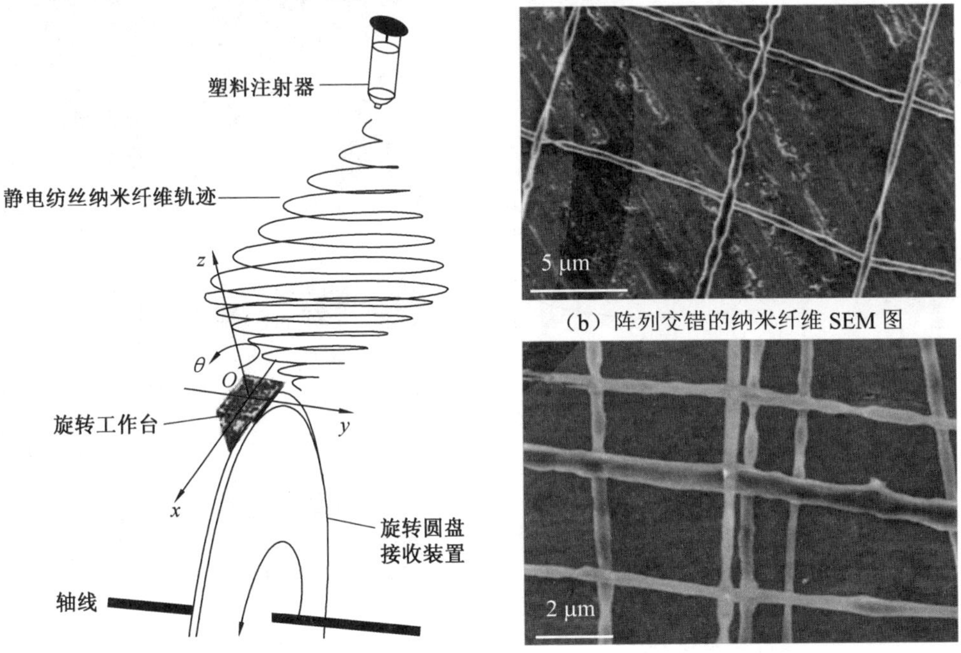

（a）圆盘接收装置示意图

（b）阵列交错的纳米纤维 SEM 图

（c）局部纳米纤维放大 SEM 图

图 1.8　圆盘接收装置及其收集的纳米纤维扫描电子显微镜（SEM）图

（2）有序排布纤维的接收装置。

为了获得单向有序纤维排布，其他接收装置也被广泛研究，例如平行板接收装置（图 1.9（a）），由两个金属接收板组成，中间由绝缘线分开。在这个装置中，静电射流在两个平板间被拉伸，所得到的纤维会单向垂直地沉积在平板上，最终两个电极间的静电作用力使得沉积的纤维发生取向排列，如图 1.9（b）和（c）所示。纳米纤维自身的有序性取决于间隙宽度和施加的电压，这些参数决定了作用在纤维上的静电场强度。在该装置上，沉积的纤维数量可以通过调节施加在两平板间的电场来控制。同时，这种装置的变形也包括两个平行的环状接收板或者两个平行的框架接收板。采用该方法制备的单向有序纳米纤维，其程度和面积均得到了明显的提高。

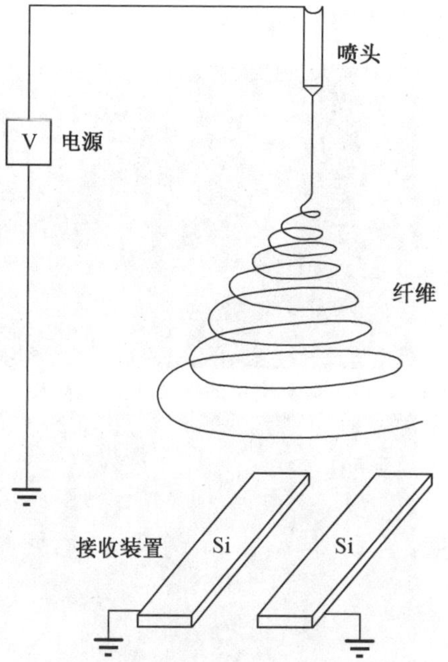

（a）平行板接收装置示意图

（b）制备的纳米纤维的暗视野光学照片

（c）单向取向纳米纤维的 SEM 图

图 1.9　平行板接收装置及其收集的纳米纤维 SEM 图

（3）图案化接收板。

除了以上沉积在金属板接收装置的无序纤维外，还可以以具有图案化的导电网格或导电纺织材料为接收装置，制备出具有图案化特点的微米或纳米纤维材料。一个带有电荷的针状物或是刀片边缘都可被用来引导纳米纤维沉积在旋转滚筒接收装置上。一种拼接网格图案同样能够以一步法，采用附加在旋转圆盘电极上的铝掩模版获得。此外，绝缘接收装置的使用也被用来生产“鸟巢状”图案，在该装置中，离子液体被掺杂到聚苯乙烯中，以增加接收装置上的电荷积聚，从而使纤维仅能在某一特定区域形成薄薄的一层。

结合多喷头、多接收装置的思路，哈尔滨工业大学（深圳）于杰教授课题组开发了多射流/多尖端接收装置，成功制备得到有序排列、可规模量产化的氮化硼（BN）纳米纤维材料，如图 1.10 所示，展现出静电纺丝在功能性纳米纤维规模化制备上的广阔应用前景。

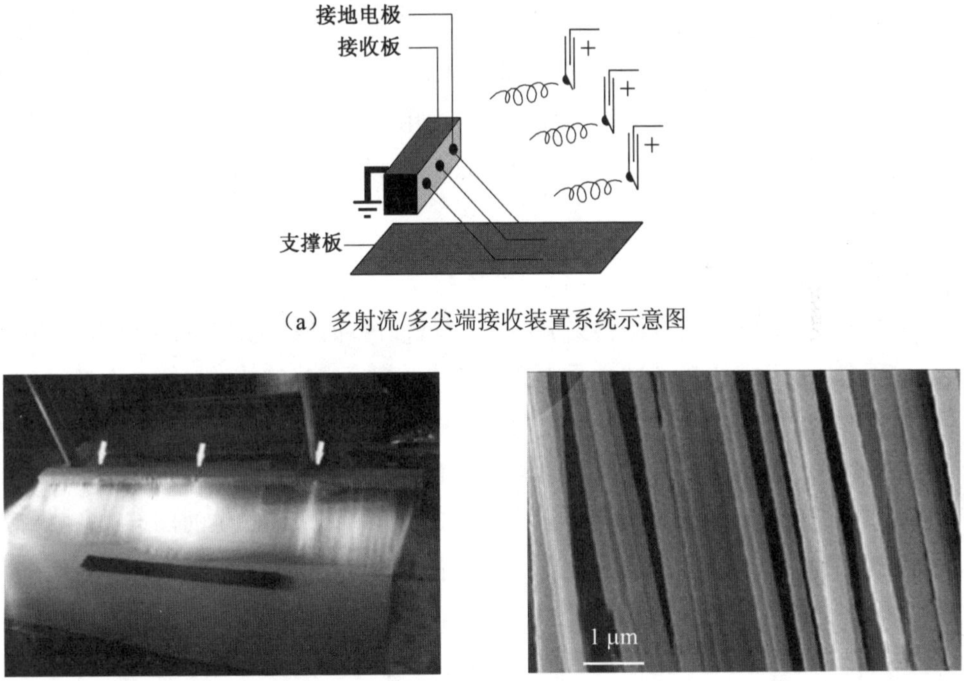

（a）多射流/多尖端接收装置系统示意图

（b）多射流/多尖端接收装置静电纺丝光学照片　　（c）前驱体聚合物纳米纤维 SEM 图

图 1.10　多射流/多尖端接收装置及其制备的纤维 SEM 图

除了以上制备过程参数和不同接收装置对纤维形貌所产生的影响，为进一步提升静电纺丝的产率、拓展纳米纤维的应用领域，研究者分别从喷头改造以及辅助场的作用等方面，对静电纺丝进行如下探讨。

1.5.2 静电纺丝的类型

1. 单喷头静电纺丝

早期的静电纺丝研究主要集中于单喷头静电纺丝。其中，所采用的喷头为金属质且端口平坦。相比于注射器所配的斜槽式端口，平坦端口能够保证形成稳定的“泰勒锥”且射流几乎没有振荡；同时，在较高的溶液供给速率条件下，平坦端口更有利于形成更多的电纺射流。

哈尔滨工业大学邱业君课题组基于单喷头静电纺丝，在实验室制得了一系列较有特色的纳米纤维材料，如氮化硼纳米纤维、超细（直径小于 20 nm）碳纳米纤维、碳化钨（WC）纳米纤维、Fe-N/C 纳米纤维、磷酸铁锂/碳（$LiFePO_4/C$）复合纳米纤维、氧化亚钛/碳（Ti_4O_7/C）复合纳米纤维等。

目前单喷头静电纺丝设备较为成熟，主要用于实验室级别的纳米纤维的制备。其不足之处在于生产效率较低，仅约为 0.02 g/h，从而使得纳米纤维生产成本较高，难以实现规模化应用。

2. 同轴静电纺丝

为进一步使纤维内部孔结构更加丰富，提升纤维的应用领域，自 20 世纪 90 年代起，研究者就逐渐开始构想关于中空结构或者核壳结构的纳米材料。而在静电纺丝领域，相比于单喷头静电纺丝，用同轴静电纺丝来设计“核壳”结构纳米纤维材料逐渐得到了广泛研究，其设备示意图如图 1.11 所示。在同轴静电纺丝中，高电压下的外部溶液相和内部溶液相在纺丝液滴形成过程中是极不稳定的，导致在喷口处形成的“泰勒锥”较为复杂。因此，在同轴静电纺丝过程中，很重要的一点就是调控内部和外部溶液的流速。在研究过程中，同轴纺丝得到的纳米纤维的形貌和均一性与内外部溶液的流速和溶液浓度密切相关；“核”和“壳”的厚度则与内外部溶液的组分和流速息息相关。

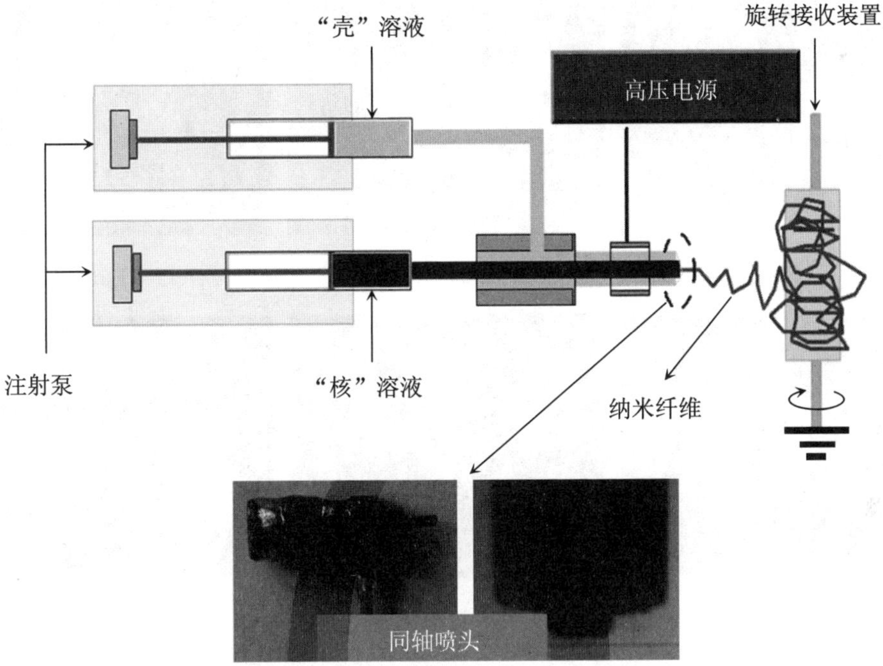

图 1.11　同轴静电纺丝设备示意图

2003 年，Yarin 等首次以同轴静电纺丝技术，采用两种不同聚合物材料，得到具有"核壳"结构的纳米纤维。他们在研究中得出在喷头到接收板之间，溶液较快的纺丝速度能够有效阻止两种扩散系数不同的聚合物溶液混合，该团队以聚十二烷基噻吩（PDT）为"核"，聚氧化乙烯（PEO）为"壳"，所得到的纳米纤维上存在着两相间明显的边界。另外，PDT 由于自身较低的分子量并不适合形成纤维状结构，因此，同轴静电纺丝有利于扩展作为"核"的聚合物材料，尤其是一些通常不能通过静电纺丝形成纤维状结构的功能性试剂。

近期，Li 等采用液态环氧树脂作为"核"，聚苯乙烯（PS）作为"壳"，采用同轴静电纺丝制备出聚苯乙烯纤维。如图 1.12 所示，随着环氧树脂作为"核"组分的增加，纤维直径在增大，同时，作为"壳"的聚苯乙烯的厚度在逐渐减小。当环氧树脂的溶液供给速率达到 9 $\mu L \cdot min^{-1}$ 时，"壳"的厚度已经很薄了。另外，在所有的"核壳"结构纤维中，均不存在裂痕和孔洞，表明该结构具有较好的稳定性。

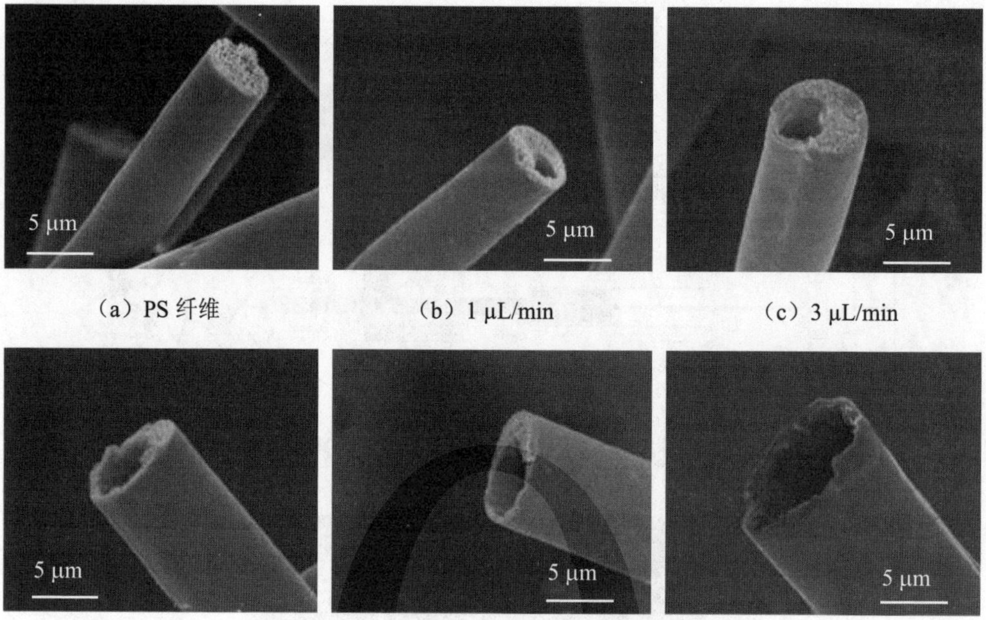

（a）PS 纤维　（b）1 μL/min　（c）3 μL/min

（d）5 μL/min　（e）7 μL/min　（f）9 μL/min

图 1.12　不同溶液供给速率下制备的纤维 SEM 图

3. 多喷头静电纺丝

相比于传统的铸造或者挤出方式，静电纺丝由于仅有一个喷头，因此沉积速度受到影响，限制了静电纺丝的实际应用。为了增加产量，众多团队探究了采用多喷头的方式同时创造出多个纤维喷流。

虽然多喷头静电纺丝技术是基于单喷头静电纺丝，但是，因为增加了喷头数量，静电纺丝设备的复杂性增加，所以，多喷头静电纺丝技术的成功需要满足一些特定的条件。有研究表明，在多喷头静电纺丝中，喷头间距必须充分，使得静电场分布更加均匀。这是因为，静电射流不仅受到施加电场和自身库仑作用力的影响，而且射流间的库仑斥力也会对射流产生影响。在库仑斥力作用下，射流会被相邻射流向外推开，而内部的射流则会沿着纺丝喷头处挤压。

另外，Chu 等通过比较分析单喷头和多喷头静电纺丝工作电场对纳米纤维制备的影响，指出了对于多喷头静电纺丝，存在着两个重要的影响因素：一是，纺丝溶液一定要能在一定压力或者流速下连续传递到每一个喷头处，使得形成的“泰勒锥”更加均一；二是，在每一个喷头尖端处液体所受的静电场力必须足够克服纺丝溶液的表面张力。第一个因素可以通过合理的机械设计来得到解决，但是第二个因素则涉及静电场强度的控制。在多喷头静电纺丝系统中，静电场强度是由施加的电场和喷头位置排布共同决定的。喷头位置的不恰当排布会导致电场的不均匀分布，将会严重弱化位于中心位置的静电场强度，因此，喷头位置的排布对于多喷头静电纺丝设备尤为重要。结合以上两点，研究者们分别从喷头数量和喷头位置排布对多喷头静电纺丝技术进行了探讨。

首先是采用线性排布的方式，研究了喷头数量对静电纺丝的影响。随着喷头数量的增加，在相同时间内，纳米纤维的产量也随之增大。但是，拉伸射流所需的电压也要随之提升，喷头间的电场干扰也变得更强。另外，位于两边的射流偏离角度增大，所获得的纳米纤维平均直径减少，且纳米纤维的不均匀程度增多。结合以上观点，在一定程度上，虽然增加喷头数量会提升产量，但是多喷头之间电场的影响将会对多喷头静电纺丝的效果产生负面影响。

在喷头位置排布方面，Tomaszewski 团队分别设计了线形、椭圆形和圆形喷头排布，如图 1.13 所示。经过比较得出，线形和椭圆形喷头排布的静电纺丝效率比较低，尤其是在线形排布过程中，纳米纤维几乎不能从喷头中心位置处喷出。三者中静电纺丝效率最高且制备的纳米纤维质量最好的喷头排布方式是圆形排布。

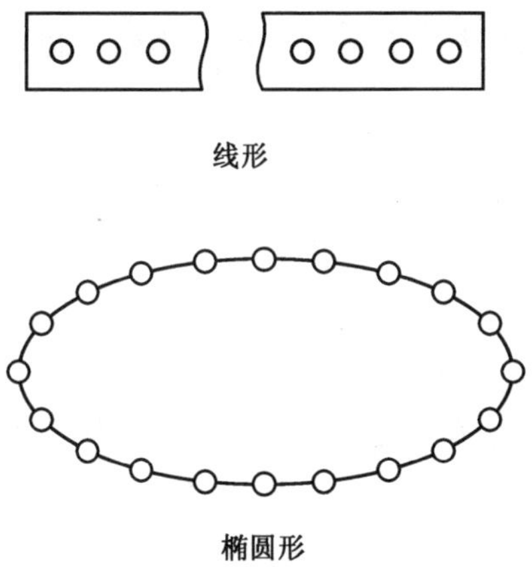

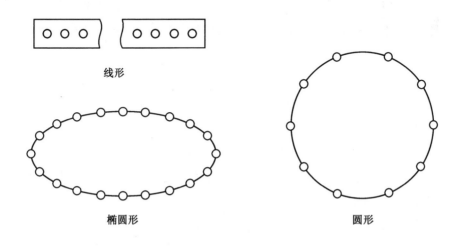

图 1.13　三种多喷头静电纺丝喷头位置排布示意图

随着纺丝喷头的增多，实际生产中喷头之间的电场会出现严重的“边缘效应”，刘延波等通过 Comsol multiphysics 软件对采用 10～200 个喷头的静电纺丝场强大小及分布进行了模拟和分析，分别对两侧单独加压和不同喷头长度这两种方式进行探讨，得出，虽然两侧单独加压能够使场强不均匀程度降低，但最佳的电压参数仍需进一步探讨，而采用不同长度的喷头，则可使场强不均匀程度低于 3%，可满足工业上对多喷头静电纺丝场强均匀性的需求。

4. 无喷头静电纺丝

虽然多喷头纺丝能够提升静电纺丝效率，但是，仍面临着诸多问题，例如，纤维产量的增加有限、喷头处聚合物会形成阻塞以及设备占地面积较大等。因而，在此基础上，研究者提出了无喷头静电纺丝技术。

自 20 世纪 70 年代，无喷头静电纺丝概念由 Simm 及其同事提出以来，在 2014—2016 年迎来了井喷式发展。无喷头静电纺丝的定义是，从一个开放的液体表面施加电场后直接电纺得到纳米纤维。同时，无喷头静电纺丝所得到的射流是连续的，完全不受到喷头所产生的毛细管作用影响。其形成机理可解释如下：导电液体在微观上形成自组装波动，最终当施加的电场强度超过阈值后形成射流。射流的形态和所得到的纳米纤维形貌受喷头处电场强度和电纺区域的影响极大，而喷头处的电场强度和电纺区域又来自于施加的电场和无喷头吐丝器形状。因此，在无喷头静电纺丝中，喷丝板在纺丝过程、纤维质量和最终产量上起到了关键性作用。

根据喷丝头的不同，无喷头静电纺丝装置分为静态无喷头静电纺丝和动态无喷头静电纺丝两种类型。

（1）静态无喷头静电纺丝。

静态无喷头静电纺丝主要是在重力、磁场作用力或者外界压力作用下来实现纺丝的。2004 年，Yarin 等开发了一种磁场作用下的无喷头静电纺丝装置，如图 1.14 所示。图 1.14 中底层 a 为磁流体溶液，上层 b 为纺丝溶液，c 为接收板（对电极），d 为浸入磁流体的电极，e 为高压电，f 为永磁体或电磁体装置。在磁场作用下，磁流体溶液会产生波动，进而带动上层溶液形成波浪状。在施加电场后，上层溶液形成的波峰处即会形成射流，进而形成纤维。虽然相比有喷头的纺丝装置，该纺丝装置可极大提升纺丝效率，但是其所制备的纤维表面粗糙，直径分布较大，且该装置结构较为复杂，不利于在实践中加以推广。

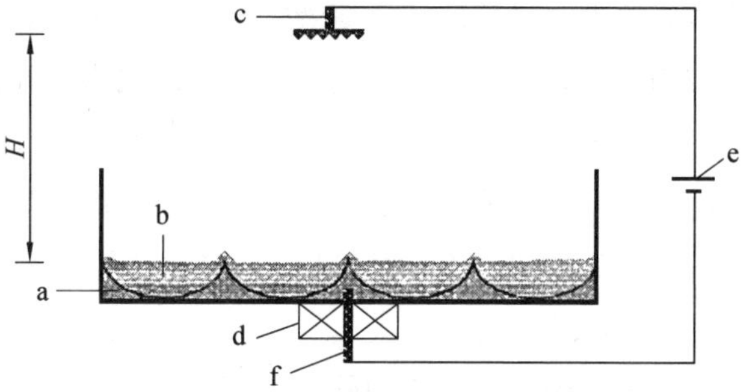

图 1.14　磁场作用下的无喷头静电纺丝装置

之后，何吉欢结合气泡动力学原理，通过模仿蜘蛛纺丝的方式，在喷气式静电纺丝装置的基础上，改进得到两款喷气式静电纺丝装置并且研发出一款新型的旋转转盘式气泡静电纺丝装置，相关装置及原理图如图 1.15 所示。

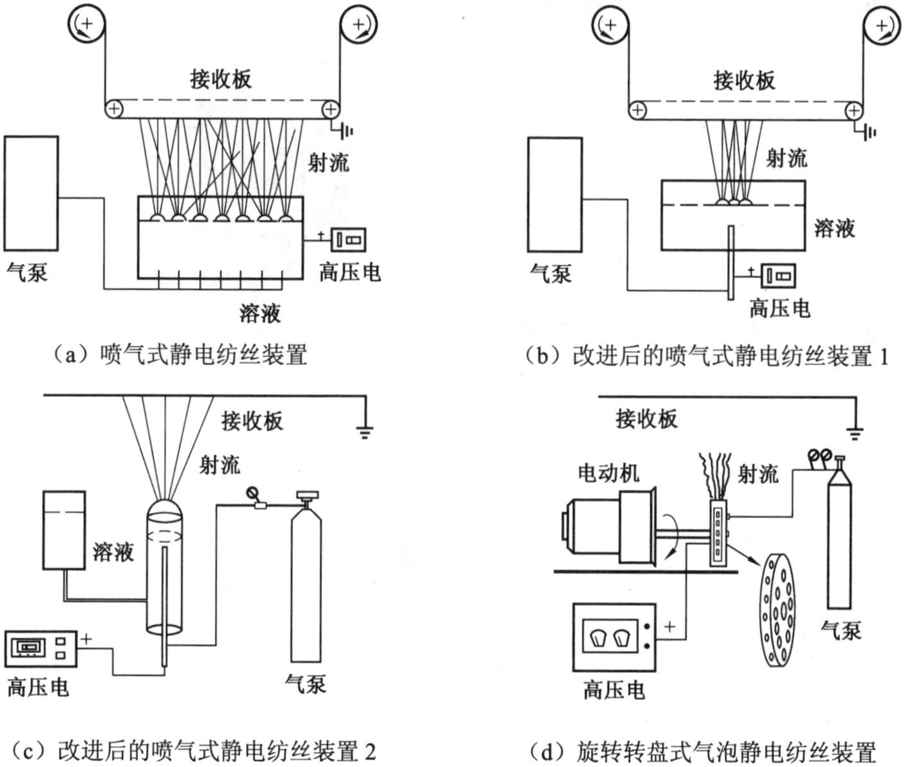

图 1.15　喷气式静电纺丝装置、旋转转盘式气泡静电纺丝装置及原理图

气泡纺丝的基本原理为，气泵使纺丝溶液表面形成连续、均匀的气泡，该气泡类似于“泰勒锥”结构。当施加电压后，气泡表面会形成均匀分布的电荷，当电场强度大于或等于气泡表面张力时，伴随着气泡破碎形成了稳定的射流。

在静电纺丝工业化过程中，气泡纺丝技术具有操作简单便捷、产率较高等优势，这主要得益于以下三点：①相比于“泰勒锥”，气泡纺丝所需的电压较低；②气泡破裂时会形成更多的射流，进而提升了纳米纤维的产量；③相比传统静电纺丝，气泡纺丝中的气泡不依赖于电场，因而在相同电压下，能形成更多射流，提升纺丝产量。

华东大学于昊楠等设计了一种新型的碟形喷头，如图 1.16 所示。相比于磁性纺丝、气泡喷丝等装置，碟形喷头通过采用圆形结构，克服了在其他无针头纺丝过程中存在的机械扰动。同时纺丝溶液在喷头边缘射出，因为边缘上电场强度均一，所形成的纳米纤维直径分布均匀，差异较小。

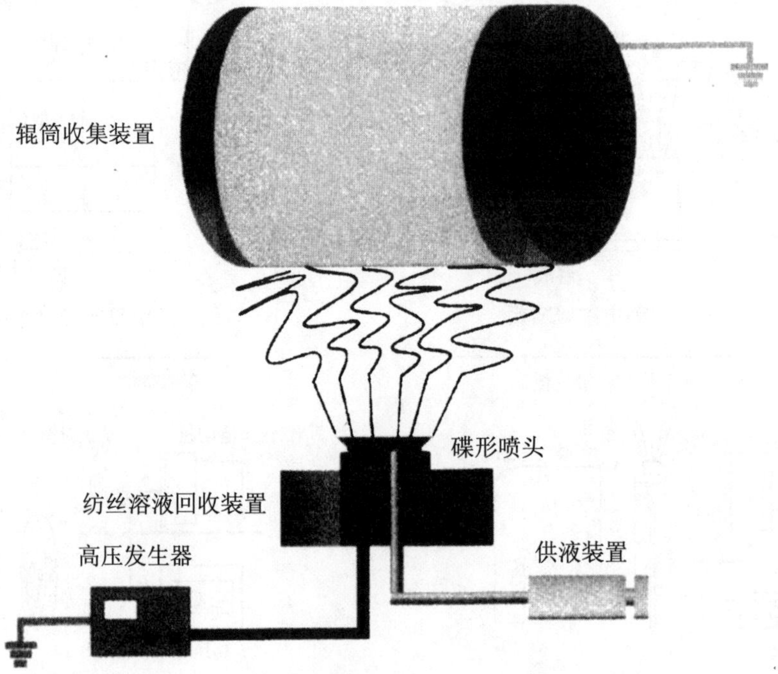

图 1.16　碟形喷头示意图

（2）动态无喷头静电纺丝。

动态无喷头静电纺丝主要是通过喷头的机械运动促进射流形成。在众多动态无喷头静电纺丝中，圆筒型已经被证实是一种有效的静电纺丝方式，并且已经被 Elmarco 公司成功商业化，该公司这款商业化的设备型号为纳米蜘蛛，如图 1.17 所示。该装置中射流的形成主要分为以下三个步骤：①通过装有聚合物溶液的储槽在钢丝导轨上以一定速率往复运动，进而在钢丝表面形成液滴；②施加外电场后，钢丝导轨上的液滴受静电力的牵引形成“泰勒锥”；③射流从“泰勒锥”拉伸出来，并不断牵伸劈裂，最终形成纳米纤维。纳米蜘蛛已经能够展示出聚合物溶液态静电纺丝和聚合物熔融态静电纺丝的生产能力。

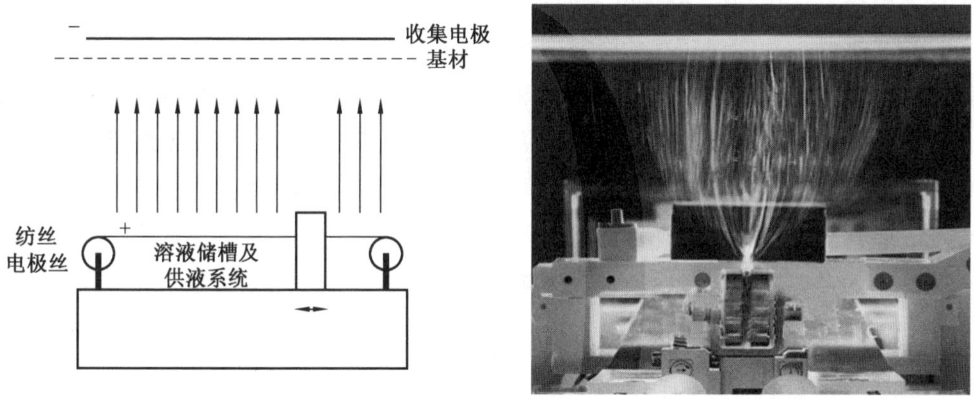

（a）纳米蜘蛛工作机理图　　（b）工作中的钢丝喷头光学照片

图 1.17　圆筒型动态无喷头静电纺丝工作原理及其光学照片

5. 近场静电纺丝

近场静电纺丝技术是用来生产图案化纤维的，图 1.18（a）所示为近场静电纺丝装置示意图。相比较于其他静电纺丝方式，首先，近场静电纺丝采用的是固体探针来取代传统的针状喷头，固体探针能够进一步增强电场强度，并增强带有电荷的射流的稳定程度；其次，该纺丝方式中，射流工作距离较短，射流在电纺过程中还未发生弯曲即在一种可控的方式上沉积到较小区域中；最后，当结合一种可移动的收集装置后，即可在预先设计好的图案上进行直写得到微米或者毫米尺寸的图案化纤维。近场静电纺丝技术可以通过显微镜技术的帮助，来增加直写纳米纤维的精细程度，并且这种可视化的操作技术也提升了静电纺丝的可兼容性，促进了电流体打印技术在微纳米一体化方面的应用。

同时，在原理上，传统的静电纺丝射流在初始时经过短暂平稳的轨道后，在静电斥力作用下会逐渐呈现出无序扰动的状态，而纳米纤维拉伸变细的过程则主要发生在这种无序状态下。由于初始时带电的射流拉伸比很小，因此传统静电纺丝过程中在较短距离下是很难获得超细纳米纤维的。而在近场静电纺丝中，电极和接收装置间的距离缩短到 500 μm～3 mm 之间，稳定的射流可以用来制备可控沉积的电纺纳米纤维。Sun 等采用直径为 25 μm 的钨探针作为喷丝装置，以替代传统静电纺丝的中空针头装置。最初喷出的射流要小于钨探针的直径，随着溶剂挥发，最终得到的纳米纤维直径在 50～300 nm 之间。

为了提升近场静电纺丝获得的纳米纤维的精密化程度，研究带电射流的可控程度是很有必要的。2010 年，Zheng 等基于麦克斯韦黏弹性理论对近场静电纺丝进行了理论模拟，研究结果表明，接收装置的移动速度是影响所沉积纳米纤维形貌的关键性因素。随着接收装置移动速度的不同，所得到的纳米纤维形貌分别为线形、波浪形、单线圈形和多线圈形。同时，因为电荷间斥力的不平衡性质是盘绕的纳米纤维形成中的主要影响因素，所以，对接收装置的电导率与纳米纤维形貌之间的关系也进行了研究。研究结果表明，通过提升接收装置的电导率，有利于增大相邻纳米纤维线圈的距离，进而使多线圈纳米纤维区域直径减小。

近年，徐方远等基于近场纺丝技术，通过工艺优化参数精确地在单根纳米纤维上采用逐层堆积的方式，制备了多个特征尺寸（小于 10 μm）、高度与宽度比在 60 左右的三维结构材料，如图 1.18（b）和（c）所示，初步表明静电纺丝法在微纳米 3D 打印方面应用的可行性。相比传统的激光 3D 打印技术，该方法拓宽了原材料选择范围、降低了生产成本且简化了生产工艺。但是，在该工艺研究中，过程参数及电场分析仍需进一步完善，且大曲率的纳米纤维堆积及实体填充则是未来要解决的重点，同时，新型喷头设计也是实现该工艺在任何接收平台上制造 3D 结构的重要因素。

2018 年，林灿然等采用 Comsol multiphysics 软件简化了近场静电纺丝装置。如图 1.18（d）所示，通过添加平行辅助电极，较为有效地抑制了射流在电场作用下的扰动，接收装置上沉积的纤维卷曲率大大降低。同时，该研究者又设计出适用于近场静电纺丝纳米纤维薄膜的数控打印系统，通过 x 轴和 y 轴的插补运动，有效实现了纤维在水平和垂直方向的堆叠。

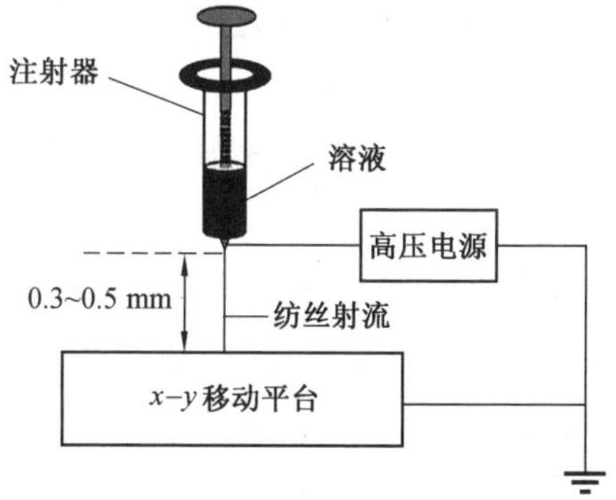

（a）近场静电纺丝装置示意图

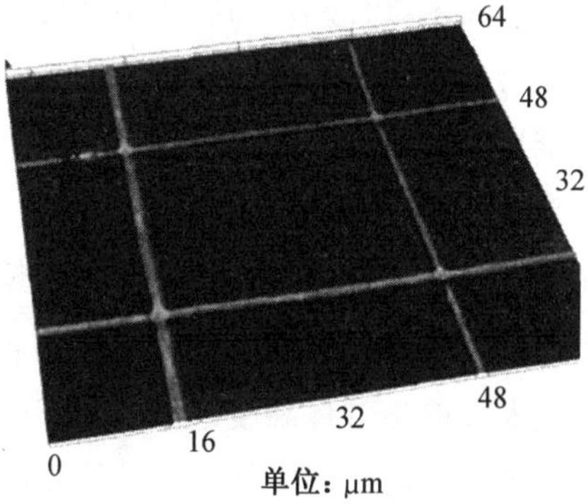

（b）二维静电纺丝网格图

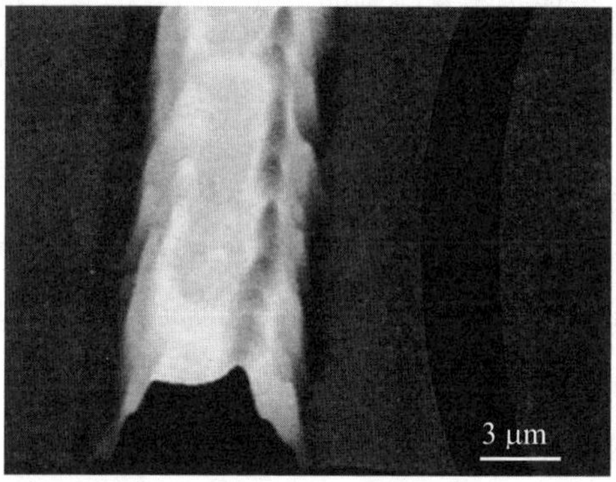

（c）二维静电纺丝纤维形貌图

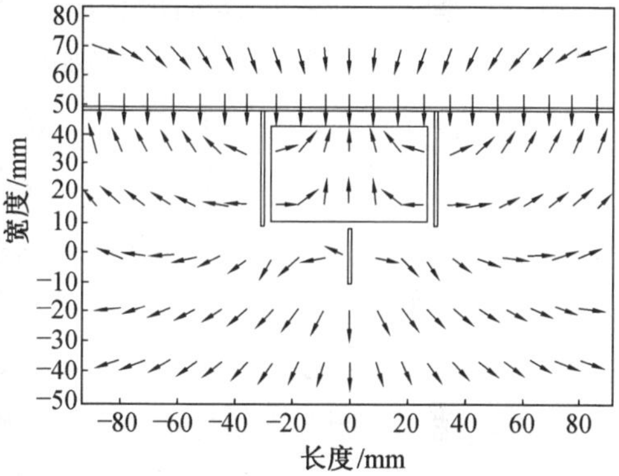

（d）平行辅助电极添加后的电场强度矢量分布图

图 1.18　近场静电纺丝

6. 熔融静电纺丝

在静电纺丝获取纳米纤维过程中，除了溶液静电纺丝外，还有熔融静电纺丝。相比溶液静电纺丝过程，由于没有溶剂的蒸发，在工业上熔融静电纺丝更为安全和环保，所制备的纳米纤维也不存在很多小孔和残留的有毒溶剂。但是，除了要考虑到开发聚合物熔融静电纺丝所需设备的复杂性之外，仍然有以下三个问题需要被考虑到：①纺丝系统中高电压的隔离和安全性能；②所制备纤维的进一步改善；③纤维产能的进一步提升。

对于溶液静电纺丝，溶液黏度通常在某一特定范围内（5～20 Pa·s）才能更顺利地进行纺丝。相似地，在熔融静电纺丝中，如果黏度太大，电场强度也是很难克服熔融体黏度来进行的，而分子量太低所导致的低黏度也会在纺丝过程中形成很多微孔球状物。在熔融静电纺丝过程中，通常的问题是纺丝黏度太大，使得所制备的纳米纤维直径过粗或者纺丝过程容易失败，因此，研究者采用了许多方式减小熔融体黏度来获得更好的纤维材料。结合不同研究者的实验参数，熔融体黏度范围主要分布在 20～200 Pa·s。多数研究者使用熔体流动指数（MFR）来表征熔融体黏度，在应用过程中主要范围是在 300～2 000 $g \cdot (10\ min)^{-1}$。一些研究者也在尝试通过改性的方式来降低熔融体黏度，获取更细的纤维直径，比如使用硬脂酸。

另外，在溶液静电纺丝中，施加的电压范围一般在 7～20 kV，在熔融静电纺丝中，其自身较大的介电常数使得要施加的电压在 20～100 kV。在熔融静电纺丝设备中，通常是选择非直接加热或者将纺丝装置与加热装置隔绝的方式，因此，除了电加热的方法外，还有激光加热、高温加热、热空气加热和微波加热等一系列方式。

在射流特征方面，不同于溶液静电纺丝存在的稳定态和不稳定态，熔融静电纺丝过程几乎不存在任何抖动。但是，在某些使用了较低聚合物黏度的熔融静电纺丝中，射流的旋转或者抖动仍然能被观察到。考虑到在纤维改善方面，溶剂的蒸发和射流的扰动是主要因素，熔融静电纺丝过程在很大程度上不需要溶剂蒸发，在熔融界面上电荷的自由程度也较低，在纺丝过程中并不容易形成不稳定的电荷扰动，因此，在静电纺丝中制备扰动且不稳定态的纳米纤维更为不易。

纤维纯度主要会影响材料的比表面积、孔隙率、抗过滤性，从而会间接影响纤维的性能和应用。在溶液静电纺丝中，所得的纳米纤维直径通常在 100～1 000 nm 之间，孔径在 2～465 μm 之间，适合应用在空气过滤、液体和颗粒阻挡等领域。相比之下，熔融静电纺丝直径通常超过 1 μm。一些研究者通过采用瞬时高温加热、添加黏度降低试剂或者无机盐的方式，使所制备的纤维直径减少到 1 μm 或者更低；还有研究者使用空气辅助的方式，使熔融静电纺丝所得纤维直径降低到 200～300 nm，但是，仅有少量材料能够达到这种直径范围，所需要的生产设备和制造流程也更为复杂。

在温度和湿度方面，对于溶液静电纺丝，环境温度和湿度对溶液黏度、溶剂蒸发速率会产生影响，进而间接影响纤维纯度、表面孔结构和纤维可纺程度。在熔融纺丝中，虽然环境湿度对纤维纺丝过程几乎没有影响，但是当环境湿度过高，也容易引起高电压下的空气击穿。Zhmayev 等研究了环境温度对纺丝过程的影响，他们发现，当环境温度增加到接近管口温度时，纤维纯度会受到很大影响。Warner 等通过观察纺丝区域中不同位置的三维温度分布，发现纺丝过程中的温度对纤维直径会产生很大影响。因此，在熔融纺丝过程中，为了避免空气击穿，环境湿度不应该过高，并且为了得到更好的纤维，纺丝过程中的温度务必精确控制。

在熔融静电纺丝过程中，最基本的参数包括熔融温度、纺丝距离、施加电压和供给速率。Liu 等在不同流体指数下采用正交实验，对熔融温度、纺丝电压、纺丝距离和 MFR 与纤维直径间的关系进行了研究。所得结果表明，各个参数在影响纤维直径重要性上，MFR>纺丝电压>纺丝距离>熔融温度。另外，Lyons 等研究表明，电场强度的增大有利于减小纤维直径，同时，随着熔融纺丝供给速率的降低，“泰勒锥”区域和纤维直径均有所减小。Zhou 等研究表明，提升熔融温度有利于减小纤维直径，并使得纤维直径分布更加密集。而对于电场强度的提升，虽然纤维直径也会降低，但是这种改变不如熔融温度的影响大。Rangkupan 和 Reneker 共同进行了真空熔融静电纺丝实验，研究了熔融纺丝黏度对可纺性和纤维直径的影响，这是因为相比溶液静电纺丝，熔融静电纺丝的黏度至少高了一个数量级。研究表明，虽然通过增加纺丝电压、提升电场强度和减小熔融纺丝供给速率等方式能够有效提升纤维纯度，但是这些需求要同时平衡纺丝的有效性。

结合以上研究，本书通过表 1.2 更直观总结出各类别静电纺丝技术在纳米纤维制备上的优缺点，为纳米纤维在不同领域中的应用提供有益的参考。

表 1.2　各类别静电纺丝技术优缺点

技术名称	优缺点
单喷头静电纺丝	操作简单，设备成熟，但生产效率较低
同轴静电纺丝	能够一步法制备连续中空多孔纳米纤维
多喷头静电纺丝	产量提升，但喷头间电场干扰仍限制其发展
无喷头静电纺丝	产率较高、操作便捷且占地面积小，目前已商业化
近场静电纺丝	能够制备图案化纳米纤维结构
熔融静电纺丝	工业生产更为环保、安全，目前已在中试阶段

1.5.3 静电纺丝产物的形貌

除了以上通过施加外场作用、改变接收装置等方式实现电纺纳米纤维有序化和图案化之外，静电纺丝技术还能通过采用不同原材料复合、与其他制备方法相结合的方式，得到具有丰富表面或/和内部结构的纳米纤维材料。

例如，Gao 等采用静电纺丝法，通过调节预氧化工艺升温速率，分别得到具有实心、中空和线-管结构的纳米纤维材料，如图 1.19 所示。

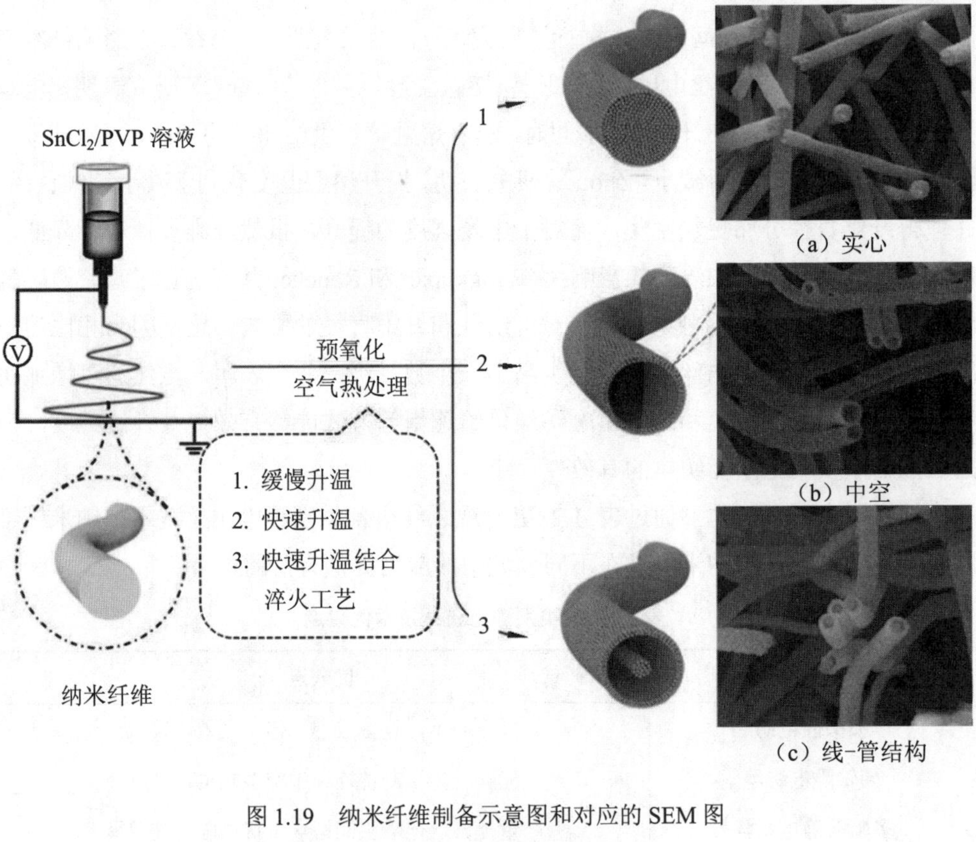

图 1.19 纳米纤维制备示意图和对应的 SEM 图

（PVP 为聚乙烯吡咯烷酮）

当升温速率维持在 2 ℃·min^{-1}，前驱体纳米纤维处于均匀加热的状态下，意味着 PVP 的分解和 SnO_2 的结晶是同步进行的，从而导致纤维收缩成一种实心结构（图 1.19（a））。当升温速率高达 250 ℃·min^{-1} 时，由于温度传导迟滞效应，因此纳米纤维内外部分的温差极大，结果就是，纳米纤维外部会立即达到 PVP 分解和 SnO_2 结晶温度，瞬间在纳米纤维外表面形成一种实心的 SnO_2 结构；随着 PVP 分解和 SnO_2 结晶逐渐从外向内蔓延，所形成的 SnO_2 纳米晶体紧密附着在外部壳体上；在热处理时间充足的情况下（500 ℃，2 min），"核"材料会逐渐外延生长，最终形成中空纳米结构（图 1.19（b））。在热处理后施加淬火工艺，如图 1.19（c）所示，外延的"核"材料会在某一位置凝固，进而在管状结构中形成了线形的实心结构。

Yuan 等结合胶体静电纺丝和水热处理的方式，大量制备了均一的、无染色剂添加的可调色电纺纤维薄膜，其制造过程示意图如图 1.20（a）所示。从图 1.20（b）～（d）可看出，通过胶体电纺的方式，聚苯乙烯-甲基丙烯酸甲酯-丙烯酸（P（St-MMA-AA））胶粒均匀分布在纤维表面，其颗粒大小分别为 220 nm、246 nm 和 280 nm。同时，所反射出的光线分别对应绿色、红色和紫红色（图 1.20（e）～（g）），且颜色在整个膜表面分布均一。该研究表明，可通过调节胶粒尺寸来进一步调节薄膜的反射颜色，且仅使用三种尺寸纳米球即可覆盖几乎所有可见光谱。在未来的研究中，通过负载高折射率对比度无机纳米颗粒，则能够进一步增强着色纤维的机械强度。

另外，2018 年，哈尔滨工业大学（深圳）于杰教授课题组结合热化学气相沉积法（CVD），以静电纺丝经 NH_3 氮化后得到的碳纳米纤维（CNF）为基础，以 CH_4 为碳源，得到具有三维结构的石墨烯纤维薄膜，如图 1.21 所示。相比于目前的三维石墨烯材料，该石墨烯纤维薄膜具有超高的电导率、优异的电磁屏蔽性能和良好的湿度可调节性（在超疏水到超亲水性之间可调节）。更重要的是，该方法可大量/大批量制备柔性石墨烯纤维薄膜，为三维石墨烯材料在催化、能源存储和电磁屏蔽等方面的应用提供了一种有效的制备方案。

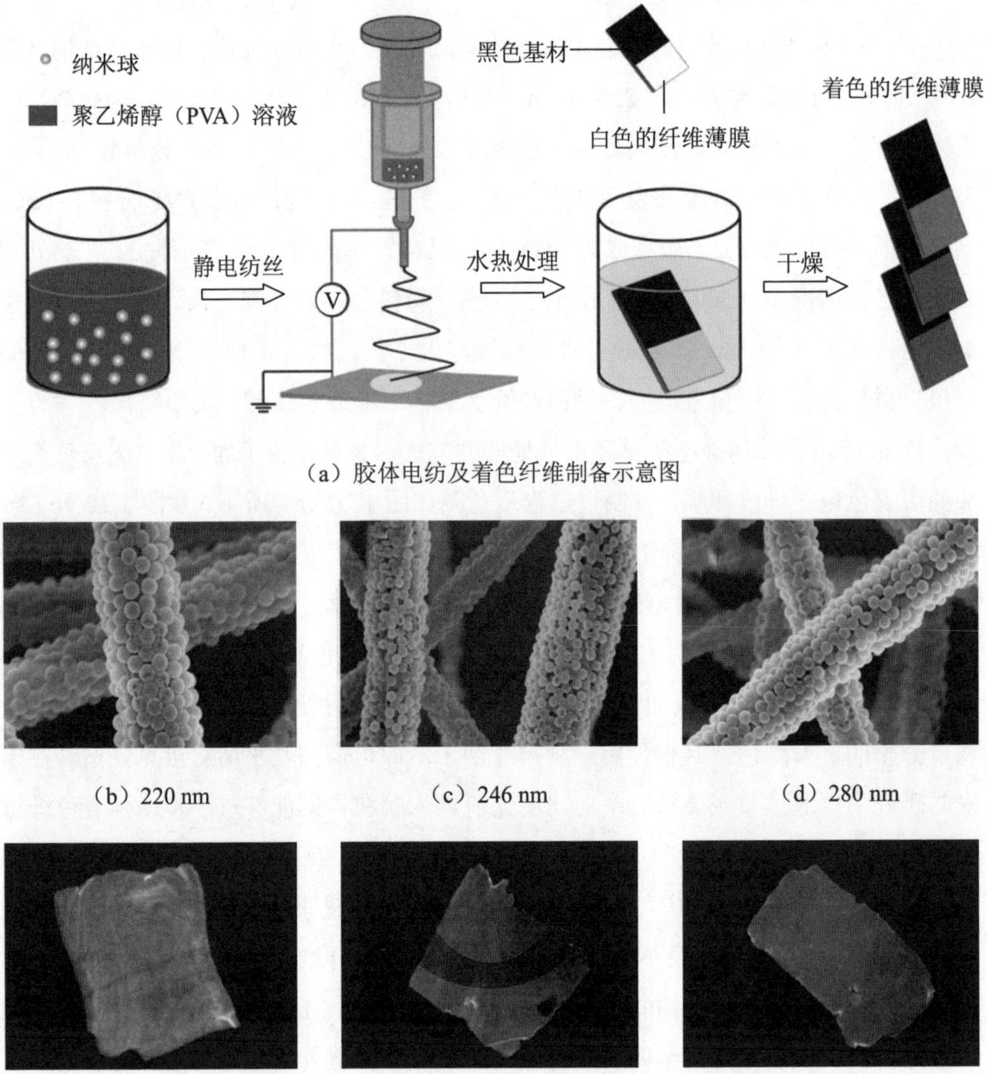

（a）胶体电纺及着色纤维制备示意图

（b）220 nm　（c）246 nm　（d）280 nm

（e）绿，220 nm　（f）红，246 nm　（g）紫红，280 nm

图 1.20　结合胶体静电纺丝和水热处理的方式制备的可调色电纺纤维薄膜

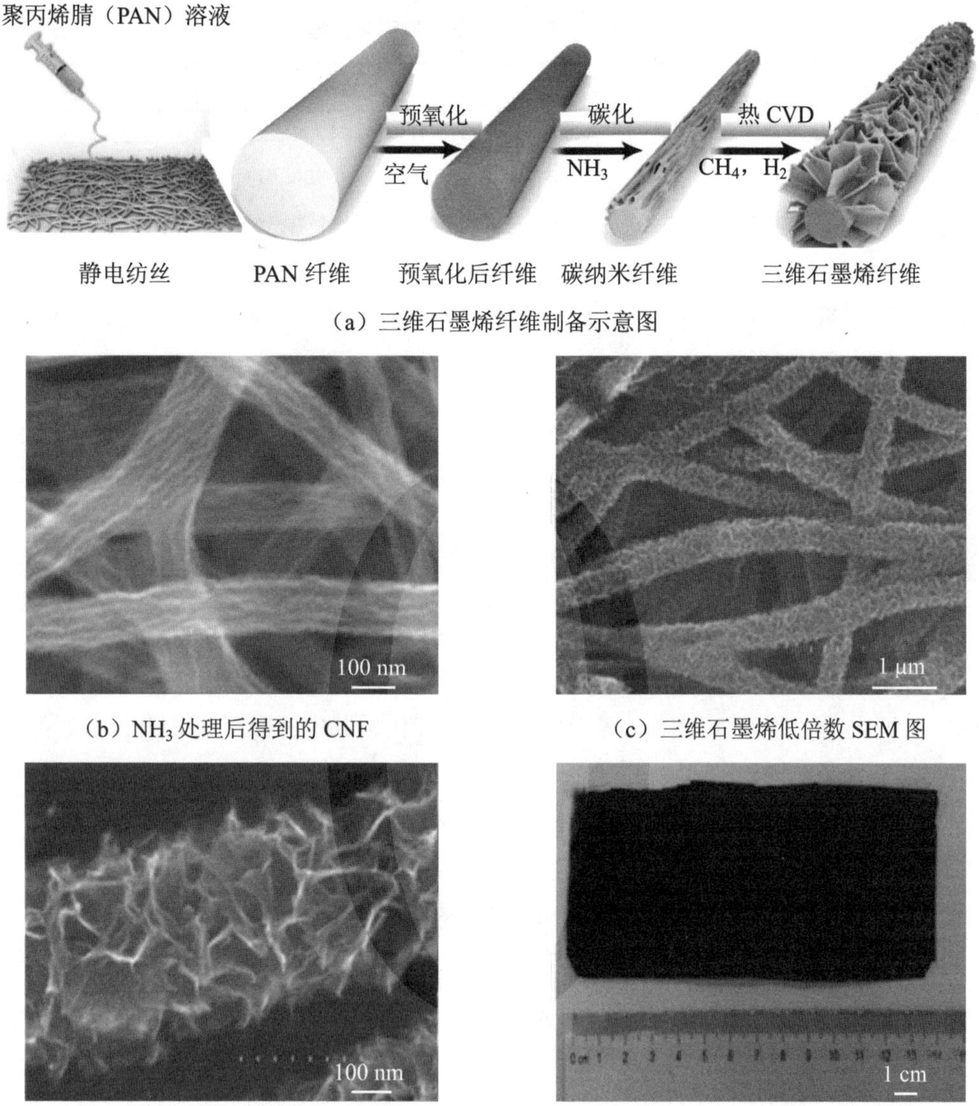

（a）三维石墨烯纤维制备示意图

（b）NH_3 处理后得到的 CNF　　（c）三维石墨烯低倍数 SEM 图

（d）三维石墨烯高倍数 SEM 图　　（e）三维石墨烯纤维薄膜光学照片

图 1.21　结合静电纺丝与热化学气相沉积法制备的三维石墨烯纤维薄膜

通过静电纺丝的方式除了能够得到以上形貌的纳米纤维材料外，包括有机多孔纳米纤维材料、单向排列纳米纤维材料和“三明治”结构纳米纤维材料等均已在众多领域展现出优异的性能，且不同结构下材料组分的多样性见表 1.3。

表 1.3　静电纺丝纳米材料组分分类

类别	实例
实心纳米纤维	富氧 CNF，石墨化碳纤维布，PAN@炭黑纤维，Ti_4O_7@CNF
空心管状纳米纤维	中空微孔 NF，pPAN/SeS_2 NFs，Cu/CuO@ZnO NF
具有核壳结构的纳米纤维	SnO_2/ZnO 中空纳米纤维，中空 SnO_2 纳米纤维
二次生长纳米纤维	气相生长碳纤维布，成卷 CNF，CNTs/GNFs，CNTs@CNFs
具有三维结构的纳米纤维	PVDF-HFP/PEDOT NFs，三维 $V_4Nb_{18}O_{55}$ NFs，WSe_2/C NFs
多层结构的纳米纤维	石墨烯纤维布，层状 $Ba_5Ta_4O_{15}$ NFs
多孔纳米纤维	多孔 CNF，类根须状多孔 CNF，Mo-W 氧化物 NFs

参考文献

[1] CHANG X L, ZU Y, ZHAO Y L. Size and structure effects in the nanotoxic response of nanomaterials[J]. Chinese Science Bulletin, 2011, 56(2): 108-118.

[2] 杜亚丹, 张璐瑶, 江鑫梅, 等. 纳米材料在催化领域中的应用研究[J]. 科教导刊-电子版（上旬）, 2018(8): 277.

[3] ONDARÇUHU T, JOACHIM C. Drawing a single nanofibre over hundreds of microns[J]. Europhysics Letters, 1998, 42(2): 215-220.

[4] 王臻, 力虎林. 模板法制备高度有序的聚苯胺纳米纤维阵列[J]. 高等学校化学学报, 2002, 23(4): 721-723.

[5] 李国炜, 林新兴, 汤祖武, 等. AAO 模板法制备纤维素纳米纤维[J]. 纤维素科学与技术, 2017, 25(2):9-14.

[6] HE C L, NIE W, FENG W. Engineering of biomimetic nanofibrous matrices for drug delivery and tissue engineering[J]. Journal of Materials Chemistry B, 2014, 2(45): 7828-7848.

[7] SONG S Y, PAN L J, LI Y, et al. Self-assembly of polyaniline: mechanism study[J]. Chinese Journal of Chemical Physics, 2008, 21(2): 187-192.

[8] FENG J T, YAN W, ZHANG L Z. Synthesis of polypyrrole micro/nanofibers via a

self-assembly process[J]. Microchim Acta, 2009, 166(3): 261-267.

[9] KAGEYAMA K, TAMAZAWA J, AIDA T. Extrusion polymerization: catalyzed synthesis of crystalline linear polyethylene nanofibers within a mesoporous silica[J]. Science, 1999, 285(5436): 2113-2115.

[10] XU T, YANG H Y, YANG D Z, et al. Polylactic acid nanofibre scaffold decorated with chitosan islandlike topography for bone tissue engineering[J]. ACS Applied Materials & Interfaces, 2017, 9(25): 21094-21104.

[11] 安林红，王跃. 纳米纤维技术的开发及应用[J]. 当代石油石化，2002，10(1): 41-45.

[12] ZHAO R J, WANG Z Z, ZOU T, et al. "Green" prepare SnO_2 nanofibers by shaddock peels: application for detection of volatile organic compound gases[J]. Journal of Materials Science, 2019, 30(3): 3032-3044.

[13] LIU X, CHEN C F, YE H W, et al. One-step hydrothermal growth of carbon nanofibers and insitu assembly of Ag nanowire@carbon nanofiber@Ag nanoparticles ternary composites for efficient photocatalytic removal of organic pollutants[J]. Carbon, 2018, 131: 213-222.

[14] SARBATLY R, KRISHNAIAH D, KAMIN Z. A review of polymer nanofibres by electrospinning and their application in oil-water separation for cleaning up marine oil spills[J]. Marine Pollution Bulletin, 2016(1/2), 106: 8-16.

[15] TEO W E. Introduction to electrospinning parameters and fiber control[M]. Singapore: World Scientific Publishing Company, 2015.

[16] BUCHKO C J, CHEN L C, SHEN Y, et al. Processing and microstructural characterization of porous biocompatible protein polymer thin films[J]. Polymer, 1999, 40(26): 7397-7407.

[17] 李欣，孟家光，门明峰. 纤维素/PVA 再生纤维静电纺丝影响因素[J]. 上海纺织科技, 2017, 45(10): 28-30.

[18] 肖婉红，曾泳春. 静电纺丝工艺参数对纤维直径影响的研究：实验及数值模拟[J]. 东华大学学报（自然科学版), 2009, 35(6): 632-638.

[19] BOSWORTH L A, DOWNES S. Acetone, a sustainable solvent for electrospinning poly(ε-caprolactone) fibres: effect of varying parameters and solution concentrations on fibre diameter[J]. Journal of Polymers and the Environment, 2012, 20(3): 879-886.

[20] 陈泉，周新，朱宇，等. 静电纺丝中用滚筒辅助动态水接收装置制备定向聚丙烯腈纳米纤维[J]. 科技创新导报，2012, 9(34): 93, 95.

[21] ZUSSMAN E, THERON A, YARIN A L. Formation of nanofiber crossbars in electrospinning[J]. Applied Physics Letters, 2003, 82(6): 973-975.

[22] LI D, WANG Y L, XIA Y N. Electrospinning of polymeric and ceramic nanofibers as uniaxially aligned arrays[J]. Nano Letters, 2003, 3(8): 1167-1171.

[23] QIU Y J, YU J, RAFIQUE J, et al. Large-scale production of aligned long boron nitride nanofibers by multijet/multicollector electrospinning[J]. Journal of Physical Chemistry C, 2009, 113(26): 11228-11234.

[24] QIU Y J, YU J, YIN J, et al. Synthesis of continuous boron nitride nanofibers by solution coating electrospun template fibers[J]. Nanotechnology, 2009, 20(34): 345603.

[25] QIU Y J, YU J, SHI T N, et al. Nitrogen-doped ultrathin carbon nanofibers derived from electrospinning: large-scale production, unique structure, and application as electrocatalysts for oxygen reduction[J]. Journal of Power Sources, 2011, 196(23): 9862-9867.

[26] ZHOU X S, QIU Y J, YU J, et al. Tungsten carbide nanofibers prepared by electrospinning with high electrocatalytic activity for oxygen reduction[J]. International Journal of Hydrogen Energy, 2011, 36(13): 7398-7404.

[27] YIN J, QIU Y J, YU J. Onion-like graphitic nanoshell structured Fe-N/C nanofibers derived from electrospinning for oxygen reduction reaction in acid media[J]. Electrochemistry Communications, 2013, 30: 1-4.

[28] QIU Y J, GENG Y H, LI N N, et al. Nonstoichiometric $LiFePO_4$/C nanofibers by electrospinning as cathode materials for lithium-ion battery[J]. Materials Chemistry and Physics, 2014, 144(3): 226-229.

[29] GUO Y, LI J, PITCHERI R, et al. Electrospun Ti_4O_7/C conductive nanofibers as interlayer for lithium-sulfur batteries with ultra long cycle life and high-rate capability[J]. Chemical Engineering Journal, 2019, 355: 390-398.

[30] LUO H J, HUANG Y, WANG D S, et al. Coaxial-electrospinning as a new method to study confined crystallization of polymer[J]. Journal of Polymer Science, Part B: Polymer Physics, 2013, 51(5): 376-383.

[31] SUN Z, ZUSSMAN E, YARIN A L, et al. Compound core-shell polymer nanofibers by co-electrospinning [J]. Advanced Materials, 2003, 15(22): 1929-1932.

[32] LI P C, SHANG Z, CUI K J, et al. Coaxial electrospinning core-shell fibers for self-healing scratch on coatings[J]. Chinese Chemical Letters, 2019, 30(1): 157-159.

[33] CHU B, HSIAO B S, FANG D F. Apparatus and methods for electrospinning polymeric fibers and membranes: US6713011[P]. 2004-03-30.

[34] TOMASZEWSKI W, SZADKOWSKI M. Investigation of electrospinning with the use of a multi-jet electrospinning head[J]. Fibres and Textiles in Eastern Europe, 2005, 13(4): 22-26.

[35] 刘延波，曹红，张立改，等. 多针头静电纺丝过程中场强的改善[J]. 成都纺织高等专科学校学报，2017, 34(2): 30-36.

[36] YARIN A L, ZUSSMAN E. Upward needleless electrospinning of multiple nanofibers [J]. Polymer, 2004, 45(9): 2977-2980.

[37] HE J H. Nano bubble dynamics in spider spinning system[J]. Journal of Animal and Veterinary Advances, 2008, 7(2): 207-209.

[38] 何吉欢，李晓霞，田丹. 气泡纺技术及其纳米纤维的工业化生产[J]. 纺织学报，2018，39(12): 175-180.

[39] 于昊楠，魏亮，杨占平，等. 无针式碟形静电纺丝喷头不同圆周倾角的模拟与试验[J]. 东华大学学报(自然科学版), 2018, 44(1): 18-27.

[40] YOU X Y, YE C C, GUO P. Electric field manipulation for deposition control in near-field electrospinning[J]. Journal of Manufacturing Processes, 2017, 30: 431-438.

[41] ZENG J, CHEN X, WANG H, et al. The direction and stability control system for

near-field electrospinning direct-writing technology[C]. //2016 IEEE International Conference on Manipulation, Manufacturing and Measurement on the Nanoscale (3M-NANO). July 18-22, 2016. Chongqing, China. IEEE, 2016: 294-297.

[42] SUN D H, CHANG C, LI S, et al. Near-field electrospinning[J]. Nano Letters, 2006, 6(4): 839-842.

[43] ZHENG G F, LI W W, WANG X, et al. Experiment and simulation of coiled nanofiber deposition behavior from near-field electrospinning[C]. //2010 IEEE 5th International Conference on Nano/Micro Engineered and Molecular Systems. January 20-23, 2010, Xiamen. IEEE, 2010: 284-288.

[44] 罗国希. 基于近场静电纺丝的聚合物微纳米结构制备及其应用关键技术研究[D]. 重庆: 重庆大学，2016.

[45] 徐方远, 贺健康, 曹毅, 等. 基于静电纺丝的微纳 3D 打印工艺研究[C]//第 16 届全国特种加工学术会议论文集(下). 厦门, 2015: 291-295.

[46] 林灿然, 王晗, 曾俊, 等. 近场直写静电纺丝电场仿真与路径规划分析[J]. 包装工程，2018, 39(15): 146-152.

[47] ZHMAYEV E, CHO D, JOO Y L. Nanofibers from gas-assisted polymer melt electrospinning[J]. Polymer, 2010, 51(18): 4140-4144.

[48] WARNER S, FOWLER A, UGBOLUE S, et al. Cost-effective nanofiber formation-mel electrospinning[C/OL]. National Textile Center Annual Report [2005-9]. http://www.ntcresearck.org/current/FY2005/F05-MD01.PDF.

[49] LIU Y, DENG R J, HAO M F, et al. Orthogonal design study on factors effecting on fibers diameter of melt electrospinning[J]. Polymer Engineering & Science, 2010, 50 (10): 2074-2078.

[50] LYONS J, LI C, KO F. Melt-electrospinning part I: processing parameters and geometric properties[J]. Polymer, 2004, 45(22): 7597-7603.

[51] ZHOU H J, GREEN T B, JOO Y L. The thermal effects on electrospinning of polylactic acid melts[J]. Polymer, 2006, 47(21): 7497-7505.

[52] RANGKUPAN R, RENEKER D H. Electrospinning process of molten polypropylene in vacuum[J]. Journal of Metals, Materials and Minerals, 2003, 12(2):

81-87.

[53] GAO S W, WANG N, LI S, et al. A multi-wall Sn/SnO_2@carbon hollow nanofiber anode material for high-rate and long-life lithium-ion batteries[J]. Angewandte Chemie International Edition, 2020, 59(6): 2465-2472.

[54] YUAN W, ZHOU N, SHI L, et al. Structural coloration of colloidal fiber by photonic band gap and resonant Mie scattering[J]. ACS Applied Materials & Interfaces, 2015, 7(25): 14064-14071.

[55] ZENG J, JI X X, MA Y H, et al. 3D graphene fibers grown by thermal chemical vapor deposition[J]. Advanced Material, 2018, 30(12): e1705380.

[56] PANDHUMAS T, PANAWONG C, LOIHA S, et al. Porous electrospun fibers as optical sensor for metal ion[J]. Chiang Mai Journal of Science, 2017, 44 (4): 1704-1713.

[57] YOO S H, JOH H I, LEE S. Synthesis of porous carbon nanofiber with bamboo-like carbon nanofiber branches by one-step carbonization process[J]. Applied Surface Science, 2017, 402: 456-462.

[58] WANG Q, YAO Q, CHANG J, et al. Enhanced thermoelectric properties of CNT/PANI composite nanofibers by highly orienting the arrangement of polymer chains[J]. Journal of Materials Chemistry, 2012, 22(34): 17612-17618.

[59] ZHENG Z, GAN L, LI H Q, et al. A fully transparent and flexible ultraviolet-visible photodetector based on controlled electrospun ZnO-CdO heterojunction nanofiber arrays[J]. Advanced Functional Materials, 2015, 25(37): 5885-5894.

[60] HUANG L, LI J J, LI Y B, et al. Lightweight and flexible hybrid film based on delicate design of electrospun nanofibers for high-performance electromagnetic interference shielding[J]. Nanoscale, 2019, 11(17): 8616-8625.

[61] LIN Y, SUN C, ZHAN S L, et al. Ultrahigh discharge efficiency and high energy density in sandwich structure $K_{0.5}Na_{0.5}NbO_3$ nanofibers/poly(vinylidene fluoride) composites[J]. Advanced Materials Interfaces, 2020, 7(9): 2000033.

[62] WU K S, HU Y, SHEN Z, et al. Highly efficient and green fabrication of a modified C nanofiber interlayer for high-performance Li-S batteries[J]. Journal of Material

Chemistry A, 2018, 6(6): 2693-2699.

[63] LEE D K, AHN C W, JEON H J. Web-structured graphitic carbon fiber felt as an interlayer for rechargeable lithium-sulfur batteries with highly improved cycling performance[J]. Journal of Power Sources, 2017, 360: 559-568.

[64] PENG Y Y, ZHANG Y Y, WANG Y H, et al. Directly coating a multifunctional interlayer on the cathode via electrospinning for advanced lithium-sulfur batteries[J]. ACS Applied Materials & Interfaces, 2017, 9(35): 29804-29811.

[65] TANG H, YAO S S, XUE S K, et al. In-situ synthesis of carbon@Ti_4O_7 non-woven fabric as a multi-functional interlayer for excellent lithium-sulfur battery[J]. Electrochimica Acta, 2018, 263: 158-167.

[66] ZHANG X Q, HE B, LI W C, et al. Hollow carbon nanofibers with dynamic adjustable pore sizes and closed ends as hosts for high-rate lithium-sulfur battery cathodes[J]. Nano Research, 2018, 11(3): 1238-1246.

[67] LI Z, ZHANG J T, LU Y, et al. A pyrolyzed polyacrylonitrile/selenium disulfide composite cathode with remarkable lithium and sodium storage performances[J]. Science Advance, 2018, 4(6): eaat1687.

[68] HWANG S-H, KIM Y K, HONG S H, et al. Cu/CuO@ZnO hollow nanofiber gas sensor: effect of hollow nanofiber structure and P-N junction on operating temperature and sensitivity[J]. Sensors, 2019, 19(14): 3151.

[69] TAN D, LEE W, KIM Y E, et al. SnO_2/ZnO composite hollow nanofiber electrocatalyst for efficient CO_2 reduction to formate[J]. ACS Sustainable Chemistry & Engineering, 2020, 8(29): 10639-10645.

[70] YI J X, ZHANG H, ZHANG Z B, et al. Hierarchical porous hollow SnO_2 nanofiber sensing electrode for high performance potentiometric H_2 sensor[J]. Sensors and Actuators B: Chemical, 2018, 268: 456-464.

[71] DENG J N, LI J, GUO J Q, et al. Vapor growth carbon fiber felt as an efficient interlayer for trapping polysulfide in lithium-sulfur battery[J]. International Journal of Electrochemical Science, 2018, 13(4): 3651-3659.

[72] THAKUR A, MANNA A, SAMIR S, et al. Polymer nanocomposite reinforced with selectively synthesized coiled carbon nanofibers[J]. Composite Interfaces, 2020, 27(2): 215-226.

[73] ZHOU Y S, JIN P, ZHOU Y T, et al. High-performance symmetric supercapacitors based on carbon nanotube/graphite nanofiber nanocomposites[J]. Scientific Reports, 2018, 8(1): 9005.

[74] KSHETRI T, THANH T D, SINGH S B, et al. Hierarchical material of carbon nanotubes grown on carbon nanofibers for high performance electrochemical capacitor[J]. Chemical Engineering Journal, 2018, 345: 39-47.

[75] KWEON O Y, LEE S J, OH J H. Wearable high-performance pressure sensors based on three-dimensional electrospun conductive nanofibers[J]. NPG Asia Materials, 2018, 10: 540-551.

[76] WU H L, FAN N R, LI J, et al. Electrospun three-dimensional $V_4Nb_{18}O_{55}$ nanofibers for advanced lithium uptake[J]. Sustainable Energy & Fuels, 2019, 3(6): 1384-1387.

[77] LI J, HAN S B, ZHANG J W, et al. Synthesis of three-dimensional free-standing WSe_2/C hybrid nanofibers as anodes for high-capacity lithium/sodium ion batteries [J]. Journal of Materials Chemistry A, 2019, 7(34): 19898-19908.

[78] LI Z, HUANG T Q, GAO W W, et al. Hydrothermally activated graphene fiber fabrics for textile electrodes of supercapacitors[J]. ACS Nano, 2017, 11(11): 11056-11065.

[79] BLOESSER A, VOEPEL P, LOEH M O, et al. Tailoring the diameter of electrospun layered perovskite nanofibers for photocatalytic water splitting[J]. Journal of Materials Chemistry A, 2018, 6(5): 1971-1978.

[80] KANG W M, FAN L L, DENG N P, et al. Sulfur-embedded porous carbon nanofiber composites for high stability lithium-sulfur batteries[J]. Chemical Engineering Journal, 2018, 333: 185-190.

[81] ZHAO X H, KIM M, LIU Y, et al. Root-like porous carbon nanofibers with high sulfur loading enabling superior areal capacity of lithium sulfur batteries[J]. Carbon, 2018, 128: 138-146.

[82] ZHANG J N, LU H, LU H B, et al. Porous bimetallic Mo-W oxide nanofibers fabricated by electrospinning with enhanced acetone sensing performances[J]. Journal of Alloys and Compounds, 2019, 779: 531-542.

第2章 纳米纤维在能源领域的应用

随着社会的快速发展，人类对能源的需求正在迅速增长，化石能源危机及化石燃料的使用带来的地球生态环境污染成为阻碍人类生活质量提高和科技进步的关键因素。目前，基于化石燃料的能源经济面临着严重的风险，能源更新的紧迫性使得寻求先进的能源储存和转换技术成为世界各地发展的重中之重。锂离子电池、超级电容器、燃料电池、染料敏化太阳能电池、有机光伏电池和锂硫电池等能量储存和转化装置被认为是缓解能源危机和环境问题最有潜力的发展方向。然而，随着生活水平的提高，能源应用范围越来越广泛，对这些设备的能量密度、功率密度、循环寿命等提出了更高的要求。

为了满足日益增长的迫切能源需求，如何降低成本，延长循环寿命，同时保持能量密度是推动储能器件发展的关键因素。静电纺丝技术具有操作简单可控、低成本效益、大规模生产的潜力，已经广泛应用于能源领域。电纺纳米纤维具有独特的一维结构、纤维的多样性、比表面积大、孔隙率高等特点，使其在能量储存和转化设备的各个元件中具有很大的应用潜力。本章将依次详细讨论静电纺丝技术在锂离子电池、超级电容器、燃料电池、染料敏化太阳能电池、有机光伏和锂硫电池等能量储存和转化装置中的应用。

2.1 锂离子电池

2.1.1 引言

近年来，锂离子电池作为便携式电子设备和车辆的电源显示出巨大的发展前景，具有高能量密度、长循环寿命、高工作电压、低自放电、无记忆效应和宽工作温度范围等优点，被认为是解决这些环境和能源问题的关键。锂离子电池是将化学能转换为电能的转换装置，反过来也可以。它的主要构成包含正极材料、负极材料、隔膜、电解液以及电池外壳。正极是锂离子源，提供充放电过程中的锂离子；负极是锂离子的受体，通过氧化还原过程进行 Li^+脱嵌；正负极材料的氧化还原过程为整个电池提供容量。相对于其他二次电池，具有无可比拟的优势。目前对锂离子电池的要求是能够在大电流密度下进行快速充放电，因此，需要在循环寿命、能量密度、功率密度、倍率性能、安全性和成本等方面进行诸多改进。

提高锂离子电池效率和循环寿命的有效方法之一是通过对电池的电极和电解质材料的纳米结构与形貌进行设计，从而提高质量和体积能量密度，并降低成本。在这方面，静电纺丝的多功能性使得它们非常适合制备正负极，电纺纤维具有更小的纳米几何形状和结构，并降低了制造成本。电纺纳米纤维具有直径可控、孔隙率高、比表面积大、孔结构相互连通等优点，在锂离子电池的应用中有利于离子传输，能够有效提高电导率，从而产生快速的传质和扩散动力学，最终能够有效提高电池的循环性能和倍率能力。本节详细介绍了电纺纳米纤维在锂离子电池中应用的最新进展。

2.1.2 锂离子电池电极

锂离子电池的电化学性能的主要取决于正极材料、负极材料和电解液。它的本质是一种嵌脱式电池，在电池充放电过程中锂离子在正负极之间可逆地嵌入与脱嵌，这一工作原理的电池被称为“摇椅式电池”。当电池充电时锂离子从正极脱嵌经过电解液和隔膜嵌入负极。当电池放电时则相反，锂离子从负极进入正极。外电路电子从负极到正极形成闭合回路，通过这种反复的形式完成能量转换。为了控制电子转移的速率，正极必须使用离子导电但电绝缘的介质（通常是液体或聚合物电解质）从物理上和电气上与负极隔离。在大多数情况下，锂离子电池是基于以下电化学反应进行工作的：

正极反应：

$$LiCoO_2 \rightleftharpoons Li_{1-x}CoO_2 + xLi^+ + xe^-$$

负极反应：

$$C + xLi^+ + xe^- \rightleftharpoons Li_xC$$

锂离子电池所采用的总反应：

$$LiCoO_2 + C \rightleftharpoons Li_{1-x}CoO_2 + Li_xC$$

要提高电池的效率，电极材料必须具有高的比容量和充电密度，产生高的电池电压，并且在两个电极上发生的电化学反应具有高度可逆性，以维持数千次充放电循环的比电荷。纳米结构材料具有许多优势，被广泛用于制备高性能锂离子电池电极。其中，纳米纤维结构被认为是电极材料有效的代替者，它不仅显著提高了锂离子的嵌入/脱嵌效率，提高了电极的倍率性能和功率密度，而且显著增强了电子在电极材料上的传输。更重要的是，纳米纤维电极的高比表面积使得电极材料和电解液间的接触面积大大提高，这增加了电池反应的活性位点，从而提高了电池的容量和倍率能力。此外，纳米纤维电极为活性材料在嵌入/脱嵌过程中的体积效应问题提供了缓冲作用，从而避免了材料结构的损坏，提高了电池的循环寿命。目前，研究者通过调整静电纺丝的物质组合和后处理，制备出了多种形态结构，如纳米纤维、纳米棒、纳米线、纳米带、中空和芯鞘双轴和三轴纤维，应用于锂离子电池电极材料，有效地解决了锂离子电池电化学过程中存在的问题。

（1）负极材料。

负极材料在锂离子电池的电化学储锂过程中起着关键作用，决定了整体电池的大部分电化学性能。因此，负极材料的研究是目前锂离子电池发展研究工作的重中之重。碳纳米纤维负极是近年来备受关注的负极材料。在众多的纳米材料中，电纺碳纳米纤维具有连续的内交联结构，可以提高电子和锂离子的导电性，用于锂离子电池无须导电添加剂，从而提高锂离子电池的整体能量密度。此外，纳米纤维还具有优良的力学性能，无须添加黏结剂就可以直接用作负极材料，简化了电极材料的制作工艺。

然而单纯的碳纳米纤维表面储锂位点有限，很难进行深度储锂，从而导致较低的比容量，为了进一步改善碳纳米纤维的电化学性能，当前主流的方法是采用杂原子掺杂和提高孔密度。研究发现，硫、硼、磷、氮等杂原子掺杂可以在碳纤维基体上产生更多的缺陷、增加储锂活性位点，从而提高锂离子插层能力；提高碳纳米纤维孔隙率能够增加活性中心的数量，促进电极材料与电解液的接触，从而提高储锂容量。Kang 等人以低成本的三聚氰胺和聚丙烯腈为前驱体，通过静电纺丝、碳化和 NH_3 处理成功制备了富氮多孔碳纳米纤维（NPCNFs）。所制备的 NPCNFs 具有相互连接的纳米纤维形态、大比表面积、发达的微孔结构和较高含量的吡啶氮掺杂。将这种材料作为锂离子电池自支撑负极，具有高容量、长循环寿命和出色的倍率性能。在 50 $mA{\cdot}g^{-1}$ 的电流密度下，具有 1 323 $mAh{\cdot}g^{-1}$ 的高可逆比容量。为了进一步提高孔隙率和改变氮元素掺杂形态，Kim 等人进一步通过使用聚丙烯腈、废聚乙烯醇缩丁醛（W-PVB）和尿素结合静电纺丝和碳化制备了多孔氮掺杂碳纳米纤维（N-CNFO）（图 2.1）。研究发现，W-PVB 的加入产生了包括各种尺寸的孔的开放通道，尿素增加了碳纤维的氮含量；通过改变尿素的加入质量比，可以改变氮元素的掺杂形态。锂离子电池负极材料具有出色的电化学性能，包括高可逆比容量（0.2 C 时为 734 $mAh{\cdot}g^{-1}$）、优异的倍率比容量（分别在 3 C 和 5 C 时为 388 $mAh{\cdot}g^{-1}$ 和 358 $mAh{\cdot}g^{-1}$）和出色的循环性能（在 1 C 下 500 个循环后为 330 $mAh{\cdot}g^{-1}$）。这些发现证实增加孔隙率和氮掺杂是增强碳纤维作为锂离子电池负极中储锂的有效策略。

此外，通过掺杂储锂活性组分来提高电极材料的容量被认为是一种有效方法。在这一方面，Yin 等通过静电纺丝过程中加入高容量单质铋制备了铋掺的碳纳米纤维，并探索了作为锂离子电池的负极工作机理。研究表明，直径约为 20 nm 的 Bi 纳米颗粒均匀分散并包覆在一维碳纳米纤维中，这种独特的嵌入式结构能有效缓解嵌锂脱锂过程中的机械应力并防止了 Bi 纳米粒子的聚集，从而达到良好的循环稳定性，在 100 $mA{\cdot}g^{-1}$ 电流密度下循环 500 次后具有 316.7 $mAh{\cdot}g^{-1}$ 的可逆比容量。他们的工作证实掺杂具有储锂活性的单质也是提高碳纳米纤维锂离子电池电化学性能的有效途径。

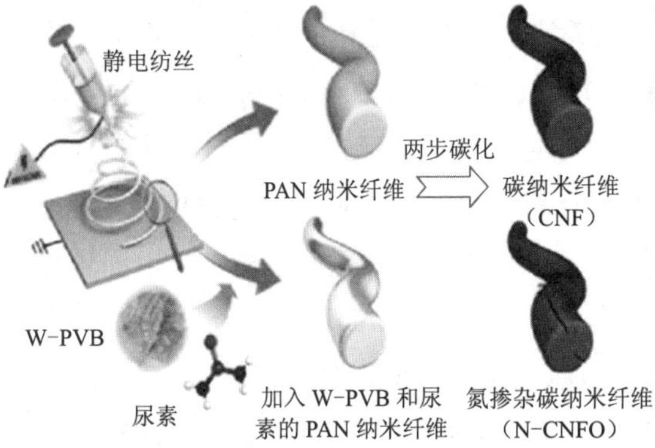

（a）CNF 和 N-CNFO 合成的示意图

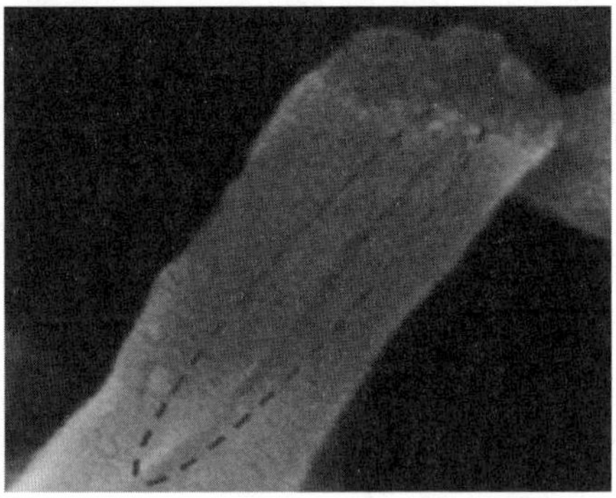

（b）高倍率的 N-CNFO 的 SEM 图像

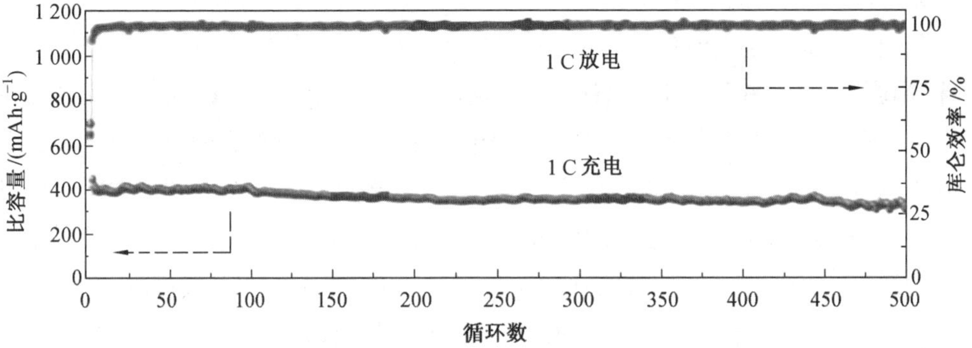

（c）N-CNFO 在 1 C 下的循环稳定性

图 2.1　N-CNFO 纳米纤维的制备、表征及其电池性能测试

尽管相当多的研究工作已经极大地改善了锂离子电池碳纳米纤维负极材料的电化学性能，但由于石墨碳有限的锂离子插入位点，理论比容量相对较低（仅为 372 $mAh\cdot g^{-1}$），并且在电化学过程中锂化电压较低，也面临着本质上的安全性问题。为了提高锂离子电池负极容量、改善负极材料的安全性，大量的非碳材料得到了发展。合金型负极材料每个金属原子可锂化储存多个锂离子，这使得合金类化合物具有最高的体积能量密度和质量能量密度；并且这些材料显示出更有利的平均脱锂电压（0.4～0.6 V），这大大增加了截止电压以避免形成锂枝晶。相比于石墨电极，这类材料具有更好的安全性。因此，这类材料被认为是大功率密度锂离子电池最具潜力的负极材料。

其中，硅（Si）负极受到人们的广泛关注，因为它是地壳中第二丰富的元素，成本低，环境友好，易于合成，有超高的理论比容量（4 200 mAh·g^{-1}）。然而，硅的本征低电导率和锂离子嵌入/脱嵌过程中（对应于 $Li_{4.4}Si$ 形式）的大体积变化（大约 400%）导致的低初始库仑效率和短暂的循环寿命，使其实际应用受到阻碍。静电纺丝技术能够将纳米活性材料进行有效包覆，从而产生一维几何结构，采用同轴静电纺丝或单喷嘴静电纺丝技术结合涂层法合成的核壳型 Si/C 纳米纤维材料具有大体积，可提供高效电子传输路径、限制电解质、在碳外形成稳定固体电解质界面膜。Si/C 复合颗粒作为核心，碳作为壳层。碳壳可以缓冲核心 Si/C 在充放电过程中产生的巨大体积膨胀和收缩，同时可以防止活性材料的机械失效引起的容量衰减。Fu 等人通过静电纺丝和化学气相沉积技术将 Si 纳米颗粒封装在碳纳米纤维的空腔中，用作柔性自支撑锂离子电池负极，由于密闭的空腔能够有效缓解 Si 的体积膨胀、抑制纳米颗粒，碳纤维有效提高了材料的导电性，制备的 Si@CNF@C 表现出优异的性能。为了增强电子传输途径，Wu 等采用静电纺丝与电沉积相结合的方法合成了还原氧化石墨烯（rGO）包覆的 Si/C 纳米纤维 rGO（Si/P−CNF@rGO)。由于 rGO 作为壳层缓解 Si 的体积变化并促进电子传输，Si/P−CNF@rGO 用作锂离子电池负极表现出优异的循环稳定性，在电流密度为 0.2 A·g^{-1} 下具有 1 851.3 mAh·g^{-1} 的高可逆比容量，在第 50 次循环时几乎没有容量衰减；即使在 4 A·g^{-1} 的大电流密度下仍具有 421.5 mAh·g^{-1} 的可逆比容量。这些工作表明自支撑的硅碳纳米纤维结构具有很高的柔韧性和良好的电化学性能，可作为锂离子电池的柔性电极。

作为硅同族，锡（Sn）基材料具有较高的初始比容量（960 mAh·g^{-1}）和安全性，被认为是锂离子电池系统中替代石墨的另一种有前途的负极。与 Si 相比，Sn 具有更高的电子传导性。然而，它们的缺点也类似，同样具有较低的锂离子反应动力学和锂嵌入/脱嵌体积变大（Sn：300%），严重损害了锂离子电池的倍率性能和循环寿命。此外，Sn 与 Li 反应过程中聚集分布不均匀导致不稳定的固体电解质界面膜（SEI 膜）的生成，造成了电池容量的衰退。为了解决这些问题，将 Sn 嵌入富碳材料的研究已经取得了一定的进展。中国科学技术大学余彦等人通过静电纺丝开发了一种竹节状中空碳纤维结构包含 Sn@C 碳纳米颗粒。中空纤维碳壳有利于调节机械应力，便于电子的传输。重要的是，碳包覆在 Sn 纳米粒子表面的碳层可以缓解 Sn 纳米粒子的团聚。材料的纳米化能够有效提高反应活性位点，从而提高活性材料的利用率。Zhang

等人通过静电纺丝将 Sn 量子点均匀分布在氮掺杂多孔碳纳米纤维上，氮掺杂多孔碳纳米纤维结构的多孔结构和高导电性不仅为电解质离子和电子提供了快速的传输途径，而且解决了电极的团聚和粉化问题。此外，由超小锡颗粒提供的电子和离子的短扩散路径进一步改善了倍率性能。除了 Si、Sn 外，其他物质（如 Ge、Sb、Al 和 Ag）也已经被静电纺丝技术制成纳米纤维用作锂离子电池负极。

过渡金属氧化物 M_xO_y（M = Mn, Fe, Co, Ni, Cu, …）为转换型负极材料，这类材料通过金属单质与氧化物之间物相转换的多电子反应提供高比容量，目前是研究者关注的热点。在电化学反应过程中，这些氧化物在储锂时，被还原成相应的金属单质同时伴随着 Li_2O 组分的生成，在脱锂化过程中可逆地氧化成相应的氧化态。然而，过渡金属氧化物的热力学限制和动力学限制，使它们通常在循环过程中具有大的体积变化、快速的容量衰减以及低的初始库仑效率。此外，过渡金属氧化物相对低的电导率、电压滞后等问题也是研究人员改善电极材料性能需要着手的地方。Chen 等报道了一种通过静电纺丝方法简单可控地制备直径为 200～300 nm 的 Co_3O_4 纳米管。所获得的中空纳米管结构和介孔壁不仅有助于快速离子传输，而且对体积变化具有良好的缓冲作用，与 Co_3O_4 纳米线负极（0.25 C/805.1 $mAh{\cdot}g^{-1}$；1 C/587.9 $mAh{\cdot}g^{-1}$）相比，Co_3O_4 纳米管具有更高的比容量和更好的循环稳定性（0.25 C/856.4 $mAh{\cdot}g^{-1}$；1 C/677.2 $mAh{\cdot}g^{-1}$）。此外，通过静电纺丝技术构建过渡金属氧化物异质结构被认为是提高反应动力学的有效方法。Wang 等人制备了一种 Co_3O_4–CeO_2 复合氧化物纳米管（CCONs）。由于 CeO_2 的协同作用提高了电极的稳定性，在电流密度为 2 $A{\cdot}g^{-1}$ 下测试，CCONs 具有 497.3 $mAh{\cdot}g^{-1}$ 的比容量，在电流密度为 1 $A{\cdot}g^{-1}$ 时可循环 1 500 次，具有良好的循环稳定性。这些研究表明，通过静电纺丝技术对电极结构和电极材料进行合理的尺寸控制有利于增强它们的倍率性能和循环稳定性。

金属负极保护是近年来的研究热点，对负极金属进行保护能够有效抑制金属枝晶的生成，减少电化学过程中副反应的发生。Ji 等采用静电纺丝结合热化学气相沉积法在碳纤维上垂直生长石墨烯片（VGSs/CFs）作为 Li/Na 金属负极的三维的稳定载体。研究结果表明，VGS 的结构对负极的电化学性能有很大的影响，紧密连接的 VGSs 为 Li/Na 吸附提供了大量的活性位点，提高了负极的导电性，增加了电流密度的均匀性，降低了循环过程中 Li/Na 成核的势垒。基于 VGSs/CFs 复合锂负极（Li/VGSs/CFs）的对称电池在 5 $mA{\cdot}cm^{-2}$ 的碳酸盐基电解液中表现出 1 200 次循环后过电位仅为 76 mV（图 2.2（a））。采用 VGSs/CFs 复合 Na 负极（Na/VGSs/CFs）的对称电池，在 10 mA·cm^{-2}下，2 000 次循环后过电位为 98 mV（图 2.2（b））。此外，在 $Li_4Ti_5O_{12}$ 或 $LiFePO_4$ 正极的全电池中，Li/VGSs/CFs 负极表现出超高的循环稳定性和倍率性能。

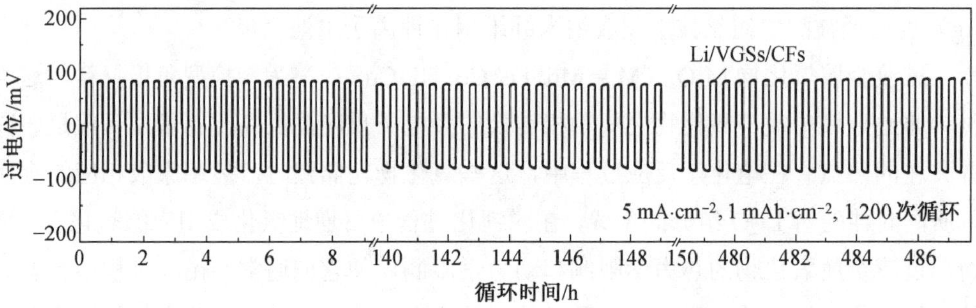

（a）Li/VGSs/CFs 对称电池在 5 mA·cm^{-2} 下的循环性能

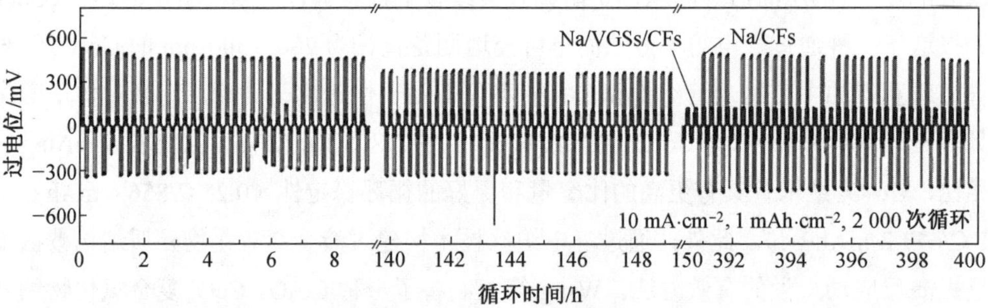

（b）NA/CFs 和 Na/VGSs/CFs 对称电池在 10 mA·cm^{-2} 下的循环性能

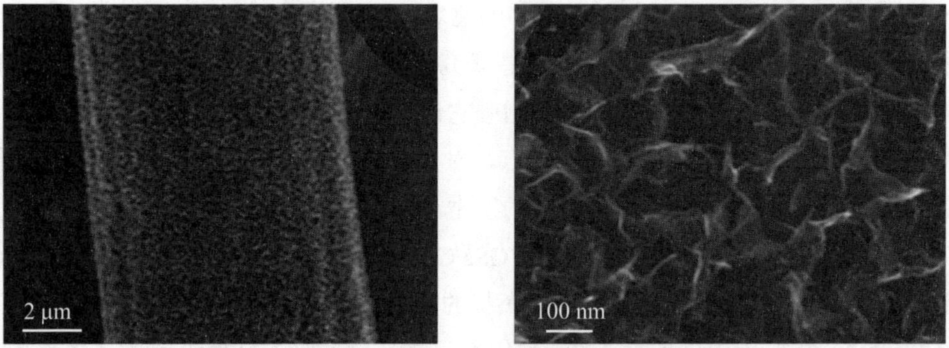

（c）不同放大倍数下生长了 10 h 的 VGSs/CFs 的表面形貌 SEM 图像

图 2.2　VGSs/CFs 基复合材料的电池循环性能和 SEM 图像

（2）正极材料。

正极材料决定着整体电池的能量密度，设计稳定的电极结构以提高正极材料的比容量是改善电池能量密度的重要方向。锂钴氧化物作为锂离子电池正极材料，具有制备简单、循环寿命长等特点，是锂离子电池工业化道路上最早使用的正极材料。当前，大多数商用锂离子电池使用 $LiCoO_2$ 作为正极材料，因为它在比能量密度和良好的循环寿命方面具有更大优势。通过静电纺丝制备了纳米纤维状 $LiCoO_2$ 正极材料，由于锂离子的扩散距离较短，可以获得快速的固态扩散速率，具有大表面积和小孔隙的电纺 $LiCoO_2$ 纤维正极的初始放电比容量为 182 $mAh{\cdot}g^{-1}$，而传统粉末和薄膜电极的初始放电比容量为 140 $mAh{\cdot}g^{-1}$。然而，由于钴离子和锂离子在电解液中形成 Li_2CO_3 和 CoF_2 等杂副反应，这种正极材料在充放电过程中也遭受容量损失。为了提高正极的稳定性，$LiFePO_4$ 被引入正极材料的研究中，它具有比容量大（170 $mAh{\cdot}g^{-1}$）、热稳定性好、放电电位高、成本低、安全等优点。

然而，$LiFePO_4$ 的电导率、锂扩散率和振实密度较低，这在电池的应用中会产生高阻抗，从而导致低倍率性能。为了克服这些缺点，传统的改性方法是在 $LiFePO_4$ 表面包覆碳等导电材料。Qiu 等通过静电纺丝和煅烧法合成了非化学计量磷酸铁锂/碳（$LiFePO_4$/C）复合纳米纤维，通过调控原料配比优化形貌和电化学性能。研究发现，原料 LiH_2PO_4 与 $FeC_6H_5O_7$ 的摩尔比为 1.3 的样品具有良好的纤维形态、多孔结构和高纯度，具有较高的锂离子电池容量和稳定性，如图 2.3（a）、图 2.3（b）所示。此外，他们进一步采用水系纺丝体系，通过静电纺丝和热处理方法将 Fe_2P 引入 $LiFePO_4$/C 纳米纤维结构中，成功地合成了磷酸铁锂/（碳和磷化亚铁）（$LiFePO_4$/（C+Fe_2P））复合纳米纤维，并探究了温度对 Fe_2P 相形成的影响。研究发现，煅烧温度在 750 ℃和 800 ℃的 $LiFePO_4$/（C+Fe_2P）复合纳米纤维显示较好的微观形态；电化学测试表明，良好的微观形貌能有效促进电化学性能的提升，Fe_2P 相的引入增强了纤维的导电性并提高了复合材料的容量，如图 2.3（c）、图 2.3（d）所示。

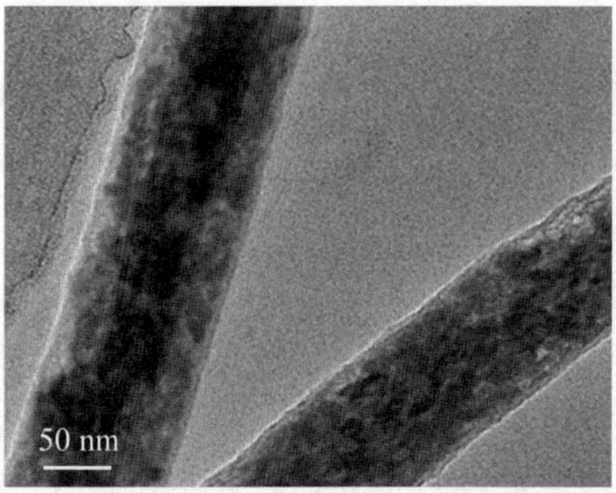

（a）$LiFePO_4/C$ 复合纳米纤维的 TEM 图

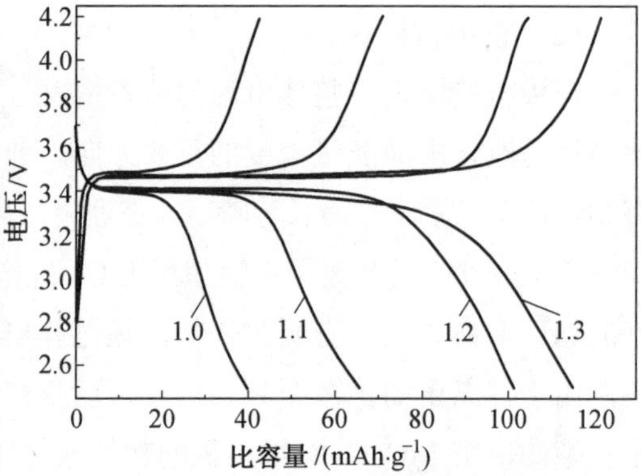

（b）$LiFePO_4/C$ 复合纳米纤维的充放电曲线图

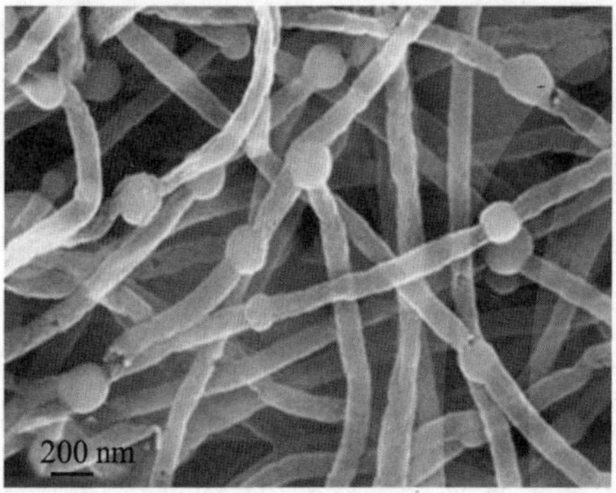

（c）$LiFePO_4/(C+Fe_2P)$ 复合纳米纤维 TEM 图

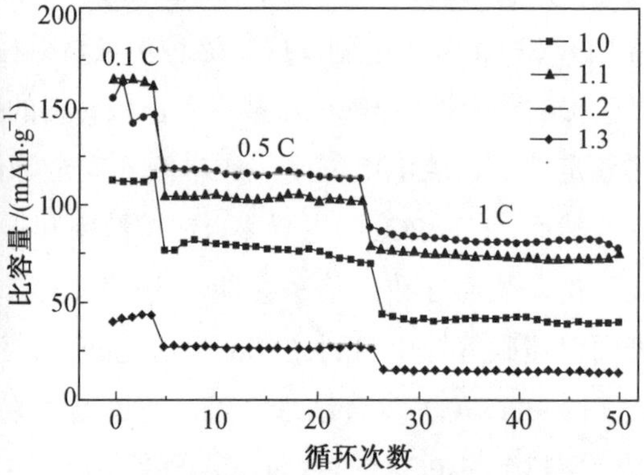

（d）$LiFePO_4/(C+Fe_2P)$复合纳米纤维倍率图

图 2.3　$LiFePO_4$ 基复合纳米纤维的透射电子显微镜（TEM）图及其电化学性能

静电纺丝技术除了用于制备 $LiFePO_4$ 纳米纤维复合材料外，还广泛应用于调控其他金属元素掺杂对 $LiFePO_4$ 形貌和性能的影响。例如，Kang 等人采用静电纺丝法制备不同锰含量掺杂的 $Li[Fe_{1-x}Mn_x]PO_4$（x=0，0.1，0.3）前驱体，然后在 800 ℃下煅烧，得到 $Li[Fe_{1-x}Mn_x]PO_4$ 纳米纤维。物理表征表明，掺杂锰含量的增加会导致 $Li[Fe_{1-x}Mn_x]PO_4$纳米纤维的形貌变得更加粗糙。结合电化学性能测试表明，由于 Mn^{2+} 离子半径比 Fe^{2+}离子半径大，Mn/Fe 含量比的提高会扩大内部空间，加速锂离子的扩散，因此促进了反应动力学。Leng 等人通过静电纺丝技术制备了 $Li_xFe_{0.2}Mn_{0.8}PO_4/C$ 纳米纤维，探究了第二碳源的种类（聚乙二醇（PEG）和柠檬酸）和热处理工艺（包括 5 ℃·min^{-1} 到 2 ℃·min^{-1} 的升温速率和热处理气氛）对纳米纤维复合材料形貌和性能的影响。将所制备的 $Li_xFe_{0.2}Mn_{0.8}PO_4$ 复合材料作为锂离子电池正极，在 0.05 C 电流下具有 174 mAh·g^{-1} 的高可逆比容量，并且表现出良好的循环稳定性；交流阻抗测试表明，随着锂的加入，$Li_xFe_{0.2}Mn_{0.8}PO_4$的电荷转移阻抗减小。此外，静电纺丝技术在开发新型正极材料和材料改性方面也得到了应用（如：$LiV_3O_8/LiV_{2.94}Nb_{0.06}O_8$等），其结构和性能如图 2.4 所示，经过静电纺丝技术合成的材料电化学性能都相应提高。

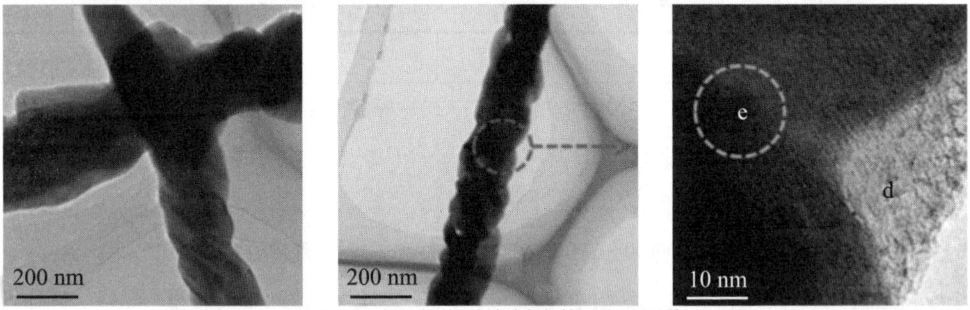

（a）LiV_3O_8复合纳米纤维的 TEM 图像

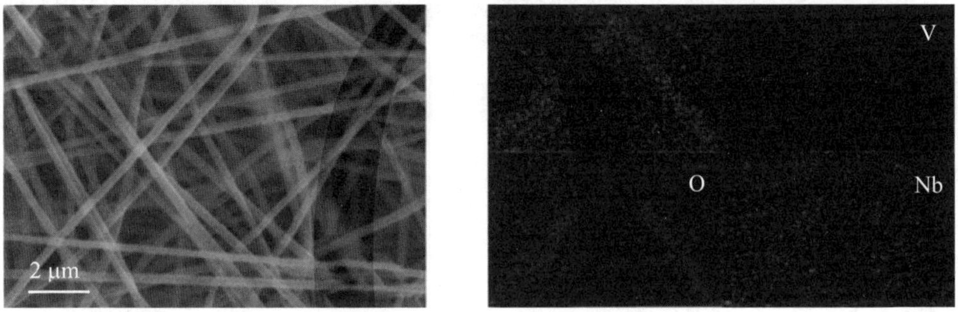

（b）$LiV_{2.94}Nb_{0.06}O_8$的 SEM 图及能谱

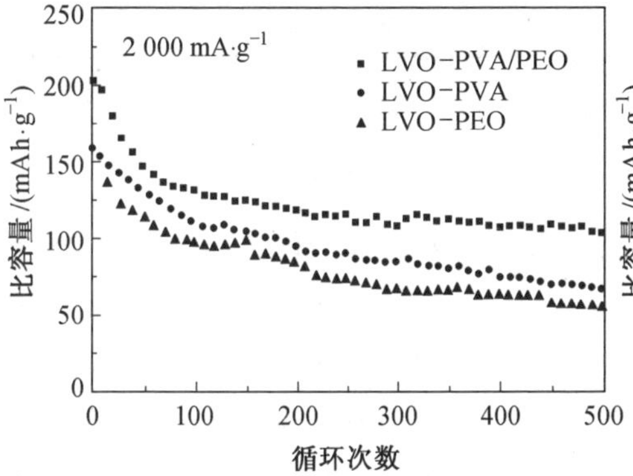

（c）LiV_3O_8复合纳米纤维的循环性能图

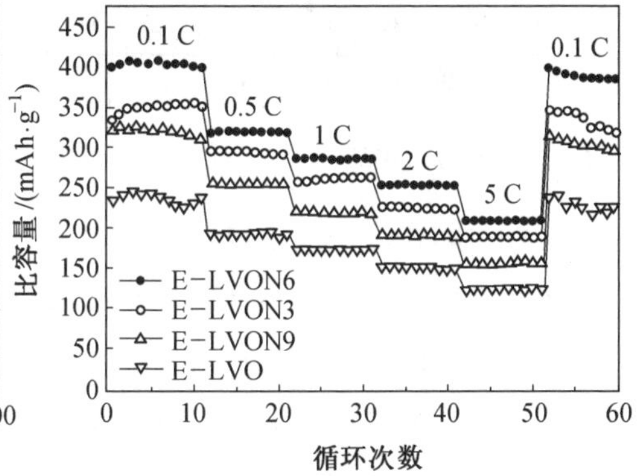

（d）$LiV_{2.94}Nb_{0.06}O_8$的倍率性能图

图 2.4　LiV_3O_8、$LiV_{3-y}Nb_yO_8$及其复合纳米纤维的 TEM 图像及其电化学性能

d—碳材料；c—LiV_3O_8纳米颗粒；LVO—LiV_3O_8；PEO—聚氧乙烯；
E-LVO—$LiV_{3-y}Nb_yO_8$（y=0）；E-LVON3—$LiV_{3-y}Nb_yO_8$（y=0.03）；
E-LVON6—$LiV_{3-y}Nb_yO_8$（y=0.06）；E-LVON9—$LiV_{3-y}Nb_yO_8$（y=0.09）；

静电纺丝技术在锂离子电池正负极材料的合成中应用非常广泛，表 2.1 展示了一些常见的电极材料，包括材料的尺寸及优势。

表 2.1 常见的锂离子电池静电纺丝电极材料

应用	材料	微观结构	优势	参考文献
负极材料	α-Fe_2O_3	200 nm 纤维	良好的循环稳定性和倍率性能	[59]
	CuO	纳米纤维	高长宽比	[60]
	$Li_4Ti_5O_{12}$	250 nm 纤维	大放电比容量	[61]
	MnO/CNFs	100～200 nm 纤维	改善接触面积	[62]
	$NiFe_2O_4$	纳米纤维	有效的离子传输途径	[63]
	NiO	100 nm 纤维	良好的循环性能	[64]
	$ZnCo_2O_4$	200～300 nm 纤维管	卓越的储锂性能	[65]
	CuO_x-Co_3O_4	多孔纤维负载氧化物	高比容量和长循环稳定性	[8]
	SnO_2-MoS_2	500 nm 纤维	快速的反应动力学	[12]
	Li_2MnSiO_4/C	70～150 nm 纤维	高比容量和高稳定性	[66]
	$Li_2Mn_{0.8}Fe_{0.2}SiO_4$/C	76～158 nm 纤维	高导电性和纯度	[67]
正极材料	$Na_3V_2(PO_4)_3$/C	200～300 nm 纤维	低成本	[68]
	V_2O_5	300～800 nm 纤维	高比容量和容量保持率	[69]
	$LiV_{2.94}Nb_{0.06}O_8$	纳米棒	高比容量	[58]

2.1.3 小结和展望

开发新的功能性纳米材料是实现锂离子电池长循环寿命和促进其进一步大规模应用的关键途径之一。一维纳米结构材料在应对这些挑战方面做出了重大贡献，因为纳米材料的结构控制与材料本身的组成同样重要。静电纺丝纳米纤维材料在能源领域具有很高的应用潜力，在电纺材料大规模应用于能源相关应用之前，仍有一些挑战有待克服，这部分是由于影响电纺技术的工艺参数尚未被完全了解。因此，静电纺丝法合成的材料种类仍有限制，有时很难完全消除珠子等缺陷，获得完全均匀的纳米纤维。此外，尽管已经在排列和桥接相邻纤维方面取得了进展，但还必须找到进一步的解决方案来缓和无机电纺纳米纤维网络的脆弱性，特别是锂离子电池电极的脆弱性。因此，在保持纤维完整性和组成的同时去除载体聚合物仍然是一个挑战。尽管存在这些缺点，但静电纺丝已被证明能够通过同轴静电纺丝制备具有更复杂结构（如核/壳和空心纳米纤维）的、均匀性优异的纳米纤维电极，以及通过共静电纺丝制备新型复合材料。成功研发出性能优异、功能齐全的锂离子电池用纳米纤维电极，将是可持续“绿色”能源走向世界所面临的挑战与机遇。

2.2　锂硫电池

2.2.1　引言

电能的储存和转换是未来低碳社会日益关注的问题之一，在过去几十年中，锂离子电池的发展极大地提高了人们的生活质量，然而当前商业化正极材料较低的理论容量限制了锂离子电池的能量密度，阻碍了锂离子电池的进一步发展。因此，需要合理设计具有更高能量密度的正极来升级电池系统。锂硫电池应运而生，与传统锂离子电池相比，锂硫电池具有无可比拟的优势（表 2.2）。硫（S：1 675 $mAh{\cdot}g^{-1}$）能够提供比传统 $LiCoO_2$ 正极高 10 倍的比容量，被认为是最有希望的正极材料，锂硫电池被认为是未来便携式电子产品和电动汽车中能源储存的有力竞争者。遗憾的是，由于多硫化锂 Li_2S_x（$4<x<8$）穿梭效应引起的容量降低和库仑效率低，缓慢的反应动力学，单质硫较差的电导率和大的体积膨胀阻碍了锂硫电池的商业应用。此外，单质硫继续被还原成短链多硫化物，最终生成绝缘且不可溶的 Li_2S_2/Li_2S，不溶的 Li_2S_2/Li_2S 沉积在金属锂电极表面，严重影响电池的库仑效率和循环寿命。目前，锂硫电池的研究工作主要从三个方面着手：提高电极材料的导电性、抑制可溶性多硫化物的穿梭效应和对锂电极进行保护。由于其卓越的纳米结构，电纺纳米纤维材料在锂硫电池的应用中显示出一些独特的特性，可以同时解决这些问题。迄今为止，电纺纤维在锂硫电池中的应用包括隔膜、中间层和正极载体。

表 2.2　锂离子电池与锂硫电池对比

电池种类	锂离子电池	锂硫电池
反应机理	离子脱嵌机理	电化学过程 S—S 键的断裂/生成
能量密度/（$Wh\cdot kg^{-1}$）	150～200	2 600
循环寿命/圈	500～2 000	300～500
负极材料	石墨（372 $mAh\cdot g^{-1}$）/硅（4 200 $mAh\cdot g^{-1}$）	金属锂（3 860 $mAh\cdot g^{-1}$）
正极材料	钴酸锂（150 $mAh\cdot g^{-1}$）/磷酸铁锂（170 $mAh\cdot g^{-1}$）	硫（1 675 $mAh\cdot g^{-1}$）
工作电压/V	3.7	2.3
电池特点	成本较高，工作电压高，工作温度范围宽，循环寿命长，安全性较低	低成本，高比能，资源丰富，环境友好，安全性较高
限制其发展的因素	石墨负极的低理论容量；硅基负极的大体积膨胀，低导电性；正极材料的低理论容量；正极材料资源短缺	单质硫的绝缘性，体积膨胀，多硫化物的穿梭效应；放电产物可逆性差，活性物质利用率低；锂负极枝晶生长问题

2. 2. 2　锂硫电池隔膜

隔膜是一种多孔膜，在作为电子绝缘的同时为离子提供传输通道，是锂硫电池的关键元件。锂硫电池的理想隔膜不仅可以在吸收电解液后获得出色的离子传导性，而且还能限制循环过程中多硫化物的穿梭效应。静电纺丝纳米膜具有突出的比表面积和较大的孔隙率，在吸收电解液和离子传输方面具有无与伦比的优势，作为锂硫电池的潜在隔膜受到了广泛的关注。此外，在纤维中引入各种有机官能团或无机介质能够有效抑制多硫化物的穿梭效应。Luo 等人通过一步静电纺丝技术合成了一种具有双重功能的羧基官能聚酰胺酸（PAA）纳米纤维隔膜，这种隔膜能够抑制多硫化物转移并促进 Li^{+}迁移。PAA 纳米纤维隔膜合成示意图、SEM 图与电化学性能图如图 2.5 所示，这种 PAA 隔膜应用在锂硫电池中，在 0.2 C 下循环 200 圈后，每个循环仅有 0.12％的超低衰减率，仅为商用 Celgard 隔膜的一半（每个循环 0.26％）（图 2.5（c））。研究表明 PAA 隔膜中的—COOH 官能团提供了负电环境，该环境不仅可有效促进 Li^{+}的迁移，同时可抑制负多硫化物阴离子的迁移。因此，PAA 纳米纤维隔膜可以充当有效的静电屏蔽，在正极侧限制多硫化物的溶解，同时有效地促进跨隔膜的 Li^+转移。

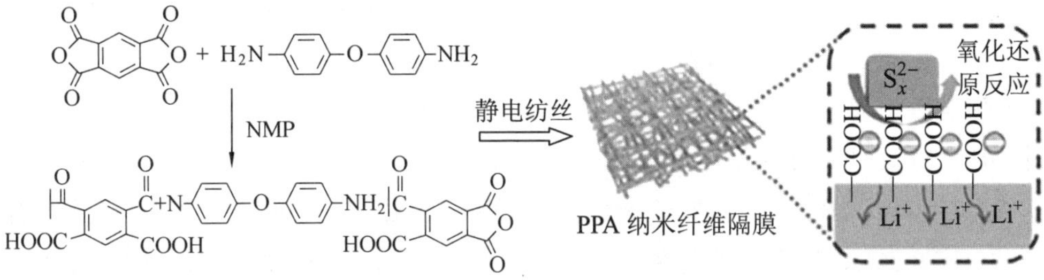

（a）PAA 纳米纤维隔膜合成示意图

（b）PAA 纳米纤维隔膜的 SEM 图

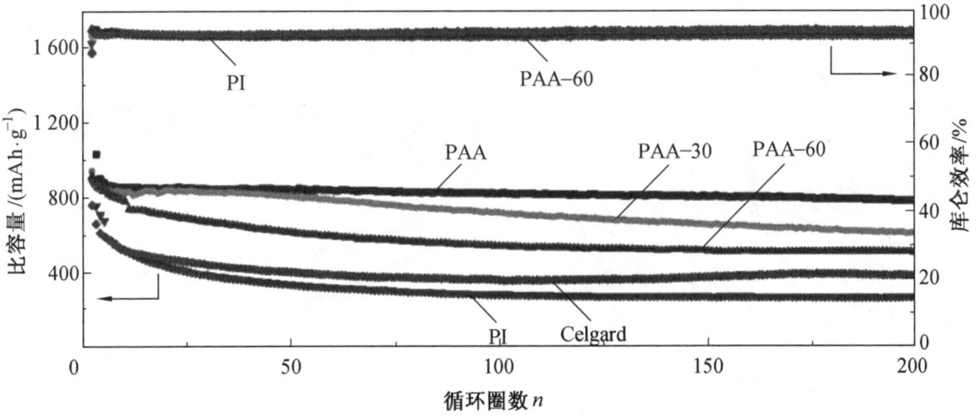

（c）电化学性能图

图 2.5　PAA 纳米纤维隔膜合成示意图、SEM 图与电化学性能图

有机和无机共掺杂隔膜被认为能够有效抑制多硫化物的穿梭效应并降低界面阻抗。因此，许多研究人员将有机物和无机物引入电纺纳米纤维膜中，以改善 Li–S 电池的循环性能。采用简单易行的方法将电纺聚偏二氟乙烯（PVDF）纳米纤维（NFs）用作成膜骨架以及电解质润湿的促进剂，并将聚（4–苯乙烯磺酸）锂盐（PSSLi）填充到空隙中，通过加热和加压来制备复合膜以获得离子交换膜（图 2.6（a））。采用 PVDF/PSSLi 为隔膜的 Li–S 电池在 0.2 C 下的首次放电比容量为 1 194 $mAh·g^{-1}$，平均库仑效率为 97%；在 0.5 C 下表现出优异的循环稳定性，200 次循环中相应的容量衰减仅为 0.26%，循环过程中库仑效率高达 97%以上，优于 PP 隔膜的 0.43%容量衰减率（图 2.6（c））。在纤维之间引入磺酸基官能团的复合膜与工业聚丙烯隔膜相比，能有效地抑制锂硫电池中多硫化物的穿梭效应，从而达到更高的放电容量和更好的循环寿命。此外，PSSLi 在分子链之间形成了一些键并增加了薄膜的结晶度，从而促进了 Li^+的运输并阻止了多硫化离子的穿梭。

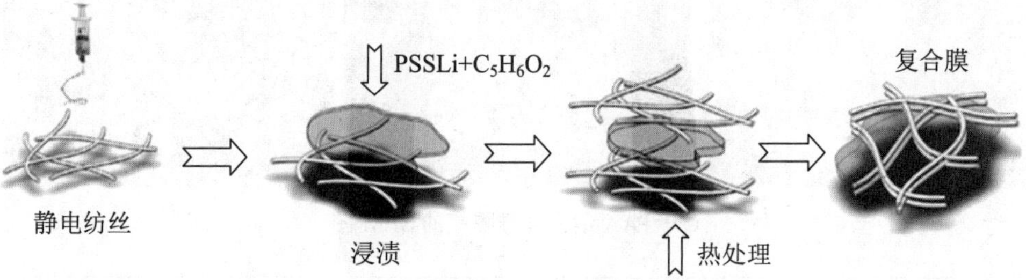

（a）PVDF/PSSLi 隔膜制造示意图

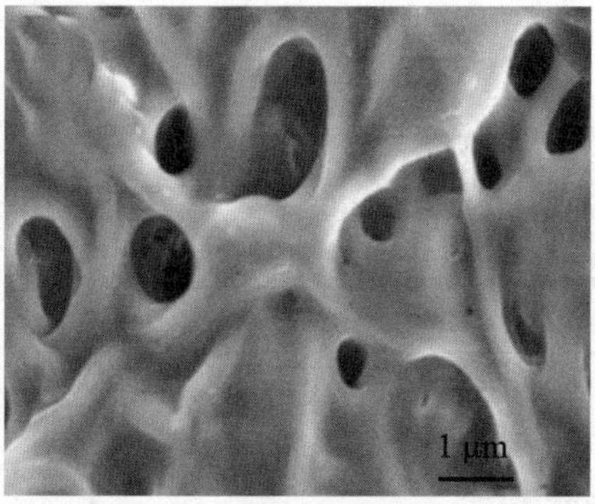

（b）PVDF/PSSLi 隔膜的 SEM 图

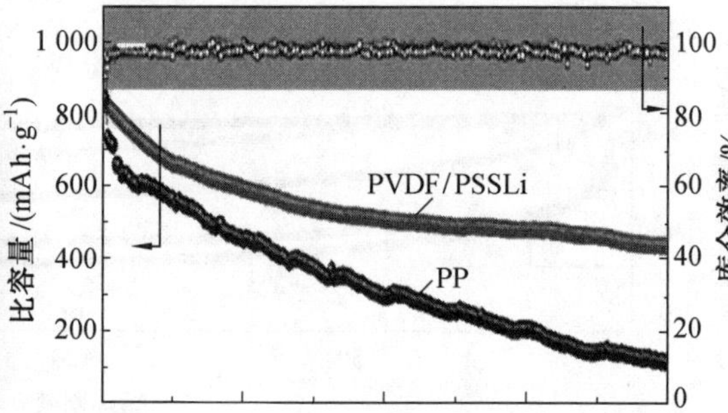

（c）电化学性能图

图 2.6　PVDF/PSSLi 隔膜制造示意图、SEM 图与电化学性能图

2.2.3　锂硫电池中间层

可溶性多硫化物（LiPSs）及其不溶性还原产物（Li_2S/Li_2S_2）沉积在锂负极表面是影响锂硫电池实际应用的主要问题。研究发现，在正极和隔膜之间引入高电导性和高比表面积的电纺纳米纤维中间层能够极大地改善锂硫电池的电化学性能。电纺纤维不仅可以提高集成电极的电导率，同时能够有效抑制电化学过程中正极多硫化物的转移。

通过原位生长开发了一种 Co 基金属有机骨架（ZIF-67）富集碳纳米纤维（CNF）复合夹层（ZIF/CNFs）。值得注意的是，ZIF/CNFs 中间层具有物理阻隔和化学捕集能力，能够有效抑制多硫化物的溶解，缓解穿梭效应。此外，三维网络结构在每个 ZIF 微反应器之间提供了互连的导电框架，以促进循环过程中的快速电子转移，从而实现出色的循环性能和倍率性能。测试结果表明，具有 ZIF/CNFs 中间层的锂硫电池在 1 C 时初始比容量高达 1 334 $mAh·g^{-1}$，并能够循环长达 300 圈（图 2.7（a）、图 2.7（b））。然而这样的中间层在促进锂硫电池反应动力学方面仍有上升空间，Guo 等采用一步法电纺丝法制备了一种柔性自支撑 Ti_4O_7/C 纳米纤维（TCNFs）中间层，微观形貌如图 2.7（c）所示。TCNFs 中层间作为物理屏障和化学吸附剂，其中比表面积大、电导率高的三维 CNFs 网络有助于 LiPSs 的转化和电子转移，而“亲硫”Ti_4O_7 可与多硫化锂产生较强化学键合，这有助于 LiPSs 的吸附和转化并达到更高的库仑效率（图 2.7（d））。在锂硫电池性能测试中，具有 TCNFs 中间层的 CMK3/S 正极比有 CNFs 中间层和无中间层的 CMK3/S 具有更好的电化学性能，在 1 C 的电流密度条件下，1 000 次循环后的比容量为 560 $mAh·g^{-1}$，在 3 C 的电流密度条件下，1 000 次循环后的比容量为 466 $mAh·g^{-1}$（图 2.7（e））。这种简单、经济、可扩展的制造方法形成的可控的独立中间层，为锂硫电池的材料结构设计和性能改进提供了新思路。

研究发现，杂原子掺杂的中间层对抑制锂硫电池的穿梭效应有更好的效果。采用简单的静电纺丝结合后续的热处理，制备了由硼氮共掺杂碳纳米纤维（BNCNF）构成的轻薄中间层，合成的中间层表面结构如图 2.7（f）所示。通过在正极和隔膜之间引入 BNCNF 薄膜，有效减小电荷转移阻抗，同时极大地抑制多硫化物的穿梭效应。通过探究发现 BNCNF 在电化学过程中与多硫化物形成 B-S 和 N-Li 化学相互作用，从而对多硫化物具有良好的吸附能力，尤其是碳纳米纤维中 N═B/N—B 结构的形成，强化了 B 原子与S_x^{2-}和 N 原子与Li^+的电性作用。BNCNF 中间层的加入使得锂硫电池的比容量和循环寿命显著提高，在 1 C 下的初始比容量为 1 054.7 mAh·g^{-1}，1 000 次循环的容量衰减率为 0.058%。即使在 5 C 的高电流密度下，初始比容量高达 612.2 mAh·g^{-1}，它在 600 次循环中的衰减率为 0.085%（图 2.7（g）、图 2.7（h））。BNCNF 中间层为提高锂硫电池的性能开辟了一条新途径，同时也为锂硫电池杂原子掺杂中间层的设计和研究提供了借鉴意义。

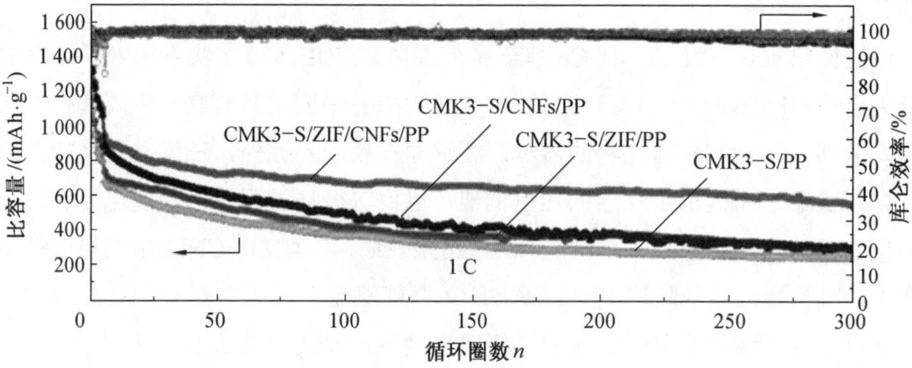

（a）ZIF/CNFs 中间层的电化学性能图

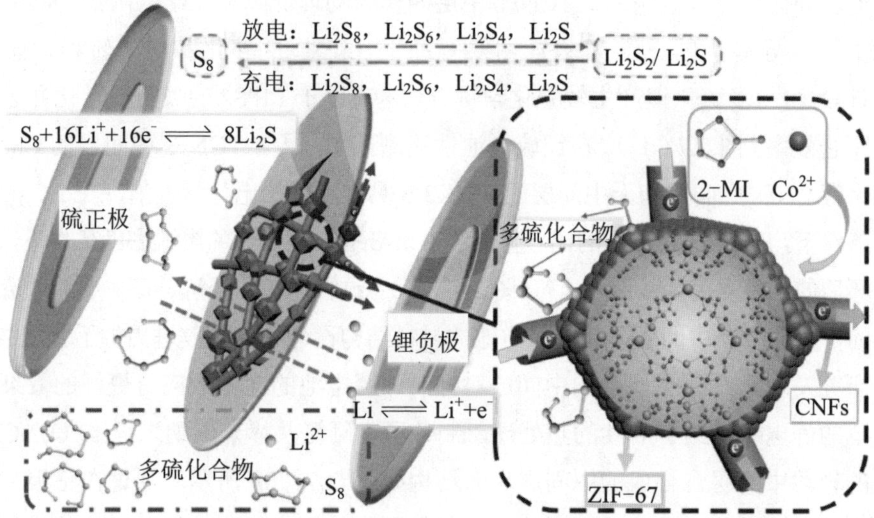

（b）ZIF/CNFs 中间层的作用机理图

图 2.7　ZIF/CNFs、TCNFs 和 BNCNF 中间层的电化学性能及其作用机理图

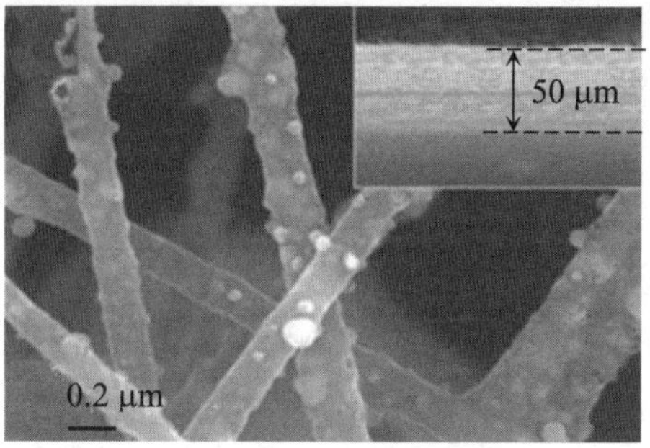

（c）TCNFs 的 SEM 图

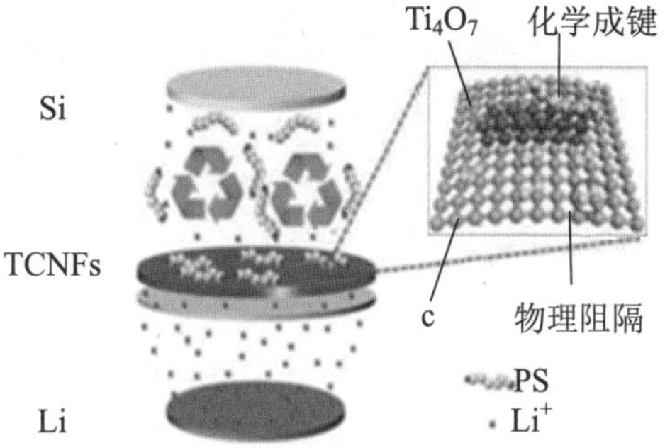

（d）TCNFs 的作用机理图

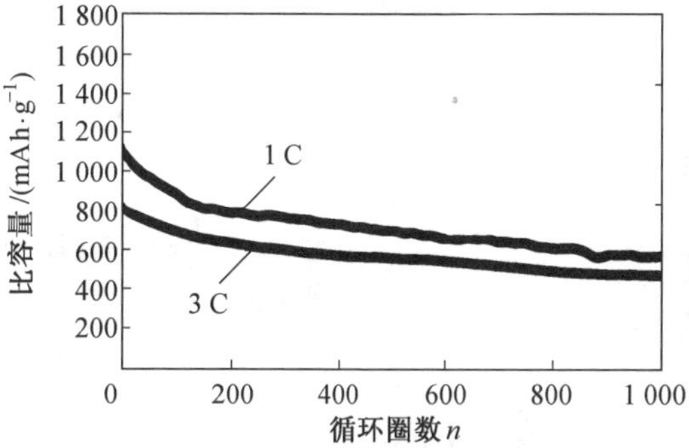

（e）TCNFs 的电化学性能图

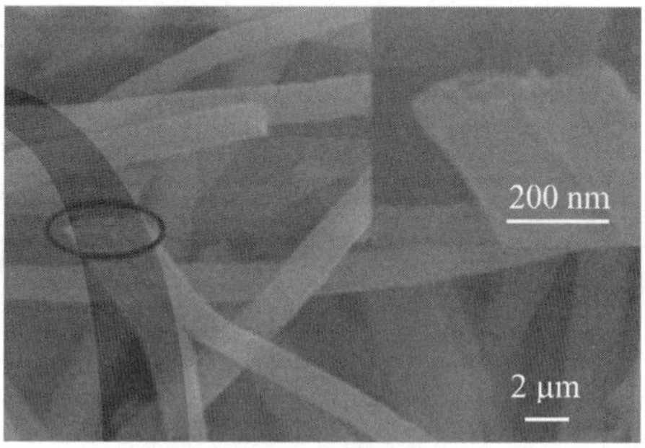

（f）BNCNF 的 SEM 图

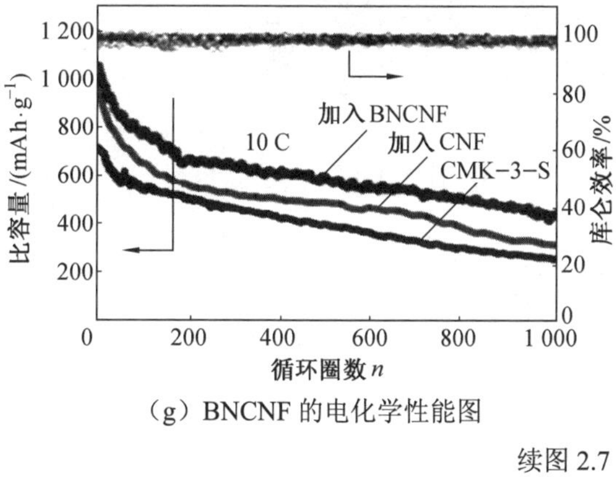

（g）BNCNF 的电化学性能图

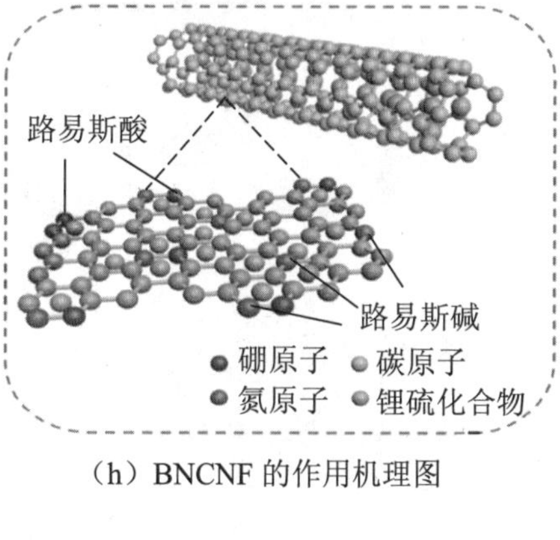

（h）BNCNF 的作用机理图

续图 2.7

多功能的中间夹层能够满足锂硫电池不同的功能需求，Li 等通过交替采用静电纺丝和喷涂技术，再经原位高温碳化还原进一步制备了具有良好的柔韧性的 Ti_4O_7-碳纳米纤维/Ti_4O_7/Ti_4O_7-碳纳米纤维（TC/T/TC）功能性夹层，材料的合成过程、微观作用机理和形貌如图 2.8（a）～（c）所示。由于中间层含有致密的极性高导电 Ti_4O_7，具有足够的物理化学位点参与 LiPSs 的捕获和转化，因此中间层对 LiPSs 具有良好的物理阻隔和化学捕集效果。此外，Ti_4O_7–纳米碳纤维的多孔表面层可以提供足够的空间来抑制体积膨胀和穿梭效应。高导电性的中间层还可作为前端集流体，以降低电阻和加速氧化还原过程。基于这些优势，把 TC/T/TC 作为锂硫电池中间层，CMK3/S 作为正极（硫面密度 2 mg·cm^{-2}），展示出良好的循环稳定性和倍率性能，在 1 C 的倍率下的初始比容量为 1 129 mAh·g^{-1}，循环 500 次后比容量保持在 686 mAh·g^{-1}；即使在 3 C 大倍率下，其初始比容量达到 738 mAh·g^{-1}，在 3 500 次超长循环后，比容量仍保持在 481 mAh·g^{-1}，高容量保留率为 65.2%，容量衰减率仅为 0.009 95%（图 2.8（d））。良好的电化学性能表明这种新型中间层在锂硫电池中具有广阔的应用前景。

（a）TC/T/TC 中间层的合成示意图

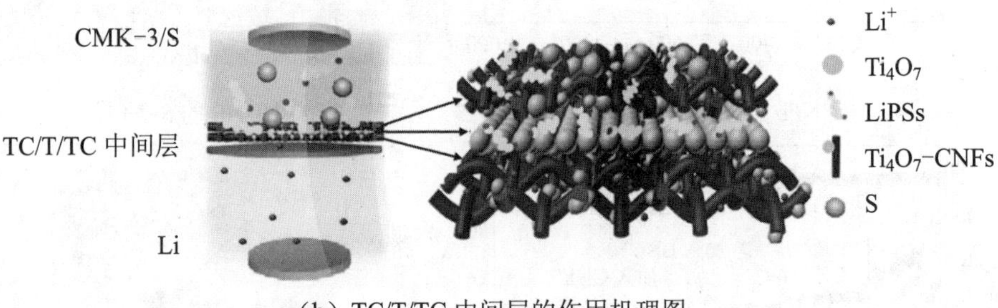

（b）TC/T/TC 中间层的作用机理图

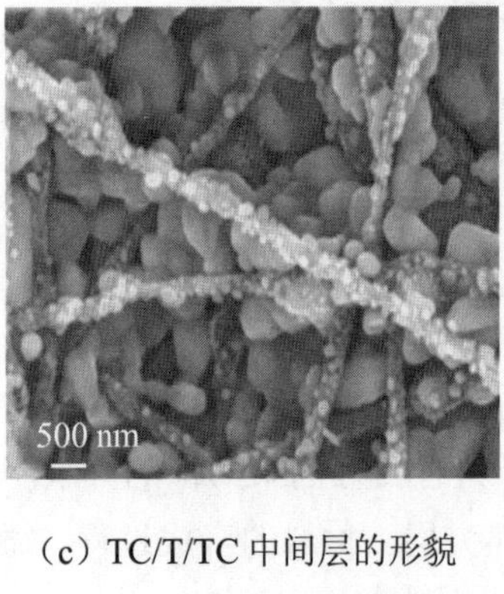

（c）TC/T/TC 中间层的形貌

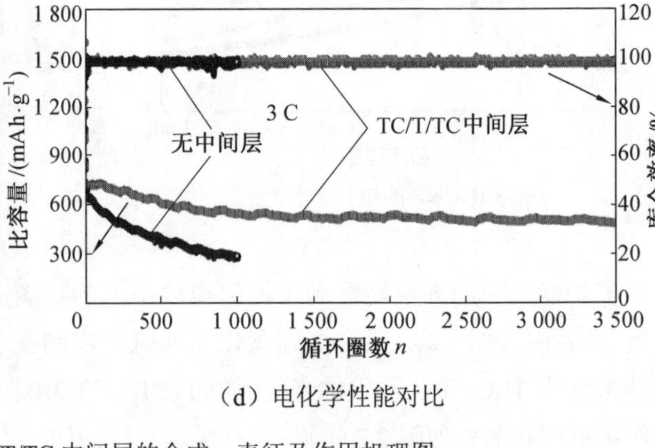

（d）电化学性能对比

图 2.8　TC/T/TC 中间层的合成、表征及作用机理图

2.2.4　锂硫电池正极载体

载硫材料的设计已经被证明是限制锂硫电池活性硫中间体、提高硫正极容量和提高循环稳定性的有效方法。理想的正极硫载体应该具有以下特性：①具有出色的导电性；②具有高孔隙率，能够负载大量活性物质；③稳定的物理和机械架构。静电纺丝技术是制造具有可调多孔结构和出色电导率的纳米纤维简单工艺。静电纺丝材料具有比表面积大、孔隙结构可控、化学性能优异等特点，已成为锂硫电池正极材料的最新研究热点。纳米纤维中的微孔和中孔可以为活性硫提供足够的反应位点，同时可吸附多硫化物，从而抑制了电化学反应中多硫化物的穿梭效应，提高活性材料的利用率；此外，还可以提供足够的空间缓解体积膨胀。多孔纳米纤维具有优异的物理和机械性能，稳定的结构以及优异的导电性，可以增强锂离子和电子传输。表 2.3 对比了目前已经报道的电纺碳基底材料的电化学性能。

表 2.3　已经报道的电纺碳基底材料的电化学性能

材料	前驱体	造孔剂	电化学性能
PCNFs	PAN	$Fe(acac)_3$	765 mAh·g^{-1} /200 圈/1 C
PCNFs	PAN	$Fe(acac)_3/HNO_3$	720 mAh·g^{-1}/200 圈/1 C
CNFs/Li_2S	PVP	Li_2SO_4	450 mAh·g^{-1}/200 圈/1 C
PCNF	PVP	P123/SiO_2	首放 1 234 mAh·g^{-1}/0.2 C
N-CNF	PAN	SiO_2	1041 mAh·g^{-1}/100 圈/0.2 C
PCNFs-CeF_3	PAN	PTFE	901.2 mAh·g^{-1}/500 圈/0.5 C
NPCF-GO	PAN	SiO_2/HF	846 mAh·g^{-1}/100 圈/0.25 C
PCNF-CNT	PAN	SiO_2(HF)	10.8 mAh·g^{-1}/50 圈/0.6 mA·cm^{-2}
CNFs-Se	PAN	—	375 mAh·g^{-1}/100 圈/0.1 A g^{-1}
CMSs-KB	PVP	PS	679 mAh·g^{-1}/1 000 圈/1 C

Li 等报道了一种通过静电纺丝、预氧化、碳化结合浸渍过程将活性硫负载在中空纳米纤维（ANHCNFs）上，其比表面积为 608 m^2·g^{-1}，孔体积为 0.8 cm^3·g^{-1}（图 2.9（c））。ANHCNFs 独特的多孔/空心结构、高比表面积、高导电性和增强离子导电性的 N 元素掺杂等特性能够提高整体电极材料的导电性，有效地缓解多硫化物的穿梭效应，

促进多硫化物的转化，使得电池的循环性能和倍率性能得到较大的改善。ANHCNFs-S 具有质量分数为 50%的高载硫量，在 0.2 C 倍率下具有 1 230 $mAh\cdot g^{-1}$ 的初始放电比容量，循环 100 圈后保持约 920 $mAh\cdot g^{-1}$ 的可逆比容量（图 2.9（d））。这些结果证实 ANHCNFs 是锂硫电池优良的正极材料载体。

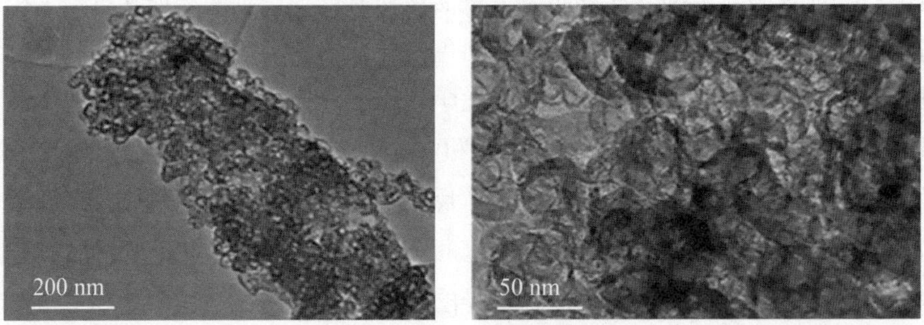

（a）ANHCNFs 的 TEM 图

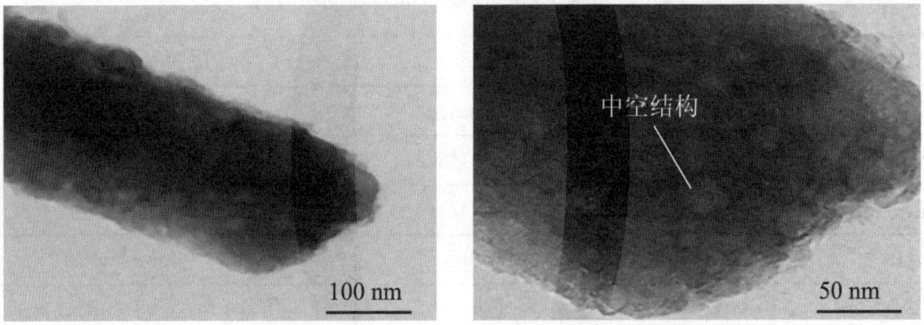

（b）ANHCNFs-S 的 TEM 图

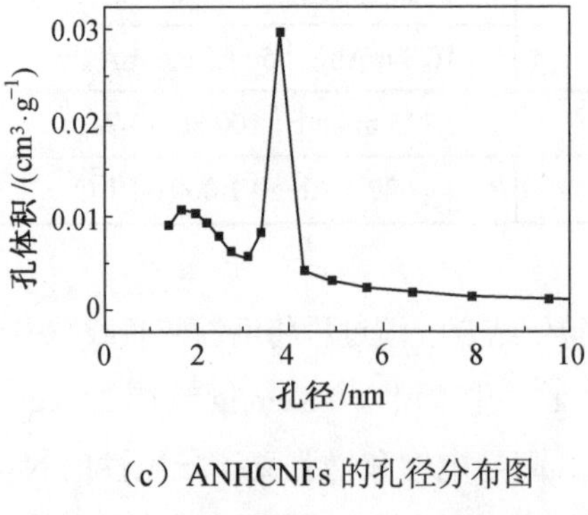

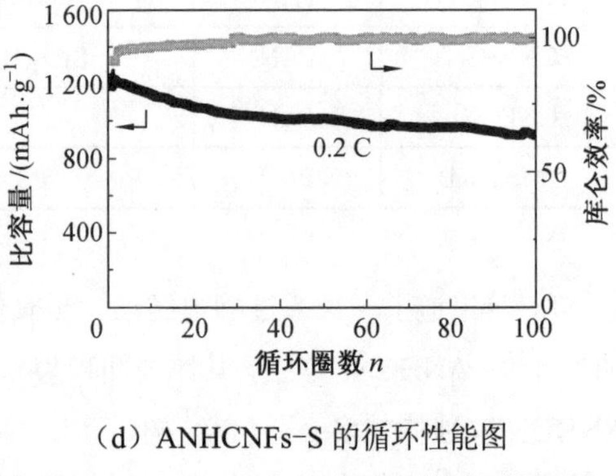

（c）ANHCNFs 的孔径分布图　　（d）ANHCNFs-S 的循环性能图

图 2.9　中空纳米纤维（ANHCNFs）正极材料载体的 TEM 图、孔径分布图及其循环性能图

通过模板法增加电纺纤维的孔隙率能够进一步提高电极材料的载硫量，Xu 等利用模板法结合静电纺丝技术制备多孔碳纤维（PCNF）载体，其微观结构如图 2.10（a）所示。他们通过优化模板剂的含量和调节活化温度来优化比表面积和孔体积。受益于载体材料的高导电性、大孔体积和具有大量锂离子渗透选择性微孔等特性，在 550 ℃的温度下制备了 PCNF 材料（命名为 PCNF/550）并且负载上硫单质得到 PNCF/A550/S，PCNF/A550/S 电极具有质量分数为 71%的高硫负载量，在 2.5 C 和 4.0 C 的倍率下经过 200 次循环后比容量分别为 625 $mAh \cdot g^{-1}$ 和 553 $mAh \cdot g^{-1}$（图 2.10（b））。通过原位透射电子显微镜发现锂化产物 Li_2S 包含在电极中，体积膨胀分数仅 35%左右，并且碳主体保持完整无断裂。这些发现为锂硫电池电极的体积膨胀与电化学性能之间的相关性提供了新的见解。Yun 等进一步通过将静电纺丝技术合成的碳纳米纤维（CNF）作为硫载体，探究了不同负载量的电化学性能，在 10.5 $mg \cdot cm^{-2}$ 的高负载量下进行电化学测试仍能提供高比容量和稳定循环（图 2.10（c）、（d））。研究表明，通过纳米结构设计进行物理封装，不仅可以解决电化学循环过程中多硫化锂的溶解问题，而且可以显著降低接触电阻，进而提高反应动力学，制备的材料显示出良好的电化学性能。此外，他们发现已反应的硫物质以特定的润湿角很好地黏附到 CNF 网络的结点上，这是由循环中捕获黏性多硫化物的基质中狭窄缝隙之间的内聚力引起的。这些研究的结果证实了电纺材料在高面积容量的锂硫电池中的潜在应用。

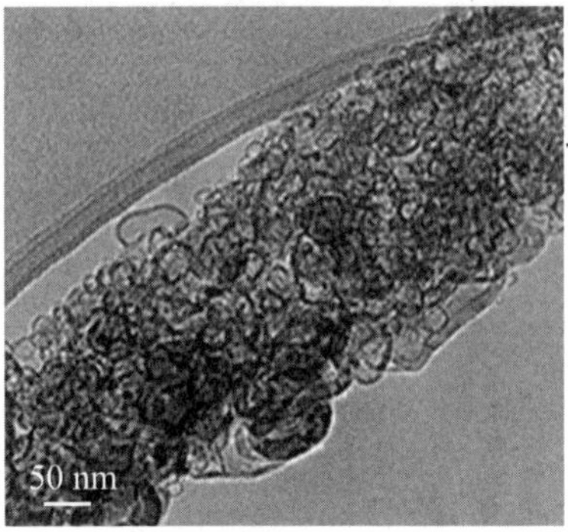

（a）PCNF 的 TEM 图

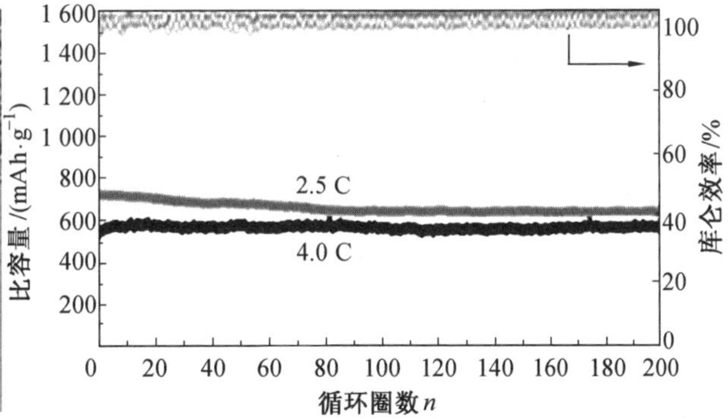

（b）PCNF/A550/S 电极的循环性能图

图 2.10　多孔碳纤维（PCNF）基正极材料的 TEM 图和循环性能图

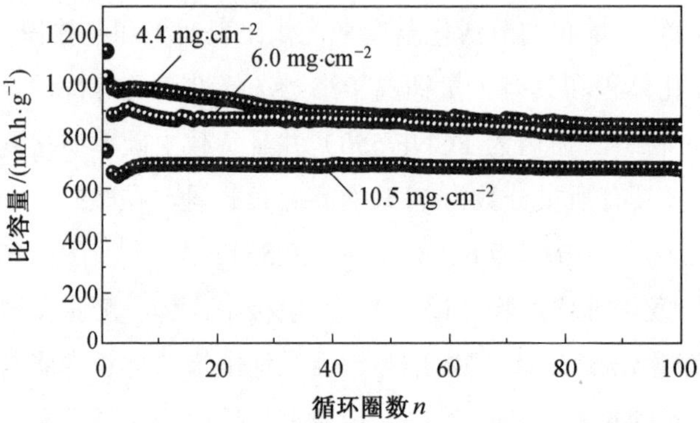

（c）CNF-S 电极不同负载量的循环性能图

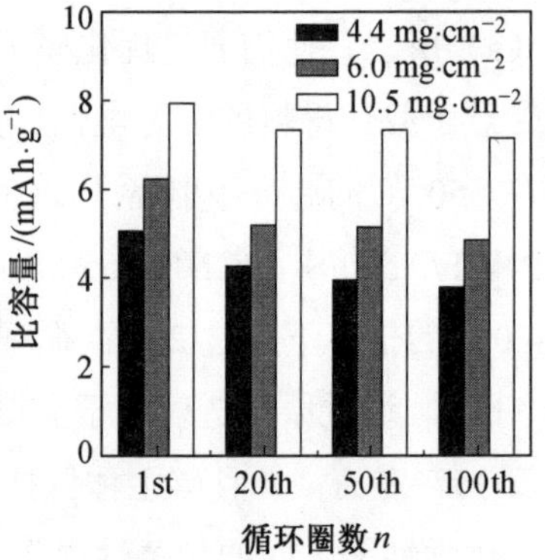

（d）不同负载量的比容量

续图 2.10

通过静电纺丝技术合成的硫载体能够有效改善硫正极的电化学性能，为了达到更佳的电化学性能，研究者试图将静电纺丝技术与其他合成手段相结合，引入其他的功能组分。Ma 等利用静电纺丝技术结合熔融扩散沉积和原子层沉积（ALD）技术制备了 SC@Se-Al_2O_3 钠硒电池正极材料。合成过程及微观结构如图 2.11（a）、（b）所示，通过调整硒含量和沉积的 Al_2O_3 层的厚度，可以实现高质量负载，提高反应动力学，有效抑制多硒化物的穿梭效应。作为自支撑超稳定和独立的碳-硒正极，电极中 Se 的质量分数高达 67%，在 0.5 $A·g^{-1}$ 电流密度下循环 1 000 次后仍然具有 503 $mAh·g^{-1}$ 的高可逆比容量（图 2.11（c））。这项工作证实了静电纺丝技术与其他制备手段相结合是一种可行的、可控制的和有效的材料制备手段，能够用于高质量负载和长循环寿命的新型储能器件。

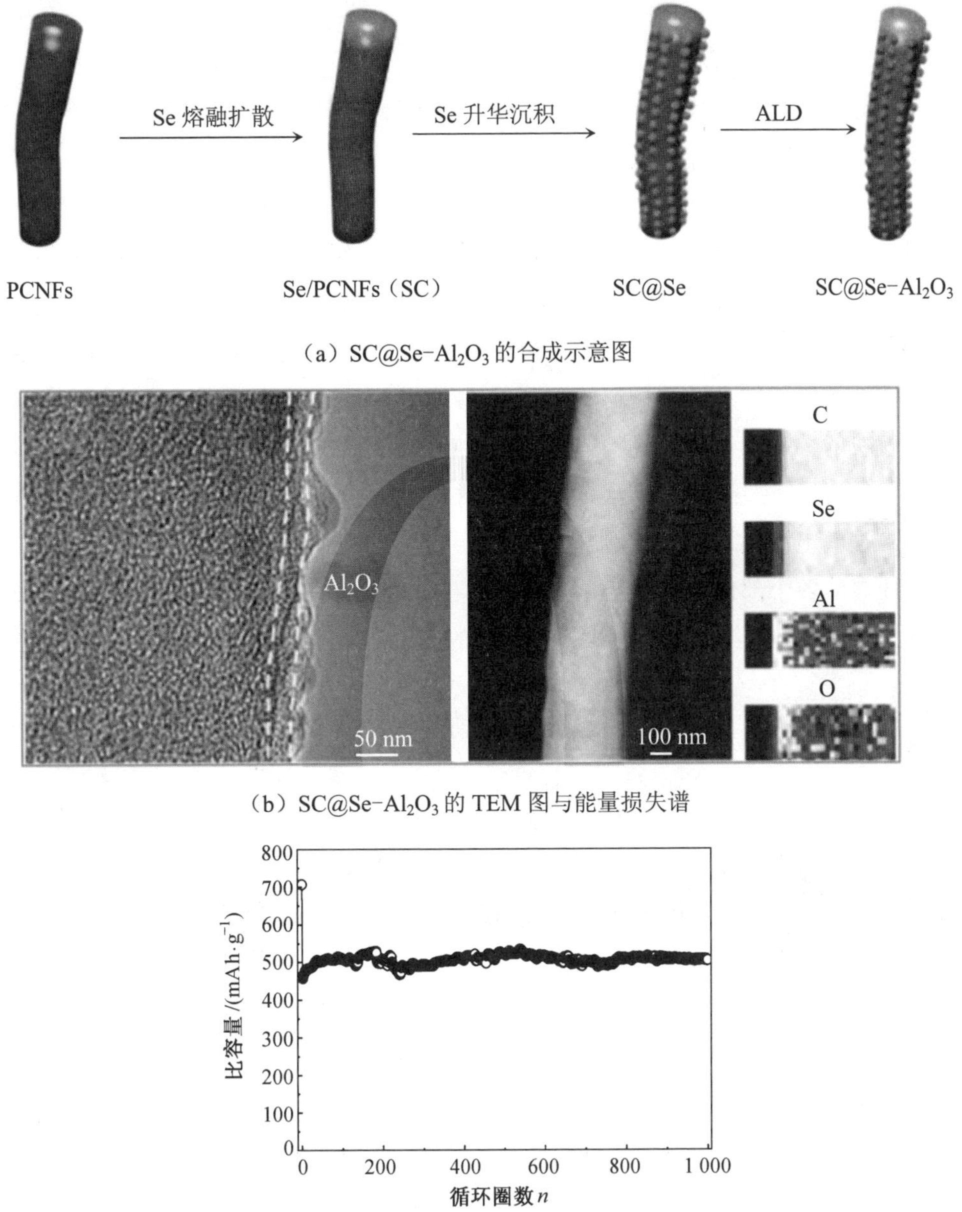

（a）SC@Se-Al$_2$O$_3$ 的合成示意图

（b）SC@Se-Al$_2$O$_3$ 的 TEM 图与能量损失谱

（c）SC@Se-Al$_2$O$_3$ 的循环性能图

图 2.11　SC@Se-Al$_2$O$_3$ 正极材料的合成、表征及循环性能图

2.2.5 小结与展望

锂硫电池的大规模应用受到许多问题的阻碍，例如硫的导电性差、活性材料利用率低、多硫化物的穿梭效应以及缓解体积膨胀等。电纺纳米纤维材料由于表面积大、结构设计性强、缺陷或活性部位丰富、导电性好、柔韧性好，能够有效提高载流量、提高硫电极的导电性和活性材料的利用率、缓解体积膨胀，并为多硫化物的穿梭效应提供物理屏障，因此电纺纳米纤维材料被认为是可用于解决锂硫电池诸多问题的候选材料之一。从工业合成和商业应用的角度来看，较低的生产率是应用于锂硫电池的电纺纳米纤维材料的主要瓶颈，如何大规模生产并最大限度降低成本是需要解决的关键。研究发现一些因素在影响电化学性能方面具有重要意义，例如纳米纤维的直径、掺杂的颗粒及其分布、孔径/分布和电导率，这些都需要进行探索。当前，锂硫电池体系的许多问题尚未完全弄清。随着电化学科学的发展，电纺材料在锂硫电池各个部件中的应用越来越广，在各个元件中的具体应用仍需研究者进行不断探索。静电纺丝技术具有大规模生产的潜力，有望促进锂硫电池的工业化进程。

2.3 超级电容器

近年来，对高速、高性能可充式储能装置需求的迅速增长，促使许多研究人员开发出新型超级电容器材料。与锂离子电池和传统的介质电容器相比，超级电容器可以提供更高的功率密度和能量密度，并提供更好的倍率性能和良好的循环稳定性。最重要的是，超级电容器可同时经受高温和低温，因此其应用范围广泛。在用于超级电容器的材料中，一维纳米纤维具有许多潜在的优越性，如固有的高孔隙率结构与优良的电子和离子导电性。本节讨论了静电纺丝纳米纤维产品在超级电容器中的创新应用，并介绍了近年来的一些研究成果。

2.3.1 引言

超级电容器是一种电化学储能装置，具有高功率、快速电荷传播和充放电过程（以 s 为单位）、长循环寿命（超过 100 000 次循环）、低维护和低自放电特性。超级电容器具有环境友好、安全性高、工作温度范围宽、循环寿命长等优点，它极快的充电速度是一个很有价值的特性。对电容器最为关注的方面是电容，主要由两种电荷储存机制决定：一种是由离子的静电吸收而产生的双电层电容器（EDLCs）；另一种是用于赝电容器（PC）的法拉第电化学存储。与锂离子电池类似，超级电容器系统是由两个电极和一个隔膜组成的，隔膜浸入电解液中。近年来，电纺碳纳米纤维及其与金属氧化物/导电聚合物复合材料应用于超级电容器的研究引起了广泛的关注。根据超级电容器储能机理和所用活性材料的不同，用于其电极的电纺纳米纤维主要可分为三类：①基于高比表面积碳纳米纤维的双电层电容器（EDLCs）；②基于导电聚合物如聚苯胺、聚吡咯和聚噻吩的赝电容器（PC）；③基于碳材料和过渡金属氧化物如 RuO_2、Fe_3O_4、NiO 和 MnO_2 的混合电容器（HCs）。表 2.4 显示了用于超级电容器电极的静电纺丝材料的直径、比表面积及其应用于超级电容器的容量、优点等。

表 2.4　用于超级电容器电极的静电纺丝材料

材料	直径/nm	比表面积/（$m^2 \cdot g^{-1}$）	容量	优点
碳纳米纤维（CNFs）	11～53	—	在 1.8 V 的工作电压下，功率密度为 450 $W \cdot kg^{-1}$ 时，能量密度为 29.1 $Wh \cdot kg^{-1}$	提高电化学活性和容量
PAN/PMMA 衍生的碳纳米纤维	215～242	248	当扫描速率为 2 $mV \cdot s^{-1}$ 时，比容量为 210 $F \cdot g^{-1}$，在 2 000 次循环后，扫描速率为 2 $mV \cdot s^{-1}$ 时，容量基本保持 100%	多孔结构，高比表面积
对齐的碳纳米纤维	441～522	635	当扫描速率为 10 $mV \cdot s^{-1}$ 时，比容量为 165 $F \cdot g^{-1}$。当扫描速率从 10 $mV \cdot s^{-1}$ 到 50 $mV \cdot s^{-1}$ 时，速率容量达到 14.9%，在电流密度为 1 $A \cdot g^{-1}$ 时，速率降低了 10.8%	提高导电性
介孔多孔氮掺杂碳纳米纤维（HPNCNF）	350	656	电流密度为 2 $A \cdot g^{-1}$ 时，比容量为 289 $F \cdot g^{-1}$，电流密度为 10 $A \cdot g^{-1}$ 时，容量在 1 000 次循环后保持 95.6%	增强界面润湿性，提高电子导电性

续表 2.4

材料	直径/nm	比表面积/（$m^2 \cdot g^{-1}$）	容量	优点
氮掺杂碳纳米纤维（N-CNFs）	300	763	电流密度为 0.1 $A \cdot g^{-1}$ 时，比容量为 251.2 $F \cdot g^{-1}$，电流密度为 20 $A \cdot g^{-1}$ 时，容量在 2 000 次循环后保持 99%	提高了孔径分布的协同效应
氮掺杂中空活性碳纳米纤维（HACNF）	内径：150 外径：300	701	电流密度为 0.2 $A \cdot g^{-1}$ 时，比容量为 197 $F \cdot g^{-1}$，电流密度为 5 $A \cdot g^{-1}$ 时，1 000 次循环后容量保持 98.6%	改善界面的离子输运性、导电性和润湿性
NiO 纳米纤维/泡沫镍（NiO-NFs/Ni）	280～400	—	电流密度为 2 $A \cdot g^{-1}$ 时，比容量为 737 $F \cdot g^{-1}$，即使在 40 $A \cdot g^{-1}$ 下，容量仍保持在 570 $F \cdot g^{-1}$	利用多孔性提高电子和离子的传输能力
Co_3O_4/CNFs	150～300	230.6	电流密度为 1 $A \cdot g^{-1}$ 时，比容量为 586 $F \cdot g^{-1}$，电流密度为 2 $A \cdot g^{-1}$ 时，容量在 2 000 次循环后保持 74%	增强导电性
$CuCo_2O_4$ 纳米线	450～500	—	电流密度为 1 $mA \cdot cm^{-1}$ 时，比容量为 443.9 $mF \cdot cm^{-1}$，经过 1 500 次循环后，容量保持 90%	提高速率能力和电导率
CuO 纳米线	30～50	—	在功率密度分别为 800、1 500、4 000 和 8 400 $W \cdot kg^{-1}$ 时，能量密度分别为 29.5、23.5、19.2 和 16.4 $Wh \cdot kg^{-1}$	提高能量密度
含 RuO_2 的中空活性碳纳米纤维（RuPM-ACNF）	180～220	约 763	电流密度为 1 $mA \cdot cm^{-1}$ 时，比容量为 188 $F \cdot g^{-1}$，在 3 000 次循环后，容量保持 75%	提高了孔/中孔体积、高速率能力、容量；高比表面积
含异质原子（B、N、O）的多孔 MnO_2/碳纳米纤维材料（MnB-CNF）	MnB(10)-CNF：730	MnB(10)-CNF：约 562	当电流密度为 1 $mA \cdot cm^{-1}$ 时，比容量高达 210 $F \cdot g^{-1}$	利用丰富的介孔提高孔隙率，增强电子和离子的传输
含中空颗粒的掺氮碳纳米纤维（HPCNFs-N）	PAN/ZIF-8：约 1.4 μm	HPCNFs-N:417.9	在电流密度为 50 $A \cdot g^{-1}$ 时，比容量仍达到 193.4 $F \cdot g^{-1}$	引入杂原子提高电化学活性

2.3.2　超级电容器电极

1. 导电聚合物纳米纤维

导电聚合物在各种有机器件中得到了广泛的研究和应用。为了提高器件的性能或扩展器件的功能，导电聚合物通常必须是纳米结构。电纺导电聚合物纳米纤维已经被用作超级电容器电极。导电聚合物聚苯胺（PANI）和聚吡咯（PPy）由于具有良好的导电性和多种本征氧化还原态，被认为是很有前途的超级电容器电极候选材料之一。Shao 等人制备了碳纳米纤维/聚苯胺复合材料并随机排列 PAN@PANI 纳米纤维分别作为超级电容器的独立电极，如图 2.12 所示。

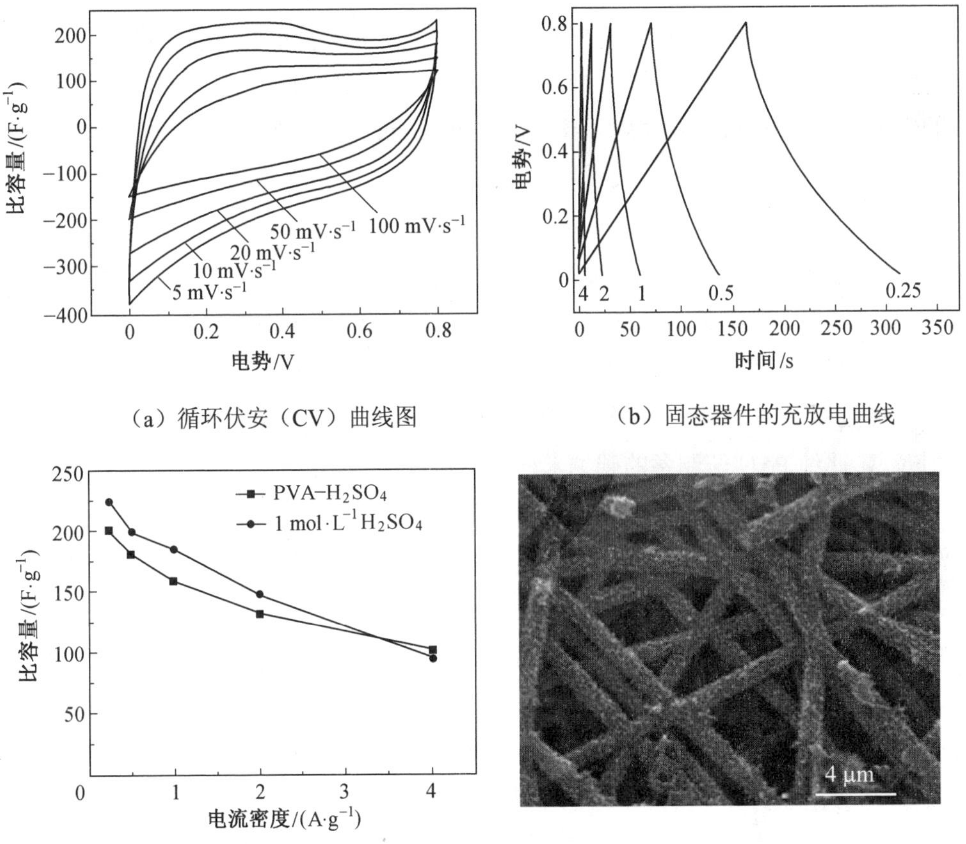

（a）循环伏安（CV）曲线图　　（b）固态器件的充放电曲线

（c）倍率性能图　　（d）CNFs/PANI 复合材料的 SEM 图像

图 2.12　碳纳米管/聚苯胺复合材料（CNFs/PANI）的 SEM 图及电化学性能

碳纳米纤维/PANI 复合材料制备的超级电容器具有优异的电化学性能，比容量为 201 $F\cdot g^{-1}$，功率密度为 103 $W\cdot kg^{-1}$ 时能量密度为 4.5 $Wh\cdot kg^{-1}$，6 000 次循环后比容量保持率为 80%，具有低泄漏电流和自放电特性。Shen 等人采用近场静电纺丝法制备了 PPy 纳米纤维，并将其应用于组装柔性微超级电容器，其比容量为 0.48 $mF\cdot cm^{-2}$。最重要的是，所制备的超级电容器可以弯曲到任何角度，而电化学性能没有明显变化。

2. 碳纳米纤维

石墨烯和碳纳米管由于具有较高的比表面积而被广泛用作超级电容器的电极。与碳纳米管相比，电纺纯碳纳米纤维作为超级电容器电极需要产生更多的纳米级孔。目前常用的聚合物有聚苯胺、聚甲基丙烯酸甲酯、聚乙烯醇等。最常用的方法是通过物理或化学方法制备碳纳米纤维。例如，Kim 等人使用旋转收集器获得了排列整齐的电纺碳纳米纤维。随着集热器转速从 250 $r\cdot min^{-1}$ 增加到 2 000 $r\cdot min^{-1}$，整齐碳纳米纤维的比表面积从 533 $m^2\cdot g^{-1}$ 增加到 635 $m^2\cdot g^{-1}$，电导率从 342 $S\cdot cm^{-1}$ 增加到 626 $S\cdot cm^{-1}$。其原因是静电纺丝使纤维与集电极表面之间的拉伸强度增加。此外，与随机碳纳米纤维相比，排列整齐的碳纳米纤维的比电容和速率容量分别提高了 35.5%和 28.4%。在碳纳米纤维中掺入其他材料可以提高电容器的性能，Lai 等人以聚苯胺和低分子量聚甲基丙烯酸甲酯为复合前驱体，通过静电纺丝法制备了碳纳米纤维。通过掺杂 PMMA，得到的碳纳米纤维具有多孔和中空结构，比电容高达 210 $F\cdot g^{-1}$，比纯 PAN 法制备的碳纳米纤维高出近 38%。

另外一种方法是通过 NH_3 活化来改善碳纳米纤维的表面化学环境。通过静电纺丝法和在不同温度下的 NH_3 碳化法制备了具有径向生长石墨烯片结构的超薄碳纳米纤维（图 2.13），发现碳纳米纤维表面存在大量的氮氧改性。用这种材料制成的超级电容器能够在 1.8 V 的高电压下工作，而 Na_2SO_4 水电解质由于水的分解，电压限制在 1.23 V 左右。更重要的是，所制备的超级电容器在 450 $W\cdot kg^{-1}$ 功率密度下具有 29.1 $Wh\cdot kg^{-1}$ 的高能量密度，并且有良好的循环性能，经过 5 000 次充放电循环后容量保持率超过 93.7%，具有很大的实际应用潜力。

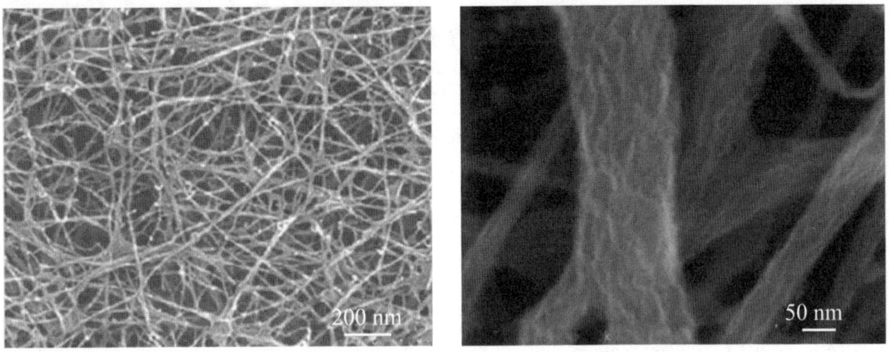

（a）在 1 100 ℃下制备的 CNF 的 SEM 图像

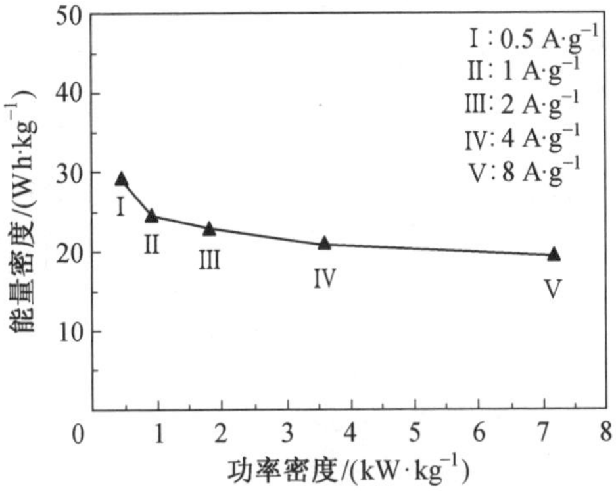

（b）拉贡（Ragone）曲线图

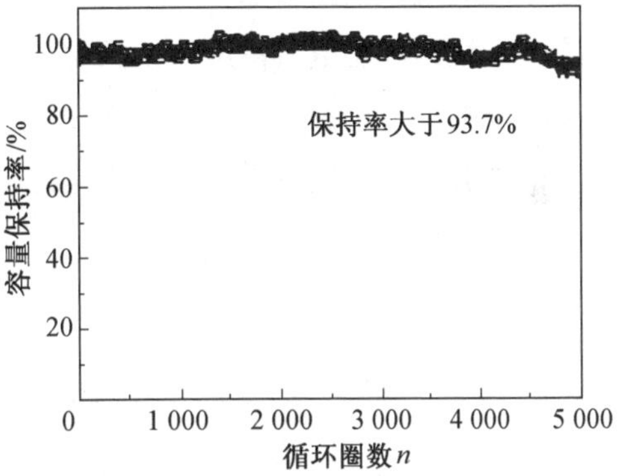

（c）1 100 CNFs 超级电容器在 1 A·g^{-1} 下的长期循环性能

图 2.13　氮掺杂碳纳米纤维的 SEM 图和电化学性能

3. 杂化电纺纳米纤维

目前，一系列过渡金属氧化物和氢氧化物由于具有较高的理论比电容和快速的多电子表面法拉第氧化还原反应，如 MnO_2、Co_3O_4、$CuCo_2O_4$、CuO、NiO 和 RuO_2，已被广泛用作超级电容器电极。在这些材料中，由于 MnO_2 在水电解质中具有较高的理论比电容（约 1 110 F·g^{-1}），成本低，环境友好，电位窗口宽，被认为是最有前途的赝电容材料。Kim 等人利用静电纺丝技术制备了掺杂硼的多孔 MnO_2/碳纳米纤维（MnB-CNF），并将其用作超级电容器电极。与不含 B 原子的普通 Mn-CNF 相比，MnB-CNF 在功率密度 400 W·kg^{-1} 下具有更高的能量密度（22.6 Wh·kg^{-1}）。

Lou 和他的同事通过将金属有机框架材料（ZIF-8）纳米粒子静电纺丝到 PAN

纳米纤维前体中，然后在氮气中碳化，合成了一种含中空颗粒的氮掺杂碳纳米纤维（HPCNFs-N），如图 2.14 所示。所制备的 HPCNFs-N 作为超级电容器电极，在功率密度为 250 W·kg^{-1} 时，具有 10.96 Wh·kg^{-1} 的高能量密度和良好的耐久性，10 d 后容量保持率为 98.2%，在电流密度为 5.0 A·g^{-1} 下进行了 10 000 次循环，其优异性能的原因在于：①交联碳空心纳米颗粒组成的层状结构可以提供丰富的电化学活性位点，提高电化学动力学和结构稳定性；②纳米纤维结构有利于离子/电子传输；③N 掺杂的碳纳米纤维可以调节带隙以利于电化学反应。除此之外，一些双金属碳化物也被用于研究并得到了较好的性能。Hao 等采用静电纺丝法成功地合成了一维 Fe_2MoC/CNFs，如图 2.15 所示。实验结果表明，Fe_2MoC/CNFs 在 800 ℃下具有完整的一维结构，在电流密度为 1 A·g^{-1} 时，高比容量为 347.8 F·g^{-1}，并且在循环容量为 100%时，电容保持率为 100%左右。在功率密度为 300 W·kg^{-1} 时，用 Fe_2MoC/CNFs 作为正极的非对称超级电容器币形电池装置的能量密度为 14.5 Wh·kg^{1}，循环 5 000 次后循环寿命可达 93%。

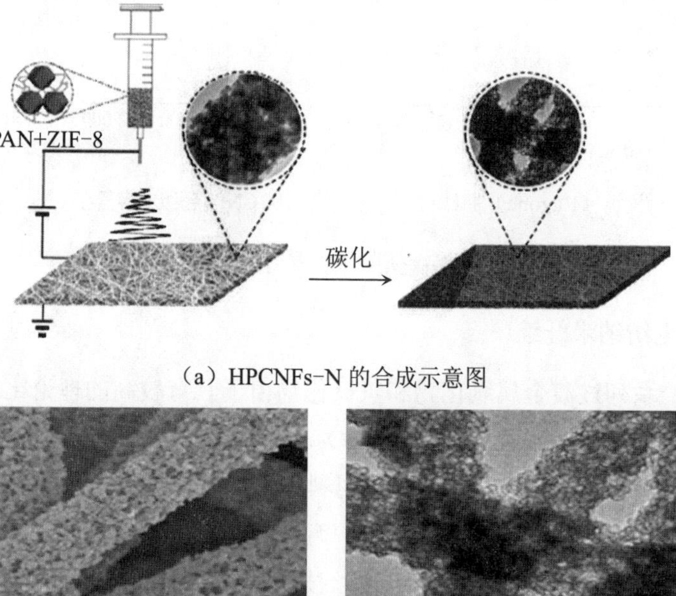

（a）HPCNFs-N 的合成示意图

（b）HPCNFs-N 样品的 SEM 图像

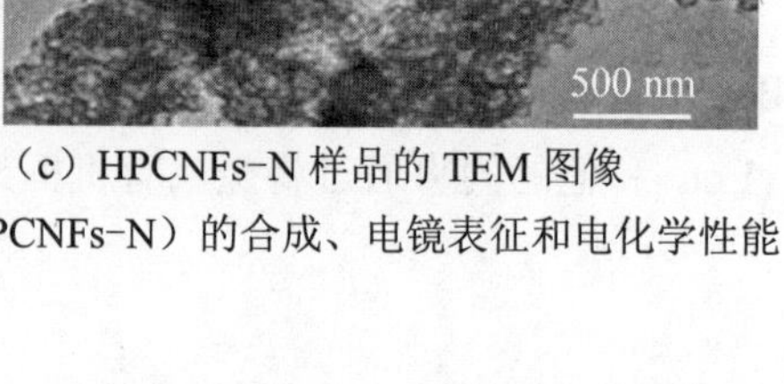

（c）HPCNFs-N 样品的 TEM 图像

图 2.14　含中空颗粒的氮掺杂碳纳米纤维（HPCNFs-N）的合成、电镜表征和电化学性能

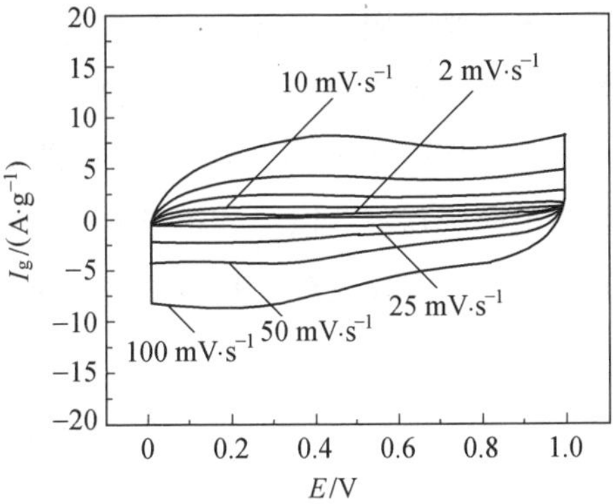

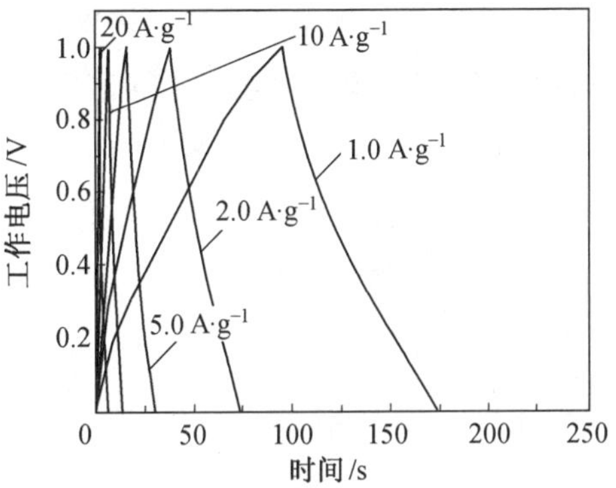

（d）恒电流不同电流密度下的充放电曲线　（e）HPCNFs-N 样品在不同扫描速率下的 CV 曲线

续图 2.14

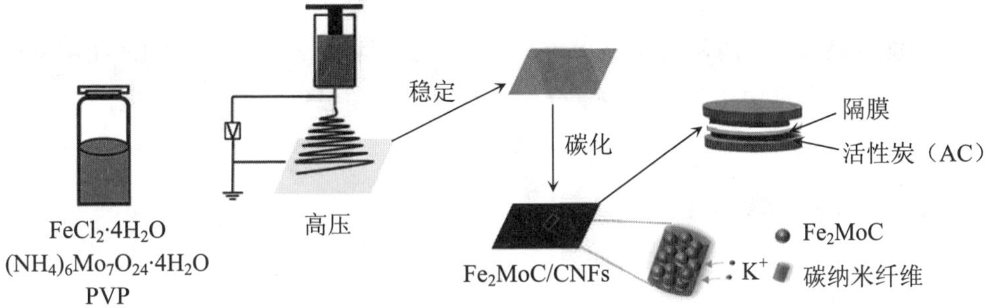

（a）电纺和非对称超级电容器纽扣式电池装置制备 $Fe_2MoC/CNFs$ 示意图

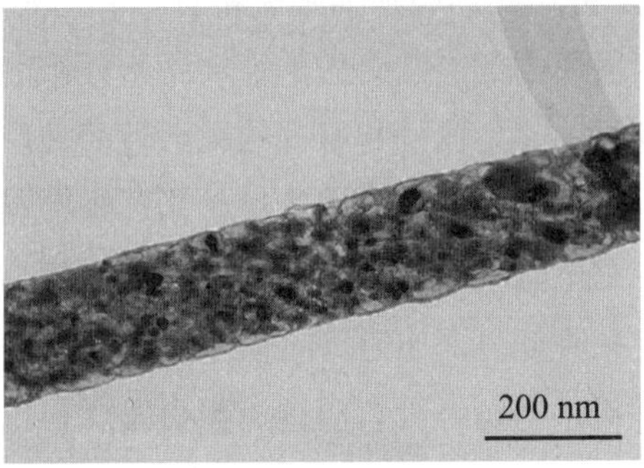

（b）$Fe_2MoC/CNFs$ 的 TEM 图像

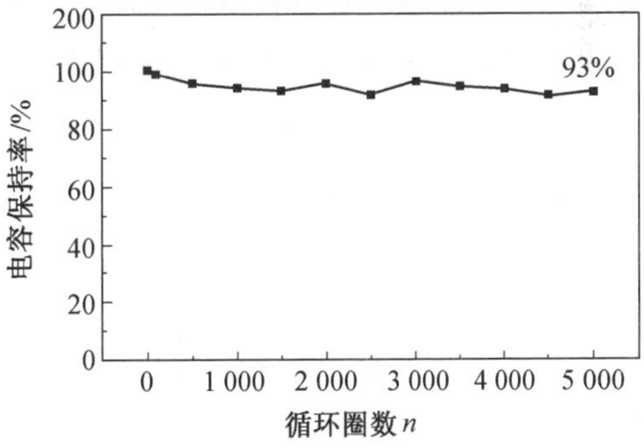

（c）电池循环 5 000 次后的稳定性

图 2.15　一维 $Fe_2MoC/CNFs$ 纳米纤维的制备、表征和电化学性能

除了用作超级电容器的电极外，一些电纺纳米纤维还应用于所有固态超级电容器的隔膜和电解液。例如，Liu 等采用电纺 PVA/PAA 膜作为隔膜和电解液的储存和运输，以提高超级电容器器件的柔韧性和可折叠性。此外，Jabbarnia 等人采用掺杂炭黑的电纺 PVDF/PVP 纳米纤维作为超级电容器的隔膜材料，以提高超级电容器的热稳定性、机械稳定性和离子导电性。结果表明，炭黑不仅改善了其润湿性，而且具有良好的稳定性和较高的介电常数。

2.3.3 小结与展望

超级电容器在解决快速增长的全球能源消耗方面有很大的作用。以上研究清楚地预示了静电纺丝法制备的一维纳米结构具有孔隙率高、团聚可能性小、长径比大等优点，有望成为制备超级电容器的新途径。特别是经过化学反应进一步改性后，超级电容器在循环寿命、工作电压、能量密度和功率密度等方面的性能都能迅速提高。因此，有理由相信电纺纳米纤维电极可以为超级电容器的发展提供助力。

2.4 燃料电池

近几十年来，针对现代社会的能源需求和新兴的生态问题，追求新颖、低成本、环保的能量储存/转换系统引起了人们极大的研究兴趣。燃料电池因具有高效率、高功率密度、低温室气体排放等优点而备受关注。静电纺丝被认为是制备有机、无机、金属或杂化组分组成的一维微观结构纳米材料最通用的方法。由于连续纳米纤维的组分以随机或定向方式排列，静电纺丝也因其尺寸、方向性和组分的灵活性而被广泛用于制备电催化剂和燃料电池电解质材料。本节详细介绍静电纺丝纳米纤维在燃料电池不同元件中的具体设计和应用，重点介绍电纺纳米纤维在提高燃料电池的功率密度、离子导电性、化学稳定性、机械强度和降低界面电阻等方面的研究进展，作为一种在能量储存和转换系统潜在应用的材料，电纺纤维有望进一步引发燃料电池的发展和演变。

2.4.1 引言

燃料电池是能量转换装置，可以通过氢或富氢燃料的电化学氧化将化学能转化为电能，主要包含质子交换膜燃料电池（PEMFC）、甲醇燃料电池（DMFC）、固态氧化物燃料电池（SOFC）、熔融碳酸盐类燃料电池（MCFC）和碱性电解质燃料电池（AFC）等，这些电池除燃料和电解质有差别外，其他组成相似。以 PEMFC 为例，氢气被送到膜电极组件（MEA）的阴极铂（Pt）侧被催化分解成质子和电子。这种负极反应表示为

$$H_2 \longrightarrow 2H^+ + 2e^- \qquad E_0(\text{vs. SHE}) = 0\ \text{V}$$

生成的质子穿过质子交换膜（PEM）进入正极。然而，电子沿着外电路移动到 MEA 的正极侧，产生电流。在正极，随着氧气输入，氧分子与渗透到质子交换膜中的质子及通过外部电路到达的电子发生反应，形成水分子。正极反应如下：

$$\frac{1}{2}O_2 + 2H^+ + 2e^- \longrightarrow H_2O \qquad E_0(\text{vs. SHE}) = 1.229\ \text{V}$$

电池总反应为

$$H_2 + \frac{1}{2}O_2 \longrightarrow H_2O \qquad E_0(\text{vs. SHE}) = 1.229\ \text{V}$$

PEMFC 的结构及原理如图 2.16 所示。

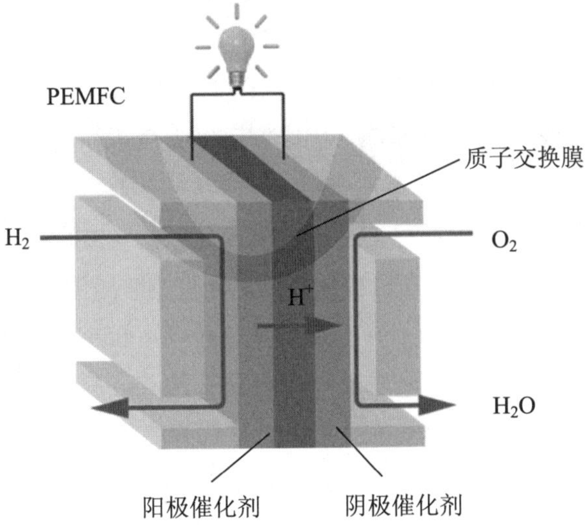

图 2.16　PEMFC 的结构及原理

目前，燃料电池的发展主要集中在固定和运输方面。在过去 20 年中，为了进一步提高能量储存和转化效率、延长燃料电池的使用寿命和降低成本，许多研究工作致力于提高性能和稳定性。这些工作的重点放在膜电极的所有组件上，包括电解质膜（在高温和低相对湿度下增加导电性，提高机械强度和化学稳定性）、电催化剂（降低催化剂负载量，提高比活性和降低成本）、催化剂载体材料（稳定不同载体材料进行氧化还原反应）。

静电纺丝技术作为制备一维纳米纤维的常用方法，在燃料电池催化剂材料和电解质膜中都有着广阔的应用前景。电纺纳米纤维具有长径比高、材料多样性、导电性好、比表面积大、力学和化学稳定性好等特点，目前正广泛应用于燃料电池膜电极组件中。

2.4.2 电纺电解质膜

静电纺丝技术作为一种制备纳米纤维膜的常用技术，已被应用于研发具有良好形态和机械性能的燃料电池复合膜，尤其是对质子交换膜进行改性。第一种方法是将不导电或电导率较低的聚合物电纺成多孔基质，当孔中填充有高质子传导性组分时，该基质起到机械增强作用。第二种方法是将高质子传导性基质电纺成多孔纳米纤维膜，使用第二聚合物能增强机械稳定性。通常，这种膜表现出高的质子传导性、低的气体渗透性、良好的耐化学性和热稳定性。例如，Nafion（一种全氟磺酸树脂）通常被用作商业聚合物电解质膜，因为它可以为质子和离子流提供连续的路径。但是，纯 Nafion 黏度低，因此不适合用于静电纺丝。通常需要在电纺丝溶液中加入高分子量聚合物，例如聚丙烯酸（PAA）、聚环氧乙烷（PEO）、聚乙烯醇（PVA）和聚乙烯吡咯烷酮（PVP）以提高机械性能。复合膜能够提供良好的机械性能和热性能。然而，与纯的 Nafion 膜相比，复合膜的质子传导率较小，这归因于纳米纤维中存在的酸性基团。此外，其他聚合物的存在会降低 Nafion 膜的质子传导性。因此，探讨合适的载体聚合物使用含量有利于制备性能优异的复合膜。

Mollá 等人将聚乙烯醇电纺多孔垫浸入 Nafion 溶液中制备复合膜（图 2.17（a）、（b））。结果表明，由于 Nafion 和 PVA 共连续相的存在，复合膜具有良好的力学性能和热性能。但是，由于纳米纤维中存在酸性基团，复合膜的质子电导率比未镀 Nafion 膜小。此外，其他聚合物的存在会降低 Nafion 膜的质子导电性，因此，在大多数情况下，许多研究试图将载体聚合物的含量降至最低。Dong 等人仅用质量分数为 0.1%的 PEO 载体聚合物通过静电纺丝制备了高纯度 Nafion 纳米纤维（图 2.17（c）、（d）），其质子电导率为 1.5 $S\cdot cm^{-1}$，而 Nafion 薄膜的质子电导率为 0.1 $S\cdot cm^{-1}$。研究还发现，随着纤维直径的减小，质子电导率急剧增加。这些结果归因于离子聚集体沿纤维轴方向的排列。采用静电纺丝法制备的磺化聚酰亚胺（PI）纳米纤维与磺化聚酰亚胺复合电解质膜，由于纳米纤维中的聚酰亚胺在静电纺丝过程中明显定向/聚集，因此获得的电解质膜比铸膜具有更高的质子电导率。因此，与商用 Nafion 膜相比，静电纺丝法制备的 Nafion 膜具有较高的质子电导率、较低的透气性、良好的耐化学性和热稳定性。

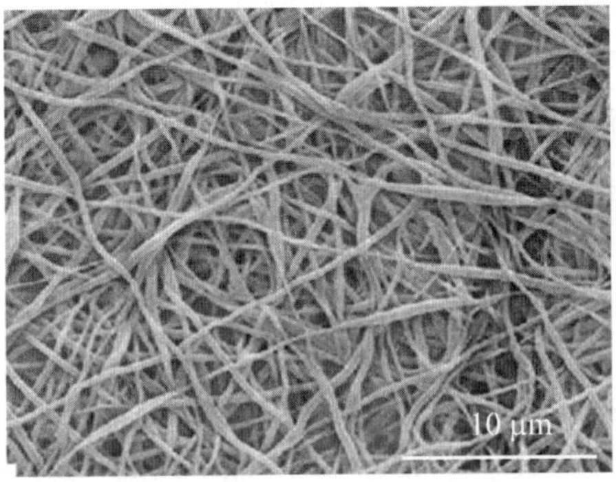

（a）在 Nafion 分散体中首次浸渍后的纳米纤维垫

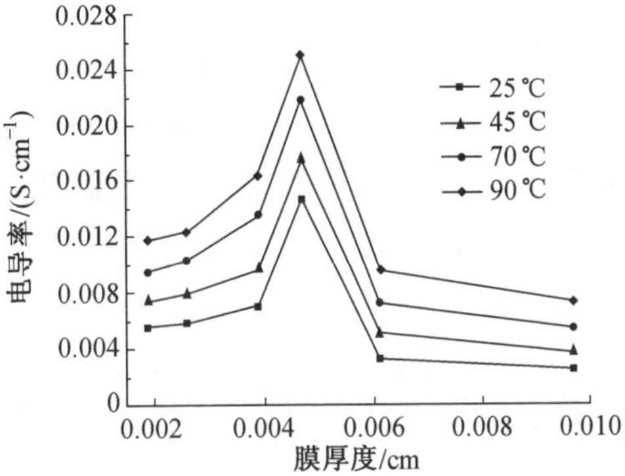

（b）Nafion/PVA 复合膜的电导率随温度和厚度的变化

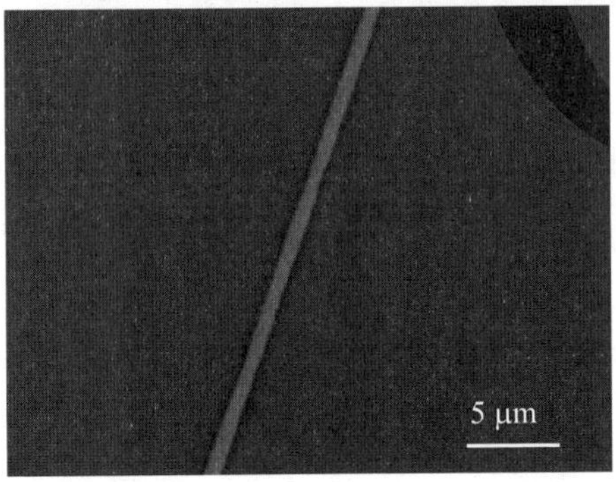

（c）高纯度 Nafion 纳米纤维的 SEM 图像

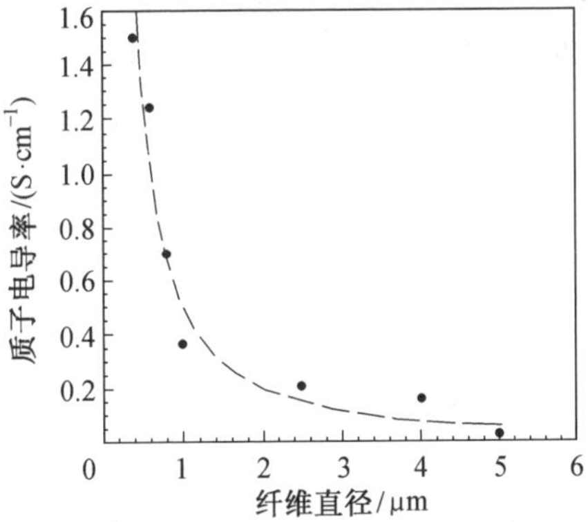

（d）高纯度 Nafion 纳米纤维的质子电导率随纤维直径的变化

图 2.17　电纺电解质膜的 SEM 图及其电导率特性

2.4.3 电催化剂

燃料电池中发生的反应为氢氧化反应（HOR）和阴极氧还原反应（ORR），其中HOR 反应的过电势较低，因此在阳极上所需要的 Pt 催化剂相对较少。而发生在阴极的 ORR 反应动力学非常缓慢，比 HOR 至少慢 5 个数量级，导致电池能量密度和循环效率降低，因此需要发展高性能的电催化剂。目前，在燃料电池中，Pt 基催化剂是最常用的阴极催化材料。但由于 Pt 成本较高、活性不稳定等缺陷，大大限制了燃料电池的发展和应用。因此，如何开发出低成本、高稳定性的新型催化剂和催化剂载体，以代替目前使用的贵金属电催化剂，在燃料电池商业化进程中至关重要。

静电纺丝技术相对较低的成本和较高的实际能耗，在相关领域的应用引起了人们极大的兴趣。静电纺丝能够很好地控制纤维的微观结构，所制备的材料尺寸、方向和组成上灵活多样。目前，利用静电纺丝技术来制造应用于燃料电池电催化剂，实现电极反应的高效进行是当前研究工作的热点。

Sung 等人利用静电纺丝法和热处理工艺，以 PVP 和六氯铂酸为原料，制备了 Pt 纳米线（Pt-NWs）。研究结果表明，与少量 Pt-NWs 混合的 Pt/C 电极表现出比纯 Pt/C 电极更好的 ORR 催化性能。Kim 等人将静电纺丝与热处理相结合，增加了氢气还原步骤，制备了纳米纤维网络结构的 Pt 催化剂，与传统的 Pt/C 阴极电催化剂相比，Pt 纳米纤维的催化活度提高了超过 50%。该纳米纤维在燃料电池中的反应机理和微观结构图如图 2.18（a）、（b）所示。

图 2.18（c）所示为以 Pt / C 和 Pt 纳米线为阴极催化剂（电池温度为 70 ℃，1 bar（1 bar=10^5 Pa）H_2、1 bar O_2）的几种结构上的 Pt 质量活度及 ORR 的 I–V 曲线。图 2.18（d）所示为用制备的纳米结构催化剂作为阴极电极制造的 PEMFC 单电池的塔费尔（Tafel）图。

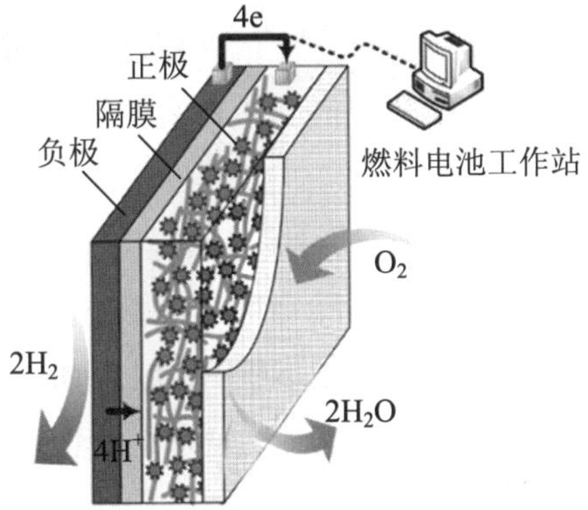

（a）反应机理图

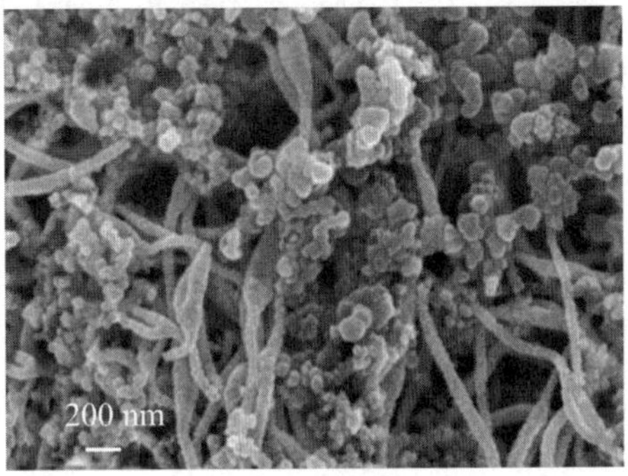

（b）微观形貌图

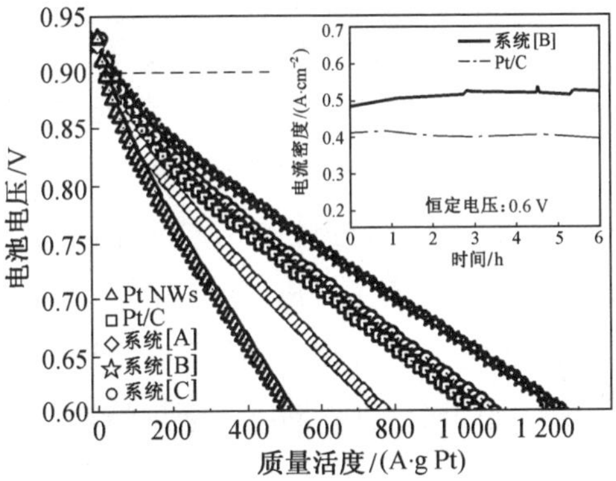

（c）质量活度及 ORR 的 I–V 曲线图

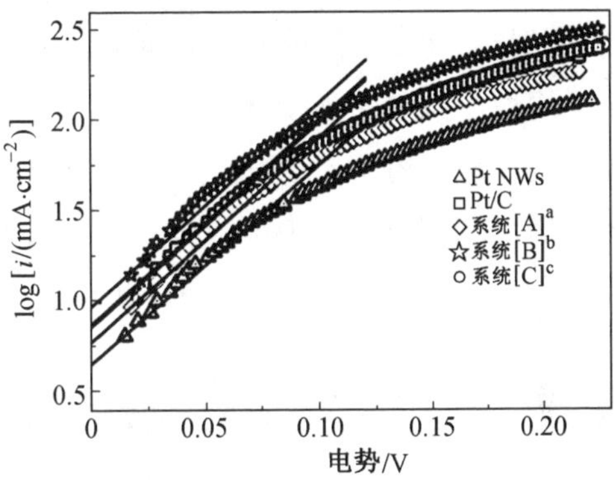

（d）Tafel 图

图 2.18　电纺电催化剂的机理、SEM 图及其电化学性能

通过静电纺丝合成的 Pt 基合金纳米线也表现出优异的 ORR 性能。Higgins 等人通过静电纺丝制备的铂钴合金纳米线（PtCoNWs）展示出优异的氧还原反应（ORR）活性，与纯 PtNWs 相比，在 0.9 V（vs. RHE）的电极电位下，电化学活性增加了 4 倍；与商业 Pt/C 催化剂相比，活性提高了近 7 倍，并且通过重复的电位循环，电化学活性表面积的保留比例提高。

非贵金属催化材料中，被研究最多的是过渡金属材料。过渡金属材料具有易合成、性能稳定、活性优异的特点，被认为是能够取代贵金属 Pt 的还原反应（ORR）催化剂。其中，金属铁和钴已被广泛研究，并且表现出优异的 ORR 电催化活性。通过静电纺丝、稳定化、浸渍和碳化等步骤，引入了 Fe 元素，制备了 Fe−N/C 纳米纤维（Fe−N/CNFs）电催化剂，铁可以促进聚合物向碳质材料的石墨化转化，增强材料对电化学氧化的抵抗力，从而增强 Fe–N/CNFs 催化剂的稳定性。在酸性和碱性溶液中，其 ORR 均显示出更好的活性和稳定性。催化剂的微观形貌图及性能如图 2.19（a）、（b）所示。此外，发现制备的 Fe−N/CNFs 催化剂在酸性电解质中不仅具有优异的 ORR 电催化活性，而且具有可观的耐久性。研究发现，电化学性能的提升归因于材料的多孔结构、高的季铵氮和吡啶氮掺杂以及高含量的弯曲状石墨化碳。Fe−NCNFs 优异的性能使其成为商业化 Pt/C 催化剂潜在的替代品。单组分的金属催化剂容易发生团聚，导致催化活性的降低。过渡金属合金催化剂（如：FeCo、FeNi 等）的特点在于不同价态和不同金属之间可以发生电子跃迁，从而大大提高了材料的导电性，为 O_2 的吸附和活化提供更多活性位点，从而提高催化剂材料的 ORR 性能。通过静电纺丝、浸渍及热处理的工艺，制备了聚丙烯腈（PAN）基氮掺杂碳纳米管/纳米纤维（NCNTs/NFs），如图 2.19（c）、（d）所示。经过 NH_3 气氛下高温处理，在 Fe−Co 双金属的作用下，以 PAN 为碳源的纤维表面生长出碳纳米管。该纳米管具有相对较高的密度，其电化学活性可与商用 20% Pt/C 催化剂相媲美，经过 12 h 测试后，其电化学活性能够保持约 94%，具有较高的稳定性。同时还通过静电纺丝法和浸渍法结合制备了三元合金 FeCoNi−N/碳纳米纤维催化剂。探讨了 Fe/Co/Ni 比例对其电化学性能的影响。当 Fe/Co/Ni 原子比为 4∶2∶1 时，该材料在 0.5 $mol \cdot L^{-1}$ H_2SO_4 中表现出良好的 ORR 电催化性能，其活性和稳定性接近于商用 20%的 Pt/C 催化剂。

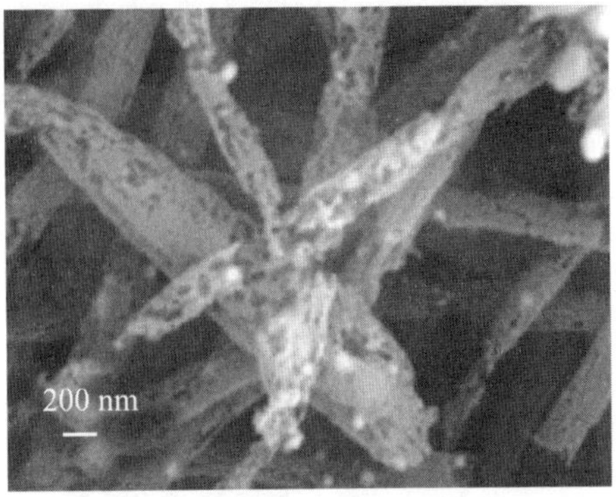

（a）Fe-N/CNFs 催化剂的 SEM 图

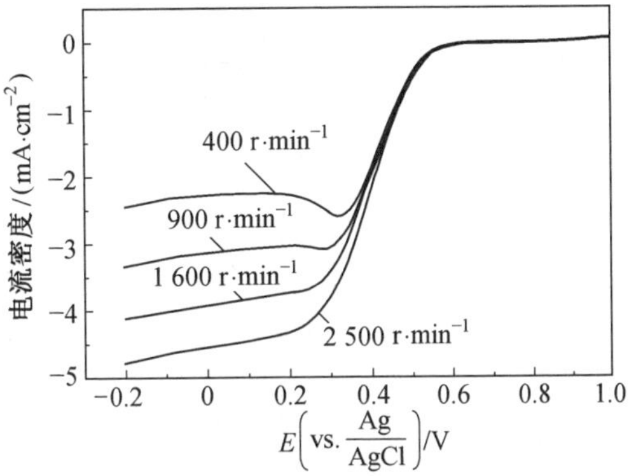

（b）旋转圆盘电极伏安图

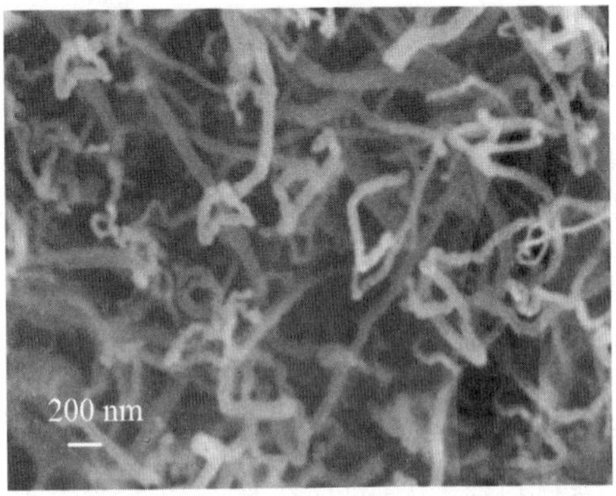

（c）FeCo-NCNTs/NFs 催化剂的 SEM 图

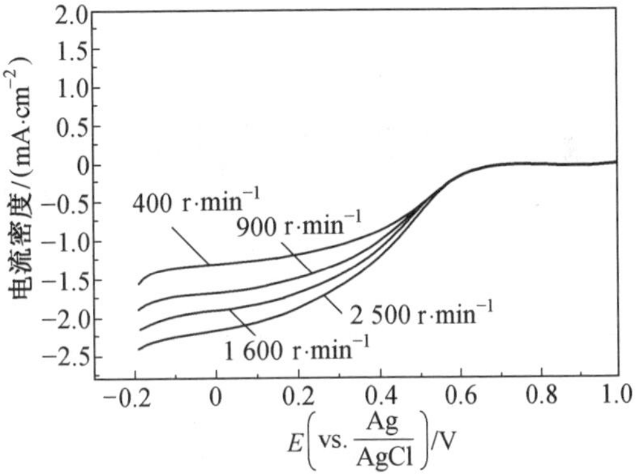

（d）FeCo-NCNTs/NFs 对应的旋转圆盘电极伏安法

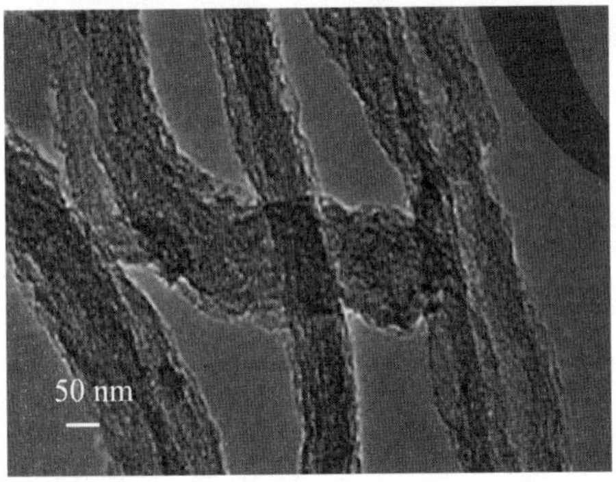

（e）酸化和 NH_3 刻蚀处理的 NCNFs 的 TEM 图

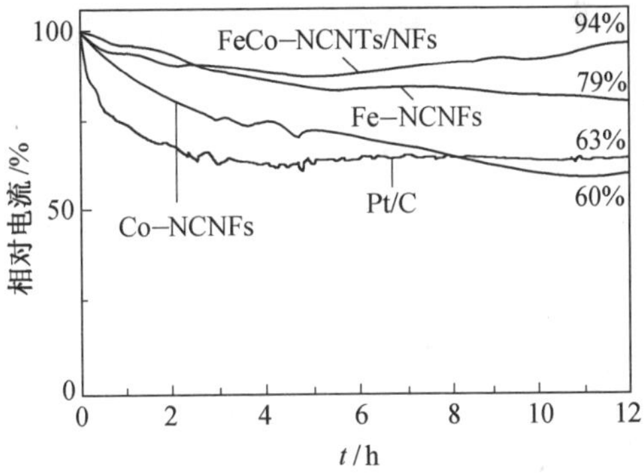

（f）FeCo-NCNTs/NFs 对应的电流计时响应图

图 2.19　电纺 ORR 电催化剂的电镜图及其电化学性能

Ⅰ—NCNFs；Ⅱ—酸化处理的 NCNFs；Ⅲ—NH_3 刻蚀处理的 NCNFs；Ⅳ—酸化和 NH_3 刻蚀处理的 NCNFs

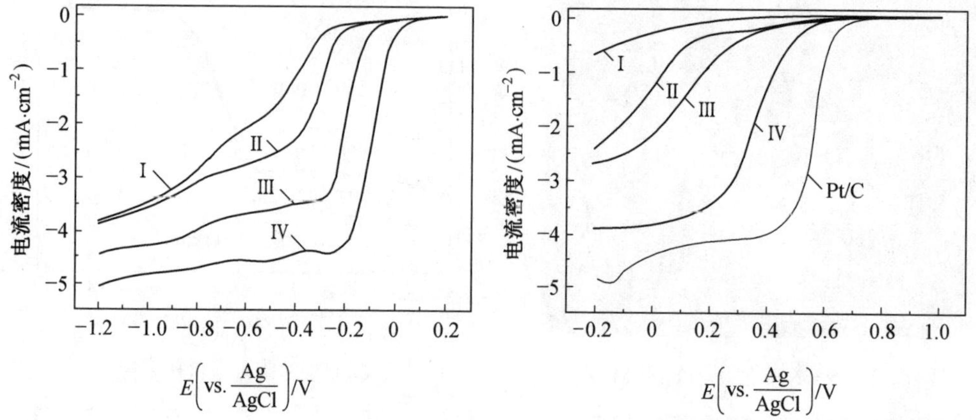

（g）处理后 NCNFs 的 RDE 曲线（$0.1\ mol\cdot L^{-1}$ KOH）

（h）处理后 NCNFs 的 RDE 曲线（$0.5\ mol\cdot L^{-1}\ H_2SO_4$）

续图 2.19

非金属元素（如 N、B、S 等）通常以掺杂的形式来实现对碳基材料的改性。尤其是氮原子在 ORR 性能的提升方面有着显著的作用，碳纳米材料中引入氮元素，不仅可以使得碳原子的电荷分布情况发生一定改变，还会将大量的缺陷位点引入碳基中，从而使其电化学性质发生变化。氮原子在碳材料结构中的存在形态也会影响氮掺杂碳基材料的氧还原催化活性。碳基材料结构上掺杂的 N 原子分为吡啶氮、吡咯氮和石墨化氮三种类型。不同类型的氮原子掺杂使得相邻碳原子的电子结构发生改变，进而影响其催化性能。Wang 等人通过静电纺丝、稳定化以及 NH_3 刻蚀的方法，以聚丙烯腈（PAN）为原料，制备了氮掺杂碳纳米纤维（NCNFs），作为 ORR 高活性电催化剂，在 KOH 电解质中的 ORR 还原电位约在 0.047 V 处，在 H_2SO_4 电解质中的 ORR 还原电位为 0.426 V。氮氧掺杂的碳基材料的 ORR 性能也有较大的提升。Zeng 等人利用静电纺丝技术结合化学气相沉积法合成 3D 石墨烯纤维（3DGFs），随后对其进行酸处理及 NH_3 气氛热处理，以获得 N、O 共掺杂的 3D 石墨烯纤维。该纳米纤维表现出优异的 ORR 电催化活性，起始电位为 1.01 V，半波电位为 0.883 V，在 50 h 后仍能保持 90%的电流。除此之外，该催化剂还具有优异的甲醇耐受性。

氮掺杂碳纳米材料在碱性溶液中对氧还原反应（ORR）表现出优异的电催化活性，而在酸性介质中的电化学活性却很差。发现通过酸化以及热处理工艺，可以提高该类催化剂在酸性和碱性溶液中的 ORR 性能。他们以聚丙烯腈（PAN）为原料，通过静电纺丝技术、酸化结合 NH_3 气氛下热处理工艺，制备了氮掺杂碳纳米纤维（NCNFs）。NCNFs 在酸性和碱性介质中均表现出优异的 ORR 电催化活性。优异的性能归因于纳米纤维较小的直径，这使得具有多孔结构的纤维暴露出更多的活性边缘。此外，较高的吡啶氮含量，也大大提高了材料的 ORR 电催化性能。其微观形貌及性能图如图 2.19（e）、（f）所示。

碳化钨（WC）催化剂也显示出优异的 ORR 催化性能。当前，WC 材料直接用于 ORR 催化的研究较少，在 ORR 中的应用主要集中在以 WC 作为贵金属催化剂的载体或高效促进剂，这是由于 WC 和贵金属组分表现出协同作用，因此增强了贵金属催化剂的 ORR 活性。以偏钨酸铵（AMT）和聚乙烯吡咯烷酮（PVP）为原料，利用静电纺丝技术和高温 NH_3 刻蚀工艺，合成了具有超细直径的碳化钨（WC）纳米纤维。经高温 NH_3 刻蚀后，纳米纤维表面覆盖的碳层被刻蚀去，WC 纳米纤维在氧还原反应中表现出良好的电催化活性和稳定性。

银（Ag）相对于贵金属（例如 Pt、Au 和 Pd）而言成本较低，ORR 电催化性能较好，并且对甲醇有着更好的耐受性，是一种代替 Pt 的理想催化剂材料。Zhang 等人通过结合静电纺丝与电镀工艺，制备了具有出色的 ORR 电催化性能的自支撑银纳米纤维（AgNFs）电催化剂，这种纳米纤维催化剂表现出优异的稳定性和极高的甲醇耐受性，相关情况如图 2.20 所示。其起始电位、半波电位和峰值电位分别为 1.041 V、0.848 V 和 0.864 V，优于商用 Pt/C 催化剂。优异的性能归因于纤维状的微观形态和晶面取向，银纳米纤维尖端上的电子发生积聚，充当电化学活性位点，而尖端上高比例的（110）晶面有利于提高催化活性。这种低成本的银纳米纤维催化剂作为氧还原反应的潜在替代催化剂具有很大的应用前景。

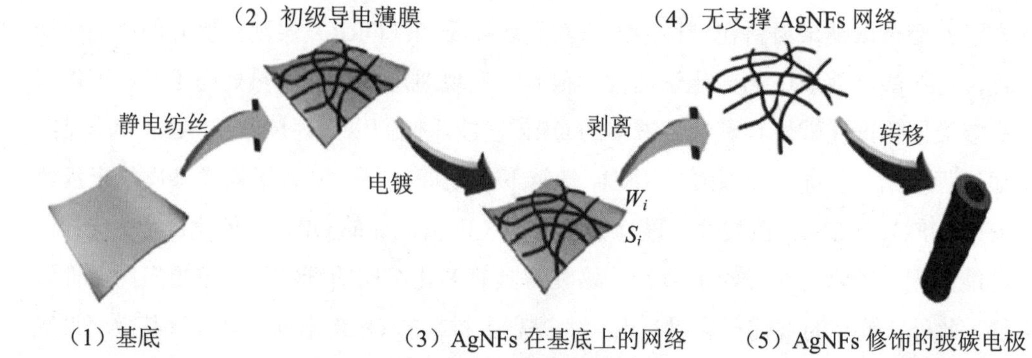

（a）自支撑银纳米纤维电催化剂制备过程

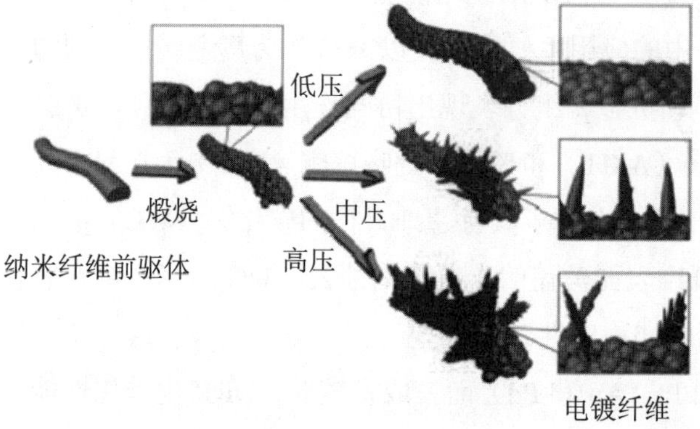

（b）不同形貌的自支撑银纳米纤维

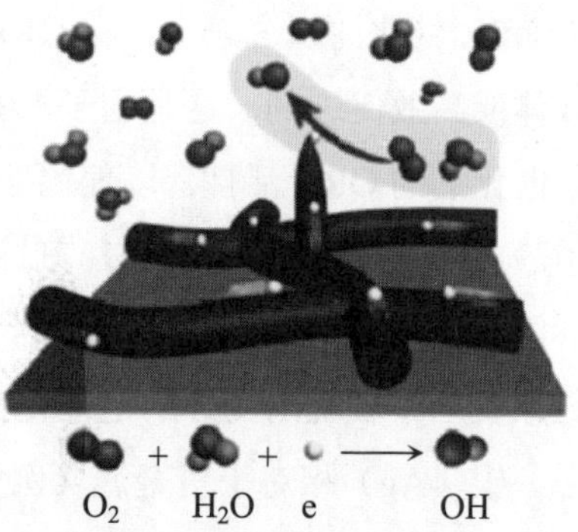

（c）氧还原催化原理

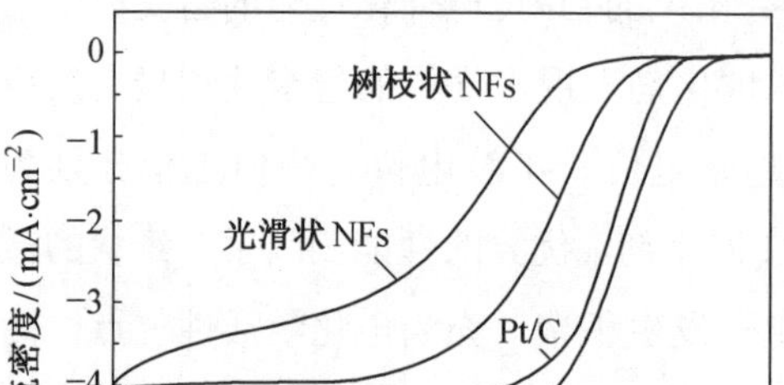

（d）催化剂在 0.1 mol·L^{-1} KOH 溶液中的线性扫描伏安（LSV）曲线

（e）多刺状银纳米线

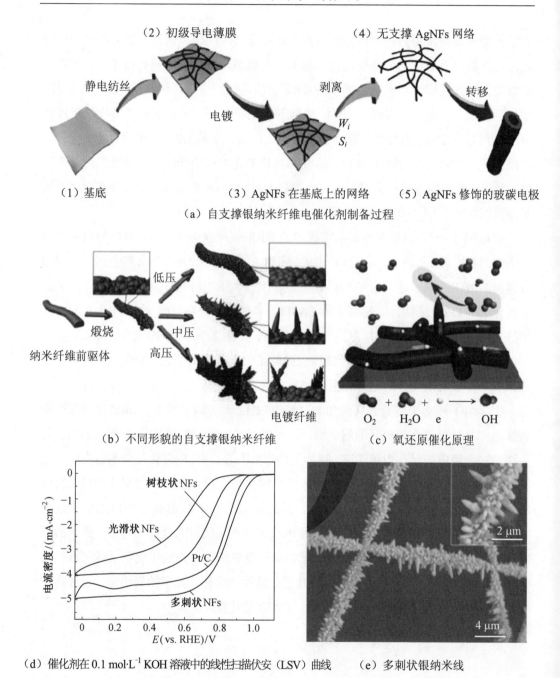

图 2.20　自支撑银纳米纤维电催化剂的制备、催化原理、SEM 图和电化学性能

2.4.4　小结与展望

静电纺丝材料在燃料电池中的大规模应用还需克服一些挑战，对于纳米纤维在催化剂和电解质膜中的应用，为了获得更好的电催化性能和更低的成本，还需要进一步探索工艺来减小纳米纤维的直径和改善直径的均一性。影响静电纺丝技术的工艺参数还不完全清楚，因此，对静电纺丝制备的催化剂和载体材料的种类仍有局限性。以上问题还有待研究者们进一步探索。

2.5　染料敏化太阳能电池

太阳不断地向地球辐射大量的能量，每秒的辐射能量超过人类每年消耗的能量的数千倍，而不涉及消耗和污染，开发廉价高效的光能转换装置已迫在眉睫。不同于硅基太阳能电池需要高纯度的原材料和复杂的制备工艺，染料敏化太阳能电池（DSSCs）具有成本低、制造工艺简单、理论能量转换效率高等优点，成为下一代太阳能电池的候选材料。在过去的几年里，静电纺纳米纤维被用于染料敏化太阳能电池中。静电纺丝法也可以用来制备聚合物凝胶电解质以代替传统的液体电解质，解决了 DSSCs 长期稳定性的问题。在这一节中，介绍了静电纺丝在设计和制造 DSSC 用纳米纤维材料中的应用。

2.5.1　引言

随着地球化石燃料的加速消耗，发展绿色可再生能源已成为世界各国的当务之急。太阳能是一种无污染的能源，被认为是最理想的能源之一。人们在开发高效的太阳能转换技术方面投入了大量的精力，最有前景的方法是制造太阳能电池。太阳能电池是将太阳能转化为电能的器件，包括硅基光伏电池、薄膜太阳能电池、聚合物太阳能电池和染料敏化太阳能电池。自 1991 年 O'Regan 和 Grätzel 的开创性工作以来，染料敏化太阳能电池因成本低、制造工艺简单、光-电转换效率高等优点，一直受到全球的关注。这些电池表现出优异的高能效，即使它们是基于廉价的原材料和简单的技术。

染料敏化太阳能电池的光-电子转换过程与 p-n 结硅太阳电池有着本质的区别。图 2.21 所示为染料敏化太阳能电池的工作原理。首先敏化剂 S^*吸收光子（式（2.1）），激发的敏化剂 S^*将电子注入半导体的导带，使敏化剂处于氧化状态 S^+（式（2.2））。注入的电子流经半导体网络后，通过外部负载到达对电极，以减少氧化还原介质（式（2.3）），后者反过来再生敏化剂（式（2.4））。在光照下，该装置构成一个再生稳定的光伏能量转换系统。

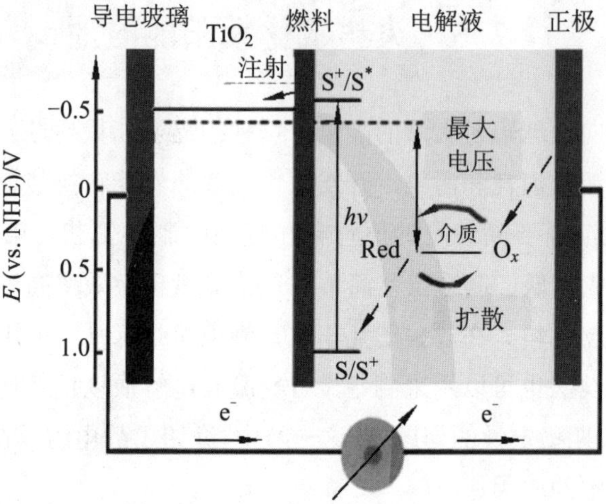

图 2.21　染料敏化太阳能电池的工作原理

$$S_{(adsorbed)} + hv \longrightarrow S^*_{(adsorbed)} \tag{2.1}$$

$$S^*_{(adsorbed)} \longrightarrow S^+_{(adsorbed)} + e^-_{(injected)} \tag{2.2}$$

$$I_3^- + 2e^-_{(cathode)} \longrightarrow 3I^-_{(cathode)} \tag{2.3}$$

$$2S^+_{(adsorbed)} + 3I^- \longrightarrow 2S_{(adsorbed)} + I_3^- \tag{2.4}$$

一些不良反应导致电池效率下降。它们是注入电子与氧化增感剂（式（2.5））或与二氧化钛表面的氧化还原介质（式（2.6））的复合。

$$S^{+}_{(adsorbed)} + e^{-}_{(titania)} \longrightarrow S_{(adsorbed)} \tag{2.5}$$

$$I_3^{-} + 2e^{-}_{(titania)} \longrightarrow 3I^{-}_{(anode)} \tag{2.6}$$

敏化剂的光激发后，电子注入半导体氧化物薄膜的导带。染料分子由氧化还原系统再生，氧化还原系统本身在对电极上由通过负载的电子再生。电位是指正常的氢电极（NHE）。

染料敏化太阳能电池的光伏性能可以用填充因子（FF）和电池效率（η）来评估，表示为

$$FF = \frac{U_{max} J_{max}}{V_{OC} J_{SC}}$$

$$\eta = \frac{U_{OC} J_{SC} FF}{P_{in}} \times 100\%$$

式中，J_{SC} 为短路电流密度，$mA \cdot cm^{-2}$；U_{OC} 为开路电压，V；FF 为填充因子；P_{in} 为入射光功率；J_{max} 和 U_{max} 定义为获得最大功率输出的电流密度和电压。FF 反映了染料敏化太阳能电池在运行过程中发生的电气和电化学损耗，主要取决于电池的串联电阻，其值通常在 0～1 之间。

染料敏化太阳能电池的最高理论 U_{OC} 为 0.95 V（TiO_2 的平带电位（−0.7 V）和 I^{-}/I_3^{-} 氧化还原电位（+0.25 V）之间的差异）。U_{OC} 还受到背电子转移反应的影响，较低的复合速率可以导致较高的 U_{OC}。J_{SC} 是染料敏化太阳能电池的一个关键参数，决定染料分子的吸附量、光阳极的集光率、光电子的扩散和聚集。

入射单色光子对电流转换效率（IPCE），有时也称为外部量子效率（EQE），是衡量器件性能的另一个基本指标。IPCE 值对应于染料敏化太阳能电池单光照明下外部电路中产生的光电流密度除以撞击电池的光子通量。根据这样的实验，IPCE 作为波长的函数可以由以下公式计算：

$$IPCE(\lambda) = 1\,240 \frac{J_{SC}(\lambda)}{\lambda P_{in}(\lambda)}$$

IPCE 值提供了关于太阳能电池单色量子效率的实用信息。

2.5.2 光阳极

染料敏化太阳能电池光阳极通常具有导电基底上的金属氧化物膜和吸附在金属氧化物表面的染料敏化剂的结构。染料敏化太阳能电池中的染料在吸收光子时产生激子（束缚电子-空穴对），并经历解离以释放自由电子和空穴。自由电子被注入金属氧化物（光阳极）中，并被传送到电极处收集。电子形成一个环路，所以产生光电流。收集的电子越多，产生的光电流和染料敏化太阳能电池的效率就越高。

首先，在最小化电荷复合的同时必须增强电荷传输，这可以通过控制光阳极材料的形貌、表面积、孔隙率和微晶尺寸来实现，因为这些特性会强烈影响电荷传输和复合过程。例如，TiO_2 或 ZnO 等光阳极材料中的电子输运与表面状态、粒子形貌和粒子间连接性密切相关。近年来，一维纳米结构材料，如纳米管、纳米棒、纳米带、纳米线和纳米纤维等，被广泛研究用作染料敏化太阳能电池的光阳极材料。与几何结构无序、电子输运中存在界面干扰的传统纳米颗粒体系相比，一维纳米纤维具有更优异的电子输运性能。人们认为一维纳米结构由于其相互连接性、细晶尺寸和高比表面积而呈现较低的晶界，因此在这种结构中电子传输速度更快。光生电子能迅速转移到光阳极材料表面并被器件收集，因此光生电子-空穴的复合较少。由纳米纤维制成的光阳极可以最大限度地提高电池器件的集光性能。对于同一种光阳极材料，光利用率越高，产生的电子越多。具有大表面积和高孔隙率的一维无机纳米纤维可以吸收最大量的染料敏化剂，促进电子的传输和光的利用。此外，与传统的由致密纳米颗粒制成的电极相比，纳米纤维的行为类似于蓬松结构，允许更好的孔填充电解质。这种将黏性电解质过滤到纳米纤维网络中的方法改善了与无机半导体的接触，并有助于氧化染料的再生，从而提高了能量转换效率。但是，光阳极材料不能太蓬松，否则光生电子的传输可能受阻或中断。为了进一步增加染料负载量和太阳能电池的整体效率，电纺纳米纤维也被缩短为纳米棒，通过单独使用或与纳米纤维/纳米颗粒结合制备。

在电纺纳米纤维的煅烧过程中，纳米纤维与导电基体的黏附性差，成为直接序列扩展器应用中的关键问题之一。为了克服这一问题，一些研究小组开发了静电纺纳米纤维的新处理方法。通过机械研磨将 TiO_2 纳米纤维研磨成纳米棒，然后在导电玻璃板上喷涂和烧结，制备出由纳米颗粒/纳米棒层组成的 TiO_2 阳极。结果表明，采用该方法可显著提高附着力，转换效率约为 5.8%。Yang 等人报道了一种由 TiO_2 纳米颗粒和多壁碳纳米管组成的光阳极。在导电玻璃板上附着力好的纳米棒中的多壁碳纳米管可以有效地收集和传输光生电子，从而减少复合，提高器件效率。热压预处理也能提高黏附性，Song 等人采用热压预处理，提高了 TiO_2 纳米纤维与基体的附着力，转换效率约为 5.02%。这些结果表明，通过提高沉积纳米纤维与基底之间的黏附力，可以改善器件的性能。

近年来，随着 TiO_2 制备方法的发展，染料敏化太阳能电池的光阳极得到了迅速发展，以提高器件性能。Zheng 等采用静电纺丝技术合成了不同相态的 TiO_2 纳米纤维作为光阳极。通过优化退火温度和阳极厚度，得到的 TiO_2 纳米纤维光阳极的功率转换效率为 6.12%。此外，他们进一步用 $TiCl_4$ 处理电纺 TiO_2，得到的阳极效率提高了 7.06%。性能的提高归因于介孔锐钛矿型/金红石型 TiO_2 复合纳米纤维的不同晶型结构具有阶梯式的能级和较高的紧密晶粒堆积比表面积。这些结果表明，静电纺纳米纤维光阳极为染料敏化太阳能电池提供了一种有效的电荷传输手段。Cao 等用分层纳米棒支化 TiO_2 纳米纤维制备了一种新型阳极。为了获得这样的结构（图 2.22），将电纺纯 TiO_2 在混合溶液中 180 ℃下进行 9 h 的水热处理，该溶液以二甘醇和去离子水为溶剂，草酸氧化钛钾为溶质。复合光阳极的转换效率提高了 6.26%，而原始的 TiO_2 纳米纤维光阳极的转换效率仅为 4.26%，这是因为该复合光阳极具有提高染料负载能力和良好的宽带光散射的协同效应。将金、银、石墨烯、氧化石墨烯（GO）等纳米粒子结合起来，是提高 TiO_2 节点性能的另一种方法，而提高转换效率的原因在于局部表面增强了集光能力。此外，粒子可以降低改性 TiO_2 光节点的表面缺陷浓度，提高电子注入和电子收集效率。Baraka 等人通过静电纺丝和水热法在 TiO_2 纳米纤维中掺杂 N 元素、氧化石墨烯（GO）和 SnO_2 制备了 GO&N@SnO_2/TiO_2 复合材料，获得了高的染料负载能力（2.164×10^{-7} mol·cm^{-2}）和较高的功率转换效率（6.18%）。这可能是因为掺杂剂通过形成宽禁带氧化物增强了电荷转移，减少了电子/空穴的复合，另外，石墨烯掺杂可能为工作电极层的电子转移开辟了新的途径。Kim 等人用石墨烯掺杂的碳纳米纤维制备了表面有 TiO_2 层的光阳极，其性能也很好。

（a）TiO_2 纤维（TF）SEM 图

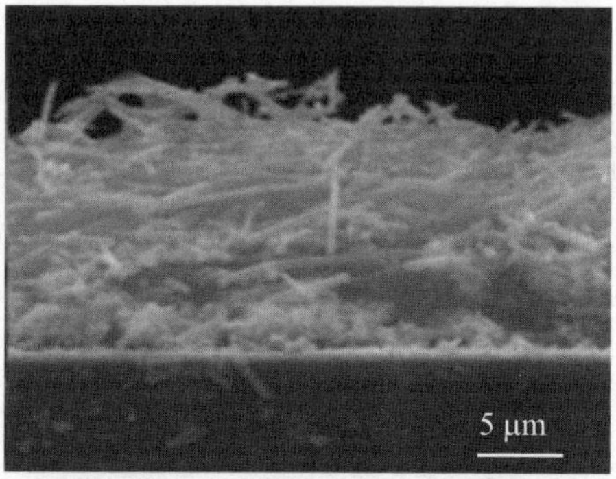

（b）TiO_2 纤维（TF）SEM 图

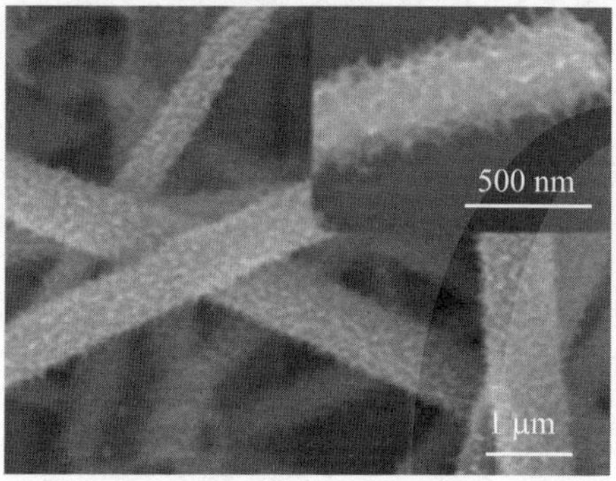

（c）分层分支 TiO_2 纤维（HBTF）SEM 图

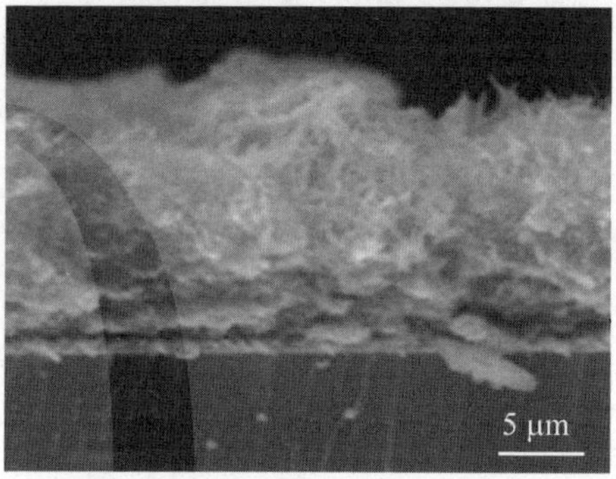

（d）分层分支 TiO_2 纤维（HBTF）SEM 图

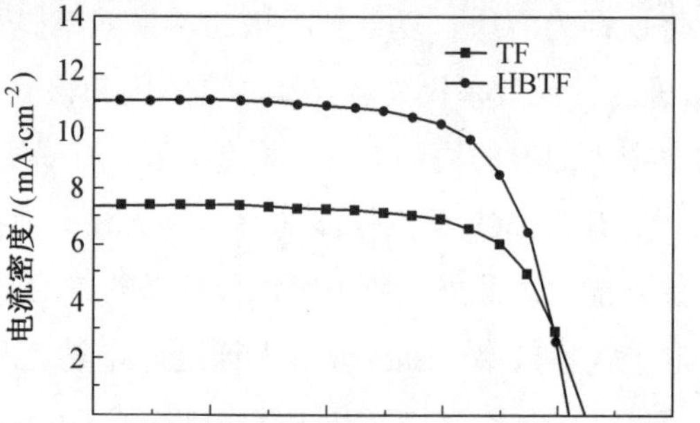

（e）厚度为 20 μm 的 TF 和 HBTF 光阳极的 DSSC 的 *J–V* 曲线

图 2.22 分层纳米棒支化 TiO_2 纳米纤维的 SEM 图和电化学性能

尽管 TiO_2 被广泛用作光阳极，但由于其与 TiO_2 类似的电子注入过程和带隙，ZnO 也被用作 DSSC 的候选材料。2007 年，静电纺 ZnO 纳米纤维首次应用于染料敏化太阳能电池，转换效率为 1.34%。Wu 等利用硫化镉（CdS）纳米晶对电纺氧化锌/聚（3-己基噻吩）（ZnO/P3HT）纳米纤维进行改性，成功制备了 CdS/ZnO 核壳纳米纤维。CdS 改性可显著提高功率转换效率 100%以上（从 0.3%提高到 0.65%）。Subramania 等人采用硒化锌量子点作为敏化剂，对 ZnO 纳米纤维进行静电纺丝并用作光阳极。ZnSe 量子点敏化 ZnO 纳米纤维的短路电流密度（6.60 $mA \cdot cm^{-2}$）有明显改善，最大功率转换效率为 1.24%，这主要是由于 ZnSe 量子点和 ZnO 纳米纤维具有更好的集光能力，以及光注入电子与聚硫电解质的复合较低。

总体来说，作为光阳极的电纺 ZnO 的转化效率远低于 TiO_2，这主要是由于与 TiO_2 材料相比，ZnO 在酸性染料溶液中的化学稳定性较差，表面积较小。到目前为止，其他金属氧化物，如 SnO_2 和 WO_3 也可以用作光阳极，但其性能仍不如 TiO_2。

2.5.3　对电极

在染料敏化太阳能电池中，对电极起着发射和收集电子的作用。染料敏化太阳能电池中常用的对电极是铂，但铂是一种昂贵的贵金属，在电解液中腐蚀性的三碘化物/碘（I_3^- / I^-）氧化还原介质会降低其电催化活性，从而影响染料敏化太阳能电池的稳定性。碳材料，如石墨、炭黑、碳纳米管等，由于成本低、易于获得，已成为取代铂作为对电极的候选材料。典型的研究是 Joshi 等的工作，其中电纺碳纳米纤维对电极具有电荷转移电阻低、电容大、I_3^- 还原速度快等特点。虽然短路电流密度和开路电压相当，但碳纳米纤维基电池的效率略低于铂基电池，这可能是高串联电阻导致填充因子较低所致。为了解决上述问题降低电阻，Park 等人采用同轴静电纺丝法，以聚丙烯腈/聚甲基丙烯酸甲酯（PAN/PMMA）共混聚合物为碳壳前驱体，PMMA 为芯前驱体，采用同轴静电纺丝法制备中空/高介孔壳结构碳纳米纤维（Meso-HACNF），并进行了稳定化、碳化和活化处理。当壳层前驱体溶液中 PMMA 的质量分数为 10%时，其能量转换效率为 7.21%，与 Pt 的 7.69%相当，并且由于其促进电子和离子转移的新颖特性而具有低电阻。Zhang 和他的同事用电纺 $Fe(NO_3)_3$/PAN 溶液合成了 α-Fe_2O_3 和 FeS 纳米棒，随后碳化得到 α-Fe_2O_3 纳米纤维，在氩气氛围下进一步硫化得到 FeS 纳米棒（图 2.23）。制备的 FeS 纳米棒对电极的光能转换效率高达 6.47%，而 Pt 对电极的光能转换效率为 3.79%，且具有比 Pt 基对电极更好的长周期性，这是由于所制备的一维纳米结构可以促进电荷输运，提高机械强度以保证稳定性催化活性中心。虽然已经取得了很大的进展，但碳基对电极取代铂仍然面临着一些挑战，如减小对电极厚度，减小总串联电阻，使其性能更具竞争力。

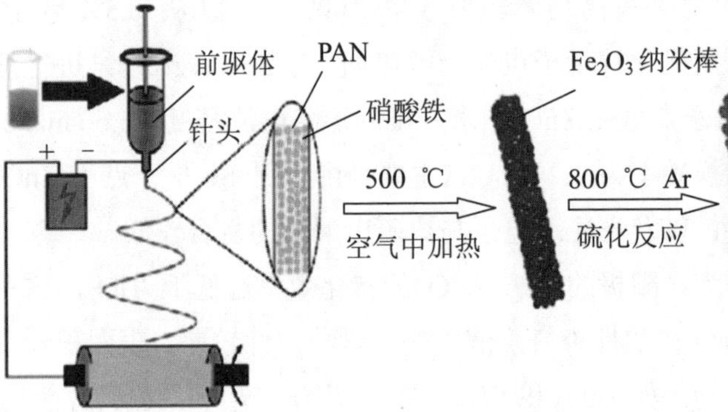

（a）α-Fe_2O_3和 FeS 纳米棒合成示意图

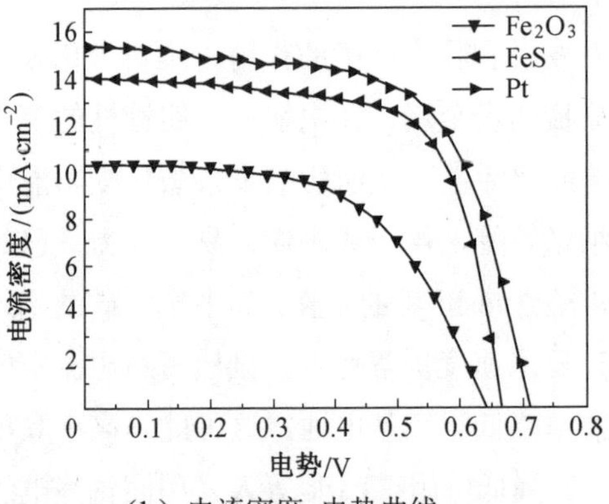

（b）电流密度-电势曲线

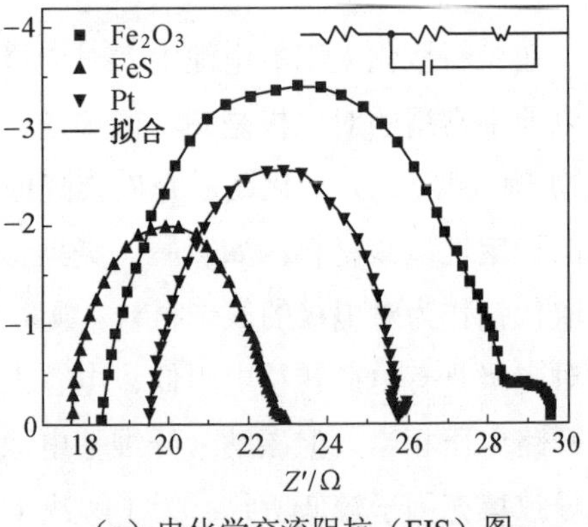

（c）电化学交流阻抗（EIS）图

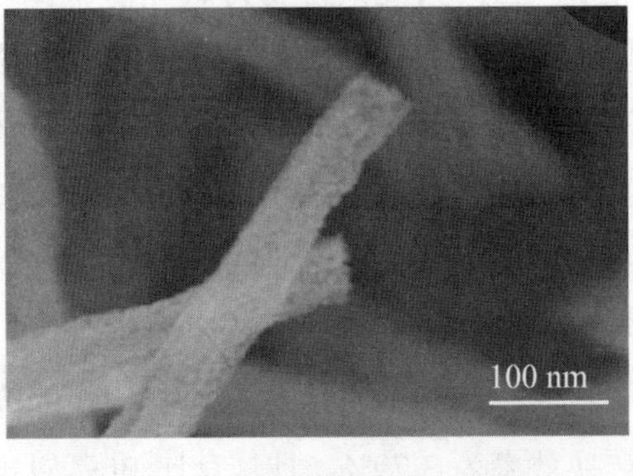

（d）α-Fe_2O_3纳米棒 SEM 图

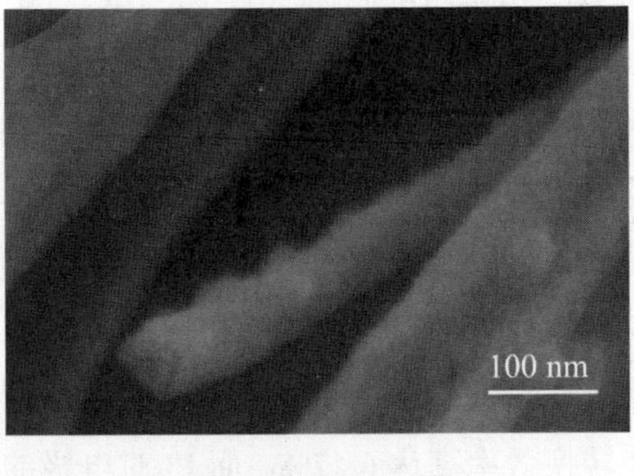

（e）FeS 纳米棒 SEM 图

图 2.23　α-Fe_2O_3和 FeS 纳米棒的合成、电化学性能和 SEM 图

2.5.4 电解质

染料敏化太阳能电池中的氧化还原电解液作为传输路径，将电子从对电极转移到氧化染料，从而补偿染料中的电子损失。染料敏化太阳能电池中的电解质可分为液体、准固态和固态。染料敏化太阳能电池中最常用的电解液是液态，通常是碘化物/三碘化物基化合物。然而，它们的使用也有缺点。液体的泄漏是一个主要的问题，因此，必须更加小心地密封太阳能电池。

另一种电解质是固态电解质，如聚合物，但是它们由于电子注入较低而效率较低。准固态电解质在染料敏化太阳能电池中有较好的发展趋势。在最近的研究中，静电纺丝纳米纤维被用作准固态电解质的替代品。研究结果表明，电纺纳米纤维作为电解质除了更稳定外，还获得了与液态电解液相近的高效率，因为它们避免了染料敏化太阳能电池的复杂的密封和不挥发等潜在问题。

2014 年，Sethupathy 等人利用基于 PVDF 和 PAN 混合的聚合物电解质制备了染料敏化太阳能电池，尽管其效率仅为 3.09%，但其实现的开路电压为 0.74 V，与之前报告的基于 PVDF 的电解质电压类似。此外，纳米纤维的孔隙率和离子电导率分别为 83.6%和 6.12×10^{-2} $S\cdot cm^{-1}$。Weerasinghe 等人使用醋酸纤维素（CA）纳米纤维作为准固态电解质，其效率为 4.0%，电池的效率进一步提高，此外，这些钙纳米纤维是可生物降解的，这一特性应在未来的环境或生物医学应用中加以考虑。Murugadoss 等人通过添加硫化钴纳米颗粒改进了 PAN 纳米纤维，使用该电解质可实现 7.41%的效率，这是由于增加了离子电导率和电极/电解质界面电荷转移的界面面积。

2.5.5 小结与展望

本节基于电纺纳米纤维对染料敏化太阳能电池组件的光阳极、电解质进行了深入的研究。然而，电池光阳极的比表面积一般不超过 100 $m^2\cdot g^{-1}$，并且纳米纤维之间的空隙过多。这两个问题都使得纳米纤维的染料吸收率仍然低于大多数纳米颗粒光阳极，从而导致静电纺丝纳米纤维直接序列太阳能电池的光电流相对较低。同时，由于纳米纤维的分层多孔性，因此阳极与基体之间的黏附能力较差，目前解决这些问题的主要方法是在静电纺丝前采用压缩层和 TiO_2 溶胶的后处理以及致密化技术。

此外，纳米纤维为固态电解质提供了越来越多的传输路径，使得无铂对电极在高效率染料敏化太阳能电池中具有竞争力。

2.6 有机光伏电池

2.6.1 引言

有机太阳能电池需要较大的界面面积才能有效分离激子，而这通常是通过热退火来实现的。然而，这种退火与低玻璃化转变温度的聚合物不相容，与柔性衬底不相容，也不与大面积器件兼容。另一种方法是使电子施主和受主与每种材料的连续相形成有效的界面。静电纺丝是一种制备导电聚合物纳米纤维的有效方法，可作为有机太阳能电池的电子供体材料。为了提高有机太阳能电池的效率并克服载流子的低迁移率问题，将纳米结构无机材料用作电子受体或电子传输层，有机和无机半导体结合作为有机-无机混合光伏电池已被广泛研究。以静电纺丝纳米纤维的形式使用无机半导体作为电子传输层是一种很有前途的替代方法，此外静电纺丝也可用于制备出具有柔性的纳米纤维基透明电极，这些电极很有希望作为有机薄膜太阳能电池和有机-无机混合太阳能电池的柔性衬底。本节重点介绍静电纺丝纳米纤维在有机薄膜太阳能电池和有机-无机混合太阳能电池方面的开发及应用。

光伏电池是一种半导体二极管，可以将太阳光转换成直流电。有机光伏是利用有机材料，特别是有机小分子或导电有机聚合物，通过光吸收和电荷输运产生光电效应的太阳能电池。有机太阳能电池中的有机小分子和导电有机聚合物的一个共同特点是它们具有大的共轭体系，碳原子与交替的单键和双键共价键形成共轭体系。当光子到达有机半导体材料时，它将被吸收，然后产生一个激发态，并限制在有机分子或导电聚合物链的一个区域内；激发态可视为静电相互作用结合在一起的电子-空穴对，即所谓的激子；激子被有效场分解成自由电子-空穴对，有效场是通过在两种不同的材料之间建立的，有效场通过使电子从吸收体（所谓的电子给体）的导带落到受主分子（所谓的电子受主）的导带来分解激子，电子受体材料的导带边缘必须低于电子供体材料的导带边缘。电子-空穴对有效分离的最有效结是电子给体和受体材料的体异质结。

有机太阳能电池与典型的无机光伏电池，即硅太阳能电池或碲化镉（CdTe）光伏电池相比，其功率转换效率仍然较低。这是由于低聚合物吸收光谱，特别是有机半导体中激发电子或空穴的低迁移率。为了改善功率转换效率，人们对有机和无机半导体作为有机-无机混合光伏电池的组合进行了大量的研究，以通过使用纳米结构无机材料作为电子传输层来克服载流子的低迁移率。

2.6.2　有机薄膜太阳能电池

有机薄膜太阳能电池中的活性层通常是基于体异质结（BHJ）系统，其中光活性层由两种不同半导体电子供体（p 型）和电子受体（n 型）的连续混合物而成。在大多数有机半导体中，激子扩散长度为 5～10 nm，光活性层的纳米级控制对于实现高功率转换效率至关重要。为了在纳米尺度上调整表面形貌，传统放入方法通常是采用热退火和添加剂处理。一维纳米材料作为一种很有前景的材料，已经在活性层中作为供体或受体广泛应用。通过调控纳米线的直径和尺寸，可以很好地匹配激子扩散长度，在纳米线太阳电池发生复合之前确保电荷分离。此外，纳米纤维的引入有效调控了器件活性层的形貌，并增加了活性层内部自由载流子的迁移率以及迁移率平衡，进而提高了器件的短路电流及填充因子。这些特性使得一维纳米材料成为 BHJ 层的一个有竞争力的应用材料。

以小分子有机半导体纳米线太阳能电池为例，C60 纳米线和 N,N′-二辛基-3,4,9,10-苝二羧酸酯（PTCDI-C_8）纳米线被报道用作体异质结纳米结构电子给体（n 型）。此外，聚合物纳米线太阳能电池的研究主要集中在 p 型纳米线太阳能电池上，如 P3ATs 及其嵌段共聚物。也有少量 n 型聚合物纳米线太阳能电池被报道，在这一方面，Briseno 等人报道了一种 n 型聚苯并双咪唑并菲咯啉（BBL）纳米带，有效提高了光电转换效率。大多数的纳米线太阳能电池都是由给体或受体纳米线制成的，但在同时包含给体和受体的纳米线 BHJ 太阳能电池也有研究。Jenekhe 及其同事制备了一系列低聚噻吩官能化萘二亚胺（NDI-nTH 和 NDI-nT）纳米线，并将其作为电子受体层应用于纳米线 BHJ 有机太阳能电池，结构由 ITO/PEDOT:PSS（40 nm）/P3HT（15 nm）/活性层（约 80 nm）/1,3,5-三（1-苯基-1H-苯并咪唑-2-基）苯（TPBI）（3 nm）/Al（80 nm）组成，P3HT 和 TPBI 薄膜分别作为电子阻挡层和空穴阻挡层，活性层由 P3HT 纳米线与 NDI 纳米线（质量比 1∶1）混合，这种纳米线基太阳能电池的功率转换效率为 1.15%，证实这种活性中间层能够有效提升功率转换效率。

周期性和垂直排列的 BHJ 太阳能电池被认为是实现高效率的理想结构。在这种器件结构中，光生激子可以很容易被分离，并且电荷载流子有它们通向每个相应电极的有效路径。图 2.24（a）所示为基于垂直排列的[6,6]-苯基-C_{61}-丁酸苯乙烯酯（PCBSD）纳米棒的有机太阳能电池。结合 PCBSD 材料的交联特性，采用 AAO 模板法实现了纳米棒的制备。有机/有机有序纳米结构异质结被认为是实现高效率的有效结构之一，Li 和同事制备了具有 P3HT∶茚-C_{60}双加合物（ICBA）BHJ 上层的有机/有机有序纳米结构异质结太阳能电池，其光电流增强，功率转换效率高达 7.3%，证实了垂直排列的纳米线系统可以促进有效的电荷输运。此外，聚合物异质结也被认为是提高太阳能电池的功率转换效率的有效手段，He 等采用双纳米印迹法制备了互穿纳米结构聚合物异质结，并将其应用于有机太阳能电池的光活性层中（图 2.24（b）），聚（(9,9-二辛基芴）-2,7-二烷基-[4,7-双（3-己基噻吩-5-基）-2,1,3-苯并噻二唑] -2'，2''-二基）（F8TBT）/P3HT 薄膜由纳米印迹法制得，形成有序的纳米阵列，间距为 20～50 nm。纳米结构太阳能电池的功率转换效率达到了 1.9%，比共混器件提高了 50%。

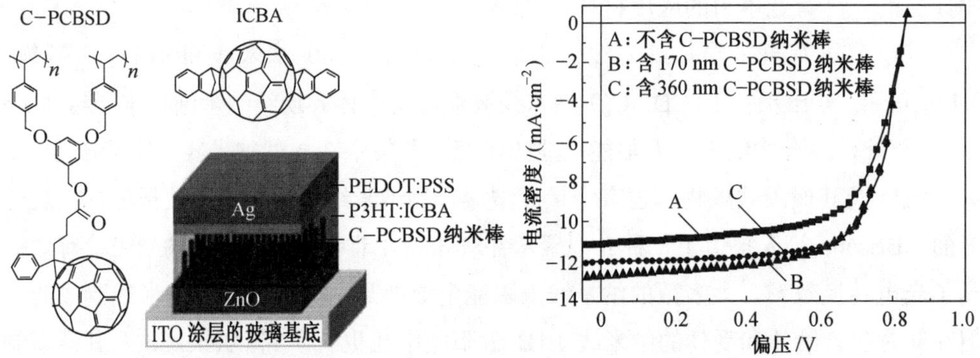

（a）以交联 PCBSD 纳米结构作为活性层制备的有机太阳能电池（左）及其光电性能（右）

图 2.24　有机薄膜太阳能电池的组装示意图、光电性能图及 SEM 图

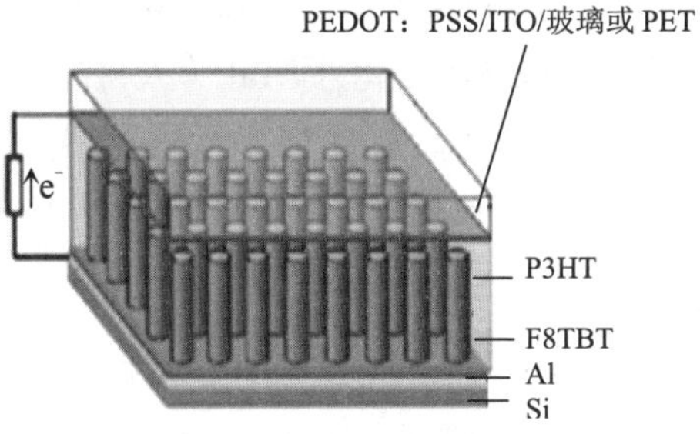

（b）纳米线-F8TBT/P3HT 有机太阳能电池原理图

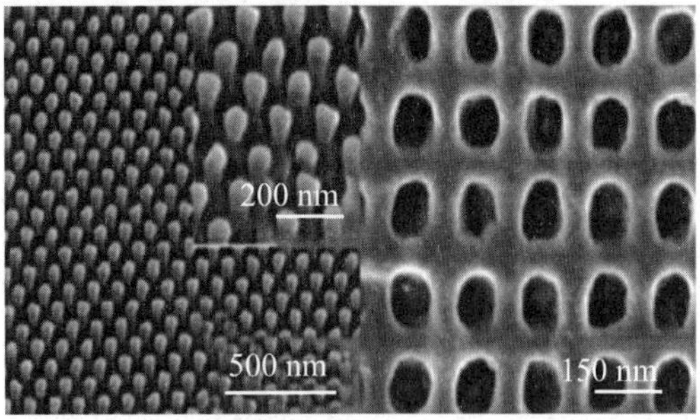

（c）相应的 SEM 图像

续图 2.24

2.6.3　有机-无机杂化太阳能电池

为了提高有机薄膜太阳能电池的效率，人们广泛研究了有机半导体和无机半导体作为有机-无机混合光伏电池的组合，以利用纳米结构无机材料作为电子传输层来克服载流子的低迁移率。静电纺丝法制备的 TiO_2 纳米纤维薄膜具有较大的孔隙，有望提高黏性聚合物凝胶电解质的渗透性。金属氧化物的一维纤维取向形态与烧结纳米颗粒相比，由于其晶界减小，因此有助于更好地进行电荷传导，烧结纳米颗粒也伴随着染料敏化剂吸附所需的高比表面积，通过纳米纤维膜的半定向电荷传输可达到高功率转换效率。Chuangchote 等通过刮刀在氟掺杂的氧化锡透明导电玻璃（FTO 玻璃）板上涂覆 TiO_2 纳米颗粒浆料，将直径为 250 nm 的 TiO_2 纳米纤维直接电纺到 TiO_2/FTO 基板上。制备出的纳米颗粒/纳米纤维复合电极经过煅烧，然后浸入钌染料中敏化。对于 0.25 cm^2 和 0.025 cm^2 两种不同的表面积，测量的转换效率分别为 8.14% 和 10.3%，有效提高了功率转换效率。基于线性一维和零维纳米粒子的组合也被证实是提高功率转换效率的有效结构，Lee 等人结合静电纺丝和溶胶-凝胶化学法制备了纳米棒和基于纳米颗粒的太阳能电池。$TiCl_4$ 处理前后的功率转换效率的测量结果是 9.52%和 11.02%。此外，Luo 等制备了线性和支链 TiO_2 纳米复合太阳能电池，结果表明，对支化纳米棒进行热处理，其效率为 0.16%，证实了线性和支链结构有利于功率转换。

将无机物纺入纤维能有效提高功率转换效率，Wu 等制备了电纺氧化锌纳米纤维基混合型太阳能电池，并对其性能进行了测试，效率为 0.51%。此外，Tai 等人将 TiO_2 的静电纺纳米纤维涂覆在导电玻璃板上，并通过使用钌染料（N719）对 TiO_2 纳米纤维表面进行改性，图 2.25（a）所示为 TiO_2/P3HT 的混合型太阳能电池的示意图，产生转化效率为 1.1%。通过静电纺丝将有机/无机混合物纺在纳米纤维中也被证实是提高电池效率的有效方法。在这方面，Olson 等人将聚 3-己基噻吩（P3HT）和[6,6]-苯基 C{61}丁酸甲酯（PCBM）混合物加入到 ZnO 纳米纤维中，图 2.25（b）所示为典型的 ITO/ZnO 纤维/P3HT/PCBM/Ag 器件和 ITO/ZnO 纤维/P3HT/PCBM 共混物/Ag 器件在强度为 100 $mW \cdot cm^{-2}$ 的 AM1.5 光照下的 *J-V* 特性，其产生的效率约为 2.03%，器件结构如图 2.25（c）所示。

为了扩大激子离解和电荷传导区域，Shim 等人引入随机取向的金属氧化物纳米纤维和排列和交叉结构。单轴排列的 TiO_2 纳米纤维多层膜和交叉排列的纳米纤维阵列如图 2.25（d）、（e）所示。研究结果表明，与随机取向的 TiO_2 纳米纤维相比，排列整齐的 TiO_2 纳米纤维阵列和聚[2-甲氧基-5-（2-乙基己氧基）-1, 4-苯撑乙烯撑]（MEH-PPV）聚合物组成的活性层由于电荷分离更好，因此具有更好的光致发光特性。测试结果如图 2.25（f）所示，电池的光诱导电流　（1.28 $mA \cdot cm^{-2}$）和功率转换效率可显著提高至少 50%，这取决于纤维中的排列程度。此外，Yu 等报道了基于具有共轭 P3HT 的有序电纺 TiO_2 和 ZnO 纳米纤维阵列的混合型太阳能电池，他们发现具有 3 个交叉排列层的有机-无机混合型太阳能电池表现出更好的性能。

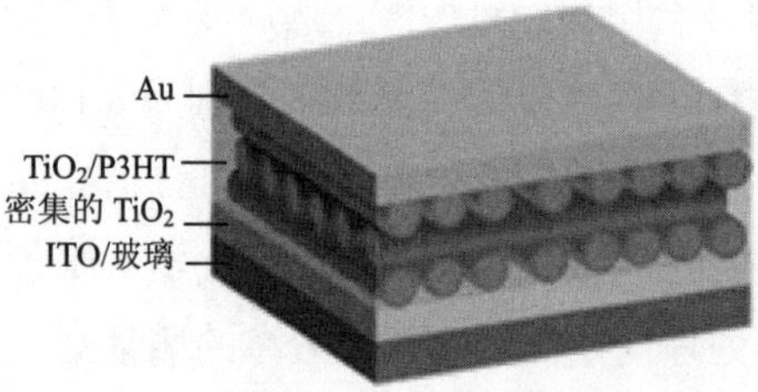

（a）基于 TiO_2 纤维/P3HT 的混合型太阳能电池组装图

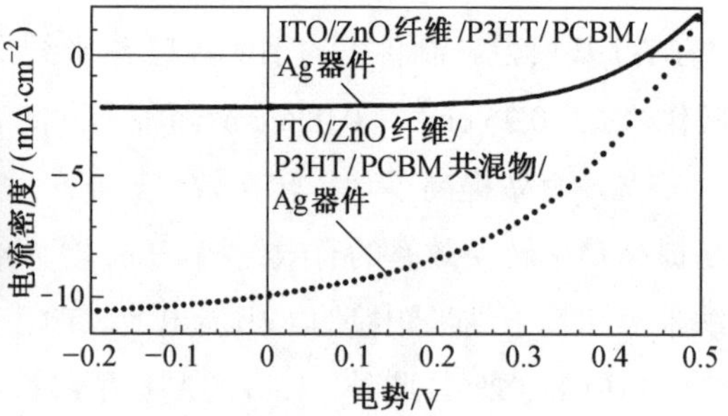

（b）光电性能图

图 2.25　有机-无机杂化太阳能电池空器件图、SEM 图及光电性能

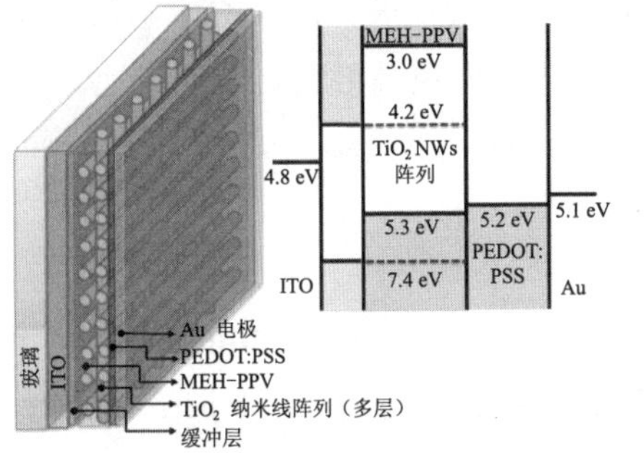

（c）TiO_2 纳米纤维的器件结构示意图和能带图

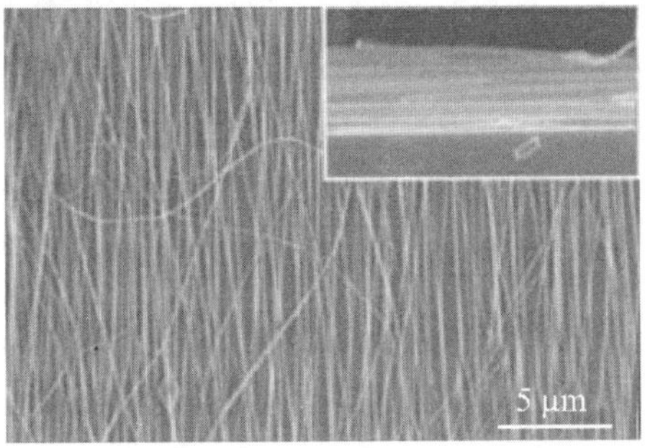

（d）单轴排列 TiO_2 纳米纤维的 SEM 图像

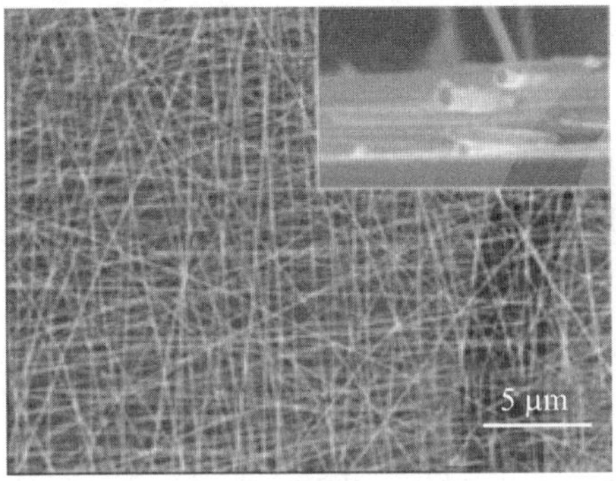

（e）交叉排列的 TiO_2 纳米纤维的 SEM 图像

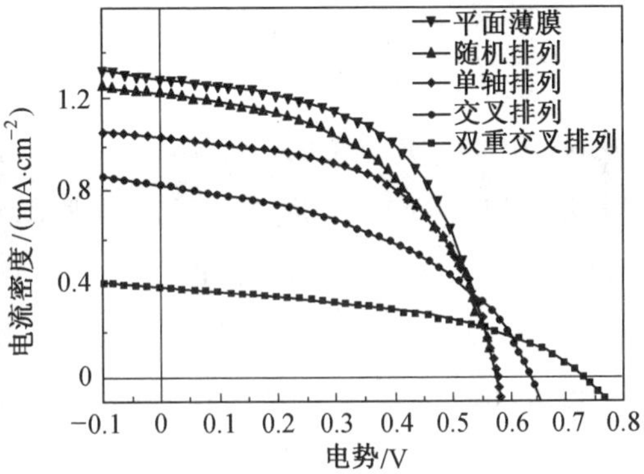

（f）MEH-PPV 的 *J*-*V* 曲线:TiO_2 与 TiO_2 相的结构变化

续图 2.25

2.6.4　小结与展望

将未来的研究方向转向可定制的纳米结构将是开发下一代有机光伏电池的一个重要阶段。虽然聚合物或混合太阳能电池由于其易于测量而越来越受到重视，但在光照和黑暗中的储存寿命测量仍需引起重视。与寿命超过 25 年的硅基太阳能电池相比，有机光伏电池的寿命通常在几分钟到几天之间。纳米纤维的多晶性和无规网状结构显示出与高效纳米颗粒电池相近的有效电子扩散系数。尽管有空间电荷复合势垒，但由于存在大量的表面陷阱，电池寿命较短。这些发现有助于制定一些在制备增强型半导体-有机混合太阳能电池时应考虑的一般规则。基于静电纺丝技术的混合型太阳能电池在学术研究和工业生产中都得到了快速的发展。潜在的成本效益和易于大规模生产使得静电纺丝技术有希望用于大规模开发更高效的太阳能电池。

参考文献

[1] WANG H G, YUAN S, MA D L, et al. Electrospun materials for lithium and sodium rechargeable batteries: from structure evolution to electrochemical performance[J]. Energy & Environmental Science, 2015, 8(6): 1660-1681.

[2] LU J, CHEN Z H, MA Z F, et al. The role of nanotechnology in the development of battery materials for electric vehicles[J]. Nature Nanotechnology, 2016, 11(12): 1031-1038.

[3] WINTER M, BRODD R J. What are batteries, fuel cells, and supercapacitors? [J]. Chemical Reviews, 2004, 104(10): 4245-4270.

[4] CABANA J, MONCONDUIT L, LARCHER D, et al. Beyond intercalation-based Li-ion batteries: the state of the art and challenges of electrode materials reacting through conversion reactions[J]. Advanced Materials, 2010, 22(35): E170-E192.

[5] EFTEKHARI A. Ordered mesoporous materials for lithium-ion batteries[J]. Microporous & Mesoporous Materials, 2017, 243: 355-369.

[6] LI L L, PENG S J, LEE J K Y, et al. Electrospun hollow nanofibers for advanced secondary batteries[J]. Nano Energy, 2017, 39: 111-139.

[7] XUE J J, WU T, DAI Y Q, et al. Electrospinning and electrospun nanofibers: methods, materials, and applications[J]. Chemical Reviews, 2019, 119(8): 5298-5415.

[8] CHEN H H, HE J, LI Y L, et al. Hierarchical CuO_x-Co_3O_4 heterostructure nanowires decorated on 3D porous nitrogen-doped carbon nanofibers as flexible and free-standing anodes for high-performance lithium-ion batteries[J]. Journal of Materials Chemistry A, 2019, 7(13): 7691-7700.

[9] ETACHERI V, MAROM R, ELAZARI R, et al. Challenges in the development of advanced Li-ion batteries: a review[J]. Energy & Environmental Science, 2011, 4(9): 3243-3262.

[10] SCROSATI B, GARCHE J. Lithium batteries: status, prospects and future[J]. Journal of Power Sources, 2010, 195(9): 2419-2430.

[11] XU B Q, LUO Y D, LIU T, et al. Ultrathin N-doped carbon-coated TiO_2 coaxial nanofibers as anodes for lithium ion batteries[J]. Journal of the American Ceramic Society, 2017,100(7): 2939-2947.

[12] CHEN H H, HE J, KE G X, et al. MoS_2 nanoflowers encapsulated into carbon nanofibers containing amorphous SnO_2 as an anode for lithium-ion batteries[J]. Nanoscale, 2019, 11(35): 16253-16261.

[13] ZHANG Y, GUO G N, CHEN C, et al. An affordable manufacturing method to boost the initial coulombic efficiency of disproportionated SiO lithium-ion battery anodes[J]. Journal of Power Sources, 2019, 426(30): 116-123.

[14] NAN D, HUANG Z H, LV R T, et al. Nitrogen-enriched electrospun porous carbon nanofiber networks as high-performance free-standing electrode materials[J]. Journal of Materials Chemistry A, 2014, 2(46): 19678-19684.

[15] PARK S W, KIM J C, AHMADDAR M, et al. Superior lithium storage in nitrogen-doped carbon nanofibers with open-channels[J]. Chemical Engineering Journal, 2017, 315: 1-9.

[16] YIN H, LI Q, CAO M, et al. Nanosized-bismuth-embedded 1D carbon nanofibers as high-performance anodes for lithium-ion and sodium-ion batteries[J]. Nano Research, 2017, 10(6): 2156-2167.

[17] HWANG J Y, MYUNG S T, SUN Y K. Sodium-ion batteries: present and future[J]. Chemical Society Reviews, 2017, 46(12): 3529-3614.

[18] TANG Y X, ZHANG Y Y, LI W L, et al. Rational material design for ultrafast rechargeable lithium-ion batteries[J]. Chemical Society Reviews, 2015, 44(17): 5926-5940.

[19] LI W H, SUN X L, YU Y. Si-, Ge-, Sn-based anode materials for lithium-ion batteries: from structure design to electrochemical performance[J]. Small Methods, 2017, 1(3): 1600037.

[20] YANG Y H, LIU S, BIAN X F, et al. Morphology- and porosity-tunable synthesis of

3D nanoporous SiGe alloy as a high-performance lithium-ion battery anode[J]. ACS Nano, 2018, 12(3): 2900-2908.

[21] STOKES K, FLYNN G, GEANEY H, et al. Axial Si-Ge heterostructure nanowires as lithium-ion battery anodes[J]. Nano Letters, 2018, 18(9): 5569-5575.

[22] WU H, CUI Y. Designing nanostructured Si anodes for high energy lithium ion batteries[J]. Nano Today, 2012, 7(5): 414-429.

[23] ZHANG X H, KONG D B, LI X L, et al. Dimensionally designed carbon-silicon hybrids for lithium storage[J]. Advanced Functional Materials, 2018, 29(2): 1806061.

[24] HWANG T H, LEE Y M, KONG B S, et al. Electrospun core-shell fibers for robust silicon nanoparticle-based lithium ion battery anodes[J]. Nano Letters, 2012, 12(2): 802-807.

[25] LEE B S, SON S B, PARK K M, et al. Fabrication of Si core/C shell nanofibers and their electrochemical performances as a lithium-ion battery anode[J]. Journal of Power Sources, 2012, 206: 267-273.

[26] ZHANG H R, QIN X Y, WU J X, et al. Electrospun core-shell silicon/carbon fibers with an internal honeycomb-like conductive carbon framework as an anode for lithium ion batteries[J]. Journal of Materials Chemistry A, 2015, 3(13): 7112-7120.

[27] FU K, LU Y, DIRICAN M, et al. Chamber-confined silicon-carbon nanofiber composites for prolonged cycling life of Li-ion batteries[J]. Nanoscale, 2014, 6(13): 7489-7495.

[28] WU C Q, LIN J, CHU R X, et al. Reduced graphene oxide as a dual-functional enhancer wrapped over silicon/porous carbon nanofibers for high-performance lithium-ion battery anodes[J]. Journal of Materials Science, 2017, 52(13): 7984-7996.

[29] XU Y H, LIU Q, ZHU Y J, et al. Uniform nano-Sn/C composite anodes for lithium ion batteries[J]. Nano Letters, 2013, 13(2): 470-474.

[30] WANG B, LUO B, LI X L, et al. The dimensionality of Sn anodes in Li-ion batteries[J]. Mater Today, 2012, 15(12): 544-552.

[31] ZHANG M, WANG T, CAO G. Promises and challenges of tin-based compounds as anode materials for lithium-ion batteries[J]. International Materials Reviews, 2015, 60(6): 330-352.

[32] YU Y, GU L, WANG C L, et al. Encapsulation of Sn@carbon nanoparticles in bamboo-like hollow carbon nanofibers as an anode material in lithium-based batteries[J]. Angewandte Chemie International Edition, 2009, 48(35): 6485-6489.

[33] ZHANG G H, ZHU J, ZENG W, et al. Tin quantum dots embedded in nitrogen-doped carbon nanofibers as excellent anode for lithium-ion batteries[J]. Nano Energy, 2014, 9: 61-70.

[34] KIM C, SONG G, LUO L L, et al. Stress-tolerant nanoporous germanium nanofibers for long cycle life lithium storage with high structural stability[J]. ACS Nano, 2018, 12(8): 8169-8176.

[35] LU X, WANG P, LIU K, et al. Encapsulating nanoparticulate Sb/MoO_x into porous carbon nanofibers via electrospinning for efficient lithium storage[J]. Chemical Engineering Journal, 2018, 336: 701-109.

[36] CHEAH Y L, ARAVINDAN V, MADHAVI S. Improved elevated temperature performance of Al-intercalated V_2O_5 electrospun nanofibers for lithium-ion batteries[J]. ACS Applied Materials & Interfaces, 2012, 4(6): 3270-3277.

[37] SHILPA, SHARMA A. Enhanced electrochemical performance of electrospun Ag/hollow glassy carbon nanofibers as free-standing Li-ion battery anode[J]. Electrochimica Acta, 2015, 176: 1266-1271.

[38] KANG D M, LIU Q L, SI R, et al. Crosslinking-derived MnO/carbon hybrid with ultrasmall nanoparticles for increasing lithium storage capacity during cycling[J]. Carbon, 2016, 99: 138-147.

[39] LIU J, XU X J, HU R Z, et al. Uniform hierarchical Fe_3O_4@polypyrrole nanocages for superior lithium ion battery anodes[J]. Advanced Energy Materials, 2016, 6(13): 1600256.

[40] YANG Y, HUANG J, ZENG J, et al. Direct electrophoretic deposition of binder-free Co_3O_4/graphene sandwich-like hybrid electrode as remarkable lithium ion battery

anode[J]. ACS Applied Materials & Interfaces, 2017, 9(38): 32801-32811.

[41] HWAN O S, PARK J S, SU J M, et al. Design and synthesis of tube-in-tube structured NiO nanobelts with superior electrochemical properties for lithium-ion storage[J]. Chemical Engineering Journal, 2018, 347: 889-899.

[42] KO S, LEE J I, YANG H S, et al. Mesoporous CuO particles threaded with CNTs for high-performance lithium-ion battery anodes[J]. Advanced materials, 2012, 24(32): 4451-4456.

[43] LIU H C, SHI L D, LI D Z, et al. Rational design of hierarchical ZnO@carbon nanoflower for high performance lithium ion battery anodes[J]. Journal of Power Sources, 2018, 387: 64-71.

[44] REDDY M V, SUBBA RAO G V, CHOWDARI B V. Metal oxides and oxysalts as anode materials for Li ion batteries[J]. Chemical Reviews, 2013, 113(7): 5364-5457.

[45] LU Y, YU L, LOU X W D. Nanostructured conversion-type anode materials for advanced lithium-ion batteries [J]. Chem, 2018, 4(5): 972-996.

[46] CHEN M H, XIA X H, YIN J H, et al. Construction of Co_3O_4 nanotubes as high-performance anode material for lithium ion batteries[J]. Electrochimica Acta, 2015, 160: 15-21.

[47] YUAN C P, WANG H G, LIU J Q, et al. Facile synthesis of Co_3O_4-CeO_2 composite oxide nanotubes and their multifunctional applications for lithium ion batteries and CO oxidation[J]. Journal of Colloid and Interface Science, 2017, 494: 274-281.

[48] JI X X, LIN Z J, ZENG J, et al. Controlling structure of vertically grown graphene sheets on carbon fibers for hosting Li and Na metals as rechargeable battery anodes[J]. Carbon, 2020, 158: 394-405.

[49] NARSIMULU D, RAO B N, SATYANARAYANA N, et al. High capacity electrospun $MgFe_2O_4$-C composite nanofibers as an anode material for lithium ion batteries[J]. Chemistry Select, 2018, 3(27): 8010-8017.

[50] CHEN L J, LIAO J D, CHUANG Y J, et al. Synthesis and characterization of PVP/$LiCoO_2$ nanofibers by electrospinning route[J]. Journal of Applied Polymer Science, 2011, 121(1): 154-160.

[51] GU Y X, CHEN D R, JIAO X L. Synthesis and electrochemical properties of nanostructured $LiCoO_2$ fibers as cathode materials for lithium-ion batteries[J]. Journal of Physical Chemistry B, 2005, 109(38): 17901-17906.

[52] ZHU C B, YU Y, GU L, et al. Electrospinning of highly electroactive carbon-coated single-crystalline $LiFePO_4$ nanowires[J]. Angewandte Chemie International Edition, 2010, 50(28): 6278-6282.

[53] QIU Y J, GENG Y H, LI N N, et al. Nonstoichiometric $LiFePO_4$/C nanofibers by electrospinning as cathode materials for lithium-ion battery[J]. Materials Chemistry & Physics, 2014, 144(3): 226-229.

[54] QIU Y J, GENG Y H, YU J, et al. High-capacity cathode for lithium-ion battery from $LiFePO_4$/(C+Fe_2P) composite nanofibers by electrospinning[J]. Journal of Materials Science, 2014, 49(2): 504-509.

[55] KANG C S, KIM C, KIM J E, et al. New observation of morphology of $Li[Fe_{1-x}Mn_x]PO_4$ nano-fibers (x=0, 0.1, 0.3) as a cathode for lithium secondary batteries by electrospinning process[J]. Journal of Physics and Chemistry of Solids, 2013, 74: 536-540.

[56] LENG F F, XU Y, JING L Y, et al. Electrospun polycrystalline $Li_xFe_{0.2}Mn_{0.8}PO_4$/carbon composite fibers for lithium-ion battery[J]. Colloids & Surfaces A Physicochemical & Engineering Aspects, 2016, 495: 54-61.

[57] REN W H, ZHENG Z P, LUO Y Z, et al. An electrospun hierarchical LiV_3O_8 nanowire-in-network for high-rate and long-life lithium batteries[J]. Journal of Materials Chemistry A, 2015, 3(39): 19850-19856.

[58] WANG L P, DENG L B, LI Y L, et al. Nb^{5+} doped LiV_3O_8 nanorods with extraordinary rate performance and cycling stability as cathodes for lithium-ion batteries[J]. Electrochimica Acta, 2018, 284: 366-375.

[59] CHAUDHARI S, SRINIVASAN M. 1D hollow α-Fe_2O_3 electrospun nanofibers as high performance anode material for lithium ion batteries[J]. Journal of Materials Chemistry, 2012, 22(43): 23049-23056.

[60] SAHAY R, SURESH K P, ARAVINDAN V, et al. High aspect ratio electrospun CuO

nanofibers as anode material for lithium-ion batteries with superior cycleability[J]. Journal of Physical Chemistry C, 2012, 116(34): 18087-18092.

[61] PARK H, SONG T, HAN H, et al. Electrospun $Li_4Ti_5O_{12}$ nanofibers sheathed with conductive TiN/TiO_xN_y layer asananode material for high power Li-ion batteries[J]. Journal of Power Sources, 2013, 244: 726-730.

[62] LIU B, HU X L, XU H H, et al. Encapsulation of MnO nanocrystals in electrospun carbon nanofibers as high-performance anode materials for lithium-ion batteries[J]. Scientific Reports, 2014, 4: 4229-4235.

[63] CHERIAN C T, SUNDARAMURTHY J, REDDY M V, et al. Morphologically robust $NiFe_2O_4$ nanofibers as high capacity li-ion battery anode material[J]. ACS Applied Materials & Interfaces, 2013, 5(20): 9957-9963.

[64] ARAVINDAN V, SURESH KUMAR P, SUNDARAMURTHY J, et al. Electrospun NiO nanofibers as high performance anode material for Li-ion batterie[J]. Journal of Power Sources, 2013, 227: 284-290.

[65] LUO W, HU X L, SUN Y M, et al. Electrospun porous $ZnCo_2O_4$ nanotubes as a high-performance anode material for lithium-ion batteries[J]. Journal of Materials Chemistry, 2012, 22(18): 8916-8921.

[66] ZHANG S, LIN Z, JI L W, et al. Cr-doped Li_2MnSiO_4/carbon composite nanofibers as high-energy cathodes for Li-ion batteries[J]. Journal of Materials Chemistry, 2012, 22(29): 14661-14666.

[67] ZHANG S, LI Y, XU G J, et al. High-capacity $Li_2Mn_{0.8}Fe_{0.2}SiO_4$/carbon composite nanofiber cathodes for lithium-ion batteries[J]. Journal of Power Sources, 2012, 213: 10-15.

[68] LIU J, TANG K, SONG K P, et al. Electrospun $Na_3V_2(PO_4)_3$/C nanofibers as stable cathode materials for sodium-ion batteries[J]. Nanoscale, 2014, 6(10): 5081-5086.

[69] CHEAH Y L, GUPTA N, PRAMANA S S, et al. Morphology, structure and electrochemical properties of single phase electrospun vanadium pentoxide nanofibers for lithium ion batteries[J]. Journal of Power Sources, 2011, 196(15): 6465-6472.

[70] YU J, XIAO J W, LI A R, et al. Enhanced multiple anchoring and catalytic conversion of polysulfides by amorphous MoS_3 nanoboxes for high-performance Li-S batteries[J]. Angewandte Chemie International Edition, 2020, 59(31): 13071-13078.

[71] LI B, LI S M, LIU J H, et al. Vertically aligned sulfur-graphene nanowalls on substrates for ultrafast lithium-sulfur batteries[J]. Nano Letters, 2015, 15(5): 3073-3079.

[72] REN J, ZHOU Y B, XIA L, et al. Rational design of a multidimensional N-doped porous carbon/MoS_2/CNT nano-architecture hybrid for high performance lithium-sulfur batteries[J]. Journal of Materials Chemistry A, 2018, 6(28): 13835-13847.

[73] ZHANG Y L, MU Z J, YANG C, et al. Rational design of MXene/1T-2H MoS_2-C nanohybrids for high-performance lithium-sulfur batteries[J]. Advanced Functional Materials, 2018, 28(38): 1707578.

[74] YANG X F, GAO X J, SUN Q, et al. Promoting the transformation of Li_2S_2 to Li_2S: significantly increasing utilization of active materials for high-sulfur-loading Li-S batteries[J]. Advanced Materials, 2019, 31(25): 1901220.

[75] LIU M, DENG N P, JU J G, et al. A review: electrospun nanofiber materials for lithium-sulfur batteries[J]. Advanced Functional Materials, 2019, 29(49):1905467.

[76] ZHOU L, DANILOV D L. Host materials anchoring polysulfides in Li-S batteries reviewed[J]. Advanced Energy Materials, 2021, 11(15): 2001304.

[77] NITTA N, WU F X, LEE J T, et al. Li-ion battery materials: present and future[J]. Materials Today, 2015, 18(5): 252-264.

[78] LI H, WANG Z X, CHEN L Q, et al. Research on advanced materials for Li-ion batteries[J]. Advanced materials, 2009, 21(45): 4593-4607.

[79] RANA M, AHAD S A, LI M, et al. Review on areal capacities and long-term cycling performances of lithium sulfur battery at high sulfur loading[J]. Advanced materials, 2019, 18: 289-310.

[80] MANTHIRAM A, FU Y Z, CHUNG S H, et al. Rechargeable lithium-sulfur batteries [J]. Chemical Reviews, 2014, 114(23): 11751-11787.

[81] MANTHIRAM A, CHUNG S H, ZU C, et al. Lithium-sulfur batteries: progress and prospects[J]. Advanced Materials, 2015, 27(12): 1980-2006.

[82] ZHU P, ZHU J, ZANG J, et al. A novel bi-functional double-layer rGO-PVDF/PVDF composite nanofiber membrane separator with enhanced thermal stability and effective polysulfide inhibition for high-performance lithium-sulfur batteries[J]. Journal of Materials Chemistry A, 2017, 5(29): 15096-15104.

[83] LUO X, LU X B, ZHOU G Y, et al. Ion-selective polyamide acid nanofiber separators for high-rate and stable lithium-sulfur batteries[J]. ACS Applied Materials & Interfaces, 2018, 10(49): 42198-42206.

[84] LIN Y S, PITCHERI R, ZHU J H, et al. Electrospun PVDF/PSSLi ionomer films as a functional separator for lithium-sulfur batteries[J]. Journal of Alloys and Compounds, 2019, 785: 627-633.

[85] LI J, JIAO C M, ZHU J H, et al. Hybrid Co-based MOF nanoboxes/CNFs interlayer as microreactors for polysulfides-trapping in lithium-sulfur batteries[J]. Journal of Energy Chemistry, 2021, 57: 469-476.

[86] GUO Y, LI J, PITCHERI R, et al. Electrospun Ti_4O_7/C conductive nanofibers as interlayer for lithium-sulfur batteries with ultra long cycle life and high-rate capability[J]. Chemical Engineering Journal, 2019, 355: 390-398.

[87] ZHU J H, PITCHERI R, KANG T, et al. A polysulfide-trapping interlayer constructed by boron and nitrogen co-doped carbon nanofibers for long-life lithium sulfur batteries[J]. Journal of Electroanalytical Chemistry, 2019, 833: 151-159.

[88] LI J, GUO Y, WEN P, et al. Constructing a sandwich-structured interlayer with strong polysulfides adsorption ability for high-performance lithium-sulfur batteries [J]. Materials Today Energy, 2019, 14: 100339.

[89] XU Z L, HUANG J Q, CHONG W G, et al. In situ TEM study of volume expansion in porous carbon nanofiber/sulfur cathodes with exceptional high-rate performance[J]. Advanced Energy Materials, 2017, 7(9): 1602078.

[90] ZHANG Y Z, ZHANG Z, LIU S, et al. Free-standing porous carbon nanofiber/carbon nanotube film as sulfur immobilizer with high areal capacity for lithium-

sulfur battery[J]. ACS Applied Materials & Interfaces, 2018, 10(10): 8749-8757.

[91] YU M L, WANG Z Y, WANG Y W, et al. Freestanding flexible Li_2S paper electrode with high mass and capacity loading for high-energy Li-S batteries[J]. Advanced Energy Materials, 2017, 7(17): 1700018.

[92] ZHAO X H, KIM M, LIU Y, et al. Root-like porous carbon nanofibers with high sulfur loading enabling superior areal capacity of lithium sulfur batteries[J]. Carbon, 2018, 128: 138-146.

[93] LIN L L, PEI F, PENG J, et al. Fiber network composed of interconnected yolk-shell carbon nanospheres for high-performance lithium-sulfur batteries[J]. Nano Energy, 2018, 54: 50-58.

[94] DENG N P, JU J G, YAN J, et al. CeF_3-doped porous carbon nanofibers as sulfur immobilizers in cathode material for high-performance lithium-sulfur batteries[J]. ACS Applied Materials & Interfaces, 2018,10(15): 12626-12638.

[95] SONG X, WANG S Q, BAO Y, et al. A high strength, free-standing cathode constructed by regulating graphitization and the pore structure in nitrogen-doped carbon nanofibers for flexible lithium-sulfur batteries[J]. Journal of Materials Chemistry A, 2017, 5(15): 6832-6839.

[96] ZHANG Y，WANG P P，TAN H, et al. Free-standing sulfur and graphitic porous carbon nanofibers composite cathode for high electrochemical performance of lithium-sulfur batteries[J]. Journal of the Electrochemical Society, 2018, 165(5): A741-A745.

[97] YAO Y, ZENG L C, HU S H, et al. Binding $S_{0.6}Se_{0.4}$ in 1D carbon nanofiber with C-S bonding for high-performance flexible Li-S batteries and Na-S batteries[J]. Small, 2017, 13(19): 1603513.

[98] QIN X Y, WU J X, XU Z L, et al. Electrosprayed multiscale porous carbon microspheres as sulfur hosts for long-life lithium-sulfur batteries[J]. Carbon, 2019, 141: 16-24.

[99] LI X Y, FU N Q, ZOU J Z, et al. Sulfur-impregnated N-doped hollow carbon nanofibers as cathode for lithium-sulfur batteries[J]. Materials Letters, 2017, 209:

505-508.

[100] YUN J H, KIM J H, KIM D K, et al. Suppressing polysulfide dissolution via cohesive forces by interwoven carbon nanofibers for high-areal-capacity lithium-sulfur batteries[J]. Nano Letters, 2018, 18(1): 475-481.

[101] MA D T, LI Y L, YANG J B, et al. Atomic layer deposition-enabled ultrastable freestanding carbon-selenium cathodes with high mass loading for sodium-selenium battery[J]. Nano Energy, 2018, 43: 317-325.

[102] CHAI D F, GÓMEZ-GARCÍA C J, LI B N, et al. Polyoxometalate-based metal-organic frameworks for boosting electrochemical capacitor performance[J]. Chemical Engineering Journal, 2019, 373: 587-597.

[103] LEE J S M, BRIGGS M E, HU C C, et al. Controlling electric double-layer capacitance and pseudocapacitance in heteroatom-doped carbons derived from hypercrosslinked microporous polymers[J]. Nano Energy, 2018, 46: 277-289.

[104] YAO L, WU Q, ZHANG P X, et al. Scalable 2D hierarchical porous carbon nanosheets for flexible supercapacitors with ultrahigh energy density[J]. Advanced Materials, 2018, 30(11): 1706054.

[105] JIA Q, YANG C, PAN Q Q, et al. High-voltage aqueous asymmetric pseudocapacitors based on methyl blue-doped polyaniline hydrogels and the derived N/S-codoped carbon aerogels[J]. Chemical Engineering Journal, 2020, 383: 123153.

[106] ZHANG M Y, SONG Y, GUO D, et al. Strongly coupled polypyrrole/molybdenum oxide hybrid films via electrochemical layer-by-layer assembly for pseudocapacitors[J]. Journal of Materials Chemistry A, 2019, 7(16): 9815-9821.

[107] WITOMSKA S, LIU Z Y, CZEPA W, et al. Graphene oxide hybrid with sulfur-nitrogen polymer for high-performance pseudocapacitors[J]. Journal of the American Chemical Society, 2019, 141(1): 482-487.

[108] FUSALBA F, GOUÉREC P, VILLERS D, et al. Electrochemical characterization of polyaniline in nonaqueous electrolyte and its evaluation as electrode material for electrochemical supercapacitors[J]. University of Cambridge, 2001, 148(1):

A1-A6.

[109] XIONG C Y, YANG Q, DANG W H, et al. Fabrication of eco-friendly carbon microtubes @ nitrogen-doped reduced graphene oxide hybrid as an excellent carbonaceous scaffold to load MnO_2 nanowall (PANI nanorod) as bifunctional material for high-performance supercapacitor and oxygen reduction reaction catalyst[J]. Journal of Power Sources, 2020, 447: 227387.

[110] MA Y, HOU C P, ZHANG H P, et al. Three-dimensional core-shell Fe_3O_4/ Polyaniline coaxial heterogeneous nanonets: preparation and high performance supercapacitor electrodes[J]. Electrochimica Acta, 2019, 315: 114-123.

[111] ZHAO P, WANG N, YAO M Q, et al. Hydrothermal electrodeposition incorporated with CVD-polymerisation to tune PPy@MnO_2 interlinked core-shell nanowires on carbon fabric for flexible solid-state asymmetric supercapacitors[J]. Chemical Engineering Journal, 2020, 380: 122488.

[112] COTTINEAU T, TOUPIN M, DELAHAYE T, et al. Nanostructured transition metal oxides for aqueous hybrid electrochemical supercapacitors[J]. Applied Physics A: Materials Science & Processing, 2006, 82(4): 599-606.

[113] ZHAO L, QIU Y J, YU J, et al. Carbon nanofibers with radially grown graphene sheets derived from electrospinning for aqueous supercapacitors with high working voltage and energy density[J]. Nanoscale, 2013, 5(11): 4902-4909.

[114] LAI C C, LO C T. Preparation of nanostructural carbon nanofibers and their electrochemical performance for supercapacitors[J]. Electrochimica Acta, 2015, 183: 85-93.

[115] KIM M, KIM Y, LEE K M, et al. Electrochemical improvement due to alignment of carbon nanofibers fabricated by electrospinning as an electrode for supercapacitor[J]. Carbon, 2016, 99: 607-618.

[116] HUANG K B, YAO Y Y, YANG X W, et al. Fabrication of flexible hierarchical porous nitrogen-doped carbon nanofiber films for application in binder-free supercapacitors[J]. Materials Chemistry & Physics, 2016, 169: 1-5.

[117] TIAN X D, ZHAO N, SONG Y, et al. Synthesis of nitrogen-doped electrospun

carbon nanofibers with superior performance as efficient supercapacitor electrodes in alkaline solution[J]. Electrochimica Acta, 2015, 185:40-51.

[118] XU Q, YU X L, LIANG Q H, et al. Nitrogen-doped hollow activated carbon nanofibers as high performance supercapacitor electrodes[J]. Journal of Electroanalytical Chemistry, 2015, 739: 84-88.

[119] KUNDU M, LIU L F. Binder-free electrodes consisting of porous NiO nanofibers directly electrospun on nickel foam for high-rate supercapacitors[J]. Materials Letters, 2015, 144: 114-118.

[120] ABOUALI S, GARAKANI M A, ZHANG B, et al. Electrospun carbon nanofibers with in situ encapsulated Co_3O_4 nanoparticles as electrodes for high-performance supercapacitors[J]. ACS Applied Materials & Interfaces, 2015, 7(24): 13503-13511.

[121] WANG Q F, CHEN D, ZHANG D H. Electrospun porous $CuCo_2O_4$ nanowire network electrode for asymmetric supercapacitors[J]. RSC Advances, 2015, 5(117): 96448-96454.

[122] VIDYADHARAN B, MISNON I I, ISMAIL J, et al. High performance asymmetric supercapacitors using electrospun copper oxide nanowires anode[J]. Journal of Alloys and Compounds, 2015, 633: 22-30.

[123] KIM B H, KIM C H, LEE D G. Mesopore-enriched activated carbon nanofiber web containing RuO_2 as electrode material for high-performance supercapacitors[J]. Journal of Electroanalytical Chemistry, 2016, 760: 64-70.

[124] LEE D G, YANG C M, KIM B H. Enhanced electrochemical properties of boron functional groups on porous carbon nanofiber/MnO_2 materials[J]. Journal of Electroanalytical Chemistry, 2017, 788: 192-197.

[125] CHEN L F, LU Y, YU L, et al. Designed formation of hollow particle-based nitrogen-doped carbon nanofibers for high-performance supercapacitors[J]. Energy & Environmental Science, 2017, 10(8): 1777-1783.

[126] LU C, CHEN X. Electrospun polyaniline nanofiber networks toward high-performance flexible supercapacitors[J]. Advanced Materials Technologies, 2019,

4(11): 1900564.

[127] MIAO F J, SHAO C L, LI X H, et al. Polyaniline-coated electrospun carbon nanofibers with high mass loading and enhanced capacitive performance as freestanding electrodes for flexible solid-state supercapacitors[J]. Energy, 2016, 95: 233-241.

[128] MIAO F J, SHAO C L, LI X H, et al. Flexible solid-state supercapacitors based on freestanding electrodes of electrospun polyacrylonitrile@polyaniline core-shell nanofibers[J]. Electrochimica Acta, 2015, 176: 293-300.

[129] SHEN C W, WANG C P, SANGHADASA M, et al. Flexible micro-supercapacitors prepared using direct-write nanofibers[J]. RSC Advances, 2017, 7(19): 11724-11731.

[130] HUANG Z D, ZHANG B, OH S W, et al. Self-assembled reduced graphene oxide/carbon nanotube thin films as electrodes for supercapacitors[J]. Journal of Materials Chemistry, 2012, 22(8): 3591-3599.

[131] HUANG Z D, ZHANG B, LIANG R, et al. Effects of reduction process and carbon nanotube content on the supercapacitive performance of flexible graphene oxide papers[J]. Carbon, 2012, 50(11): 4239-4251.

[132] XU Q, YU X L, LIANG Q H, et al. Nitrogen-doped hollow activated carbon nanofibers as high performance supercapacitor electrodes[J]. Journal of Electroanalytical Chemistry, 2015, 739: 84-88.

[133] JEONG J H, KIM Y A, KIM B H. Electrospun polyacrylonitrile/cyclodextrin-derived hierarchical porous carbon nanofiber/MnO_2 composites for supercapacitor applications[J]. Carbon, 2020, 164: 296-304.

[134] LI W Z, YANG W W, WANG N, et al. Optimization of blocked channel design for a proton exchange membrane fuel cell by coupled genetic algorithm and three-dimensional CFD modeling[J]. International Journal of Hydrogen Energy, 2020, 45(35): 17759-17770.

[135] ZHOU L J, ZHU J Y, LIN M J, et al. Tetra-alkylsulfonate functionalized poly(aryl ether) membranes with nanosized hydrophilic channels for efficient proton

conduction[J]. Journal of Energy Chemistry, 2020, 40: 57-64.
[136] WANG S Y, JIANG S P. Prospects of fuel cell technologies[J]. National Science Review, 2017, 4(2): 163-166.
[137] JIANG C R, MA J J, CORRE G, et al. Challenges in developing direct carbon fuel cells[J]. Chemical Society Reviews, 2017, 46(10): 2889-2912.
[138] BREITWIESER M, KLOSE C, KLINGELE M, et al. Simple fabrication of 12 μm thin nanocomposite fuel cell membranes by direct electrospinning and printing[J]. Journal of Power Sources, 2017, 337: 137-144.
[139] CAVALIERE S, SUBIANTO S, SAVYCH I, et al. Electrospinning: designed architectures for energy conversion and storage devices[J]. Energy & Environmental Science, 2011, 4(12): 4761-4785.
[140] MOLLÁ S, COMPAÑ V. Polyvinyl alcohol nanofiber reinforced Nafion membranes for fuel cell applications[J]. Journal of Membrane Science, 2011, 372(1-2): 191- 200.
[141] DONG B, CHEN H, SNYDER J, et al. Super proton conductive nafion nanofibers: discovery, fabrication, properties, and fuel cell performance[J]. ECS Transactions, 2011, 41(1): 1503-1506.
[142] TAMURA T, KAWAKAMI H. Aligned electrospun nanofiber composite membranes for fuel cell electrolytes[J]. Nano Letters, 2010, 10(4): 1324-1328.
[143] SHENG W C, GASTEIGER H A, SHAO-HORN Y. Hydrogen oxidation and evolution reaction kinetics on platinum: acid vs alkaline electrolytes[J]. Acta Crystallographica, 2010, 157(11): B1529.
[144] GASTEIGER H A, PANELS J E, YAN S G. Dependence of PEM fuel cell performance on catalyst loading[J]. Journal of Power Sources, 2004, 127(1-2): 162-171.
[145] SUNG M T, CHANG M H, HO M H. Investigation of cathode electrocatalysts composed of electrospun Pt nanowires and Pt/C for proton exchange membrane fuel cells[J]. Journal of Power Sources, 2014, 249: 320-326.
[146] KIM H J, KIM Y S, SEO M H, et al. Highly improved oxygen reduction

performance over Pt/C-dispersed nanowire network catalysts[J]. Electrochemistry Communications, 2010, 12(1):32-35.

[147] HIGGINS D C, WANG R Y, HOQUE M A, et al. Morphology and composition controlled platinum-cobalt alloy nanowires prepared by electrospinning as oxygen reduction catalyst[J]. Nano Energy, 2014, 10: 135-143.

[148] QIU Y J, YU J, WU W H, et al. Fe-N/C nanofiber electrocatalysts with improved activity and stability for oxygen reduction in alkaline and acid solutions[J]. Journal of Solid State Electrochemistry, 2013, 17(3): 565-573.

[149] YIN J, QIU Y, YU J. Enhanced electrochemical activity for oxygen reduction reaction from nitrogen-doped carbon nanofibers by iron doping[J]. ECS Solid State Letters, 2013, 2(5): M37-M39.

[150] FU Y, YU H Y, JIANG C, et al. NiCo alloy nanoparticles decorated on N-doped carbon nanofibers as highly active and durable oxygen electrocatalyst[J]. Advanced Functional Materials, 2018, 28(9): 1705094.

[151] LIU Q, CAO S B, QIU Y J, et al. Bimetallic Fe-Co promoting one-step growth of hierarchical nitrogen-doped carbon nanotubes/nanofibers for highly efficient oxygen reduction reaction[J]. Materials Science and Engineering B, 2017, 223: 159-166.

[152] LIU Q, CAO S B, FU Y, et al. Trimetallic FeCoNi-N/C nanofibers with high electrocatalytic activity for oxygen reduction reaction in sulfuric acid solution[J]. Journal of Electroanalytical Chemistry, 2018, 813: 52-57.

[153] SHI H, SHEN Y F, HE F, et al. Recent advances of doped carbon as non-precious catalysts for oxygen reduction reaction[J]. Journal of Materials Chemistry A, 2014, 2(38): 15704-15716.

[154] WANG H B, MAIYALAGAN T, WANG X. Review on recent progress in nitrogen-doped graphene: synthesis, characterization, and its potential applications[J]. ACS Catalysis, 2012, 2(5): 781-794.

[155] MASA J, XIA W, MUHLER M, et al. On the role of metals in nitrogen-doped carbon electrocatalysts for oxygen reduction[J]. Angewandte Chemie, 2015, 54(35):

10102-10120.

[156] WANG S G, DAI C L, LI J P, et al. The effect of different nitrogen sources on the electrocatalytic properties of nitrogen-doped electrospun carbon nanofibers for the oxygen reduction reaction[J]. International Journal of Hydrogen Energy, 2015, 40(13): 4673-4682.

[157] ZENG J, MU Y B, JI X X, et al. N,O-codoped 3D graphene fibers with densely arranged sharp edges as highly efficient electrocatalyst for oxygen reduction reaction[J]. Journal of Materials Science, 2019, 54(23): 14495-14503.

[158] YIN J, QIU Y J, YU J, et al. Enhancement of electrocatalytic activity for oxygen reduction reaction in alkaline and acid media from electrospun nitrogen-doped carbon nanofibers by surface modification[J]. RSC Advances, 2013, 3(36): 15655-15663.

[159] WEIGERT E C, STOTTLEMYER A L, ZELLNER M B, et al. Tungsten monocarbide as potential replacement of platinum for methanol electrooxidation[J]. The Journal of Physical Chemistry C, 2007, 111(40): 14617-14620.

[160] ZHOU X S, QIU Y J, YU J, et al. Tungsten carbide nanofibers prepared by electrospinning with high electrocatalytic activity for oxygen reduction[J]. International Journal of Hydrogen Energy, 2011, 36(13): 7398-7404.

[161] CHENG Y H, LI W Y, FAN X Z, et al. Modified multi-walled carbon nanotube/Ag nanoparticle composite catalyst for the oxygen reduction reaction in alkaline solution[J]. Electrochimica Acta, 2013, 111: 635-641.

[162] ZHANG L W, GUO Q Q, PITCHERI R, et al. Silver nanofibers with controllable microstructure and crystal facet as highly efficient and methanol-tolerant oxygen reduction electrocatalyst[J]. Journal of Power Sources, 2019, 413: 233-240.

[163] O'REGAN B, GRÄTZEL M. A low cost, high efficiency solar cell based on dye sensitized colloidal TiO_2 films[J]. Nature, 1991, 353(6346): 737-740.

[164] KIM G H, PARK S H, BIRAJDAR M S, et al. Core/shell structured carbon nanofiber/platinum nanoparticle hybrid web as a counter electrode for dye-sensitized solar cell[J]. Journal of Industrial and Engineering Chemistry, 2017,

52: 211-217.

[165] YUN S N, QIN Y, UHL A R, et al. New-generation integrated devices based on dye-sensitized and perovskite solar cells[J]. Energy & Environmental Science, 2018, 11(3): 476-526.

[166] TÉTREAULT N, GRÄTZEL M. Novel nanostructures for next generation dye-sensitized solar cells[J]. Energy & Environmental Science, 2012, 5(9): 8506-8516.

[167] GRÄTZEL M. Solar energy conversion by dye-sensitized photovoltaic cells[J]. Inorganic Chemistry, 2005, 44(20): 6841-6851.

[168] MADDAH H A, BERRY V, BEHURA S K. Biomolecular photosensitizers for dye-sensitized solar cells: recent developments and critical insights[J]. Renewable & Sustainable Energy Reviews, 2020, 121: 109678.

[169] BABAR F, MEHMOOD U, ASGHAR H, et al. Nanostructured photoanode materials and their deposition methods for efficient and economical third generation dye-sensitized solar cells: a comprehensive review[J]. Renewable & Sustainable Energy Reviews, 2020, 129: 109919.

[170] BISQUERT J, CAHEN D, HODES G, et al. Physical chemical principles of photovoltaic conversion with nanoparticulate, mesoporous dye-sensitized solar cells[J]. ChemInform, 2004, 108(24): 8106-8118.

[171] LU W L, JIANG R, YIN X, et al. Porous N-doped-carbon coated $CoSe_2$ anchored on carbon cloth as 3D photocathode for dye-sensitized solar cell with efficiency and stability outperforming Pt[J]. Nano Research, 2019, 12(1): 159-163.

[172] LIU X Q, IOCOZZIA J, WANG Y, et al. Noble metal-metal oxide nanohybrids with tailored nanostructures for efficient solar energy conversion, photocatalysis and environmental remediation[J]. Energy & Environmental Science, 2017, 10(2): 402-434.

[173] JO M S, CHO J S, WANG X L, et al. Improving of the photovoltaic characteristics of dye-sensitized solar cells using a photoelectrode with electrospun porous TiO_2 nanofibers[J]. Nanomaterials, 2019, 9(1): 95.

[174] ONOZUKA K, DING B, TSUGE Y, et al. Electrospinning processed nanofibrous TiO_2 membranes for photovoltaic applications[J]. Nanotechnology, 2006, 17(4): 1026-1031.

[175] SHAHZAD N, SHAH Z, SHAHZAD M I, et al. Effect of seed layer on the performance of ZnO nanorods-based photoanodes for dye-sensitized solar cells[J]. Materials Research Express, 2019, 6(10): 105523.

[176] FUJIHARA K, KUMAR A, JOSE R, et al. Spray deposition of electrospun TiO_2 nanorods for dye-sensitized solar cell[J]. Nanotechnology, 2007, 18(36): 365709.

[177] YANG L J, LEUNG W W F. Electrospun TiO_2 nanorods with carbon nanotubes for efficient electron collection in dye-sensitized solar cells[J]. Advanced Materials, 2013, 25(12): 1792-1795.

[178] SONG M Y, KIM D K, IHN K J, et al. Electrospun TiO_2 electrodes for dye-sensitized solar cells[J]. Nanotechnology, 2004, 15(12): 1861.

[179] ZHENG D, XIONG J, GUO P, et al. Fabrication of improved dye-sensitized solar cells with anatase/rutile TiO_2 nanofibers[J]. Journal of Nanoscience and Nanotechnology, 2016, 16(1): 613-618.

[180] CAO Y, DONG Y J, FENG H L, et al. Electrospun TiO_2 nanofiber based hierarchical photoanode for efficient dye-sensitized solar cells[J]. Electrochimica Acta, 2016, 189: 259-264.

[181] RHO W Y, KIM H S, CHUNG W J, et al. Enhancement of power conversion efficiency with TiO_2 nanoparticles/nanotubes-silver nanoparticles composites in dye-sensitized solar cells[J]. Applied Surface Science, 2018, 429: 23-28.

[182] LI R, ZHAO Y, HOU R E, et al. Enhancement of power conversion efficiency of dye sensitized solar cells by modifying mesoporous TiO_2 photoanode with Al-doped TiO_2 layer[J]. Journal of Photochemistry and Photobiology A: Chemistry, 2016, 319: 62-69.

[183] MOHAMED I M A, DAO V D, YASIN A S, et al. Design of an efficient photoanode for dye-sensitized solar cells using electrospun one-dimensional GO/N-doped nanocomposite SnO_2/TiO_2[J]. Applied Surface Science, 2017, 400: 355-364.

[184] LEE H, NAGAISHI T, PHAN D N, et al. Effect of graphene incorporation in carbon nanofiber decorated with TiO_2 for photoanode applications[J]. RSC Advances, 2017, 7(4): 6574-6582.

[185] BAUER C, BOSCHLOO G, MUKHTAR E, et al. Electron injection and recombination in $Ru(dcbpy)_2(NCS)_2$ sensitized nanostructured ZnO[J]. The Journal of Physical Chemistry B, 2001, 105(24): 5585-5588.

[186] CHEN H W, LIN C Y, LAI Y H, et al. Electrophoretic deposition of ZnO film and its compression for a plastic based flexible dye-sensitized solar cell[J]. Journal of Power Sources, 2011, 196(10): 4859-4864.

[187] KIM I D, HONG J M, LEE B H, et al. Dye-sensitized solar cells using network structure of electrospun ZnO nanofiber mats[J]. Applied Physics Letters, 2007, 91(16): 163109.

[188] WU S, LI J, LO S C, et al. Enhanced performance of hybrid solar cells based on ordered electrospun ZnO nanofibers modified with CdS on the surface[J]. Organic Electronics, 2012, 13(9): 1569-1575.

[189] DAS A K, SRINIVASAN A. Band gap tuning and defects suppression upon Mg doping in electrospun ZnO nanowires[J]. Journal of Materials Science: Materials in Electronics, 2017, 28(9): 6488-6492.

[190] LAW M, GREENE L E, JOHNSON J C, et al. Nanowire dye-sensitized solar cells[J]. Nature Materials, 2005, 4: 455-459.

[191] KRISHNAMOORTHY T, TANG M Z, VERMA A, et al. A facile route to vertically aligned electrospun SnO_2 nanowires on a transparent conducting oxide substrate for dye-sensitized solar cells[J]. Journal of Materials Chemistry, 2012, 22(5): 2166-2172.

[192] CHEN Z L, ZHAO J X, YANG X X, et al. Fabrication of TiO_2/WO_3 composite nanofibers by electrospinning and photocatalystic performance of the resultant fabrics[J]. Industrial & Engineering Chemistry Research, 2016, 55(1): 80-85.

[193] JOSHI P, ZHANG L F, CHEN Q L, et al. Electrospun carbon nanofibers as low-cost counter electrode for dye-sensitized solar cells[J]. ACS applied materials &

interfaces, 2010, 2(12): 3572-3577.

[194] PARK S H, KIM B K, LEE W J. Electrospun activated carbon nanofibers with hollow core/highly mesoporous shell structure as counter electrodes for dye-sensitized solar cells[J]. Journal of Power Sources, 2013, 239: 122-127.

[195] ZHANG C L, DENG L B, ZHANG P X, et al. Electrospun FeS nanorods with enhanced stability as counter electrodes for dye-sensitized solar cells[J]. Electrochimica Acta, 2017, 229: 229-238.

[196] CAVALIERE S, SUBIANTO S, SAVYCH I, et al. Electrospinning: designed architectures for energy conversion and storage devices[J]. Energy & Environmental Science, 2011, 4(12):4761-4785.

[197] DISSANAYAKE S S, DISSANAYAKE M A K L, SENEVIRATNE V A, et al. Performance of dye sensitized solar cells fabricated with electrospun polymer nanofiber based electrolyte[J]. Materials Today Proceedings, 2016, 3: S104-S111.

[198] SETHUPATHY M, PANDEY P, MANISANKAR P. Development of quasi-solid-state dye-sensitized solar cell based on an electrospun polyvinylidene fluoride-polyacrylonitrile membrane electrolyte[J]. Journal of Applied Polymer Science, 2014, 131(6): 596-602.

[199] WEERASINGHE A M J S, DISSANAYAKE M A K L, SENADEERA G K R, et al. Application of electrospun cellulose acetate nanofibre membrane based quasi-solid state electrolyte for dye sensitized solar cells[J]. Ceylon Journal of Science, 2017, 46(2): 93.

[200] MURUGADOSS V, ARUNACHALAM S, ELAYAPPAN V, et al. Development of electrospun PAN/CoS nanocomposite membrane electrolyte for high-performance DSSC[J]. Ionics, 2018, 24: 4071-4080.

[201] PARK Y, VANDEWAL, LEO K. Optical in-coupling in organic solar cells[J]. Small Methods, 2018, 2(10): 1800123.

[202] ZHANG H, WANG X K, SUN Y S, et al. Effect of IT-M doping on charge transfer and ultrafast carrier dynamics of ternary organic solar cell materials[J]. Journal of Physics D: Applied Physics, 2020, 53(9): 095103.

[203] XUE R M, ZHANG J W, LI Y W, et al. Organic solar cell materials toward commercialization[J]. Small, 2018, 14(41): e1801793.

[204] ZOU X. Investigation of the photovoltaic performance of n-ZnO/n-CdS/p-Cu_2ZnSnS_4 solar cell[J]. International Journal of Modern Physics B, 2019, 34(1-3): 2040010.

[205] NGUYEN L H, HOPPE H, ERB T, et al. Effects of annealing on the nanomorphology and performance of poly (alkylthiophene): fullerene bulk-heterojunction solar cells[J]. Advanced Functional Materials, 2007, 17(7): 1071-1078.

[206] SHENOUDA A Y, RASHAD M M. Controlling the performance of photovoltaic cell based on nanostructured $Cd_{1-x}Zn_xTe$ semiconductors[J]. Materials Research Innovations, 2019, 23(6): 363-368.

[207] REZAEI B, SHOUSHTARI A M, RABIEE M, et al. Multifactorial modeling and optimization of solution and electrospinning parameters to generate superfine polystyrene nanofibers[J]. Advances in Polymer Technology, 2018, 37(8): 2743-2755.

[208] PEET J. Bulk heterojunction solar cells: the role of processing in the fabrication and optimization of plastic solar cells [J]. Advanced Materials, 2009, 21(14/15): 1521-1527.

[209] TAVAKKOL E, TAVANAI H, ABDOLMALEKI A, et al. Production of conductive electrospun polypyrrole/poly(vinyl pyrrolidone) nanofibers[J]. Synthetic Metals, 2017, 231: 95-106.

[210] XIN H, GUO X, REN G, et al. Efficient phthalimide copolymer-based bulk heterojunction solar cells: how the processing additive influences nanoscale morphology and photovoltaic properties[J]. Advanced Energy Materials, 2012, 2(5): 575-582.

[211] HUANG H J N, NAN H R, YANG Y G, et al. Improved performance of thick films based binary and ternary bulk heterojunction organic photovoltaic devices incorporated with electrospinning processed nanofibers[J]. Advanced Materials

Interfaces, 2018, 5(20): 1800914.

[212] SALIM T, SUN S Y, WONG L H, et al. The role of poly(3-hexylthiophene) nanofibers in an all-polymer blend with a polyfluorene copolymer for solar cell applications[J]. Journal of Physical Chemistry C, 2010, 114(20):9459-9468.

[213] XIN H, REID O G, REN G Q, et al. Polymer nanowire/fullerene bulk heterojunction solar cells: how nanostructure determines photovoltaic properties[J]. ACS Nano, 2010, 4(4): 1861-1872.

[214] NOLDE F, PISULA W, MÜLLER S, et al. Synthesis and self-organization of core-extended perylene tetracarboxdiimides with branched alkyl substituents[J]. Chemistry of Materials, 2006, 18(16): 3715-3725.

[215] MAEYOSHI Y, SAEKI A, SUWA S, et al. Fullerene nanowires as a versatile platform for organic electronics[J]. Scientific Reports, 2012, 2: 600.

[216] KARAK S, RAY S K, DHAR A. Photoinduced charge transfer and photovoltaic energy conversion in self-assembled N,N'-dioctyl-3,4,9,10-perylenedicarboximide nanoribbons[J]. Applied Physics Letters, 2010, 97(4): 043306.

[217] PRADHAN S, DHAR A. Synthesis of vertically grown N,N'-dioctyl-3,4,9,10-perylenedicarboximide nanostructure for photovoltaic application[J]. Journal of Renewable & Sustainable Energy, 2013, 5(3): 031611.

[218] BRISENO A L, MANNSFELD S C B, SHAMBERGER P J, et al. Self-assembly, molecular packing, and electron transport in n-type polymer semiconductor nanobelts[J]. Chemistry of Materials, 2008, 20(14): 4712-4719.

[219] REN G Q, AHMED E, JENEKHE S A. Nanowires of oligothiophene-functionalized naphthalene diimides: self assembly, morphology, and all-nanowire bulk heterojunction solar cells[J]. Journal of Materials Chemistry, 2012, 22(46): 24373-24379.

[220] CHANG C Y, WU C E, CHEN S Y, et al. Enhanced performance and stability of a polymer solar cell by incorporation of vertically aligned, cross-linked fullerene nanorods[J]. Angewandte Chemie International Edition, 2011, 50(40): 9386-9390.

[221] HE X M, GAO F, TU G L, et al. Formation of nanopatterned polymer blends in

photovoltaic devices[J]. Nano Letters, 2010, 10(4): 1302-1307.

[222] KIM J, RHEE S, LEE H, et al. Universal elaboration of Al-doped TiO_2 as an electron extraction layer in inorganic-organic hybrid perovskite and organic solar cells[J]. Advanced Materials Interfaces, 2020, 7(10): 1902003.

[223] LIU J W, KUO Y T, KLABUNDE K J, et al. Novel dye-sensitized solar cell architecture using TiO_2-coated vertically aligned carbon nanofiber arrays[J]. ACS Applied Materials & Interfaces, 2009, 1(8): 1645-1649.

[224] CHUANGCHOTE S, SAGAWA T, YOSHIKAWA S. Efficient dye-sensitized solar cells using electrospun TiO_2 nanofibers as a light harvesting layer[J]. Applied Physics Letters, 2008, 93(3): 033310.

[225] LEE Y J, LLOYD M T, OLSON D C, et al. Optimization of ZnO nanorod array morphology for hybrid photovoltaic devices[J]. The Journal of Physical Chemistry C, 2009, 113(35): 15778-15782.

[226] LUO J, LIU C M, YANG S H, et al. Hybrid solar cells based on blends of poly(3-hexylthiophene) and surface dye-modified, ultrathin linear- and branched-TiO_2 nanorods[J]. Solar Energy Materials and Solar Cells, 2010, 94(3): 501-508.

[227] WU Y, ZHANG G Q. Performance enhancement of hybrid solar cells through chemical vapor annealing[J]. Nano Letters, 2010, 10(5): 1628-1631.

[228] TAI Q D, ZHAO X Z, YAN F. Hybrid solar cells based on poly(3-hexylthiophene) and electrospun TiO_2 nanofibers with effective interface modification[J]. Journal of Materials Chemistry, 2010, 20(35): 7366-7371.

[229] OLSON D C, PIRIS J, COLLINS R T, et al. Hybrid photovoltaic devices of polymer and ZnO nanofiber composites[J]. Thin Solid Films, 2006, 496(1): 26-29.

[230] SHIM H S, NA S I, NAM S H, et al. Efficient photovoltaic device fashioned of highly aligned multilayers of electrospun TiO_2 nanowire array with conjugated polymer[J]. Applied Physics Letters, 2008, 92(18): 183107.

[231] YU M, LONG Y Z, SUN B, et al. Recent advances in solar cells based on one-dimensional nanostructure arrays[J]. Nanoscale, 2012, 4(9): 2783-2796.

第3章　纳米纤维在催化领域的应用

催化反应是在不影响化学平衡的前提下，改变化学反应的速率，是自然界中普遍存在的一种现象，在人类文明进步与世界经济发展中扮演着非常重要的角色。催化在现实生活中发挥着不可或缺的作用，它能够将原料高效、绿色地转变为所需要的化工产品。在工业领域，大约90%的化工制品是借助于催化来生产的，在能源制备领域也具有重要的地位；在生态领域，催化反应能够处理各种废弃物品，优化产业结构，如通过光催化降解各种有机污染物，以实现环境的有效治理。

静电纺丝是一种简单、实用，可实现大规模生产应用的技术，可以制备直径从微米到纳米的超细纤维，具有大的比表面积、大孔隙率、高长径比等等特征。因其独有的特征和优异的性能，可以将静电纺丝技术与催化相结合，已广泛应用于电催化、光催化、生物酶固定等催化领域。

3.1　电催化

3.1.1　引言

能源作为社会和经济发展的动力，与人们的日常生活和国家的稳固发展密切相关。面向国家在可再生能源的开发和能源高效优化清洁利用方面的需求，针对我国可再生清洁能源地域分布及产出不均衡的问题，电化学能量存储与物质转化有望提供有效的解决方案,氢能是其中的关键部分。

氢气因其良好的环境兼容性、可再生性以及高的能量密度，是21世纪最清洁的绿色能源之一，在工业生产、航空航天等领域有不可替代的作用。

氢气的制备手段有多种，电解水是有潜力实现大规模应用的制氢方式。目前，电解水制氢工艺成熟，制得的产品氢气纯度高，杂质含量少，可应用于各类能源消耗的场所。除此之外，电解水的原材料是水，资源丰富，地球表面 70%被水覆盖，足以可见电解水工艺广阔的发展空间。

电化学析氢反应（HER）是一种简单的二电子转移过程，只有一种中间产物，机理图如图 3.1 所示。一个完整的析氢反应分为两个连续的步骤：首先通过沃尔默反应（Volmer 反应），氢原子吸附在催化剂表面，并伴随一个电子的转移，形成了吸附氢。随后生成氢气有两种途径：途径一，塔费尔反应（Tafel 反应），两个相邻的吸附氢原子的催化剂结合，形成一个氢分子；途径二，海洛夫斯基反应（Heyrovsky 反应），即一个吸附氢原子与一个质子结合，同时转移一个电子即生成一个氢分子。在酸性、中性和碱性溶液中的反应方程式如下所示。M 代表阴极电极上的催化剂，MH 代表阴极催化剂表面吸附的氢原子。

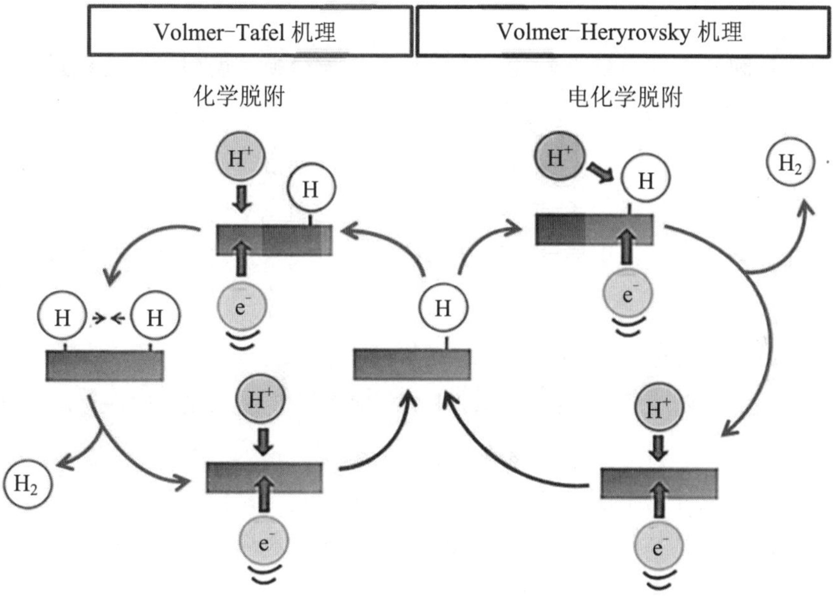

图 3.1　HER 反应机理图

（1）电化学氢吸附步骤（Volmer 反应）。

$$H_3O^+ + e^- \longrightarrow MH + H_2O \quad （酸性介质） \tag{3.1}$$

$$H_2O + e^- \longrightarrow MH + OH^- \quad （酸性或中性介质） \tag{3.2}$$

（2）电化学脱附步骤（Heyrovsky 反应）。

$$H_3O^+ + MH + e^- \longrightarrow M + H_2 + H_2O \quad (\text{酸性介质}) \tag{3.3}$$

$$H_2O + MH + e^- \longrightarrow M + H_2 + H_2O \quad (\text{碱性或中性介质}) \tag{3.4}$$

（3）复合脱附步骤（Tafel 反应）。

$$MH + MH \longrightarrow M + H_2 \tag{3.5}$$

然而，电解水缓慢的动力学因素导致了其过电位较大、能量转换效率较低等问题。因此，研究制备高效电催化剂，加快反应进程对于减小过电位，降低能耗来说至关重要。电催化剂的难点一旦实现突破，将颠覆人类的能源结构，成为生态文明建设和人类可持续发展的基础。

火山图反映了不同金属吸附能与析氢反应电流交换密度对数值之间的关系，如图 3.2 所示，贵金属 Pt、Pd 等的吸附和脱附自由能适中，位于火山图的顶端，表示其电催化活性较好，但是由于贵金属的储量较少，价格昂贵，限制了其在商业上的大规模使用。过渡金属 Ni、Cu、Co 等元素位于火山图中间部位，吸附和脱附自由能比贵金属大，但也表现出较好的催化活性，并且因其价格低廉、导电性好等优点吸引了广泛关注。

静电纺丝法为各种材料成功制备纤维状纳米结构提供了一种简单通用的技术。通过静电纺丝法可以制备内部为实心或中空的纳米纤维，其长度特别长，直径均匀，并且成分多样化，制备的纳米材料比表面积比一般材料大得多，可以提供更多的活性位点，表现出优异的电解质扩散和电荷传输能力。与其他制备方法不同，通过静电纺丝制备薄纤维是基于凝胶状聚合物溶液或熔体在高压静电场的作用下的单轴拉伸，从针头处喷射出去最终形成纤维状材料。与机械拉伸相比，静电纺丝更适合于产生直径更细的纤维，可以通过施加外部电场的非接触方案来实现纤维的拉伸。与机械拉伸一样，静电纺丝也是一个连续的过程，因此该技术适用于大批量生产。毫无疑问，静电纺丝将成为未来生成一维纳米结构的首选技术，在电催化领域有着广阔的应用前景。

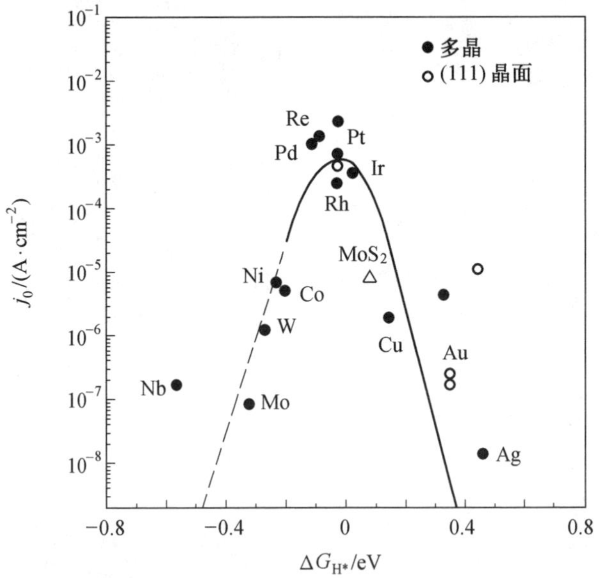

图 3.2　各金属交换电流密度对数与吉布斯自由能之间关系的火山图

3.1.2　电催化析氢催化剂

过渡金属 Ni、Cu、Co、Fe 等元素位于火山图中间部位，吸附和脱附自由能较为适中，表现出较好的催化活性，并且因其较为低廉的价格以及较好的使用寿命逐渐引起了广泛关注。此外，在过渡族非贵金属材料中掺杂非金属元素（如：N、S、P、Se 等）能够改变其微观形貌，增加活性位点，扩大催化剂的电化学活性面积，进而提高催化活性，延长电极的使用寿命。Wang 等人利用静电纺丝技术制备了磷和氮双掺杂钴基碳纳米纤维（Co-N-P-CNFs），表现出出色的催化析氢（HER）性能，如图 3.3 所示，在 0.249 V 的过电势下即可达到 10 $mA\cdot cm^{-2}$ 的电流密度。

Lin 等人通过静电纺丝及技术结合热处理工艺，制备了氮掺杂的碳化铁-碳化钼纳米纤维（Fe_3C-Mo_2C/NC），如图 3.4 所示。由于 Mo_2C 上的氢键很强，而 Fe_3C 上的氢键相对较弱，Fe_3C-Mo_2C 的异质界面能够显著提高 HER 动力学和本征活性，进而提高其 HER 性能。在 0.5 $mol\cdot L^{-1}$ H_2SO_4 的介质中，仅需 116 mV 的过电势即可达到 10 $mA\cdot cm^{-2}$ 的电流密度，Tafel 斜率仅为 43 $mV\cdot dec^{-1}$。

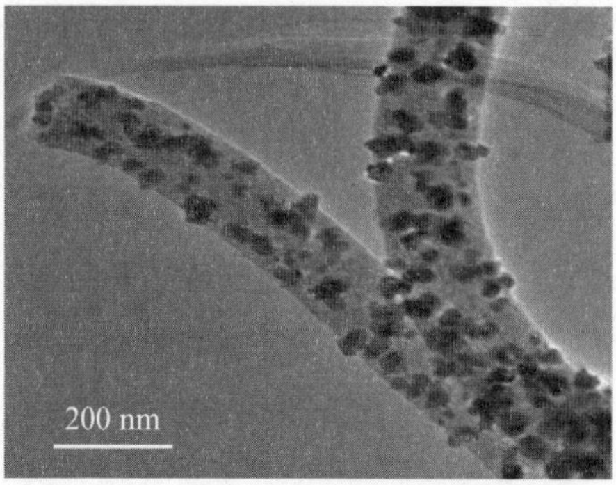

（a）Co-N-P-CNFs 纳米纤维的 TEM 图

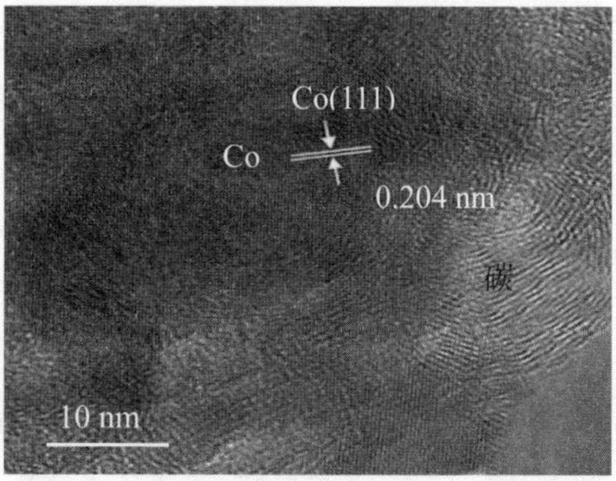

（b）Co-N-P-CNFs 纳米纤维的 HR-TEM 图

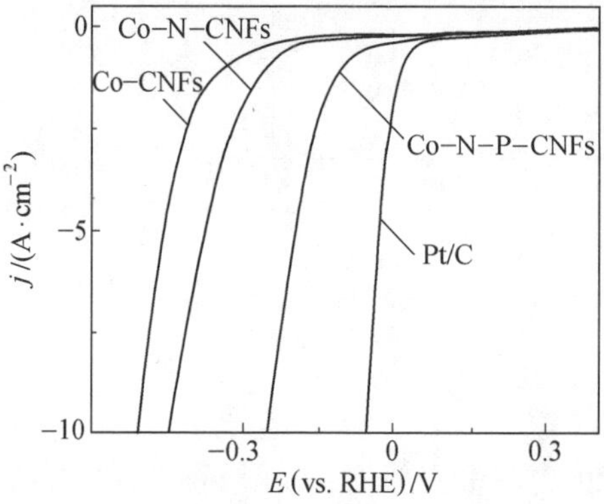

（c）各催化剂的 LSV 对比图

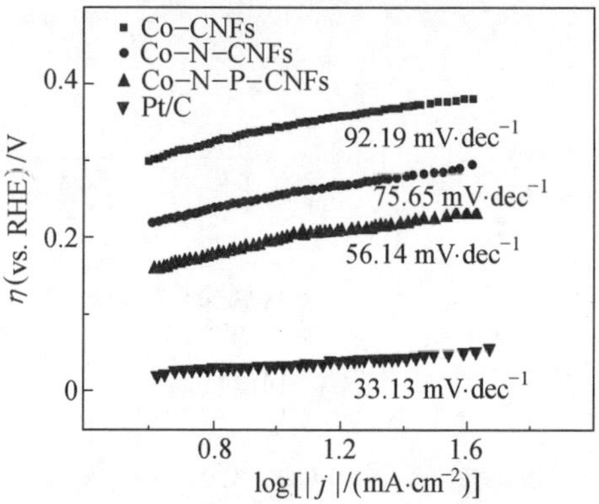

（d）各催化剂的 Tafel 斜率对比图

图 3.3　磷和氮双掺杂钴基碳纳米纤维(Co-N-P-CNFs)

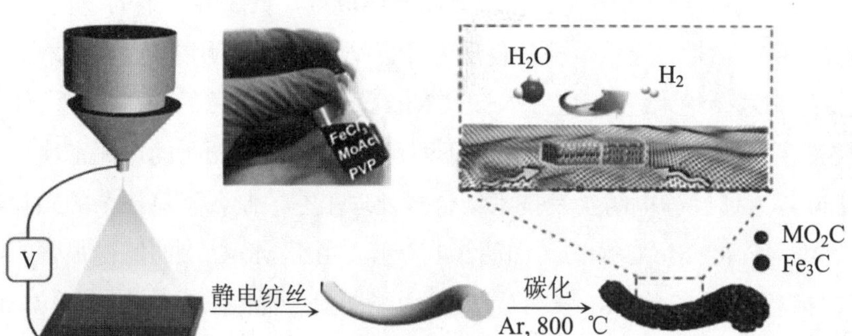

（a）Fe_3C-Mo_2C / NC 的制备机理图

图 3.4　氮掺杂的碳化铁-碳化钼纳米纤维（Fe_3C-Mo_2C/NC）

（b）Fe_3C-Mo_2C/NC 的 SEM 图

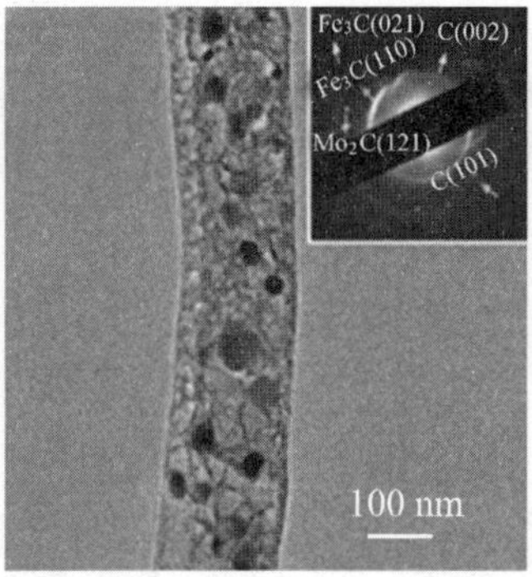

（c）Fe_3C-Mo_2C/NC 的 TEM 图

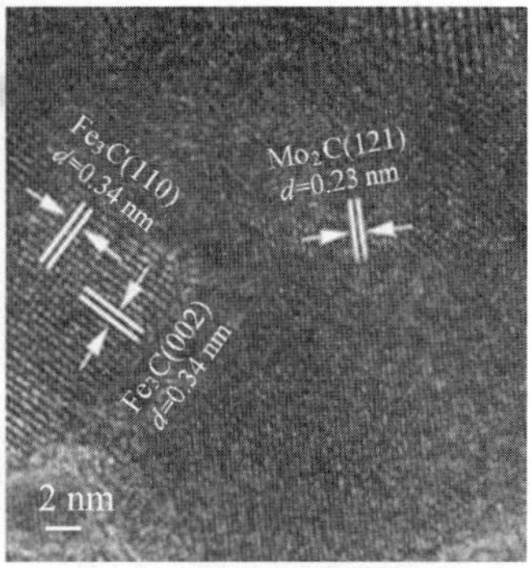

（d）Fe_3C-Mo_2C/NC 的 TEM 图

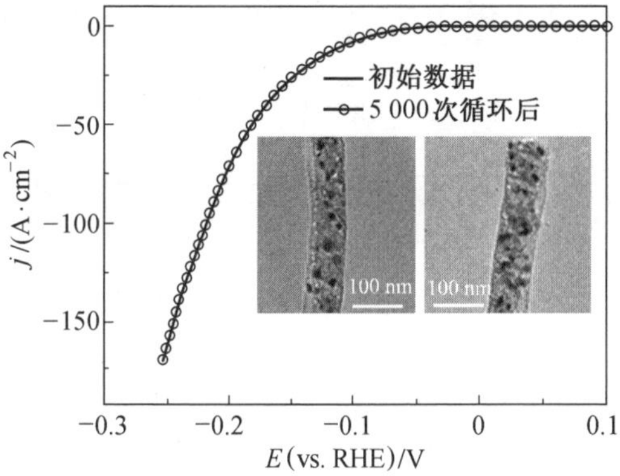

（e）经 5 000 次循环后 LSV 对比图

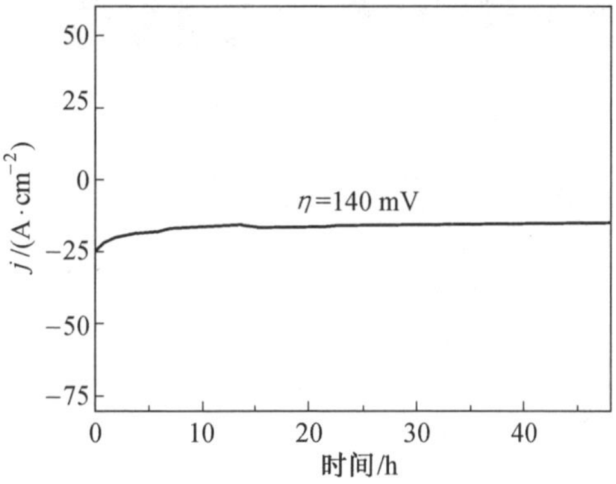

（f）Fe_3C-Mo_2C/NC 的长时间稳定性测试图

续图 3.4

大多数 HER 催化剂仅仅能够在酸性条件下发挥其催化作用，大多数 OER 催化剂都可以在碱性和中性条件下发挥其催化作用。而为了实现电解水，HER 催化剂必须在较宽 pH 范围内工作，这样才能和 OER 催化剂相匹配。目前，已研发出许多在全 pH 范围内工作的电化学催化剂。Yang 等人通过静电纺丝技术结合热处理工艺，制备了 N、P 掺杂的 Mo_2C/碳纳米纤维，具有优异的催化析氢（HER）性能。在酸性（0.5 $mol{\cdot}L^{-1}$ H_2SO_4）和碱性（1 $mol{\cdot}L^{-1}$ KOH）介质中分别仅需 197 mV 和 107 mV 的过电势即可达到 10 $mA{\cdot}cm^{-2}$ 的电流密度，Tafel 斜率分别为 81.3 $mV{\cdot}dec^{-1}$ 和 67.5 $mV{\cdot}dec^{-1}$。该催化剂在酸性和碱性条件下均表现出优异的稳定性，其微观形貌如图 3.5 所示。

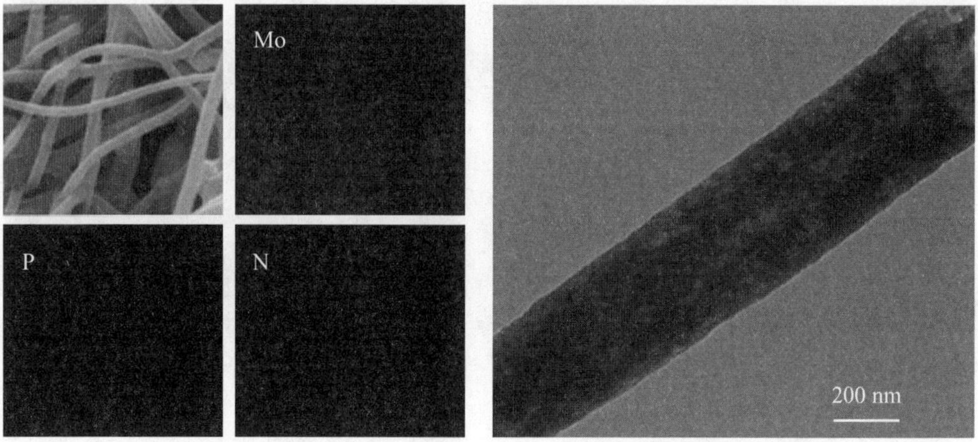

（a）4%-PMoHN-CNFs 的 SEM 和 EDS 元素分布图　　（b）4%-PMoHN-CNFs 的 TEM 图

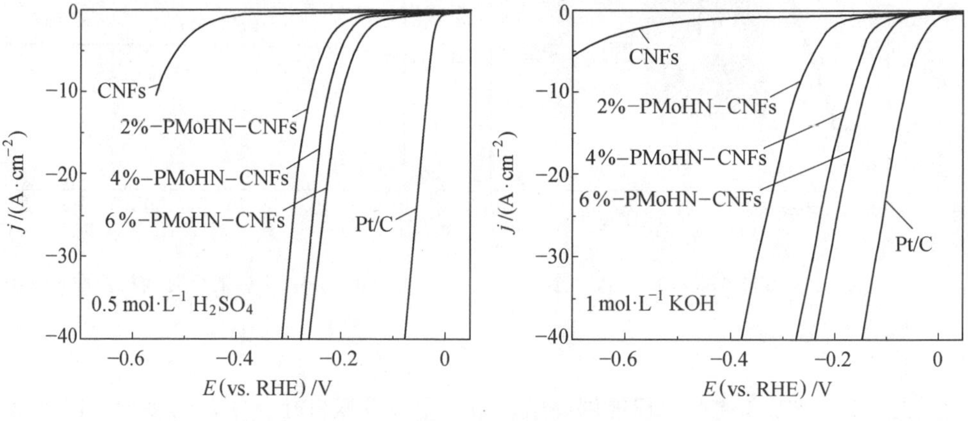

（c）0.5 mol·L^{-1} H_2SO_4 和 1 mol·L^{-1} KOH 介质中的 LSV 曲线图

图 3.5　N、P 掺杂的 Mo_2C/碳纳米纤维

双功能电催化剂也逐渐引起了人们的兴趣，开发出一种新型催化剂，其在相同的 pH 范围内可以实现 HER 和 OER 的高活性和稳定性，进而实现电解水。Tong 等人以聚丙烯腈（PAN）为原料，通过静电纺丝技术结合热处理磷化工艺，制备了氮掺杂的 CoP/NCF 纳米纤维，具有优异的 HER 和 OER 活性和稳定性，在碱性介质中达到 10 mA·cm^{-2} 的电流密度分别需要 141 mV 和 288 mV 的过电势，Tafel 斜率分别为 84 mV·dec^{-1}和 60 mV·dec^{-1}。对于整体水的分解也表现出优异的催化活性，仅需

1.64 V 的电压即可达到 10 mA·cm^{-2}的电流密度，表现出优异的全水分解性能。密度泛函理论计算表面，CoP 和 N 之间的协同作用使得 HER 和 OER 性能大大提升。其微观形貌如图 3.6 所示。

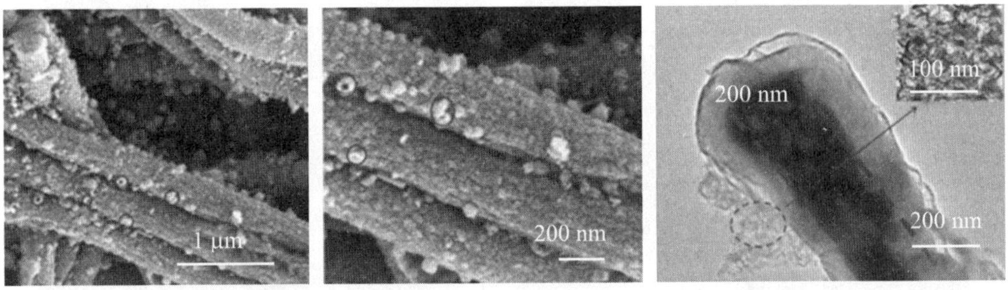

（a）CoP/NCF 的 SEM 图　（b）CoP/NCF 的 SEM 图　（c）CoP/NCF 的 TEM 图

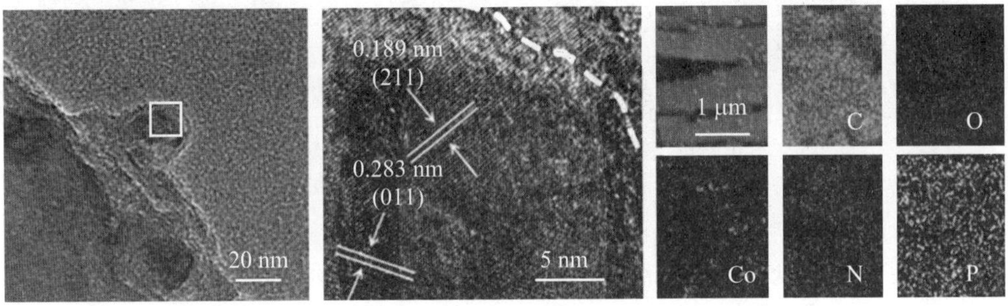

（d）CoP/NCF 的 HRTEM 图像　（e）选定区域 HRTEM 放大图（f）CoP/NCF 的 SEM 及 EDS 元素分布图

图 3.6　氮掺杂的 CoP/NCF 纳米纤维

通式为 ABO_3 的钙钛矿氧化物（其中 A 通常是稀土金属或碱土金属，B 是过渡金属）因其优异的物理、化学性质而逐渐引起了大家的关注，尤其是在碱性溶液中表现出优异的 OER 和 ORR 性能。Zhu 等人利用静电纺丝技术结合热处理工艺，制备了 $SrNb_{0.1}Co_{0.7}Fe_{0.2}O_{3-\delta}$ 钙钛矿纳米纤维。如图 3.7 所示，在碱性溶液中显示出优异的 OER 和 HER 活性和稳定性，可用于整体水的分解，仅需 1.68 V 电压即可得到 10 mA·cm^{-2}的电流密度，是一种高效的贵金属催化剂替代品。

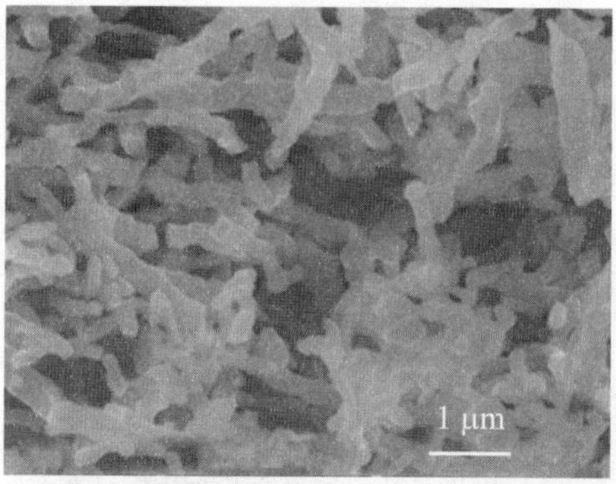

（a）钙钛矿纳米纤维的 SEM 图

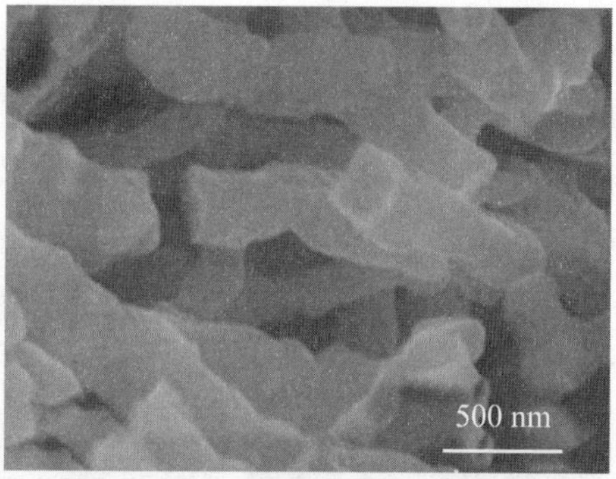

（b）钙钛矿纳米纤维的 SEM 图

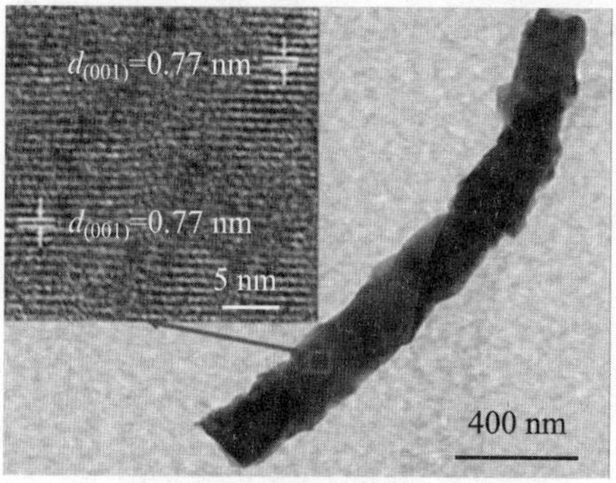

（c）钙钛矿纳米纤维的 TEM 图

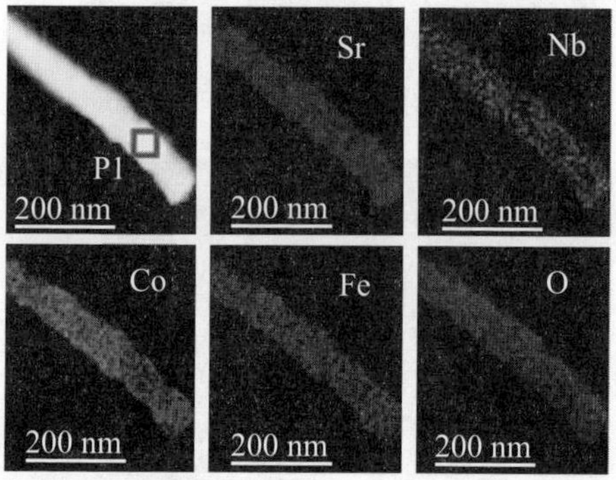

（d）钙钛矿纳米纤维的 SEM 及 EDS 元素分布图

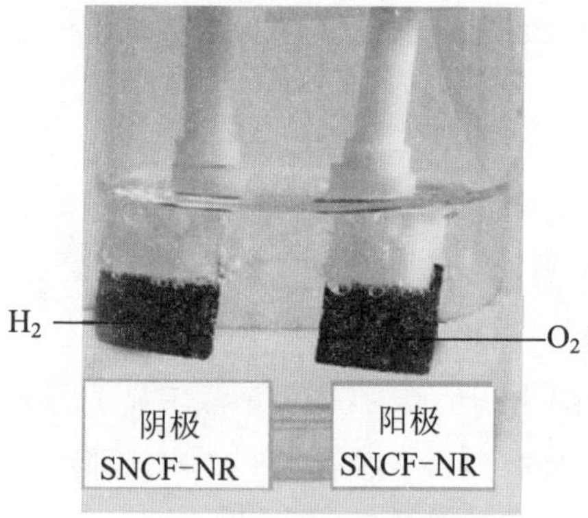

（e）全水分解器件实物图

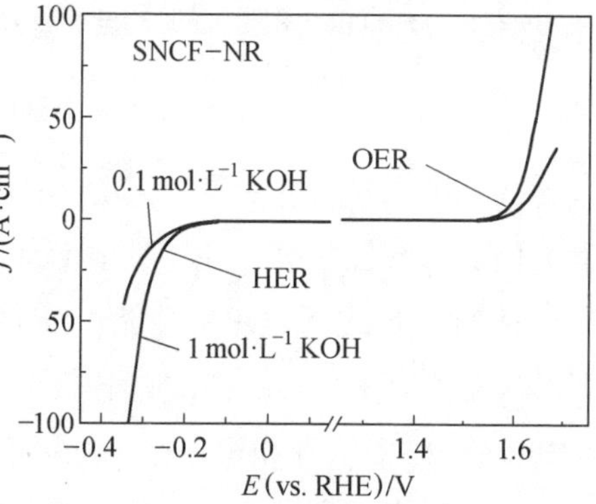

（f）钙钛矿纳米纤维的 LSV 图

图 3.7　$SrNb_{0.1}Co_{0.7}Fe_{0.2}O_{3-\delta}$钙钛矿纳米纤维

2017～2020 年 HER 催化剂研究现状调研结果见表 3.1，通过静电纺丝技术制备的纳米纤维膜在酸性和碱性介质中均具有较好的催化析氢（HER）性能，较小的塔费尔斜率也意味着纳米纤维催化剂具有更有利的动力学因素。

表 3.1　HER 催化剂研究现状调研结果

电极材料名称	电解质	达到 10 mA • cm^{-2} 电流密度所需过电势（vs. RHE）/mV	塔费尔斜率 /（mV • dec^{-1}）
Co_6W_6C–N@CNFs	0.5 $mol·L^{-1}$ H_2SO_4	86	85
	1 $mol·L^{-1}$ KOH	116	101
MoO_2/MoS_2	0.5 $mol·L^{-1}$ H_2SO_4	340	59
Ni_3Fe@N–CNT/NFs	1 $mol·L^{-1}$ KOH	72	98
CoS_2–C@MoS_2	0.5 $mol·L^{-1}$ H_2SO_4	173	61
Ni–NCNFs–Pt	0.5 $mol·L^{-1}$ H_2SO_4	47	31
Co/CoP@NC	0.5 $mol·L^{-1}$ H_2SO_4	117	60.3
CoP/PCNF	0.5 $mol·L^{-1}$ H_2SO_4	83	62
Co–PAN(5%)–800	0.5 $mol·L^{-1}$ H_2SO_4	159	—
MoC–MoP/BCNC NFs	0.5 $mol·L^{-1}$ H_2SO_4	158	58
P–$NiCo_2S_4$@CNT/CNF	0.5 $mol·L^{-1}$ H_2SO_4	74	65.9
CNFs/$CoSe_2$	0.5 $mol·L^{-1}$ H_2SO_4	119	54
C@NiO/Ni	1 $mol·L^{-1}$ KOH	395	152
N,P–Mo_xCNF	0.5 $mol·L^{-1}$ H_2SO_4	107	65.1
NiOOH/$Ni(OH)_2$	0.1 $mol·L^{-1}$ KOH	147	41
Ni/Mo_2C–NCNFs	1 $mol·L^{-1}$ KOH	143	57.8
NCNFs–MoS_2|P	0.5 $mol·L^{-1}$ H_2SO_4	98	66
MWCNT	0.5 $mol·L^{-1}$ H_2SO_4	295	42

3.1.3 电催化析氧催化剂

水的分解包括了另一个半反应，即析氧反应（OER），该反应涉及 O—H 键的断裂和 O—O 键的重组，动力学因素较为缓慢，析氧电位较高，整体水分解的能耗较大，效率较低，被认为是水分解反应的瓶颈。因此，研发出优异的 OER 催化剂是至关重要的。目前，IrO_2 和 RuO_x 在 OER 反应中具有优异的活性，但是由于贵金属的价格较为昂贵，成本较高，且耐久性差，在实际的大规模生产应用中受到了一定的限制。正因如此，研发 OER 活性高，稳定性好的非贵金属基新型催化剂是至关重要的。其火山图和 OER 反应机理图如图 3.8 所示。

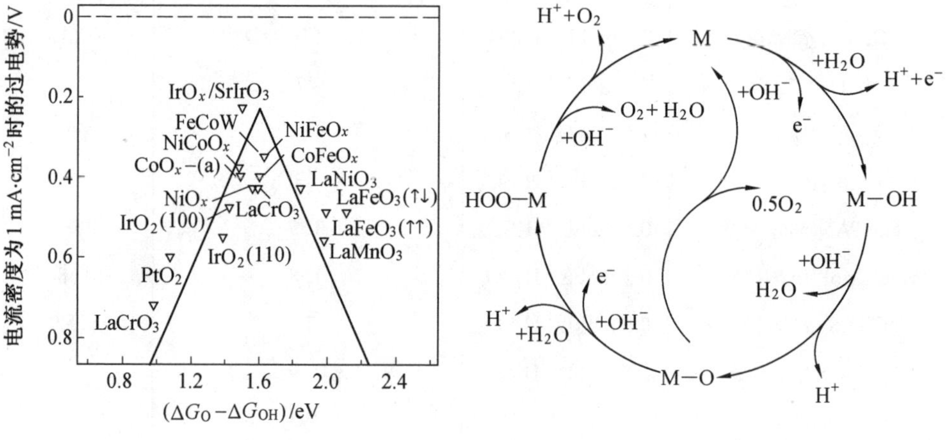

（a）电流密度对数与吉布斯自由能关系图　　（b）OER 反应机理图

图 3.8　析氧电催化反应

电化学析氧反应（OER）机理的研究比较成熟，受电极表面的影响，不同材料的 OER 反应机理不同，四电子途径是被大家广泛接受的理论。酸性条件下与碱性条件下的中间体相同，如 MOH 和 MO，不同之处在于氧气的生成过程。从 MO 中间体生成氧气的路径主要有两个：一个是 2MO 直接结合生成 O_2；另一个是涉及 MOOH 中间体的形成，随后释放出 O_2。OER 反应是非均相催化反应，中间体（MOH、MO、MOOH）对整个过程至关重要。

酸性条件：

$$2H_2O(l) \longrightarrow 2O_2(g) + 4H^+ + 4e^- \tag{3.6}$$

反应机理：

$$M + H_2O(l) \longrightarrow MOH + H^+ + e^- \tag{3.7}$$

$$MOH + OH^-(l) \longrightarrow MO + H_2O(l) + e^- \tag{3.8}$$

$$2MO \longrightarrow 2M + O_2(g) \tag{3.9}$$

$$MO + H_2O(l) \longrightarrow MOOH + H^+ + e^- \tag{3.10}$$

$$MOOH + H_2O(l) \longrightarrow M + O_2(g) + H^+ + e^- \tag{3.11}$$

碱性条件：

$$4OH^- \longrightarrow 2O_2(g) + 2H_2O(l) + 4e^- \tag{3.12}$$

反应机理：

$$M + OH^- \longrightarrow MOH \tag{3.13}$$

$$MOH + OH^- \longrightarrow MO + H_2O(l) \tag{3.14}$$

$$2MO \longrightarrow 2M + O_2(g) \tag{3.15}$$

$$MO + OH^- \longrightarrow MOOH + e^- \tag{3.16}$$

$$MOOH + OH^- \longrightarrow M + O_2(g) + H_2O(l) \tag{3.17}$$

钴基材料因其在电催化反应中较高的稳定性以及非常规的 3d 电子组态而引起了广泛的关注。S. Sam Sankar 等人通过静电纺丝技术结合热处理工艺，制备了 Co-ZIF 超细纤维，其表现出优异的 OER 性能。在 1 $mol \cdot L^{-1}$ KOH 介质中，Co-ZIF-350-Air 仅需 370 mV 过电势即可达到 10 $mA \cdot cm^{-2}$ 的电流密度，Tafel 斜率仅为 55 $mV \cdot dec^{-1}$。Guo 等人以泡沫镍为基底，利用静电纺丝技术制备了无定形非晶

Co-Fe 氢氧化物纳米纤维（NF-PVP/$CoFe_{1.3}$），如图 3.9 所示，在碱性环境中表现出卓越的 OER 性能，仅需 0.267 V 的过电势即可达到 100 $mA\cdot cm^{-2}$ 的电流密度，Tafel 斜率仅为 47.43 $mV\cdot dec^{-1}$。

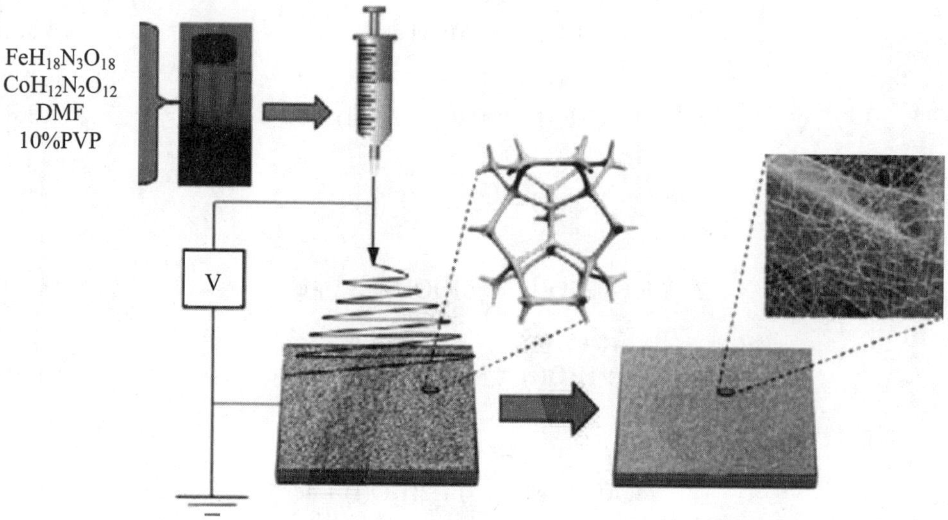

（a）NF-PVP/$CoFe_{1.3}$ 纳米纤维制备流程图

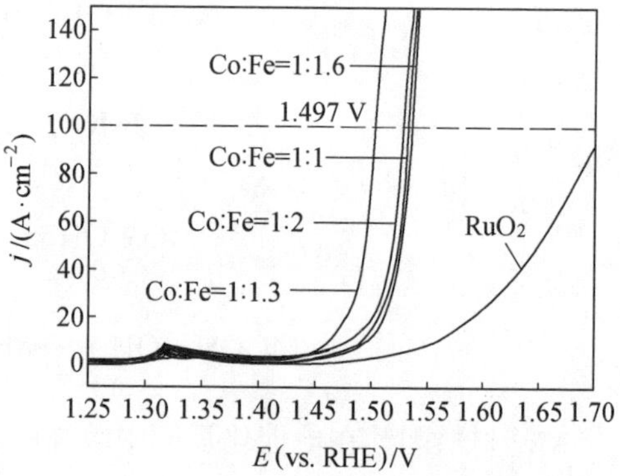

（b）NF-PVP/$CoFe_{1.3}$ 纳米纤维的 TEM 图　（c）不同 Co、Fe 比的样品与 RuO_x 的 LSV 曲线对比图

图 3.9　无定形非晶 Co-Fe 氢氧化物纳米纤维（NF-PVP/$CoFe_{1.3}$）

Wang 等人通过同轴静电纺丝技术，制备了具有氮掺杂纳米管的介孔薄壁 $CuCo_2O_4@C$ 纳米纤维，其具有优异的 ORR（327 mV@10 mA·cm^{-2}）和 OER（327 mV@10 mA·cm^{-2}）性能。用作 Zn−空气电池阴极催化剂时，表现出低的充放电电压间隙（0.79 V@10 mA·cm^{-2}）以及长的循环寿命（80 h 最多 160 个循环），具有优出色的电催化性能。Zhao 等人利用静电纺丝技术结合热处理和酸化工艺，制备了嵌入钴纳米颗粒的多孔氮掺杂纳米纤维（Co−PNCNFs），表现出优异的 OER 性能，仅需 285 mV 的过电势即可达到 10 mA·cm^{-2} 的电流密度。除此之外，发现该纳米纤维还具有优异的 HER 性能。将其用于整体水的电解，仅需 1.66 V 的电压即可达到 10 mA·cm^{-2} 的电流密度。其微观形貌图及 EDS 图如图 3.10 所示。

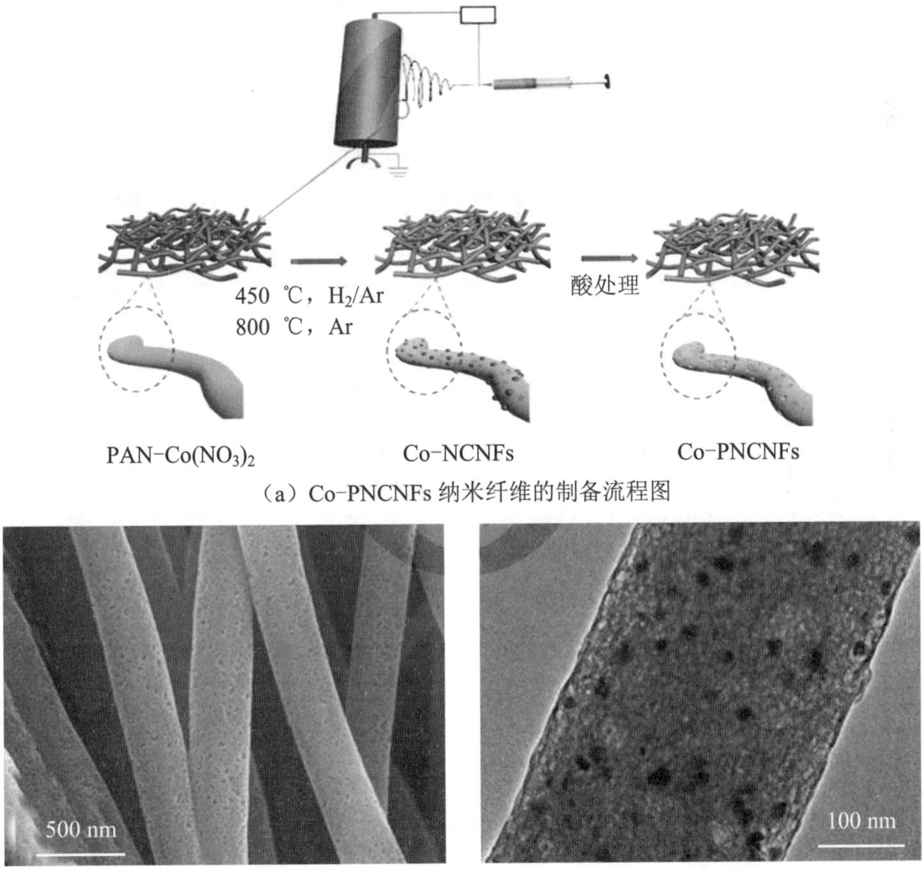

（a）Co−PNCNFs 纳米纤维的制备流程图

（b）SEM 图　　（c）TEM 图

图 3.10　嵌入钴纳米颗粒的多孔氮掺杂纳米纤维（Co−PNCNFs）

目前，Ni 基金属 OER 材料已得到广泛研究，并且发现其具有作为高效 OER 催化剂的巨大潜力。Bhushan Patil 等人利用原子层沉积法（ALD）结合静电纺丝技术，将非贵金属催化剂 NiOOH/Ni（OH）$_2$ 沉积在碳纳米纤维上，制得的纳米纤维具有优异的 OER 性能，仅需 290 mV 过电势即可达到 10 mA·cm^{-2} 的电流密度。Chen 等人通过静电纺丝技术结合热处理工艺，制备了 Ni−Fe_2O_3/CNF 复合纤维，具有优异的 OER 性能，仅需 310 mV 的过电势即可达到 10 mA·cm^{-2} 的电流密度，能够连续工作 15 h 并保持活性，如图 3.11 所示。其优异的 OER 性能归结于纳米纤维较大的孔隙率提供了更多的活性位点，铁镍氧化物的形成与碳载体之间形成良好的耦合，以及通过调节 Ni/Fe 比可以改变活性位点的化学环境以增强 OER 性能。

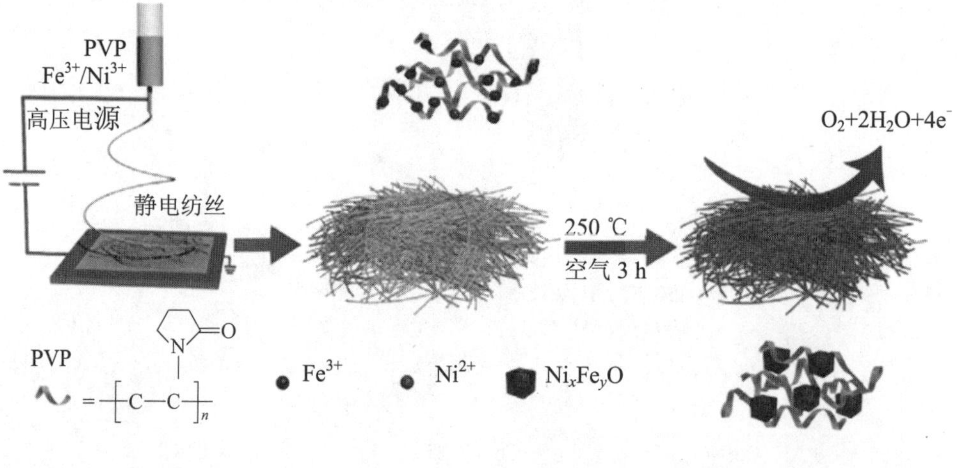

（a）Ni−Fe_2O_3/CNF 复合纤维的制备机理图（彩图见附录）

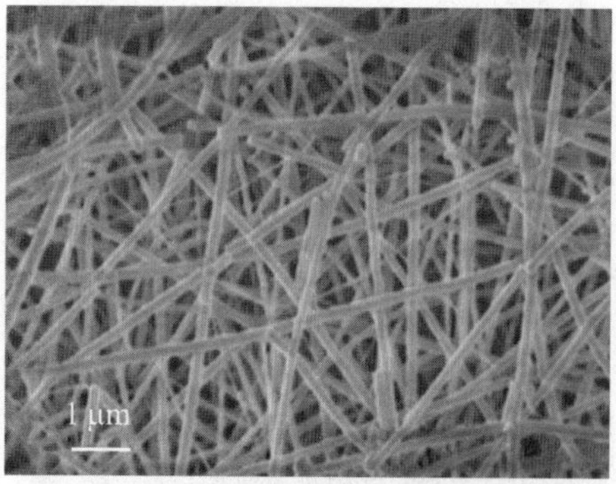

（b）Ni−Fe_2O_3/CNF 复合纤维的 SEM 图

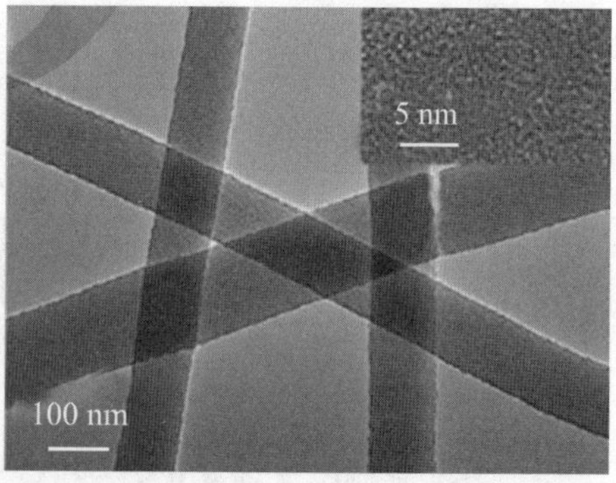

（c）Ni−Fe_2O_3/CNF 复合纤维的 TEM 图

图 3.11　Ni−Fe_2O_3/CNF 复合纤维

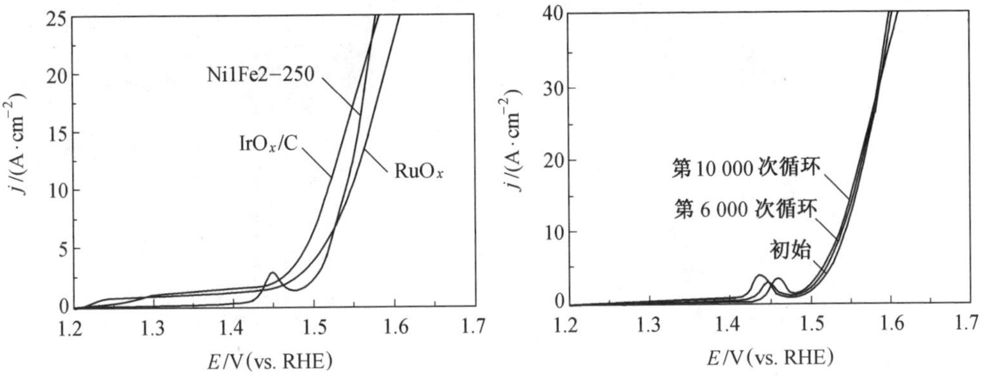

（d）Ni-Fe_2O_3/CNF 复合纤维的 LSV 曲线图

续图 3.11

Feng Cao 等人利用静电纺丝技术，制备了金属盐比例为 1∶1∶1 的氮掺杂、碳包封的镍 NiCoFe 双功能催化剂（NiCoFe@N-CNFs），其具有优异的 OER 和 ORR 活性。仅需 270 mV 过电势即可达到 10 $mA·cm^{-2}$ 的电流密度，Tafel 斜率仅为 72 $mV·dec^{-1}$，具有较好的动力学因素。Wei Peng 等人利用静电纺丝技术，制备了多孔氮掺杂的碳纳米纤维（NiFe@NCNFs），其中镍铁合金纳米粒子（NiFe）被封装在纳米纤维中，是一种高效的 OER 电催化剂。如图 3.12 所示，在 1 $mol·L^{-1}$ KOH 溶液中，仅需 294 mV 的过电势，即可达到 10 $mA·cm^{-2}$ 的电流密度。除此之外，该纳米纤维在碱性溶液中表现出优异的稳定性。

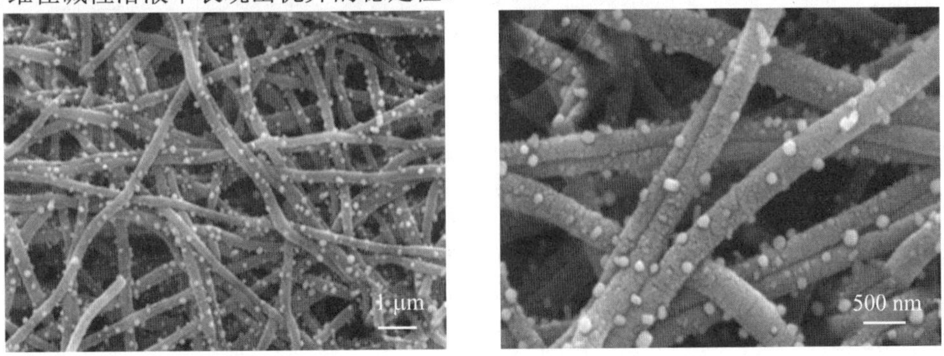

（a）NiFe@NCNFs 在不同放大倍数下的 SEM 图

图 3.12　多孔氮掺杂的碳纳米纤维（NiFe@NCNFs）

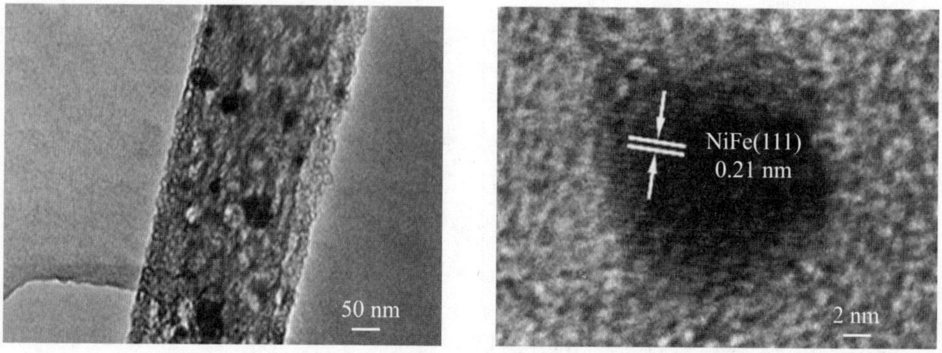

（b）不同放大倍数下的 TEM 图

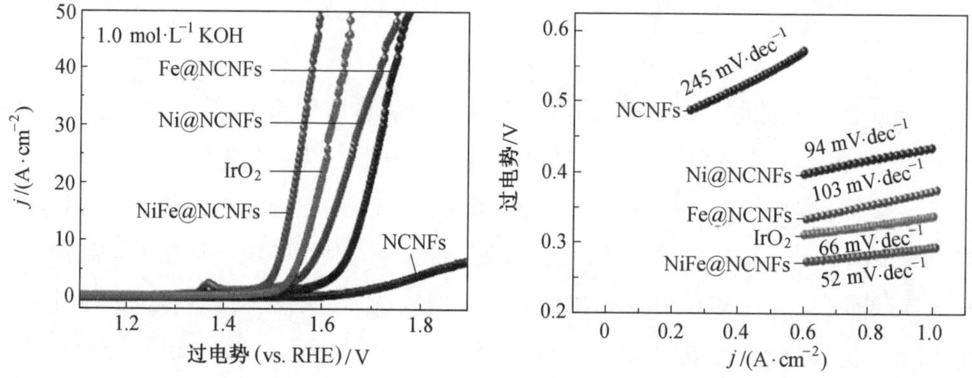

（c）1 mol·L^{-1} KOH 溶液中的 LSV 曲线图　　（d）LSV 曲线图对应的 Tafel 斜率图

续图 3.12

钙钛矿氧化物也具有优异的 OER 性能。Zhen 等人通过静电纺丝技术合成了多孔 $La_{0.6}Sr_{0.4}Co_{0.6}Fe_{0.4}O_{3-\delta}$ 纳米纤维，其具有优异的 OER 性能，仅需 647 mV(vs Ag/AgCl) 过电势即可达到 10 mA·cm^{-2} 的电流密度，比商用催化剂 LSCF（786 mV）、IrO_2（660 mV）小。除此之外，该纳米纤维还具有出色的稳定性，在 10 mA·cm^{-2} 的电流密度下 3 h 后并没有发现其性能的下降。其微观形貌图、LSV 和 Tafel 斜率图如图 3.13 所示。

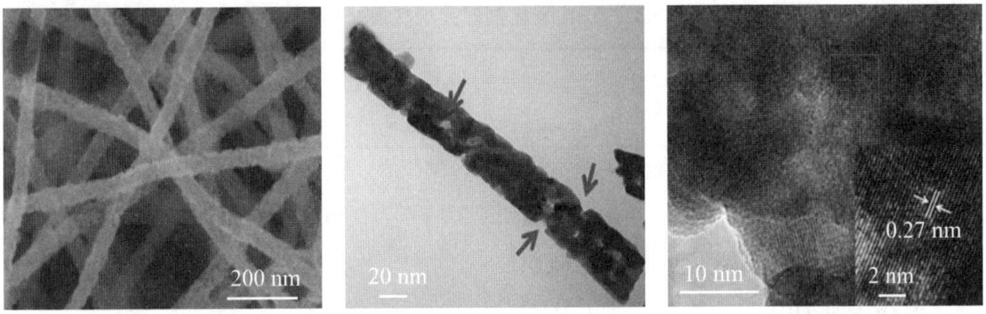

（a）SEM 图　　　　（b）TEM 图　　　　（c）HRTEM 图

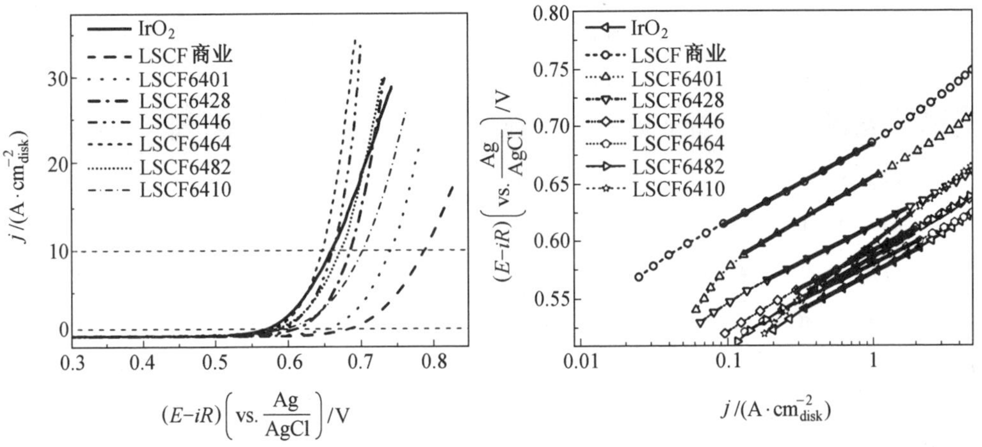

（d）0.1 mol·L^{-1}溶液中的 LSV 对比图　　　　（e）Tafel 斜率图

图 3.13　多孔 $La_{0.6}Sr_{0.4}Co_{0.6}Fe_{0.4}O_{3-\delta}$（LSCF）纳米纤维

2014～2020 年 OER 催化剂研究现状调研结果见表 3.2，通过静电纺丝技术制备的纳米纤维膜在酸性和碱性介质中均具有较好的催化析氧（OER）性能，其较小的析氧电位以及较好的动力学因素有利于促进整体水的分解，以实现其大规模的应用发展。

表 3.2 OER 催化剂研究现状调研结果

电极材料名称	电解质	达到 10 mA・cm^{-2} 电流密度所需过电势（vs. RHE）/mV	塔费尔斜率 /（mV・dec^{-1}）
$CoFe_2O_4$@N–CNFs	0.1 mol·L^{-1} KOH	349	80
CoS_2 C@MoS_2	0.5 mol·L^{-1} H_2SO_4	391	46
NiCoFe@N–CNFs	0.1 mol·L^{-1} KOH	270	72
NiOOH/Ni$(OH)_2$	0.1 mol·L^{-1} KOH	390.5	50
FeNi–NCFs	1 mol·L^{-1} KOH	317	49
NiCoP/CNF	1 mol·L^{-1} KOH	268	83
Ni/Mo_2C–NCNFs	1 mol·L^{-1} KOH	288	78.4
Co_3O_4 NF	1 mol·L^{-1} KOH	293	60.5
SFCNF	0.5 mol·L^{-1} H_2SO_4	180	54.4
FeCoNi–CNF	1 mol·L^{-1} KOH	220	57
CQDs@BSCF–NFs	1 mol·L^{-1} KOH	350	66
h–Co_3O_4/CeO_2@N–CNFs	0.1 mol·L^{-1} KOH	310	89
Fe–CoO/C	1 mol·L^{-1} KOH	362	60
Fe@Ni NFs	1 mol·L^{-1} KOH	230	10
$CoFe_2O_4$–$CoFe_x$/C	1 mol·L^{-1} KOH	350	—
Co–MnO_2|OV	1 mol·L^{-1} KOH	279	75
NiO/$NiCo_2O_4$	0.1 mol·L^{-1} KOH	357	130
Fe_3C@NCNTs–NCNFs	1 mol·L^{-1} KOH	284	56
Mo_2C–MCNFs	1 mol·L^{-1} KOH	320	68

3.1.4 小结与展望

氢气是一种绿色环保且能量密度高的理想能源载体，是 21 世纪最清洁的能源之一。而电解水是制备氢气的有效途径。由于水的氧化还原反应的动力学缓慢，极大地阻碍了电解水商业化的进程。HER/OER 催化剂的重要性毋庸置疑，迫切需要寻找高效、稳定，具有成本效益和可大规模商业化生产的材料。静电纺丝是制备纳米纤维最简便的方法之一，制得的纳米纤维比表面积大、孔隙率高，可以提供大量的活性位点，表现出优异的电解质扩散和电荷传输能力，是一种极具潜力的电催化材料。尽管目前通过静电纺丝制备高性能的 HER/OER 催化剂已经取得了一定进展，但在制备和合成方面仍需进一步努力。具体如下：

（1）在催化材料的功能设计和精准合成方面还很欠缺，尤其是兼顾高效催化效率与长时间稳定性的材料仍然缺乏。如何将静电纺丝技术与其他制备工艺相结合，以实现高活性和高稳定性的新型纳米纤维电催化材料的大规模制备，仍是我们需要努力的方向。

（2）开发具有多功能，同时具有较好析氢能力和析氧能力的纳米纤维电极是至关重要的。

（3）目前对于催化剂表界面的反应动力学和机制认识有限，缺乏一定的理论模拟。需要进一步借助理论模拟和高通量的实验筛选优化，以实现通过静电纺丝技术，制备高活性、稳定性好的纳米纤维催化剂材料。

3.2　光催化

3.2.1　引言

随着工业和科技的飞速发展，各个国家的生产力水平也在不断提高。世界上的不可再生资源也逐渐走向枯竭，这些资源的消耗不仅会带来能源的短缺，也会引发一系列环境污染问题。而这些问题能否妥善处理，成为未来人类社会可持续发展能否正常进行的关键。

光催化在环境治理，新能源开发等方面逐渐受到了人们的特别重视，通过太阳能制备氢气是解决能源和环境问题的最佳方案之一。半导体催化剂是目前世界上使用最多，应用最广的光催化剂，它能够利用可再生的太阳能除去污染物，将之分解成对环境无污染的 CO_2 和 H_2O 等小分子，以实现环境净化；也可以将水分解为氢气，实现太阳能向氢能的转换，进而解决当今所面临的能源枯竭和环境污染问题。

3.2.2 光催化剂在环境治理中的应用

半导体材料在光照下，若禁带宽度小于入射光能量，价带电子将激发至导带，产生空穴。光生电子和空穴会向表面迁移，催化剂的部分电子和空穴能够复合，剩余未发生复合的电子和空穴将会与催化剂表面的反应物发生氧化还原反应，使有机物分解，以达到光催化分解污染物的目的。光催化机理图如图 3.14 所示。

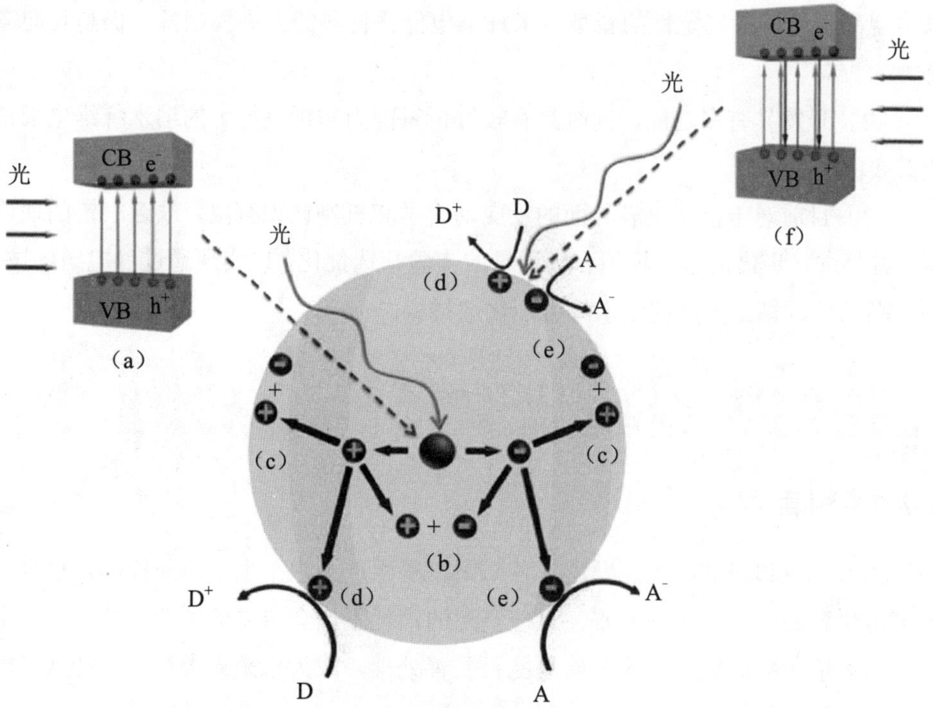

图 3.14　光催化机理图

（a）光子中的电子激发（CB 为导带，VB 为价带）；（b）电子和空穴的复合；（c）晶粒的电子和空穴的复合表面；（d）与吸附在晶粒表面上的电子供体（D）的空穴反应；（e）与吸附在晶粒表面上的电子受体（A）的电子反应；（f）分子光在光催化剂的晶粒表面上进行局部的电子激发重组

TiO_2 因其出色的光催化性能、稳定的物理化学性质、成本低、适用范围广等优势，一直是人们研究光催化剂的热点。然而，目前 TiO_2 催化剂多以纳米颗粒的形式存在，晶界和表面上存在大量缺陷，不利于光生载流子的快速转移；纳米粒子容易

发生团聚，影响使用性能等等。因此，如何提高 TiO_2 的光催化性能逐渐引起了大家的关注。

Lee 等人利用静电纺丝技术，以 PVP 和 PVDF 为原料，将 P25-TiO_2 固定在的纳米纤维上。通过溶解 PVP 引入了孔结构，以增加其比表面积，为 TiO_2 提供进入的通道。使用 2∶1 的 PVDF 与 PVP 质量比可获得最高的光催化活性。通过将 TiO_2 有效地固定到 PVDF 上，如图 3.15 所示，该纳米纤维既可以提高光催化水处理的效率，又可以降低水处理的能耗需求。

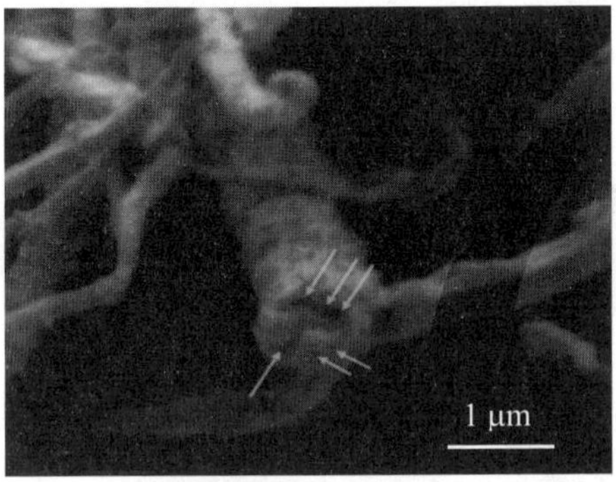

（a）纳米纤维的 SEM 图

（b）纳米纤维的 TEM 图

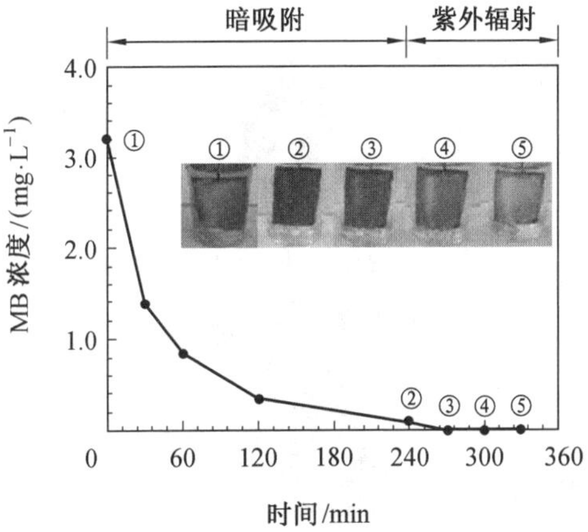

（c）亚甲基蓝（MB）在黑暗条件下解偶联吸附后通过紫外照射降解

图 3.15　P25-TiO_2 掺杂的 PVP-PVDF 纳米纤维

SnO_2是用作透明导体、电极和传感器的常见半导体材料，许多研究表明，TiO_2和 SnO_2 结合能够有效提高其光催化活性，这是因为其能带差能够使得光生电子在SnO_2上聚集，空穴在 TiO_2上聚集，在界面上形成异质结构。Peng 等人利用同轴静电纺丝技术结合热处理工艺，制备了 SnO_2/TiO_2中空纳米纤维。如图 3.16 所示，通过改变锡前驱体溶液的浓度，可以改变纳米纤维内部的形貌，随着溶液浓度的降低，纳米纤维从填充的固体到豆荚状到空心管。该纳米纤维具有比商用 TiO_2光催化剂更高的光催化活性。

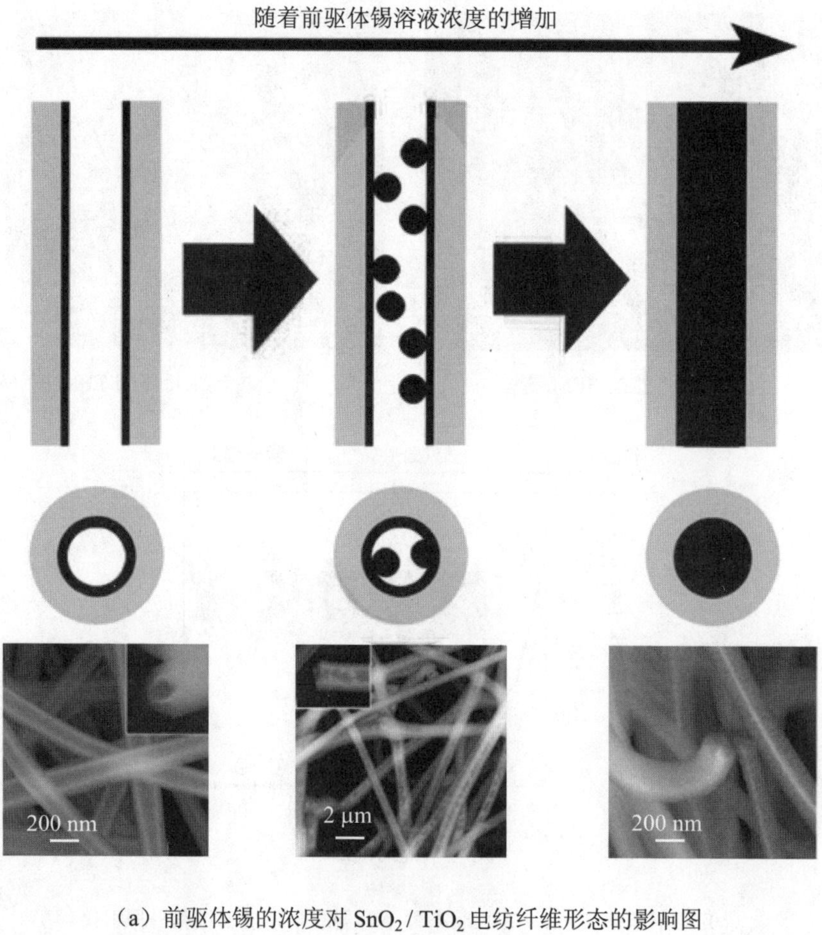

（a）前驱体锡的浓度对 SnO_2 / TiO_2 电纺纤维形态的影响图

图 3.16　SnO_2/TiO_2 中空纳米纤维

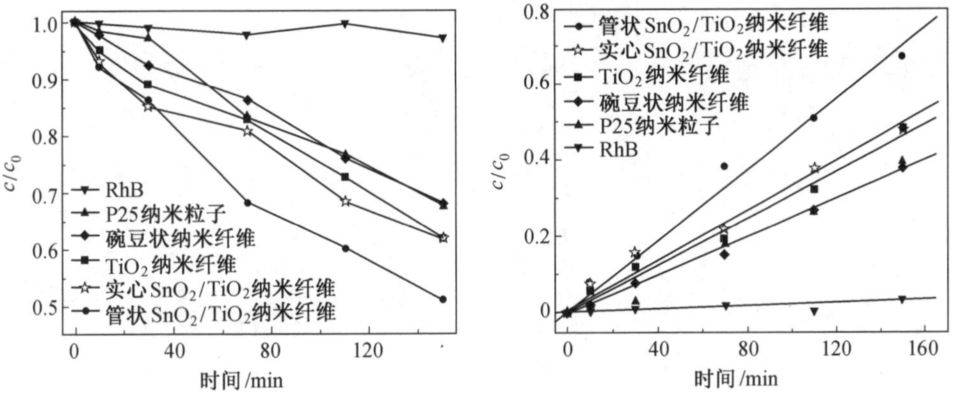

（b）不同光催化剂在紫外光的降解曲线图　（c）不同光催化剂在 RhB 光中一级动力学常数图

续图 3.16

石墨烯（GO）由碳元素组成，是呈六角形蜂窝状结构的单质材料，具有优异的导电性能和导热性能、大的比表面积以及出色的载流子迁移率。使用石墨烯或氧化石墨烯对 TiO_2 材料进行改性处理，以提高其光催化性能，逐渐引起了广泛的关注。Zhang 等人通过静电纺丝技术，制备了二氧化钛（TiO_2）/氧化石墨烯（GO）复合纳米纤维，与 TiO_2 纳米纤维相比，具有更高的光催化活性。Nasr 等人利用静电纺丝技术，制备了氧化石墨烯（GO）/二氧化钛（TiO_2）复合纳米纤维，具有优异的光催化性能。在甲基橙（MO）的光降解中，如图 3.17 所示，与商用 TiO_2-P25 相比，复合纤维的降解动力学速率比商用 TiO_2 高 6 倍，具有优异的光催化性能。

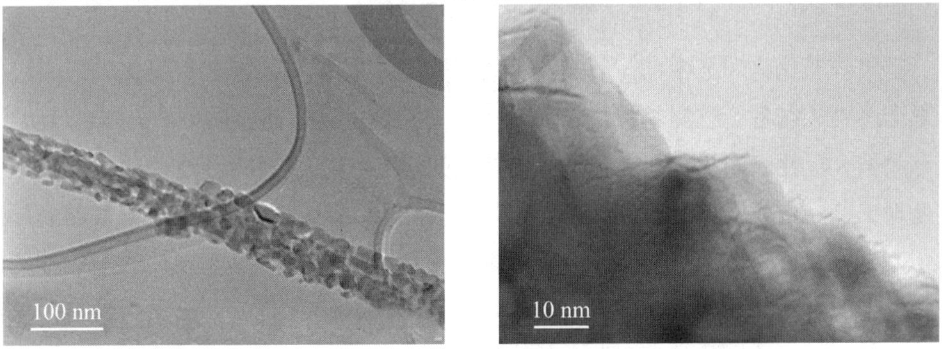

（a）GO（质量分数为 0%）/TiO_2 复合纳米纤维的 TEM 图

图 3.17　氧化石墨烯（GO）/二氧化钛（TiO_2）复合纳米纤维

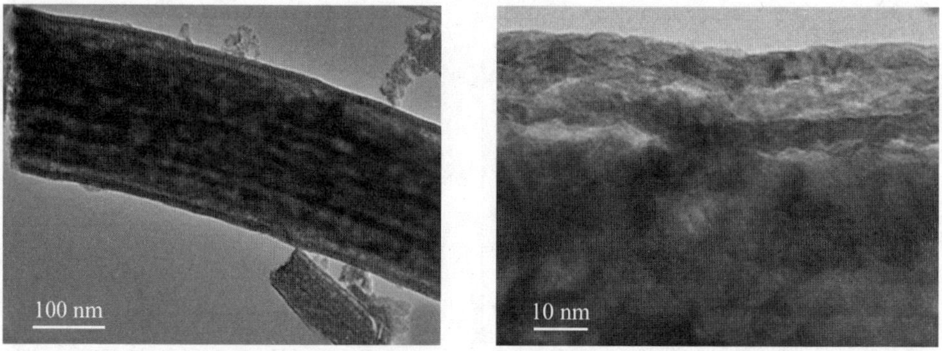

（b）GO（质量分数为 2%）/TiO_2 复合纳米纤维的 TEM 图

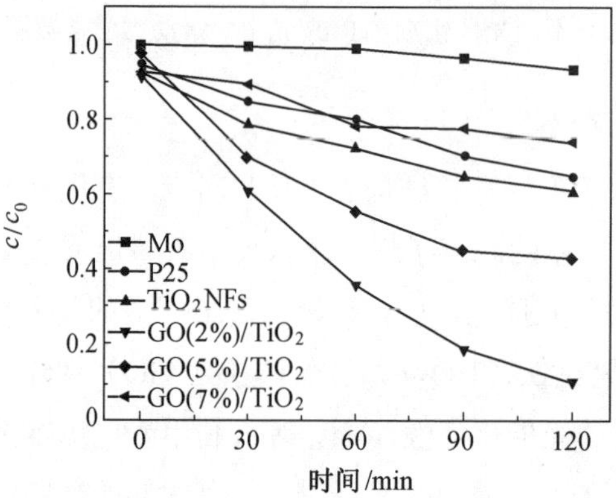

（c）复合纳米纤维对甲基橙的光降解图

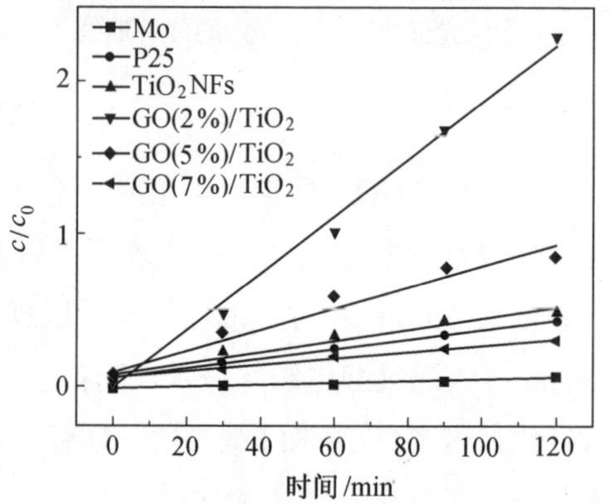

（d）复合纳米纤维对甲基橙的光降解动力学图

续图 3.17

石墨碳氮化物（g-C_3N_4）作为一种非金属的光催化剂，因其可见光响应、稳定无毒无污染、制备工艺简单等优势逐渐引起了广泛关注。将 TiO_2 和 g-C_3N_4 复合可以有效扩大活性表面积，以实现光催化性能的提高。Wang 等人通过静电纺丝技术和热处理工艺，合成了具有不同 TiO_2 含量的电纺纳米纤维 TiO_2/g-C_3N_4 异质结光催化剂。DRS 测试表明，TiO_2/g-C_3N_4 异质结在紫外和可见光区域均表现出强吸收催化性能和优异的循环光催化性能。Tang 等人以正丁醇钛（TNBT）和尿素为原料，通过一步静电纺丝法结合热处理工艺，合成了 TiO_2/g-C_3N_4 复合多孔纳米纤维，如图 3.18 所示。研究表明，光催化活性的增强主要归因于两种物质之间的异质结，其能够促进载体的转移并阻止它们的重组。

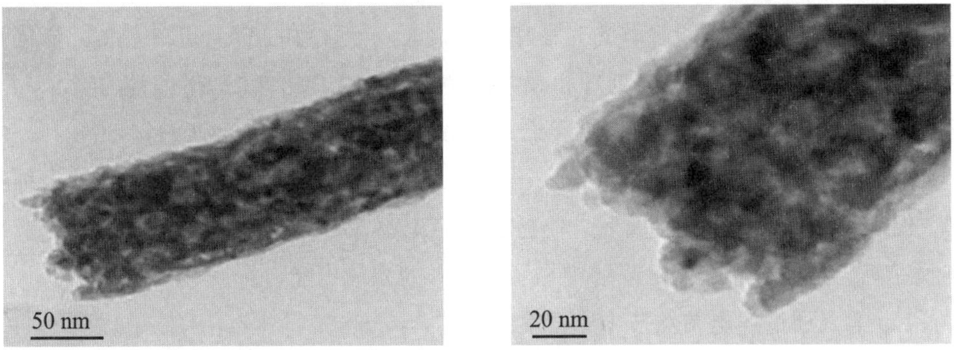

（a）复合多孔纳米纤维的 TEM 图

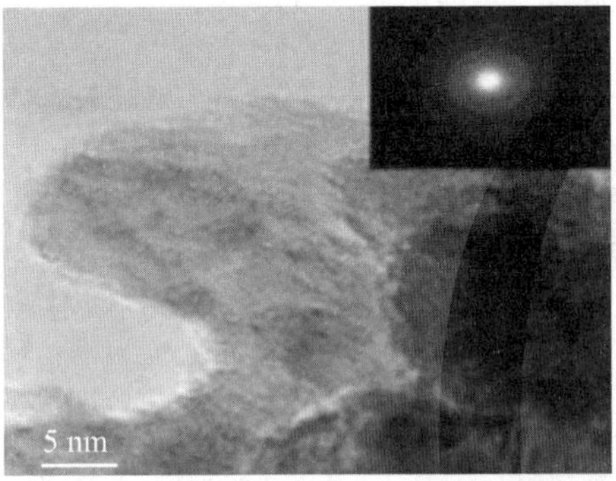

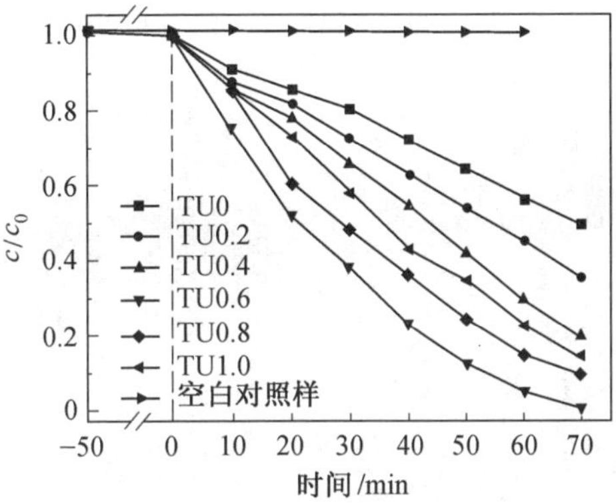

（b）复合多孔纳米纤维的 TEM 及 SAED 图　　（c）不同催化剂对 RhB 的降解曲线图

图 3.18　TiO_2/g−C_3N_4 复合多孔纳米纤维

元素掺杂是一种实现半导体功能化的重要方式之一，通过掺杂可以改变半导体的晶体结构、带隙结构，进而对其性能进行调整。Ma 等人通过静电纺丝技术，制备了 S 掺杂的 TiO_2 纳米纤维。与未掺杂的纳米纤维相比，具有更优异的光降解作用。Narendra 等人通过静电纺丝技术制备了钽（Ta）掺杂 TiO_2 介孔纳米纤维（TNF）。发现 5%Ta 掺杂样品在所有光催化剂中显示出更高的光催化活性。在紫外线和日光照射下，5%Ta 掺杂的 TNF 分别显示出比 TNF 高 5.1 和 2.2 倍的光催化活性。

银纳米颗粒（AgNPs）负载在 TiO_2 表面上是一种提高 TiO_2 光催化性能的有利方式，一维介孔 TiO_2 纳米纤维和 Ag 纳米颗粒结合，能够形成许多通道，有利于光催化性能的提高。除此之外，银纳米粒子具有很高的抗菌活性，在消毒抗菌方面能够发挥很大的作用。Michael 等人通过静电纺丝技术结合热处理工艺，制备了 Ag/TiO_2

纳米纤维。发现其对苯酚降解的光催化活性最大，与商用的 TiO_2–P25 相比，催化活性约高 3 倍。Liu 等人通过热处理和静电纺丝技术制备了银纳米粒子改性的聚乙烯醇/聚丙烯酸/羧基官能化氧化石墨烯复合纤维膜（PVA/PAA/GO–COOH@AgNPs）。由于 GO–COOH 片材具有强大的 $\pi—\pi$ 作用力和强静电作用，能够吸附水中的多种染料，促进目标染料的扩散和富集。制备的 PVA/PAA/GO–COOH@AgNPs 纳米复合纤维膜在有机染料溶液的催化降解中显示出优异的光催化能力。

Liu 等人通过静电纺丝技术结合多元醇的合成方法，将 Ag 纳米粒子沉积在 TiO_2 纳米纤维上，制备了 Ag/TiO_2 纳米纤维。与市售的 TiO_2–P25 相比，Ag/TiO_2 纳米纤维膜具有更高的渗透通量，在 30 min 的阳光照射下，能够实现 99.9%的细菌灭活和 80.0%的染料降解。其微观形貌图如图 3.19 所示。

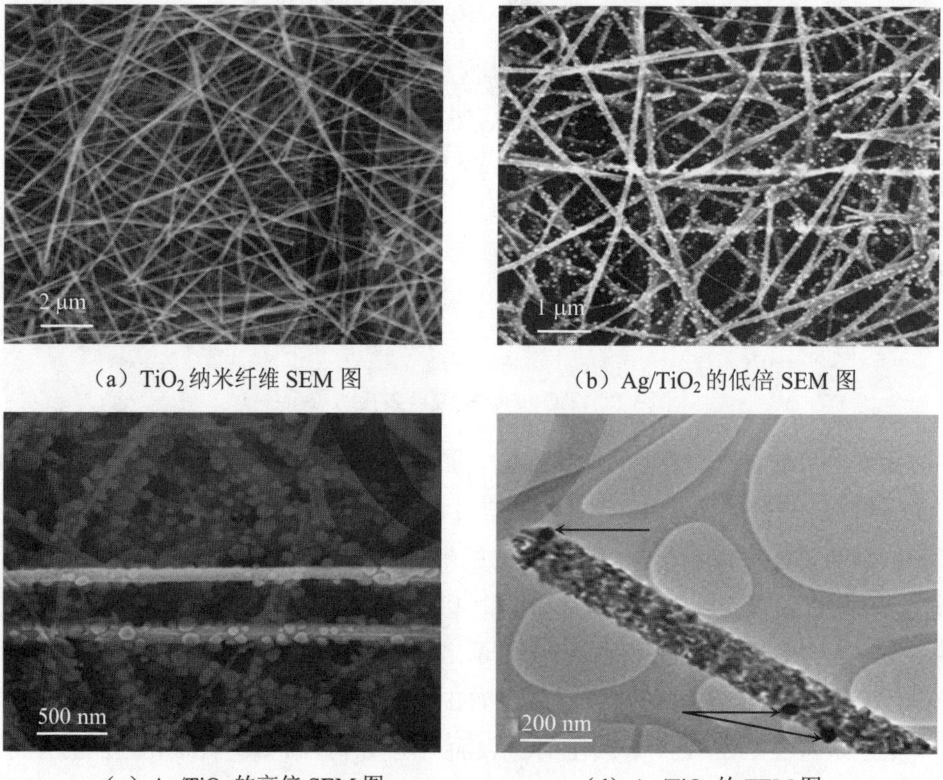

（a）TiO_2 纳米纤维 SEM 图　（b）Ag/TiO_2 的低倍 SEM 图

（c）Ag/TiO_2 的高倍 SEM 图　（d）Ag/TiO_2 的 TEM 图

图 3.19　Ag/TiO_2 纳米纤维

金纳米颗粒（AuNPs）也具有优异的光催化活性，负载在 TiO_2 上能够显著提高其光催化性能。Labeesh 等人利用同轴静电纺丝技术结合热处理工艺，制备了包封金纳米颗粒（AuNPs）的中空多孔二氧化钛（TiO_2）纳米纤维（Au@TiO_2），对于还原 4-硝基苯酚和刚果红染料，表现出优异的催化活性，并且在多次重复使用后纳米纤维仍能保持原来的形貌，具有回收使用性。其制备和在催化还原 4-NP 中的应用示意图及微观形貌图如图 3.20 所示，由 HAADF-STEM 图中可以看出，AuNPs 存在于 TiO_2 纳米纤维中。

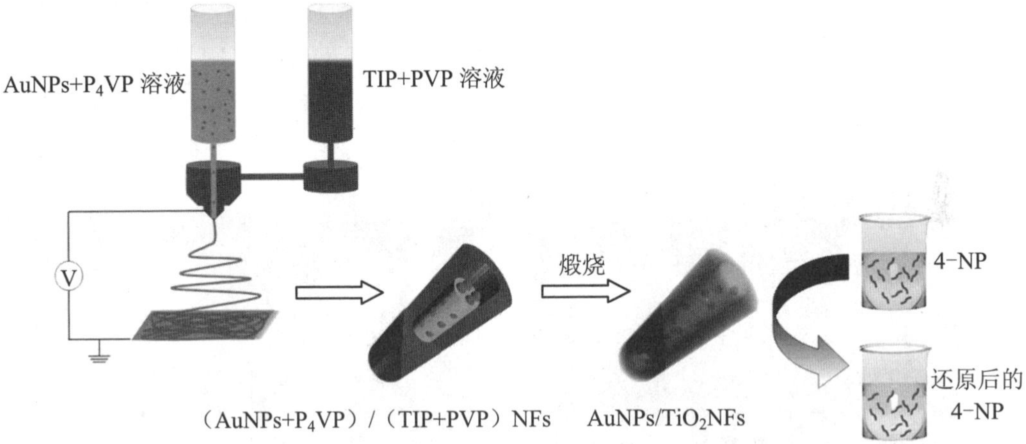

（a）Au @ TiO_2 纳米纤维的制备及其在催化还原 4-NP 中的应用示意图

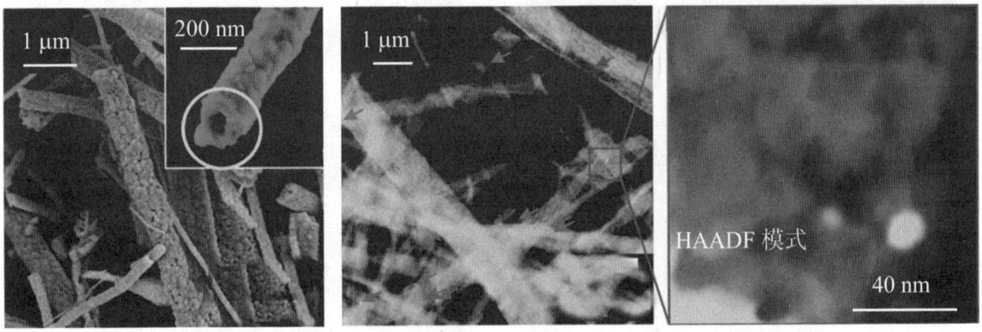

（b）SEM 图　　　　（c）Au@TiO_2 的 HAADF-STEM 图

图 3.20　包封金纳米颗粒的中空多孔二氧化钛纳米纤维（Au@TiO_2）

二维氮化硼（BN）纳米片具有高温稳定性、化学稳定性、高电阻和宽能带等性质，在纳米光电器件领域中有着巨大的潜力。Maryline 等人利用静电纺丝技术制备了氮化硼/二氧化钛（BN/TiO_2）纳米纤维光催化剂。与 TiO_2 纳米纤维相比，片状 BN 的存在改善了 TiO_2 中光致电子-空穴对的分离，并增加了带隙能量和比表面积，与商用光催化剂相比，光催化性能显著提高。

氧化锌(ZnO)是第三代半导体材料的典型采标，具有物理化学性质稳定、无毒、价格低廉的特征，与 TiO_2 一样，氧化还原能力很强，是一种高效的光催化剂。Fatma 等人通过静电纺丝和原子层沉积，结合热处理工艺，制备了由 ZnO 和 TiO_2 组成的核-壳异质结（CSHJ）纳米纤维，光催化性能极其优异。Adem 等人利用电弧放电工艺结合静电纺丝技术制备了 TiO_2/ZnO/PAN 复合纤维。如图 3.21 所示，TiO_2 纳米颗粒发生聚集，增加了表面粗糙度。在紫外光照射下，表现出优异的光催化效率，其反应速率是纯 PAN 纳米纤维的两倍。

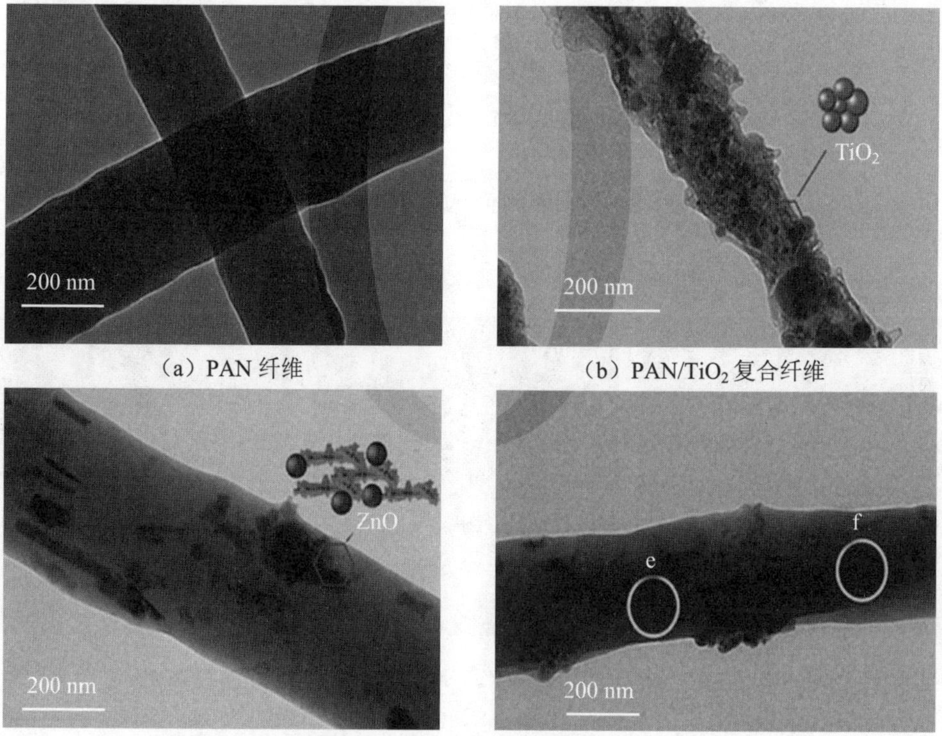

（a）PAN 纤维　　（b）PAN/TiO_2 复合纤维

（c）PAN/ZnO 复合纤维　　（d）TiO_2 / ZnO / PAN 复合纤维的 SEM 图

图 3.21　TiO_2/ZnO/PAN 复合纤维

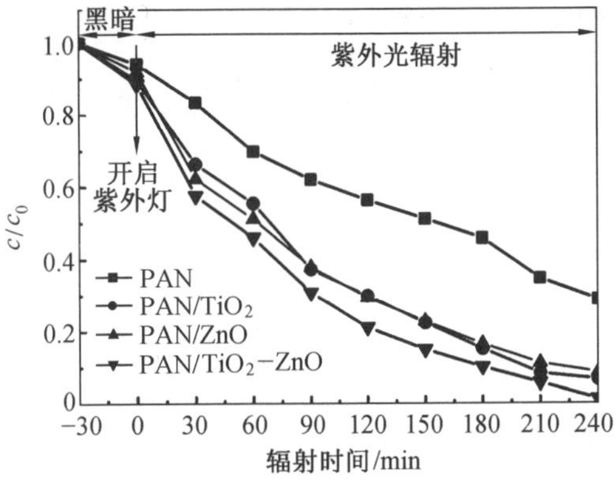

（e）不同光催化剂在紫外光下对孔雀绿试剂（MG）的催化能力比较图

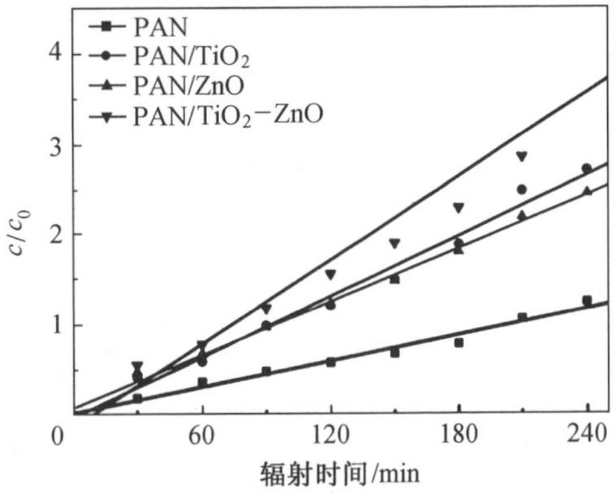

（f）不同光催化剂的表观速率常数比较图

续图 3.21

Seongpil 等人通过静电纺丝结合热处理工艺制备了石墨烯修饰（质量分数）的氧化锌（GO-ZnO）纳米纤维。如图 3.22 所示，相对于纯 ZnO 纤维膜，该纳米纤维表现出更优异的光催化活性。在紫外线照射 4 h 后，0.5%GO-ZnO 纤维膜具有最高的光催化活性（80%降解）。

（a）ZnO 纳米纤维的 SEM 图

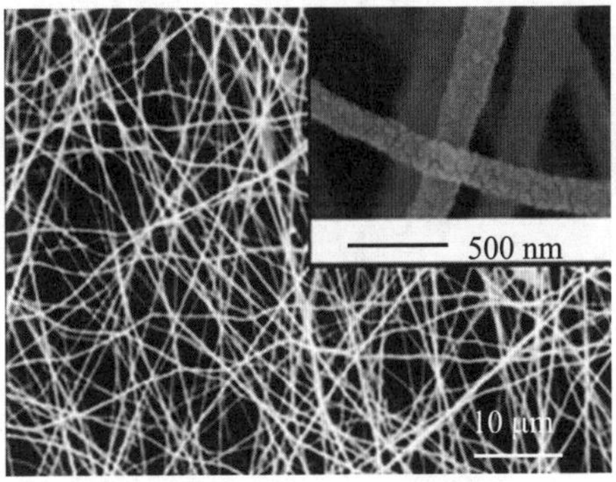

（b）0.5%GO-ZnO 纳米纤维的 SEM 图

图 3.22　石墨烯修饰的氧化锌（GO-ZnO）纳米纤维

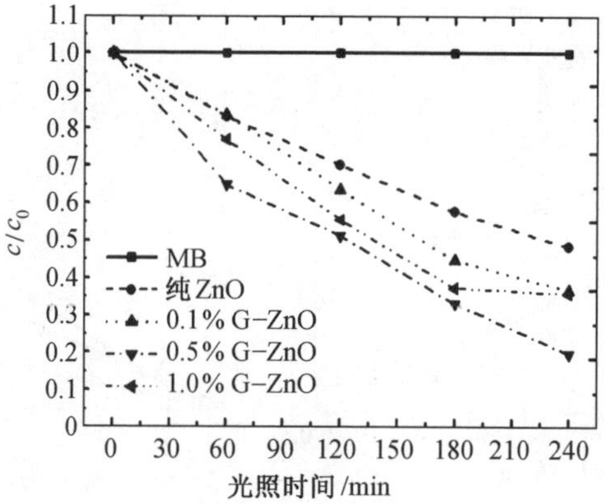

（c）纯 ZnO 纳米纤维和不同浓度掺杂的 G-ZnO 纳米纤维对于 MB 的降解随时间变化图

续图 3.22

Das 等人通过静电纺丝技术结合热处理工艺，制备了 $ZnO-SnO_2-Zn_2SnO_4$ 纳米纤维，其具有优异的光诱导电荷分离能力，并能生成大量的活性羟基（OH^-）。与纯 Zn_2SnO_4 相比，在紫外线照射下对于亚甲基蓝、刚果红以及非吸收性有机污染物（如苯酚和双酚 A）进行光催化性能测试，$ZnO-SnO_2-Zn_2SnO_4$ 复合纤维具有更优异的光催化性能。其微观形貌如图 3.23 所示。

（a）$ZnO-SnO_2-Zn_2SnO_4$ 纳米纤维的 SEM 图像

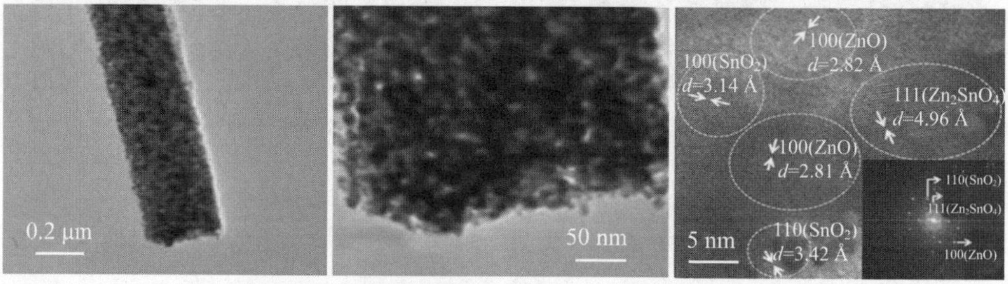

（b）单根纤维的 TEM 图像　　　　（c）HRTEM 图像

图 3.23　$ZnO-SnO_2-Zn_2SnO_4$ 纳米纤维

Ren 等人利用静电纺丝技术，结合水热和热处理工艺，制备了 ZnO 纳米颗粒高度分散的 $ZnO/BaTiO_3$ 纳米纤维异质结构。如图 3.24 所示，与 ZnO 粉末相比，纳米纤维在对甲基橙（MO）染料的降解反应中表现出更优异的光催化活性，这可能是因为光生电子与空穴对分离的协同效应以及纳米纤维的高比表面积。

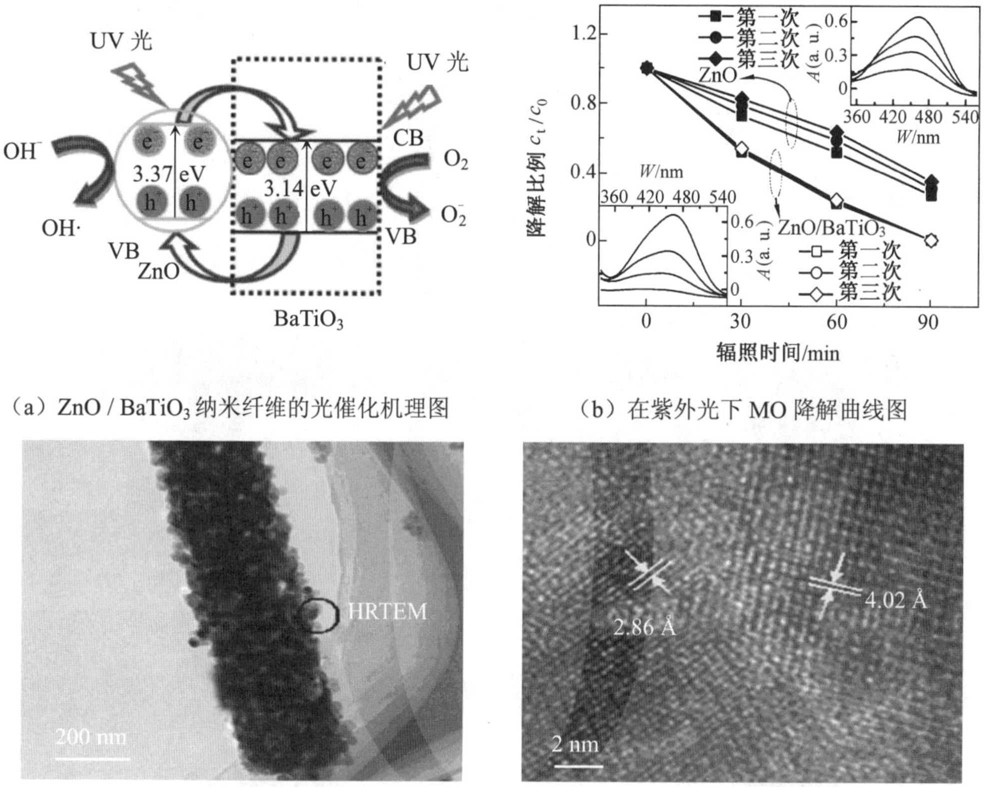

（a）$ZnO / BaTiO_3$ 纳米纤维的光催化机理图　（b）在紫外光下 MO 降解曲线图

（c）$ZnO / BaTiO_3$ 纳米纤维的 TEM 图　（d）$ZnO / BaTiO_3$ 纳米纤维的 HRTEM 图

图 3.24　$ZnO/BaTiO_3$ 纳米纤维

甲基对硫磷（MP）是一种有机磷农药和杀虫剂，广泛应用于控制棉花等农作物中的害虫，但排放到水体中会引起一系列的环境问题，需要从废水中去除。目前，已经发现二氧化钛是出色的吸附剂和光催化剂，能够降解 MP 及其有毒代谢物。Krishnasamy 等人以聚丙烯腈（PAN）为原料，利用静电纺丝技术和水热反应，制备了（La-ZnO/PAN）纳米纤维。如图 3.25 所示，该纳米纤维是一种降解 MP 的有效材料，光催化条件下 150 min 内可以实现 MP 的完全降解，且具有较好的稳定性和可回收性。

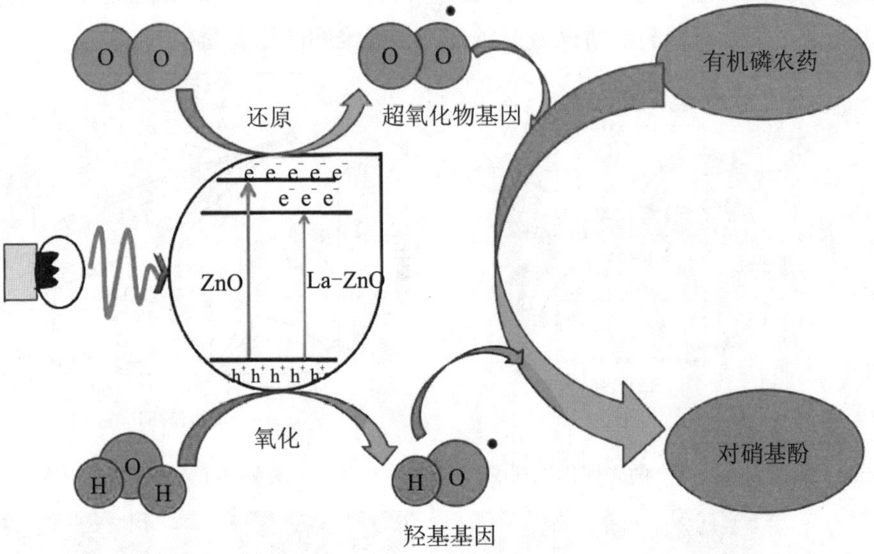

（a）光催化降解 MP 机理图

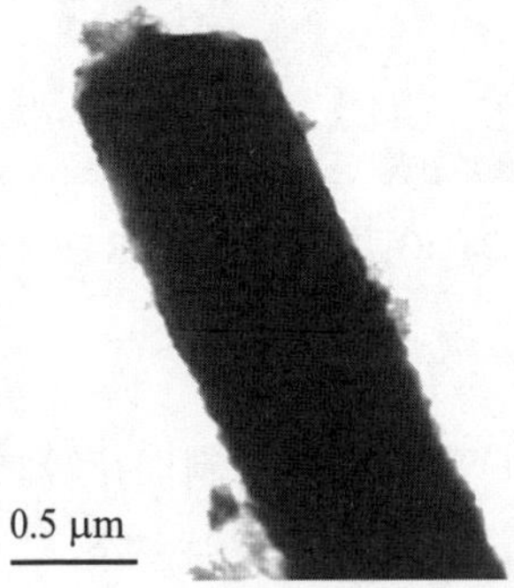

（b）La 掺杂的 ZnO/PAN 纳米纤维的 HRTEM 图

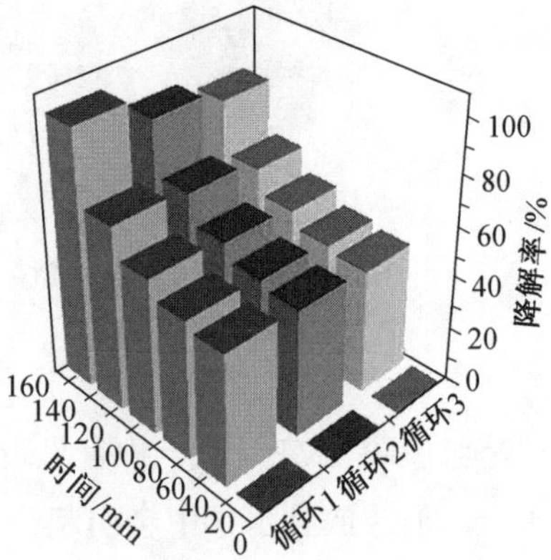

（c）光催化剂可回收性和稳定性图

图 3.25　La 掺杂的 ZnO/PAN 纳米纤维

近年来，Bi 系复合物由于其独特的结构和优异的光吸收能力引起了大量关注。Zhou 等人通过静电纺丝技术结合浸渍法，制备了 PAN/g-C_3N_4/BiOI 纳米纤维，其具有优异的光催化性能。电子-空穴对的有效分离及其在可见光区域的强吸收使该光催化剂在可见光照射下的光催化性能优于 PAN/g-C_3N_4。Bi 的碘氧化物也有着优异的光催化性能。Wang 等人通过静电纺丝技术和溶剂热法，制备了 p 型 BiOCl 纳米/n 型 TiO_2 纳米纤维（p-BiOCl/n-TiO_2HHs）的分层异质结构。由于 p-n 异质结和高表面积的影响，p-BiOCl/n-TiO_2HHs 表现出优异的紫外光催化活性。

3.2.3　光催化剂在产氢领域中的应用

半导体材料光催化分解水的基本原理和过程如图 3.26 所示。当光子的能量大于等于禁带宽度时，价带中的电子将会被激发跃迁至导带，并且在价带中留下空穴。而产生的空穴和光电子将会分别引发水发生氧化反应和还原反应，以完成水的分解。

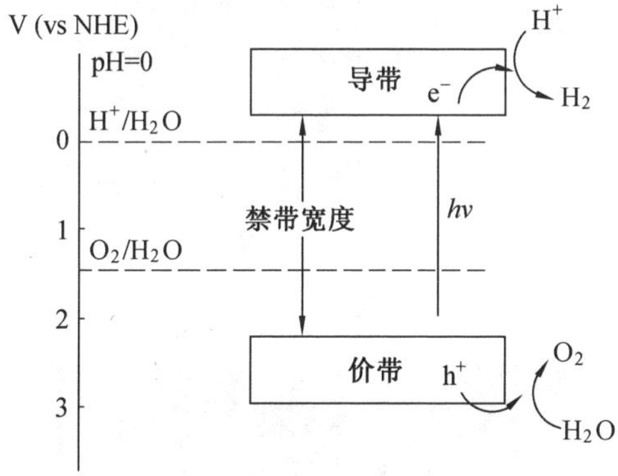

（a）光催化分解水基本原理图

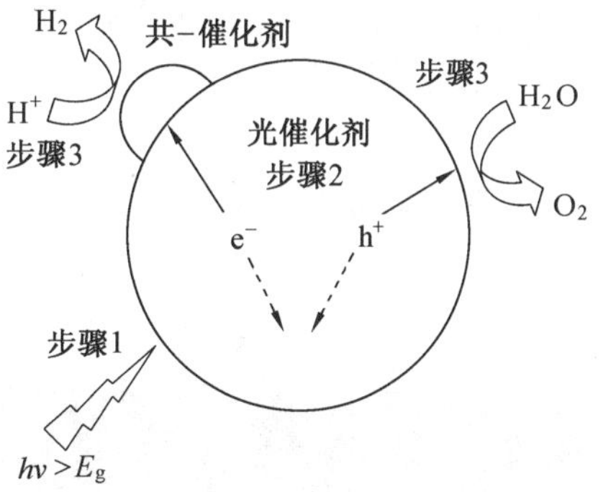

（b）光催化分解水的过程图

图 3.26　半导体材料光催化分解水的基本原理和过程

TiO_2 除了在环境治理方面有着巨大的应用，在光催化产氢领域也发挥着不可替代的作用。S. K. Choi 等人通过静电纺丝技术制备了 TiO_2 纳米纤维，与 TiO_2 纳米颗粒相比，光催化产生电流的活性高出 3 倍，产氢活性高出 7 倍。这可能是因为电子通过纳米纤维框架转移能够导致电荷的有效分离，进而提高了其光催化活性。Hou 等人通过发泡辅助静电纺丝的方法，制备了超细多孔 TiO_2 纳米纤维。如图 3.27 所示，该纳米纤维具有优异的光催化产氢速率，产氢速率为 710.8 $\mu mol \cdot g^{-1} \cdot h^{-1}$，具有巨大的应用前景。

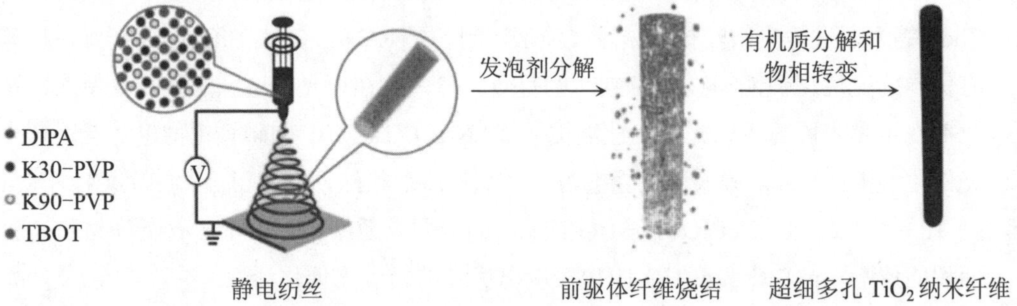

（a）制备机理图（彩图见附录）

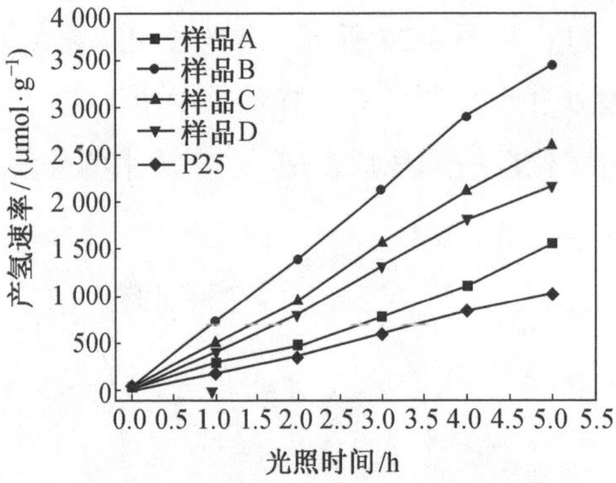

（b）不同样品随时间变化的产氢图

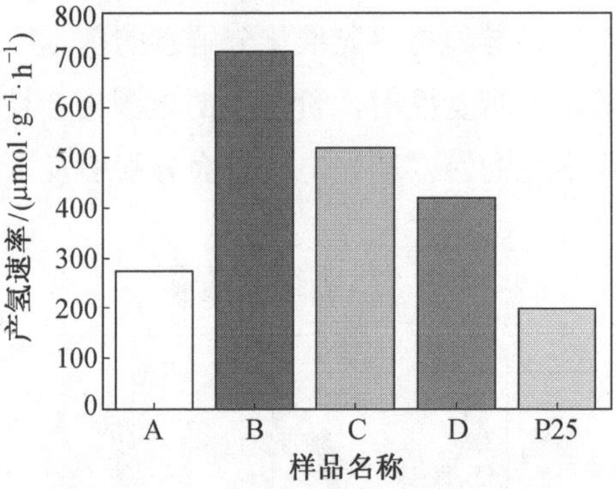

（c）不同样品产氢速率图

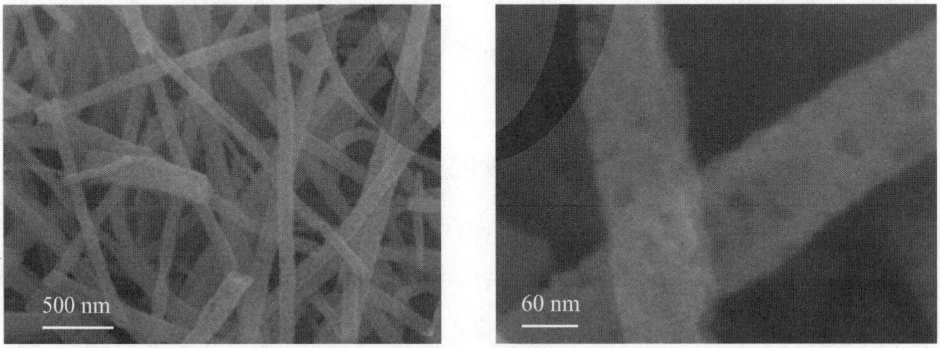

（d）TiO_2 纳米纤维的 SEM 图

图 3.27　超细多孔 TiO_2 纳米纤维

将 TiO_2 与其他物质相结合，形成异质结构，可大幅提高其光催化产氢速率。Liu 等人利用静电纺丝技术结合水热法和热处理工艺，制备了 $Cu/Cu_2O/CuO/TiO_2$ 多异质结纳米纤维，具有优异的光催化产氢性能，产氢速率为 10.04 $\mu mol\cdot h^{-1}$，是纯 TiO_2 纳米纤维的 17.3 倍。Nasir 利用静电纺丝技术制备了 TiO_2 纳米纤维，并通过水热法结合化学气相沉积，如图 3.28 所示，在 TiO_2 纳米纤维上负载 $g\text{-}C_3N_4$ 以及 $SnSe_2$ 纳米片，形成异质结构。该纳米纤维具有优异的光催化产氢速率，产氢速率为 2 375 $\mu mol\cdot g^{-1}\cdot h^{-1}$，具有广阔的应用前景。

（a）TiO_2 纳米纤维　（b）水热处理后的 TiO_2 纳米纤维

（c）CVD 负载了 $g\text{-}C_3N_4$ 的 TiO_2 纳米纤维　（d）TiO_2 纳米纤维的 SEM 图

图 3.28　负载 $g\text{-}C_3N_4$ 以及 $SnSe_2$ 纳米片的 TiO_2 纳米纤维

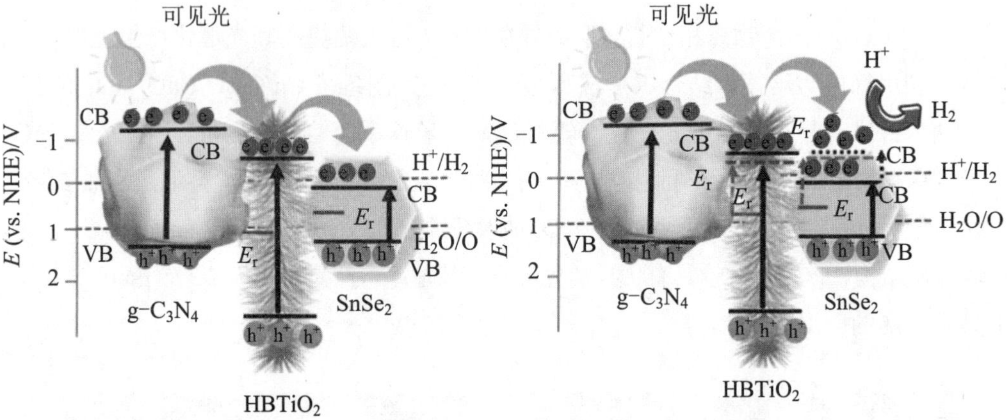

（e）可能的光催化机理图

续图 3.28

ZnO 由于其导带（CB）边缘电势比 H_2O/H_2 还原电势更负，因此可以促进质子向 H_2 还原，有着较好的光催化产氢性能。Zhou 等人利用静电纺丝技术结合热处理工艺，制备了 ZnO 纳米纤维，该纳米纤维由紧密堆积的颗粒组成，有效地抑制了纳米颗粒的聚集，同时也促进了颗粒间的电荷转移。ZnO 纳米纤维相对于纳米粒子而言具有更高的光催化产氢活性，产氢速率为 1 031 μmol·g^{-1}·h^{-1}，比纳米粒子高约 1.5 倍。ZnO 纳米纤维相对于纳米颗粒而言，具有更高的稳定性。其微观形貌图如图 3.29 所示。

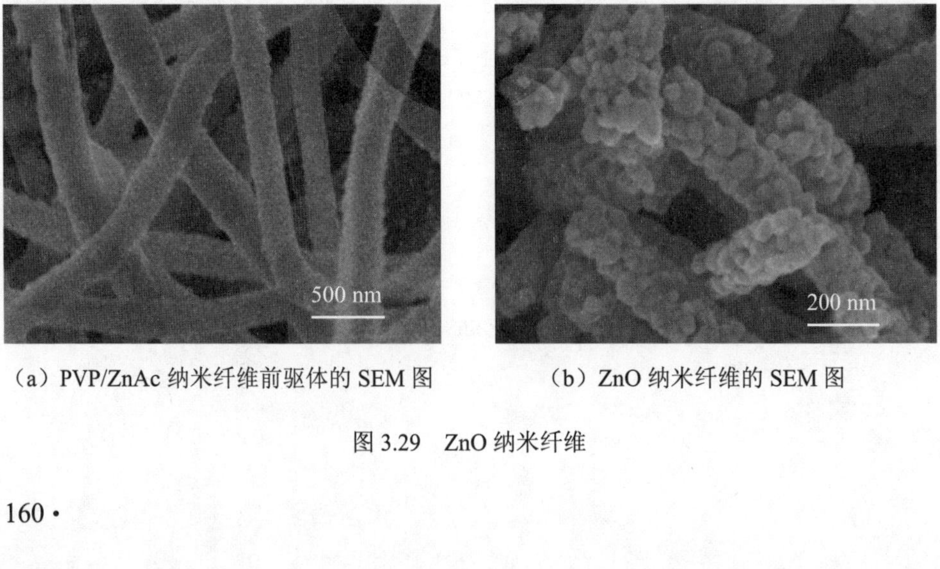

（a）PVP/ZnAc 纳米纤维前驱体的 SEM 图　　（b）ZnO 纳米纤维的 SEM 图

图 3.29　ZnO 纳米纤维

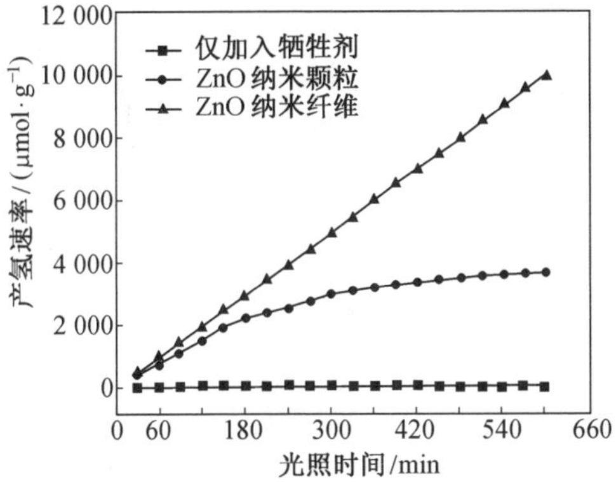

（c）随着时间变化的光催化产氢性能图

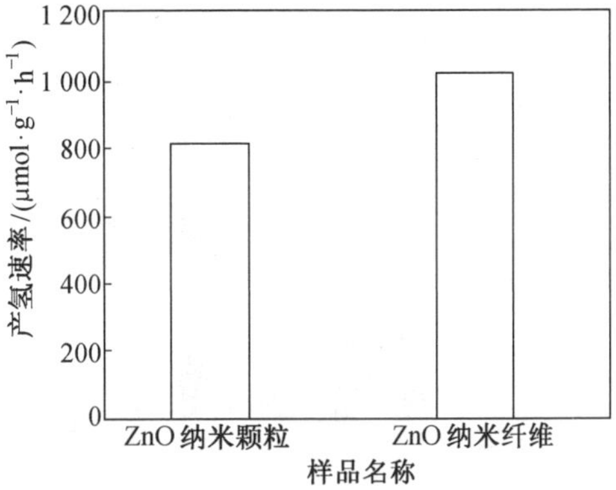

（d）ZnO 纳米纤维和颗粒的光催化产氢速率图

续图 3.29

Sun 等人通过静电纺丝技术制备了 ZnO 纤维，水热法原位硫化制备了 ZnS/ZnO，最后在其表面沉积 CdS 量子点，得到 CdS/ZnS/ZnO 三元异质结构纳米纤维。其具有优异的光催化产氢速率，产氢速率为 51.45 $mmol\cdot h^{-1}\cdot g^{-1}$，分别比 ZnO 和 ZnS/ZnO 纳米纤维高出 93.54、2.28 倍。其微观形貌如图 3.30 所示。

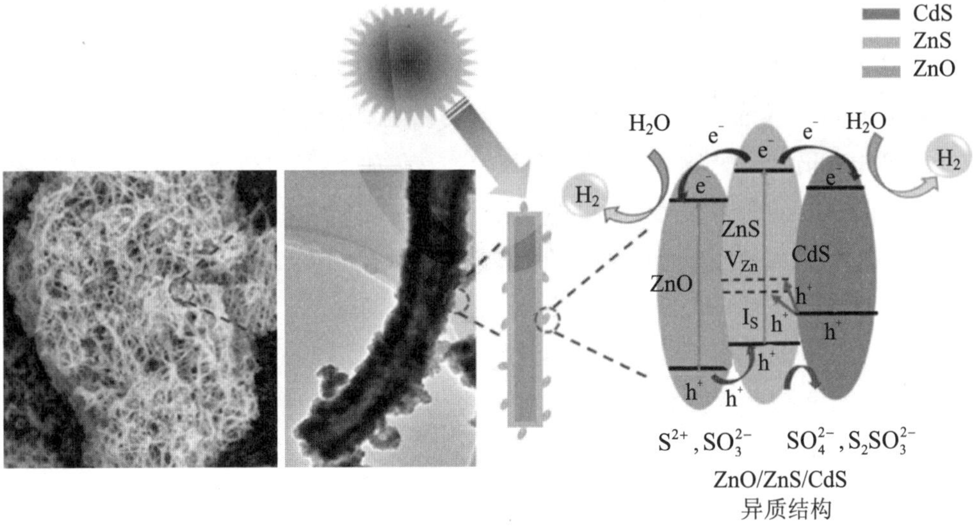

（a）CdS/ZnS/ZnO 三元异质结构纳米纤维光催化产氢机理图（彩图见附录）

图 3.30　CdS/ZnS/ZnO 三元异质结构纳米纤维

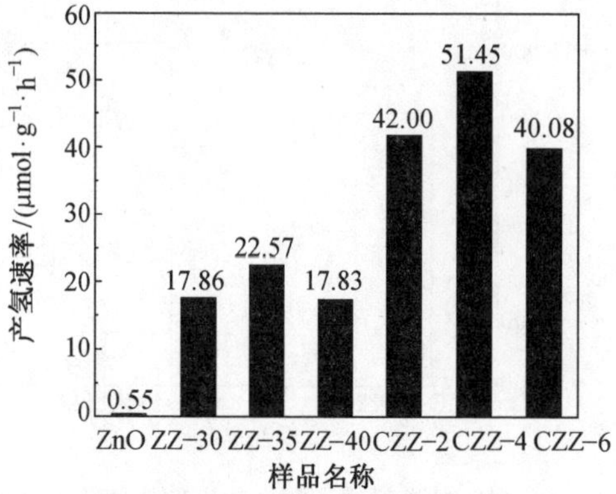

（b）各样品的产氢速率图

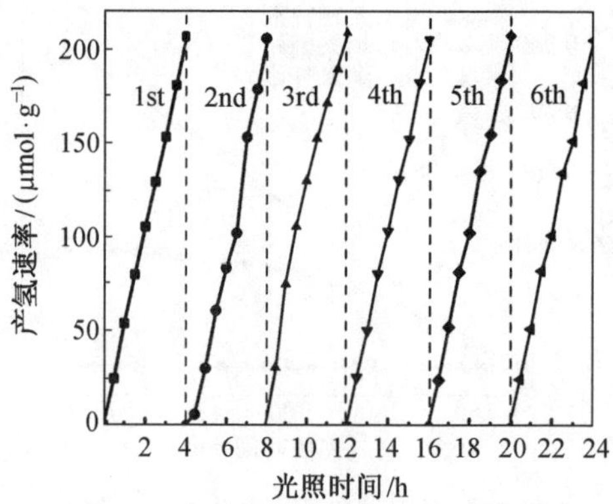

（c）光催化产氢的生产循环运行图

续图 3.30

3.2.4 小结与展望

静电纺丝技术是一种用于制造无机和聚合物纳米纤维的简单方法。如本章所述，已讨论了静电纺纳米纤维在光催化领域的一些最新成果，特别是无机半导体纳米纤维材料，其作为光催化剂具有很大的优势，例如大的比表面积和高的孔隙率。通过与金属纳米颗粒结合，与其他半导体材料偶联，掺杂非金属元素，将增强纳米纤维的光催化性能。除此之外，它们还具有优异的循环利用性能。然而，在纳米纤维光催化领域中，仍有一些问题需要进一步研究。具体如下：

（1）首先，对于多相光催化剂的机理还需进一步研究和探索。

（2）其次，如何实现金属或半导体材料的有效掺杂以实现其与光催化剂的高效偶合，进一步提高光催化剂的催化活性，仍是一大问题。

（3）最后，目前对于静电纺丝制备的纳米纤维作为光催化剂，仍主要受限于实验室，距离大规模产业化生产仍有一段距离，仍需进一步研究和探索。

3.3　酶催化

3.3.1　引言

酶本质上是具有特殊构象的蛋白质，具有高效、专一等特点，广泛应用于生物、食品、医药、环境等领域。然而，酶也存在许多缺陷。在游离状态下，生物酶虽然能够表现出更高的活性，但是长时间稳定保存较为困难，并且极易受到环境变化的影响，尤其是在强酸强碱环境或是高温情况下。除此之外，酶在游离的状态下难以彻底从反应体系中分离，易造成反应体系的二次污染，在使用过程中不能多次重复使用，造成很大的浪费。为了解决这些问题，固定化酶技术应运而生。

酶的固定化最早于 1916 年由 Nelosn 和 Griffin 两位学者发现，之后研究学者开始展开对固定化酶研究的工作。固定酶技术是指采用特定的载体，通过一定的方法将酶固定在特定载体上，这不仅能够很好地保留酶的催化特性，还能够使酶具有良好的稳定性和重复使用性。这种方法能够大大减少酶的使用量，以降低其工业成本。

酶是一种特殊的催化剂，在反应中通过降低活化能，以达到催化的目的。而选择合适的固定酶的方法是至关重要的。目前，酶的固定方法有许多种，包括物理吸附、包埋、化学交联、共价结合以及配位结合法等。固定酶方法示意图如图 3.31 所示。

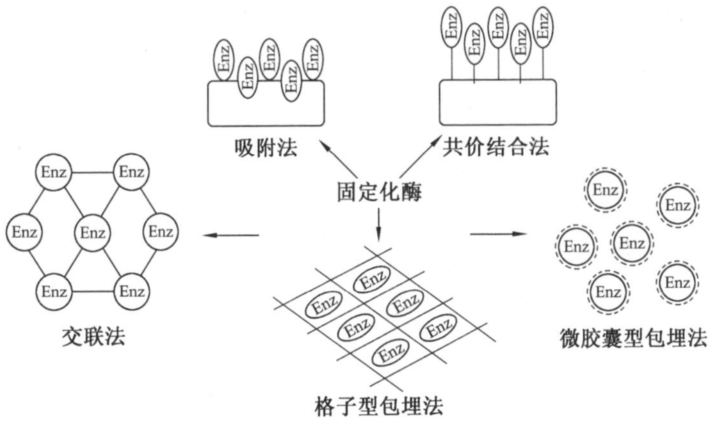

图 3.31　固定酶方法示意图

在固定酶的过程中，载体的选择是至关重要的。载体的物理化学性质直接影响着酶的固载量、固定化后的活性以及稳定性等等。酶固定化的趋势现已转向使用纳米结构材料作为酶载体，其中包括纳米多孔材料、纳米颗粒和纳米纤维。与常规载体相比，纳米结构材料提供了极高的比表面积，这对于提高固定效率是理想的。利用静电纺丝制备纳米纤维以实现酶的固定，逐渐引起了广泛关注。制备得到的具有高孔隙率和互连性的纳米纤维易于回收，可多次重复使用，具有巨大的应用前景。

3.3.2 酶固定催化剂

聚丙烯腈（PAN）具有优异的高热稳定性、良好的机械性能和优异的耐化学性，是一种广泛用于制造碳材料的聚合物。此外，PAN 纳米纤维的表面腈基团（—C≡N）很容易与水溶液中的羟胺（NH_2—OH）进行反应；并且所得的 AOPAN 纳米纤维在表面上具有偕胺肟（—C（NH_2）═ NOH）基团，该基团由羟基（—OH）和氨基（—NH_2）基团组成，可通过物理/化学相互作用进行固定化酶。

Li 等人利用静电纺丝技术，以聚丙烯腈为载体，利用酰胺化反应，将脂肪酶共价固定在 PAN 纳米纤维上。固定化脂肪酶的活性保留率为游离酶的 81%，且储存稳定性有所提高，可多次重复使用。Shinji 等人利用静电纺丝技术，制备了能够固定脂肪酶的聚丙烯腈纳米纤维，其具有优异的催化活性，且稳定性大大提高。

Li 等人利用静电纺丝技术制备了聚丙烯腈纳米纤维，用以固定过氧化氢酶。发现过氧化氢酶固定后可保留 50%以上的催化活性，其热稳定性能也有所提高。Wu 等人利用静电纺丝技术，以聚氨酯（PU）、偕胺肟聚芳腈（AOPAN）和 β-环糊精（β-CD）、氯化铁为原料，制备了复合纳米纤维膜（Fe（Ⅲ）-PU / AOPAN /β-CD），如图 3.32 所示。所制备的纳米纤维膜可以有效地固定漆酶，并且具有优异的力学性能和可再生能力。漆酶的膜固定量高达 186.34 $mg \cdot g^{-1}$，与游离漆酶相比，表现出相当高的催化活性，并显著提高了对温度和 pH 变化的抵抗力。另外，固定的漆酶的热稳定性、储存稳定性和可重复使用性大大提高。

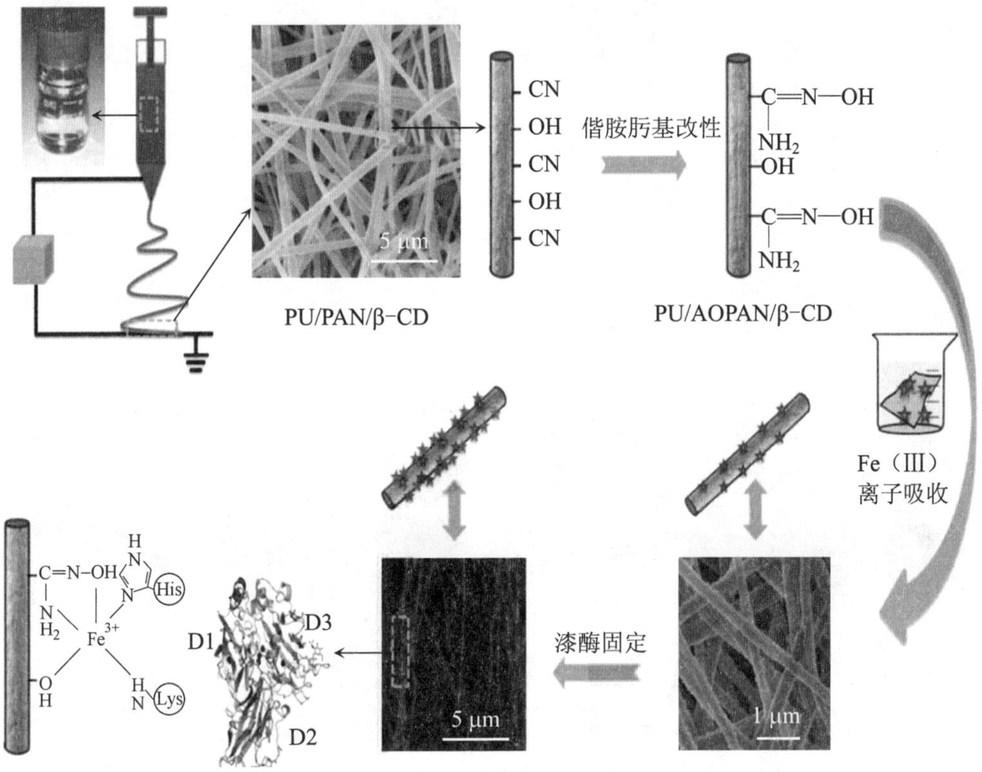

图 3.32　静电纺丝制备 Fe（Ⅲ）-PU / AOPAN /β-CD 示意图

聚乙烯醇（PVA）是一种聚合物，具有无毒、可生物降解、生物相容性好等优点，被用作固定化酶的载体，具有出色的化学和热稳定性，可利用静电纺丝技术制备成纳米纤维，在固定酶领域发挥着巨大的作用。

Porto 等人利用静电纺丝技术，将 α-淀粉酶固定在聚乙烯醇纳米纤维上。酶的掺入导致纤维形态的轻微变化，随着酶浓度的增加，纤维变得更扁平、更粗。研究发现，酶固定化可在较大的温度和 pH 范围内提高 α-淀粉酶的稳定性和活性。Rona 等人利用静电纺丝技术，制备了包封 β-葡萄糖醛酸糖苷酶的聚乙烯醇纳米纤维，如图 3.33 所示。该纳米纤维大大提高了该酶活性的保留时间到至少 7 周，能够有效地抑制细胞的增殖，预期可用于心血管移植的设计。

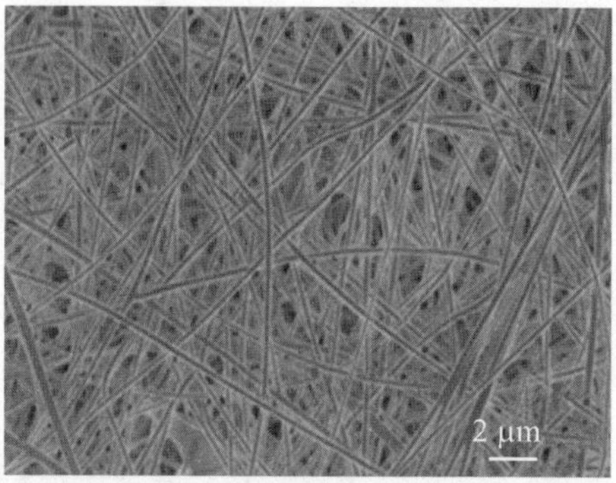

（a）PVA 纳米纤维包埋酶包埋前的 SEM 图

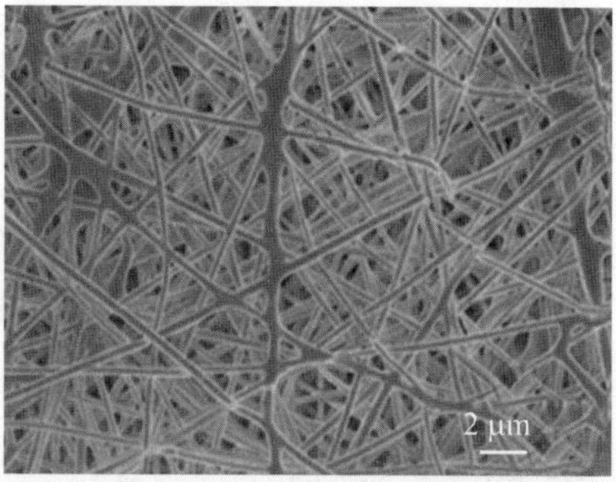

（b）PVA 纳米纤维包埋酶包埋后的 SEM 图

（c）含有体积分数为 1%罗丹明标记的包埋酶后的图像

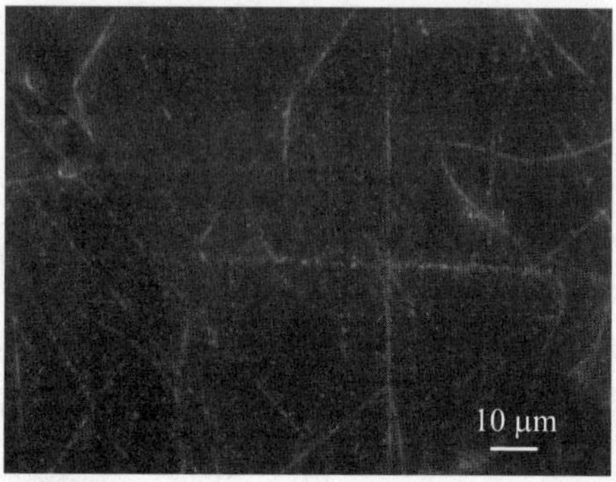

（d）含有体积分数为 5%罗丹明标记的包埋酶后的图像

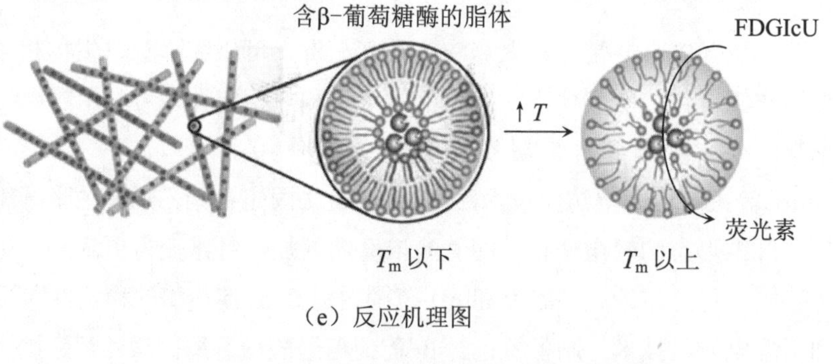

（e）反应机理图

图 3.33　包封 β-葡萄糖醛酸糖苷酶的聚乙烯醇纳米纤维

Santos 等人通过静电纺丝技术，将木聚糖酶固定在聚乙烯醇上，制备了具有酶催化性能的纳米纤维。与游离木聚糖酶相比，固定在 PVA 纤维中的木聚糖酶在极端 pH 条件（4、5、7 和 8）下表现出更高的催化活性。Saallah 等人利用静电纺丝技术，制备了平均直径为（176±46）nm 的 CGTase / PVA 纳米纤维，然后进行戊二醛气相交联，将工业酶环糊精葡聚糖转移酶（CGTase）固定在聚乙烯醇纳米纤维膜中。在进行静电纺丝和交联反应之后，酶没有发生结构变化。与由相同起始溶液（对照）制成的 CGTase / PVA 膜相比，固定化酶显示出优异的催化效率，酶负载量高出 3.6 倍，活性提高 25%，且具有好的可重复利用性。固定酶前后纤维的微观形貌图如图 3.34 所示。

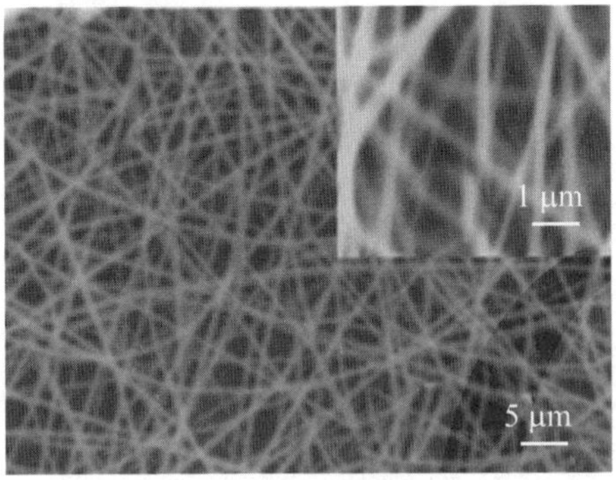

（a）PVA 纤维的 SEM 图

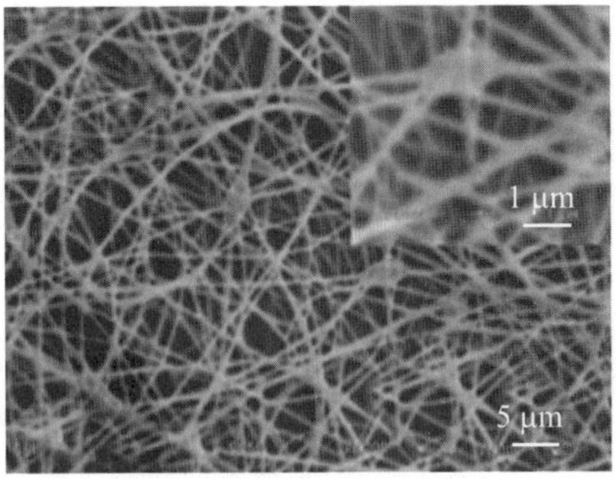

（b）CGTase/PVA 纤维的 SEM 图

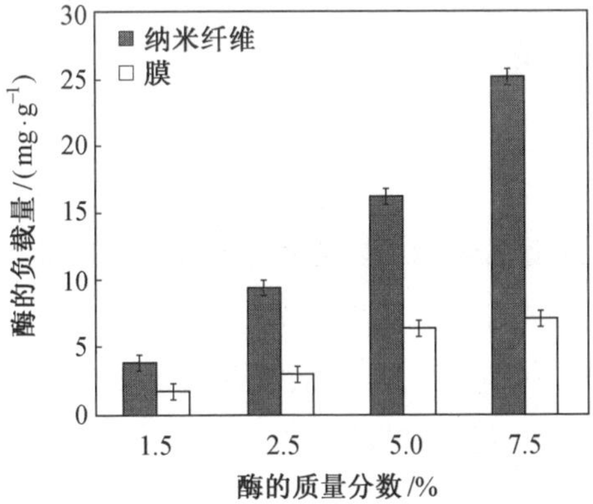

（c）不同酶浓度的 CGTase/PVA 的负载量对比图

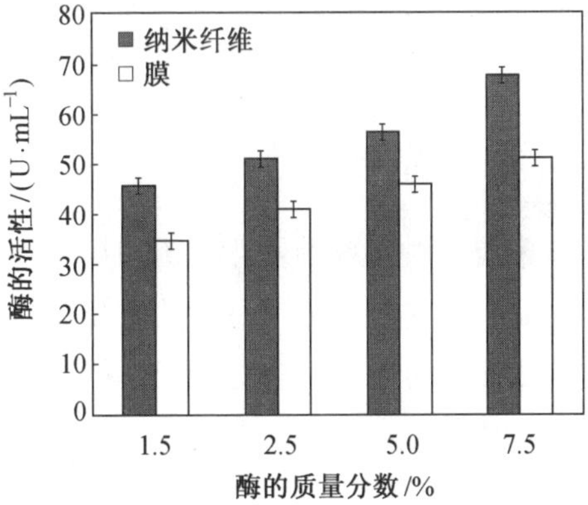

（d）不同酶浓度的 CGTase/PVA 酶活性的对比图

图 3.34　CGTase / PVA 纳米纤维膜

聚苯乙烯（PS）由于其廉价和高可加工性而被广泛用于蛋白质/酶的固定化应用中。Jia 等人通过静电纺丝技术制备了聚苯乙烯纳米纤维，用以固定 R-胰凝乳蛋白酶。相对于其他形式的固定化酶方式，其催化效率高，在有机溶剂中的活性能够提高三个数量级，且更加稳定。Martrou 等人利用静电纺丝技术，将基于表面改性的 PS 纳米纤维用于蛋白质/酶的固定，如图 3.35 所示。利用纤维表面的氨基来固定过氧化物酶（HRP），与游离酶相比，固定在纤维上酶的稳定性大大提高。当使用游离态的过氧化物酶时，在紫光条件下将导致 HRP 严重失活，严重影响了降解 2，4-二氯苯酚（2，4-DCP）的效率，而将 HRP 封装在纳米管内则可有效避免紫外线引起的酶失活。

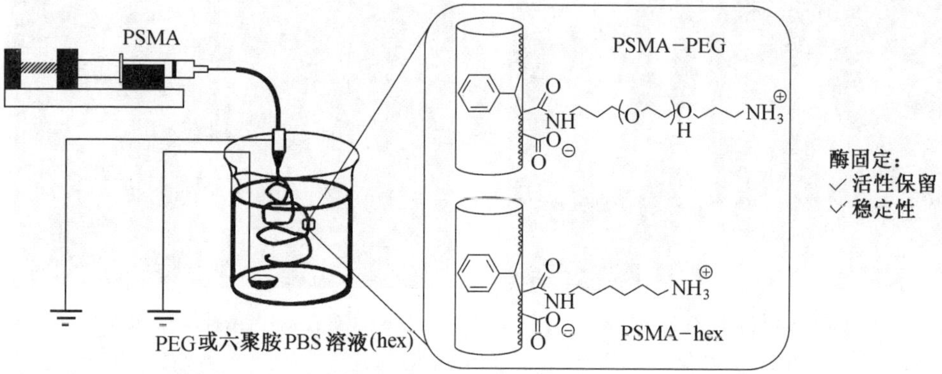

图 3.35　固定过氧化氢酶的 PS 纳米纤维制备机理图

3.3.3　小结与展望

酶是一种特殊的催化剂，在反应中通过降低活化能，以达到催化的目的。而选择合适的固定酶的方法是至关重要的。在固定酶的过程中，载体的选择是至关重要的。载体的物理化学性质直接影响着酶的固载量、固定化后的活性以及稳定性等等。通过静电纺丝制备纳米纤维作为酶的载体，可有效提高酶的催化效率、稳定性和可重复使用性，是一种具有广阔应用前景的酶固定方法。然而，在使用纳米纤维固定化酶的过程中仍有一些问题需要进一步研究。具体如下：

（1）首先，需要进一步研究和揭示通过共混静电纺丝包埋法固定酶的机理。

（2）其次，使用单一方法或技术固定化酶存在一定的局限性，限制广泛采用的关键问题之一是如何结合多种方法或技术的优势来提高固定化酶的催化效率和稳定性。

（3）最后，目前对用于酶固定化的静电纺丝纳米纤维的研究仍主要限于实验室，与实际的大规模生产还有较大距离。因此，如何实现用于酶固定化的纳米纤维的工业化大规模生产，仍是一大难题。

3.4　其他催化

3.4.1　甲醛氧化催化剂

甲醛（HCHO）是室内空气中常见的挥发性有机化合物，长期接触 HCHO 可能会导致严重的健康问题。通过纳米催化剂上的催化氧化反应除去 HCHO 受到越来越多的关注。与传统的吸附方法相比，该技术将有毒的 HCHO 氧化成无毒的 CO_2 和 H_2O，可以消除 HCHO，而不是将 HCHO 从气态转移到吸附态。

目前，已证明在金属氧化物上掺杂贵金属可以提供更多的活性位点以氧化 HCHO，包括 Pd、Au 和 Pt 对室温 HCHO 氧化反应具有较高的活性。其中，Pt/TiO_2 催化剂由于 Pt 和 TiO_2 之间的独特相互作用以及其良好的化学稳定性、生物相容性和较长的稳定性而引起了广泛关注。

Xu 等人利用静电纺丝技术结合热处理工艺和 $NaBH_4$ 还原工艺，将 Pt 纳米颗粒（NPs）沉积在 TiO_2，制备了具有铂（Pt）和不同相组成的二氧化钛（TiO_2）纳米纤维的纳米复合材料。通过改变煅烧温度可以控制 Pt/TiO_2 的相组成和结构。与包含锐钛矿相或混合相的 Pt/TiO_2 相比，纯金红石相的 HCHO 催化活性更高，周转频率（TOF）为 16.6 min^{-1}。Pt/TiO_2 在不同煅烧温度下的微观形貌图如图 3.36 所示。

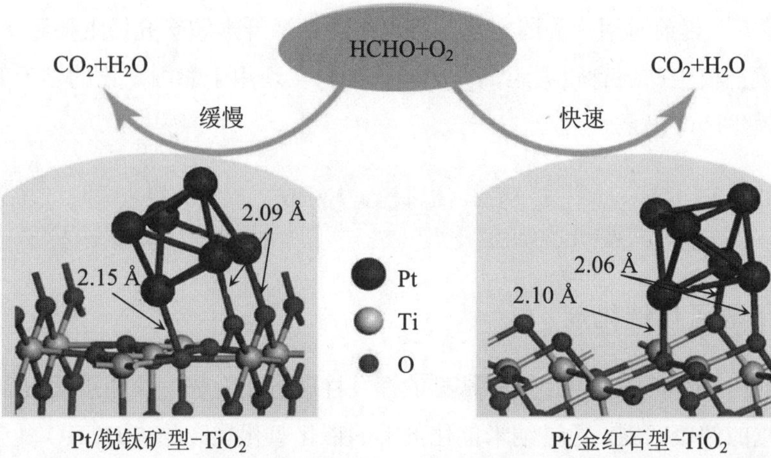

（a）催化反应机理图

（b）800 ℃煅烧后的 SEM 图

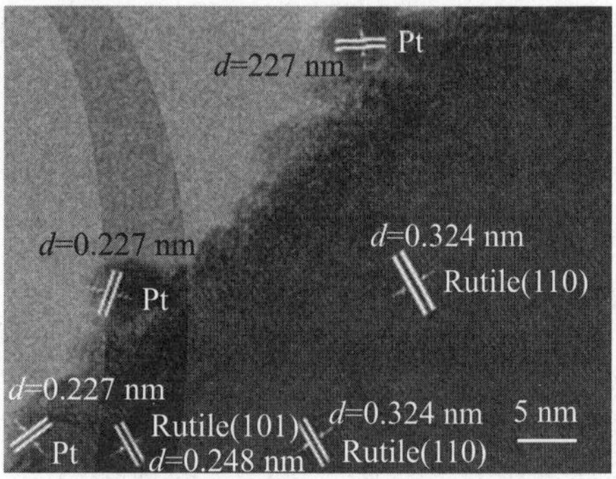

（c）800 ℃煅烧后的 HRTEM 图

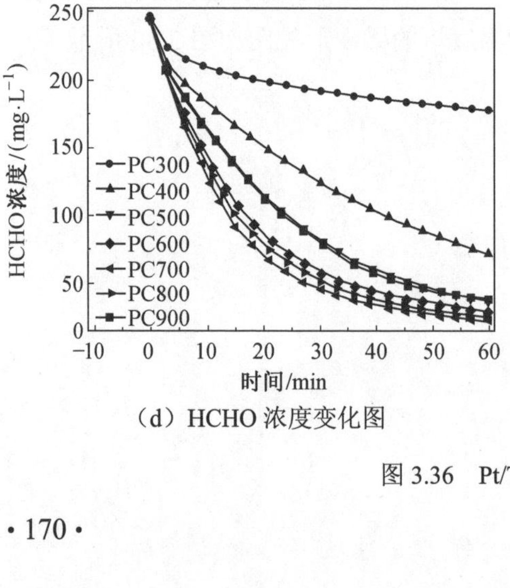

（d）HCHO 浓度变化图

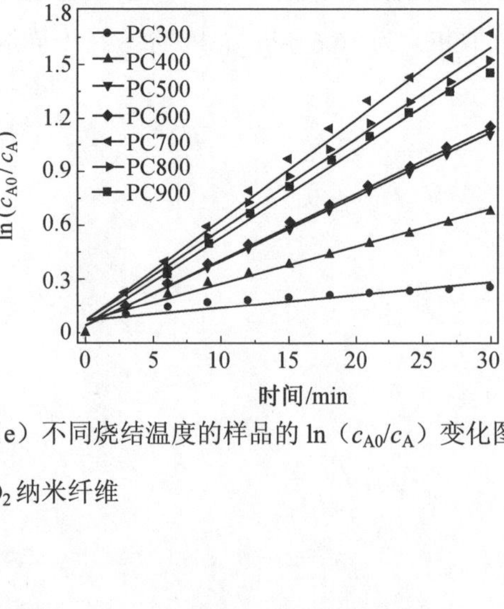

（e）不同烧结温度的样品的 ln（c_{A0}/c_A）变化图

图 3.36 Pt/TiO_2 纳米纤维

Nie 等人通过静电纺丝制备了 Pt 纳米颗粒（NPs）修饰的多孔 TiO_2 纳米纤维（Pt/TF）。如图 3.37 所示，与 Pt-NP 修饰的商用 TiO_2（P25）催化剂相比，在室温下甲醛的分解中，制得的复合纳米纤维有着更高的 HCHO 催化活性。这主要是由于其分层的宏观/介孔结构有利于反应物和产物的扩散，进而提高了催化活性。

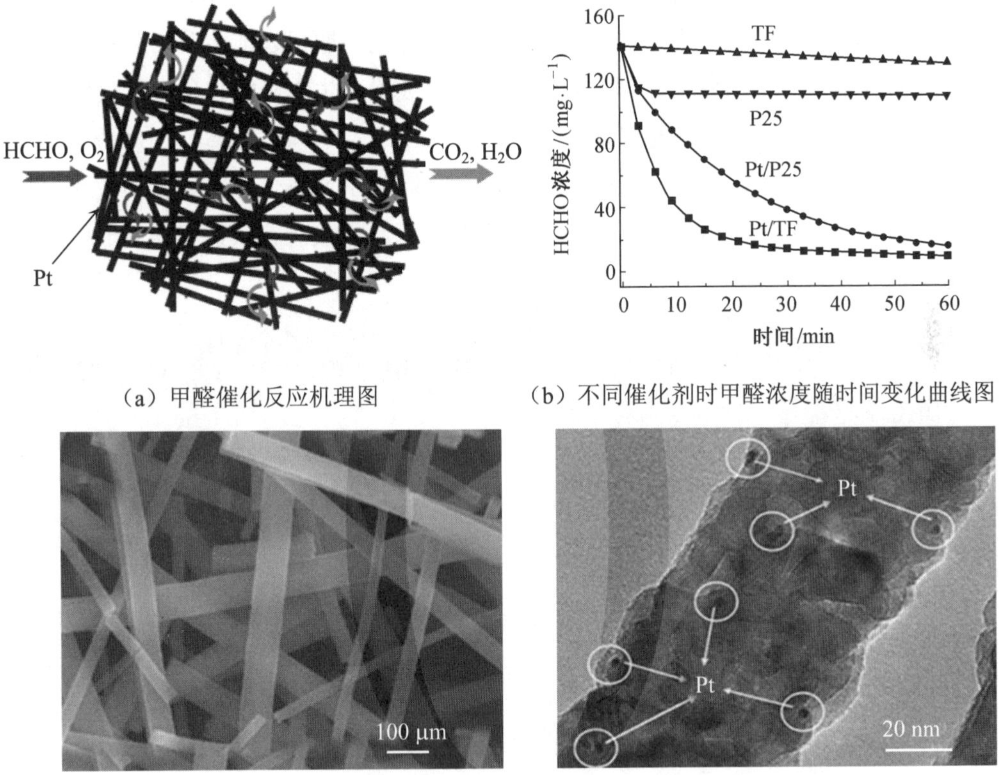

（a）甲醛催化反应机理图　　（b）不同催化剂时甲醛浓度随时间变化曲线图

（c）Pt/TF 的 SEM 图　　（d）Pt/TF 的 TEM 图

图 3.37　Pt 纳米颗粒（NPs）修饰的多孔 TiO_2 纳米纤维（Pt/TF）

然而，负载的贵金属由于其较高的成本，限制了在商业上的广泛应用。另一种提高 HCHO 催化活性的方法是改变纳米结构和形态，并使更多的活性位点暴露于反应物以增强催化活性。Wu 等人通过静电纺丝和热处理工艺相结合，制备了多孔的钴氧化物（Co_3O_4）纳米纤维。该纳米纤维由具有项链状排列的纯 Co_3O_4 纳米颗粒组成，对甲醛的氧化表现出高的催化活性（98 ℃，100%转化率）和催化稳定性（160 h，近 100%转化率）。

3.4.2 尿素氧化催化剂

尿素（$CO(NH_2)_2$）是一种常见的化学燃料，大量富含尿素的废液每天排出。若不进行任何处理，除了会促进酸雨的形成，还容易被氧化成硝酸盐等污染物，对人类健康、水源等造成严重的伤害。

因此，合理有效地处理尿素，不仅能够解决环境问题，还能将其转化成有对我们有利的资源。传统的尿素处理方式包括水解、吸附、生物降解和化学氧化，但这些方法所需设备成本较高，能量消耗较大。

尿素的电催化氧化（Urea Oxidation Reaction，UOR）不仅有着较高的效率，且产物（CO_2、N_2和H_2）稳定、无毒，逐渐引起了广泛的关注。尿素燃料电池机理图如图 3.38 所示，其反应方程式如下所示：

总反应：

$$CONH_2 + H_2O \longrightarrow N_2 + 3H_2 + CO_2 \tag{3.18}$$

阳极：

$$CONH_2 + 6OH^- \longrightarrow N_2 + 5H_2O + CO_2 + 6e^- \tag{3.19}$$

阴极：

$$2H_2O + 2e^- \longrightarrow H_2 + 2OH^- \tag{3.20}$$

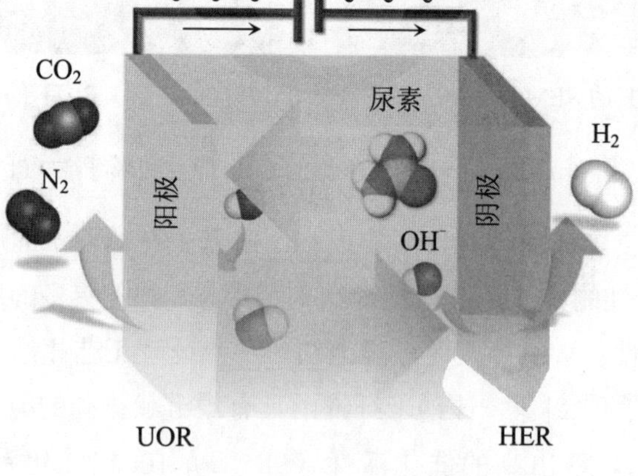

图 3.38 尿素燃料电池机理图

然而，开发具有所需起始电位的合适阳极材料并非易事，因为大多数报道的材料（包括贵金属）的高过电位会导致能耗较大。对于贵金属催化剂而言，铂和铑对尿素氧化反应（UOR）具有很高的催化能力。但是，昂贵的价格限制了其在商业上大规模应用。因此，研发出一种价格低廉、催化活性好的新型催化剂是至关重要的。

镍是一种具有优异 UOR 催化性能的非贵金属材料，但是由于其过电位高、稳定性差等问题，限制了其在 UOR 反应中的应用。因此，许多人通过制备不同形貌和不同结构的镍来提高其 UOR 催化性能。

除了通过改变形貌结构来提高催化活性之外，选用碳质纳米结构作为载体也逐渐引起关注。碳材料能够提高电子传输能力进而提高其催化 UOR 性能。目前，石墨烯、石墨、碳纳米管是使用最广泛的载体材料。在报道的碳质载体中，纳米纤维由于其大的轴向比率，具有最低的电子转移阻力。通常，碳纳米纤维的大轴向比率能够消除颗粒间出现的界面阻力。而静电纺丝技术因其简单、高产量、低成本以及对各种材料的良好的适用性，也逐渐成为工业和研究中最广泛使用的纳米纤维合成工艺。

Barakat 等人通过静电纺丝技术结合热处理工艺，制备了氮掺杂镍修饰的碳纳米纤维。氮掺杂促进了金属纳米颗粒表面上 NiOOH 活性层的形成，能够明显改善对尿素氧化的电催化活性。其微观形貌图如图 3.39 所示。

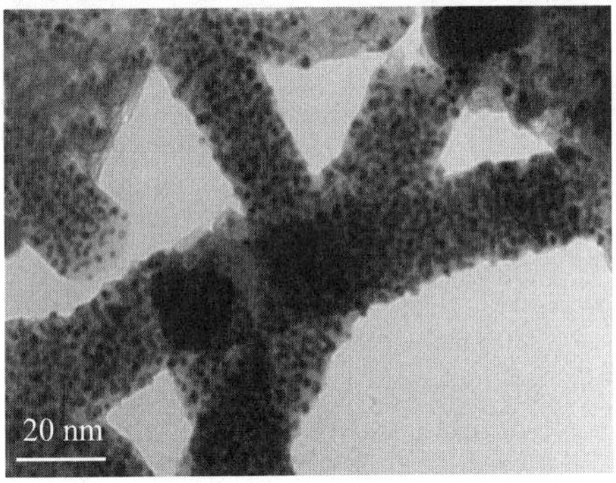

（a）Ni–C 纳米纤维的 TEM 图

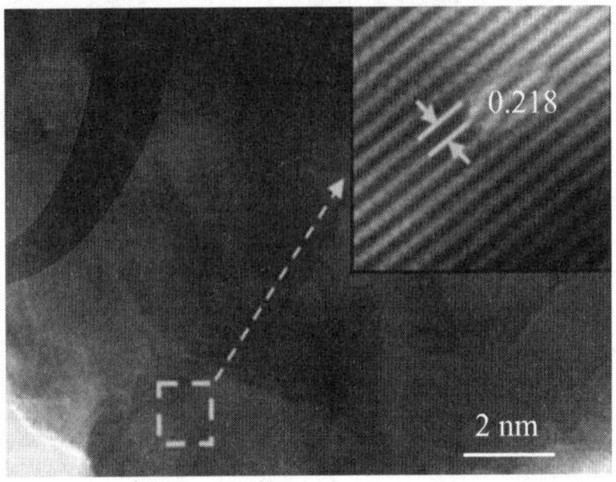

（b）Ni–C 纳米纤维的高分辨 TEM 图

图 3.39　氮掺杂镍修饰的碳纳米纤维（Ni–C）

Liu 等人利用静电纺丝技术结合碳化工艺，制备了一种镍纳米线包埋、氮硫双掺杂的碳纳米纤维（Ni/N、S–CNFs），具有优异的尿素氧化反应（UOR）电催化性能。

如图 3.40 所示，Ni/N、S-CNFs 在 0.42 V（vs.SCE）时能够获得 37.0 $mA{\cdot}mg^{-1}$ 的高电流密度，是 Ni/CNFs 的 2 倍。此外，Ni / N、S-CNFs 具有优异的催化稳定性。

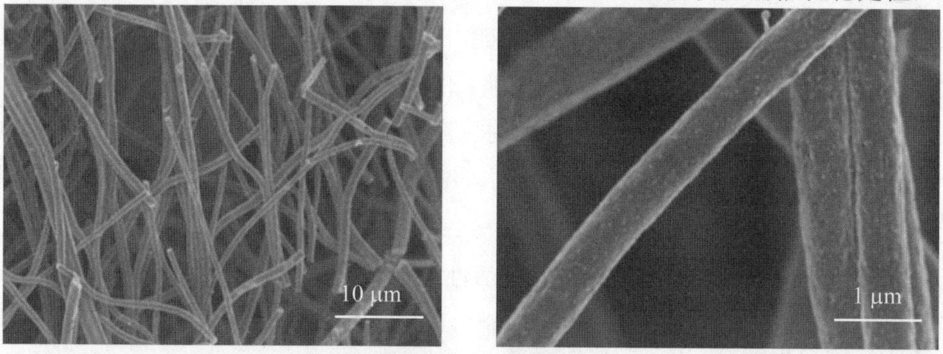

（a）Ni/CNFs 的 SEM 图

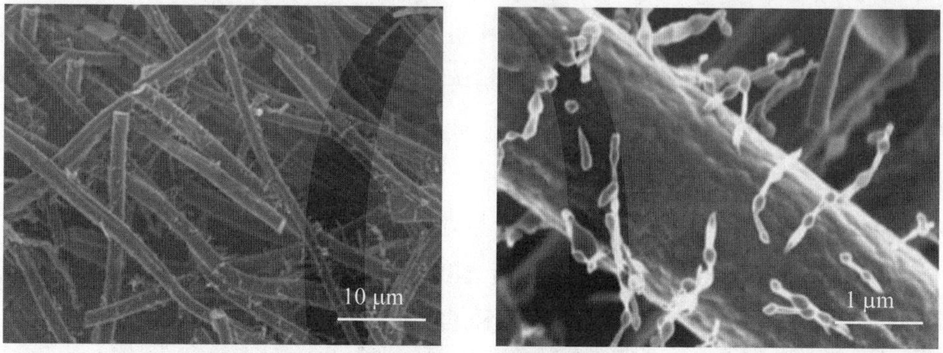

（b）Ni / N、S-CNFs 的 SEM 图

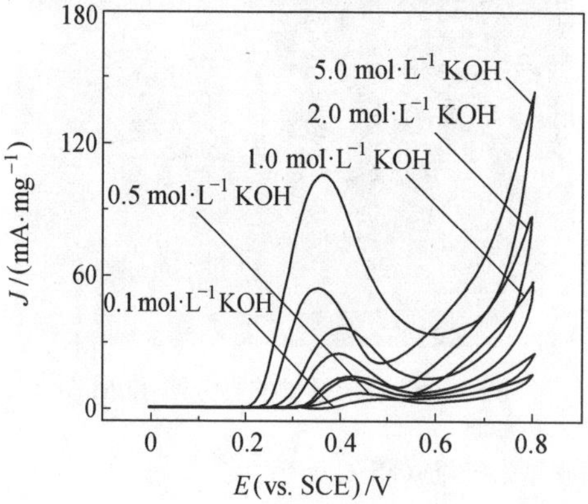

（c）Ni/CNFs 的 CV 图

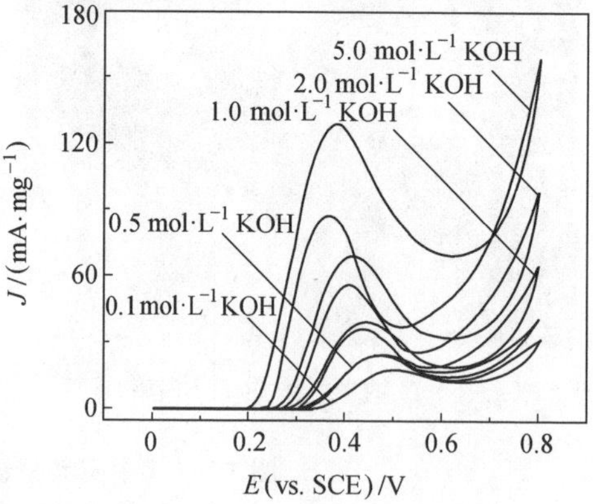

（d）Ni / N、S-CNFs 的 CV 图

图 3.40　包埋镍纳米线、氮硫双掺杂的碳纳米纤维（Ni/N、S-CNFs）

将一些本身不具活性或者活性很小的物质加入到催化剂中，能够显著提高其催化性能，这种物质称为助催化剂。锡可以与包括镍在内的许多金属形成有用的合金。在能源设备中，锡镍合金电极在锂离子电池中表现出良好的性能。Barakat 等人利用静电纺丝技术结合热处理工艺，制备了掺杂 NiSn 合金纳米颗粒的碳纳米纤维，发现 Sn 是一种优异的助催化剂。对于尿素氧化反应（UOR），如图 3.41 所示，在 55 ℃时电流密度能够达到 175 mA·cm^{-2}（vs. Ag/AgCl），是具有优异催化性能的尿素燃料电池的阳极材料。

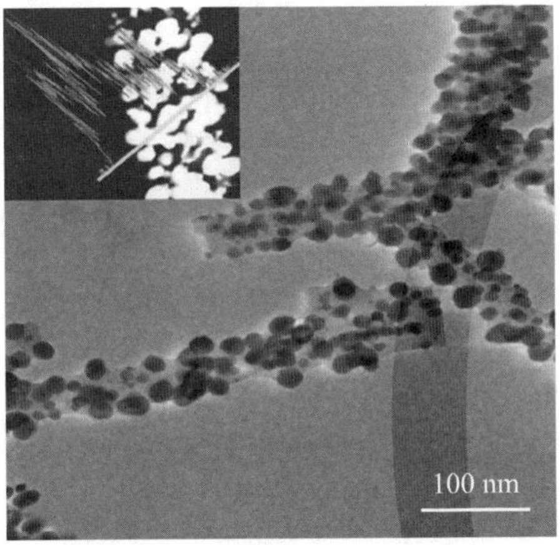

（a）NiSn 合金纳米颗粒的碳纳米纤维的 TEM 图

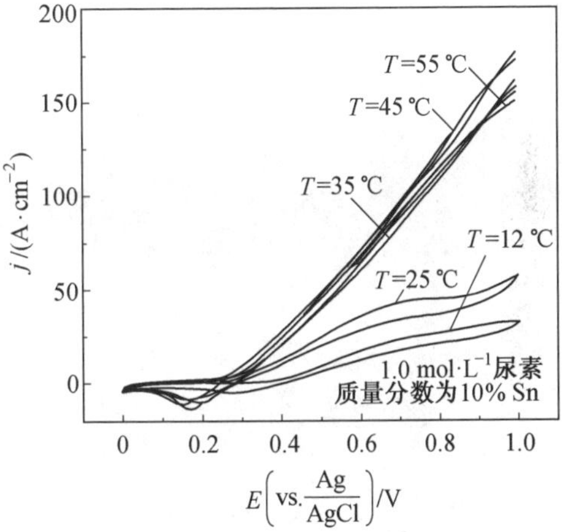

（b）掺入 NiSn 颗粒纳米纤维的 CV 图

图 3.41　掺杂 NiSn 合金纳米颗粒的碳纳米纤维

除此之外，Mn 也是一种良好的助催化剂。Baraka 等人通过静电纺丝技术结合热处理工艺，制备了 Ni、Mn 纳米粒子掺杂的碳纳米纤维，是一种高效稳定的尿素氧化电催化剂，如图 3.42 所示。与纳米粒子相比，具有更好的电催化活性。虽然纳米粒子的比表面积要大于纳米纤维，但是由于纳米纤维较大的轴向比，能够增强电子在催化剂内的转移能力，进而提高了其电催化性能。

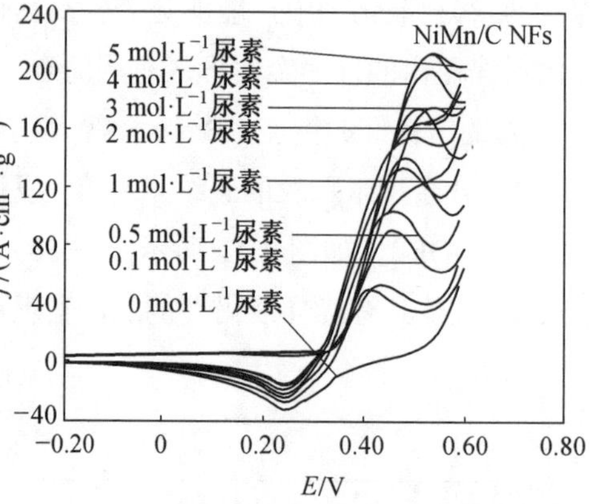

（a）NiMn/C 纳米纤维的尿素点催化活性图

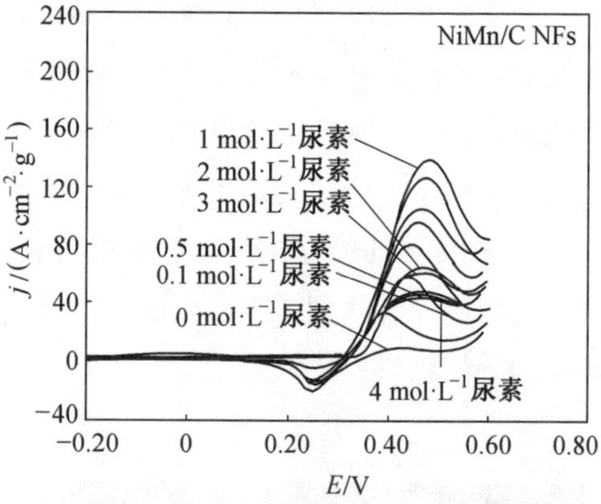

（b）NiMn/C 纳米颗粒的尿素点催化活性图

（c）NiMn/C 纳米颗粒的 SEM 图

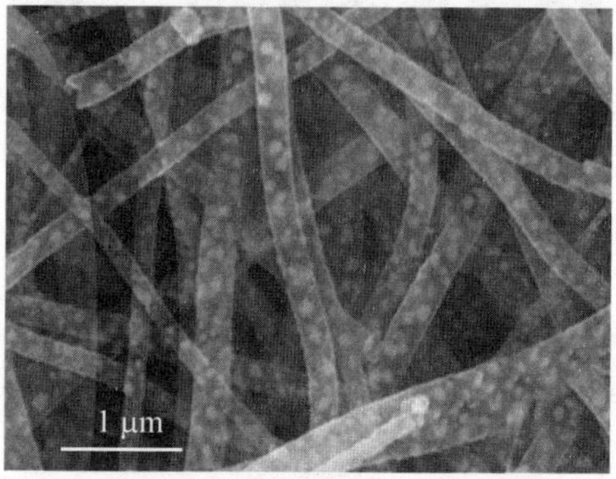

（d）NiMn/C 纳米纤维的 SEM 图

图 3.42　Ni 纳米颗粒掺杂的碳纳米纤维

3.4.3　甲醇氧化催化剂

随着人类社会的不断发展和进步，能源逐渐成为我们生存发展的物质基础。随着经济的快速发展，对能源的需求也在不断加大。而传统的化石燃料被不断开发和利用，其有限的储量造成了化石燃料资源的枯竭。化石燃料的使用也对环境造成了不可逆的损伤。因此，寻找一种绿色、高效的新型能源是至关重要的。

燃料电池可以将化学能转化为电能。其中，直接甲醇燃料电池（Direct Methanol Fuel Cells, DMFCs）具有高能量转换效率、低污染物排放、低工作温度、易于处理和液体燃料等特点，被认为是解决未来能源危机的最有前途的选择之一。

直接甲醇燃料电池的电极反应方程式如下所示：

阳极：

$$CH_3OH + H_2O \longrightarrow CO_2 + 6H^+ + 6e^- \tag{3.21}$$

阴极：

$$\frac{3}{2}O_2 + 6H^+ + 6e^- \longrightarrow 3H_2O \tag{3.22}$$

总反应：

$$CH_3OH + \frac{3}{2}O_2 \longrightarrow CO_2 + 2H_2O \tag{3.23}$$

铂（Pt）基催化具有良好的甲醇氧化催化活性，但是由于其储量较少、成本较高，限制了其在商业领域的发展。除此之外，甲醛会在 Pt 表面发生氧化，生成 CO 等中间产物，吸附在 Pt 的表面，活性位点被占据，引起催化剂中毒。因此，制备高活性、高稳定性和抗中毒能力的 Pt 基催化剂，成为一大研究热点。

Kim 等人利用了静电纺丝技术结合热处理工艺，制备了 Pt 纳米纤维，是一种优异的甲醇氧化催化剂。在氮气中经热处理然后暴露于空气中可增加 Pt 纳米线的表面积，进而提高甲醇氧化反应（MOR）的电化学活性。其微观形貌图如图 3.43 所示。

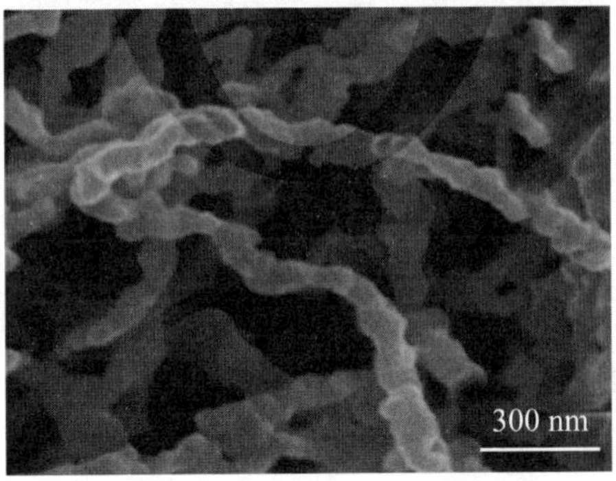

（a）Pt 纳米纤维微观形貌图

图 3.43　Pt 纳米纤维

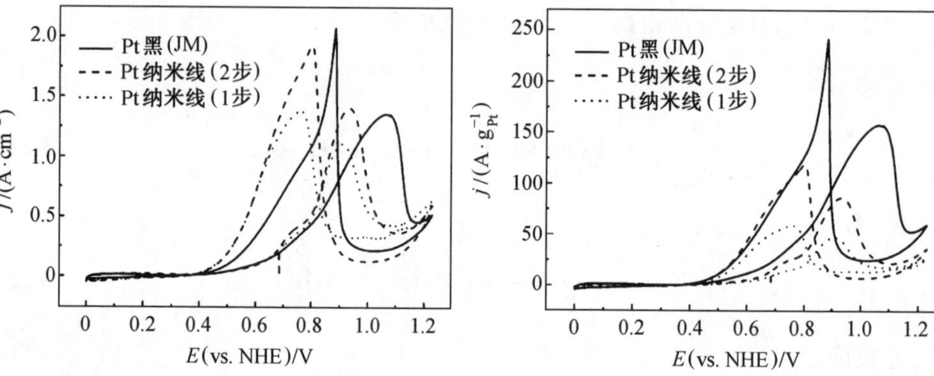

（b）Pt 纳米纤维的 MOR 比活性　　（c）Pt 纳米纤维的质量活性

续图 3.43

TiO_2 在过渡金属氧化物中，无毒且具有良好的耐腐蚀性能，在全 pH 环境中有较高的稳定性。TiO_2 载体可以降低 CO 中间体的吸附能，增加 CO 基团在 Pt 纳米结构上的迁移率。且研究表明，TiO_2 在 DMFC 中可作为助催化剂，有利于催化性能的提升。Formo 等人通过静电纺丝技术制备了 TiO_2 纳米纤维，然后用 Pt 纳米颗粒和 Pt 纳米线修饰了其表面。这些纳米纤维因为负载了 Pt 纳米结构，成为甲醇氧化的有效电催化剂，催化剂的电化学活性和耐久性均得到改善。这是因为 Pt 纳米颗粒的尺寸的减小可以使催化剂的活性表面积大大增加。也可以通过将 Pt 纳米结构的形态从纳米颗粒更改为纳米线来提高甲醇氧化（MOR）的效率，纳米线的大侧面可以提供更多的催化活性面。

氧化物（WO_3）是燃料电池的载体材料，负载在 WO_3 催化剂上的 Pt 纳米颗粒表现出优异的 CO 耐受性和更高的催化活性。Zhao 等人将静电纺丝工艺与电置换反应相结合，将 Pt 纳米粒子原位锚定在电纺 WO_3 纳米纤维表面上，制备了 Pt/WO_3 纳米纤维。这种与碳纳米管相连的纳米结构与商用 WO_3 上负载的 Pt 相比，具有出色的甲醇氧化催化活性和稳定性。

研究表明，甲醇氧化可被认为是阳极表面吸附和电化学反应的结合，为了利用其吸附能力，碳材料已被结合到许多最近报道的电催化材料中。多孔碳纳米纤维（pCNF）具有高比表面积、机械强度、电导率和化学稳定性等特点，是一种优异的电极载体材料。

聚丙烯腈具有良好的可纺性和高碳含量，可以将其与其他前体材料混合制备良好的多孔碳纳米材料，是一种常用的载体材料。Li 等人通过静电纺丝技术制备了聚丙烯腈纳米纤维（CFM），在其表面负载商用 Pt/C 催化剂，与商用碳纸（CPs）对比，其表现出更高的电催化活性，更高的稳定性。其微观形貌图如图 3.44 所示。

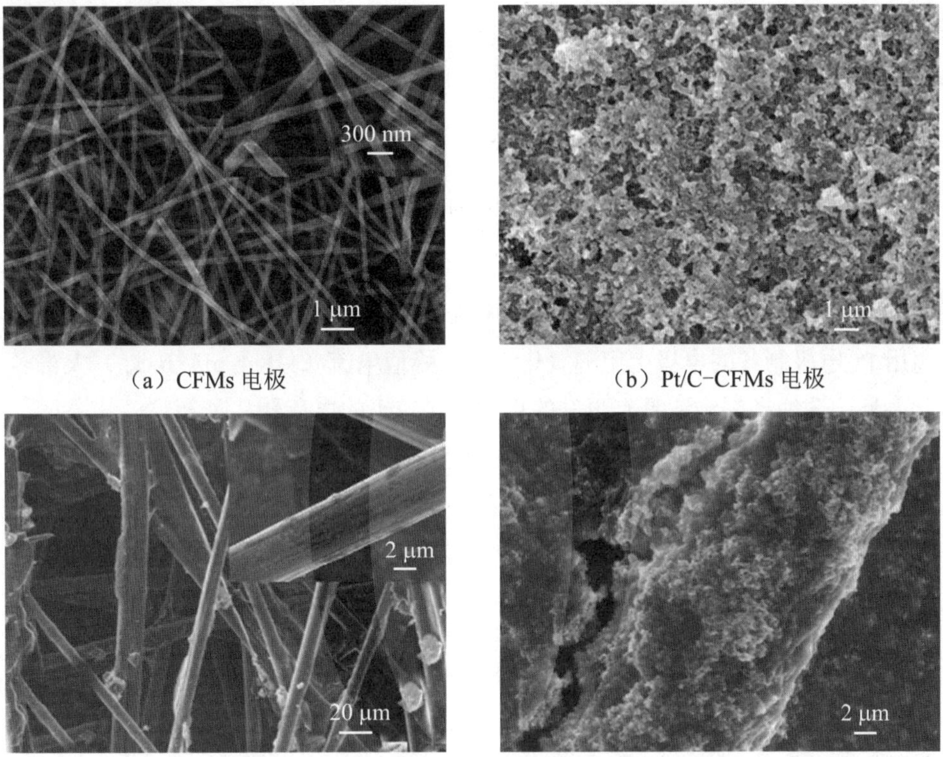

（a）CFMs 电极　（b）Pt/C-CFMs 电极

（c）商用碳纸（CPs）电极的 SEM 图　（d）Pt/C-CP 电极的 SEM 图

图 3.44　负载商用 Pt/C 催化剂的聚丙烯腈纳米纤维

Li 等人利用静电纺丝技术结合热处理，以聚丙烯腈为原料，制备了碳纳米纤维。通过循环伏安法将铂电沉积在碳纳米纤维垫上，负载在碳纳米纤维垫上的铂在电催化活性和对甲醇氧化的稳定性方面表现出高性能。其微观形貌图如图 3.45 所示。

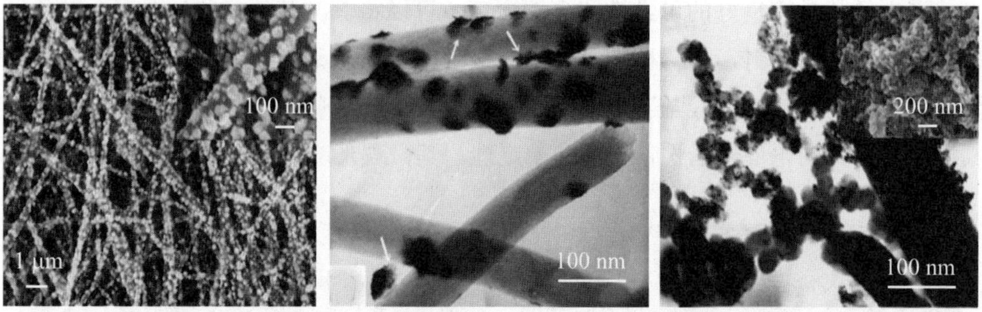

（a）Pt/CFM 的 SEM 图　（b）Pt/CFM 的 TEM 图　（c）Pt/C 电极的 TEM 和 SEM 图

图 3.45　负载 Pt 纳米颗粒的碳纳米纤维垫（CFM）

Liu 等人结合静电纺丝和化学镀工艺制备了聚丙烯腈纳米纤维（Pt/Au-PAN）。Pt/Au-PAN 电极上甲醇氧化的催化峰值电流可以达到约 450 $mA\cdot mg_{Pt}^{-1}$，这比通过加载商用 Pt 电极制备的电极上甲醇氧化的催化峰值电流（118.4 $mA\cdot mg_{Pt}^{-1}$）大得多。除此之外，该纳米纤维还具有更好的稳定性和较小的电荷转移阻力。

Pd 基电催化剂对甲醇表现出高的电催化活性，且具有耐 CO 中毒能力。Guo 等人利用静电纺丝技术结合碳化工艺，制备了钯纳米颗粒负载的碳纳米纤维复合材料（Pd/CNF）。如图 3.46 所示，Pd 纳米颗粒牢固地嵌入具有面心立方结构的 CNF 中。与市售的 Pd/C 催化剂相比，所制备的 Pd/CNF 具有更优异的电催化活性和对甲醇氧化的稳定性。

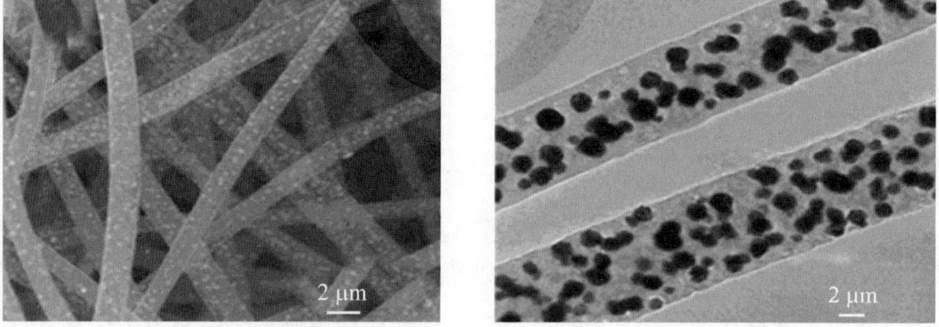

（a）Pd/CNF 纳米纤维 SEM 图　（b）Pd/CNF 纳米纤维 TEM 图

图 3.46　钯纳米颗粒负载的碳纳米纤维复合材料（Pd/CNF）

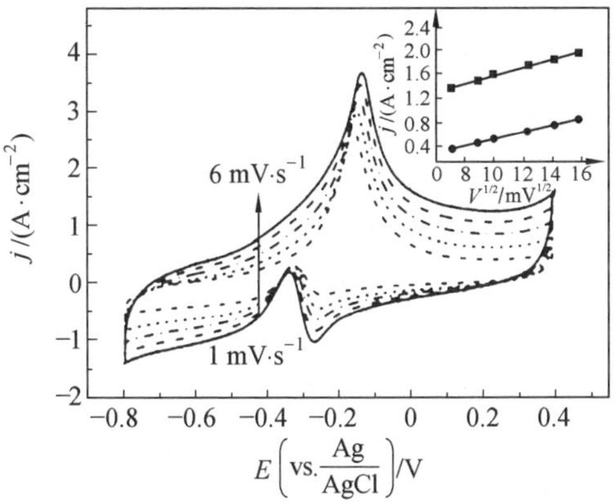

（c）Pd/CNF 纳米纤维在含 1 mol·L^{-1} CH_3OH 的 1 mol·L^{-1} KOH 溶液中不同扫速下的 CV 曲线图

续图 3.46

近年来，非贵金属催化剂因其优异的催化活性和较低的成本引起了大家的兴趣。金属镍及其合金由于低成本和良好的甲醇氧化活性而备受关注，已有研究显示，Ni/MWCNT、CoNi-CNF 和 CoNi-石墨烯对醇类氧化表现出良好的电催化作用。其中，氮掺杂的碳纳米结构具有高的表面成核位点，可以使催化剂纳米颗粒在载体表面材料上锚定和高度分散，从而导致氮掺杂的碳质载体与催化金属之间发生高相互作用。Abdullah 等人以聚丙烯腈、聚苯胺和石墨烯溶液混合物为原料，利用静电纺丝技术制备了高度掺杂的氮碳纳米纤维（N-CNF），集合化学沉淀工艺将 NiO 纳米颗粒负载在 N-CNF 表面，制备了 NiO/N-CNFs 纳米纤维。与其他基于镍和 NiO 的催化剂相比，NiO/N-CNFs 杂化物对碱性介质中的甲醇电氧化表现出优异的电催化活性。NiO 纳米颗粒负载前后的微观形貌图和不同 NiO 负载量的循环伏安图如图 3.47 所示。

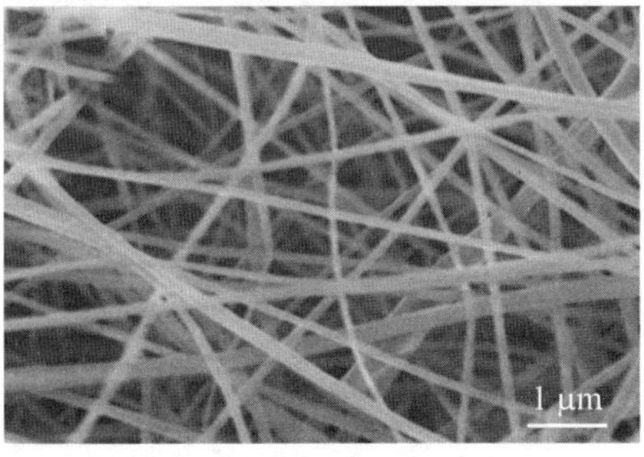

（a）负载 NiO 前的碳纳米纤维 SEM 图

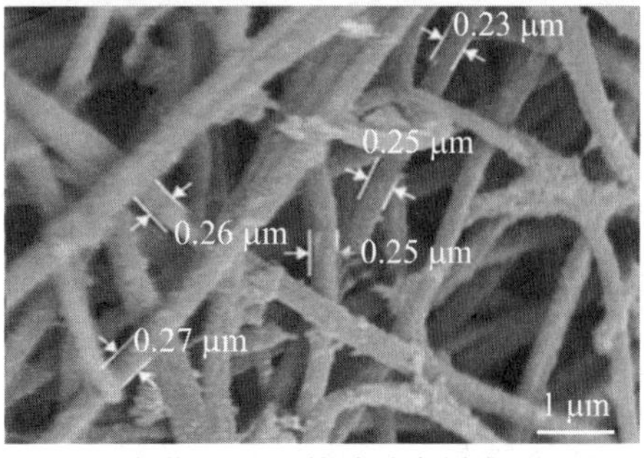

（b）负载 NiO 后的碳纳米纤维 SEM 图

图 3.47　负载 NiO 的氮掺杂碳纳米纤维（NiO/N-CNFs）

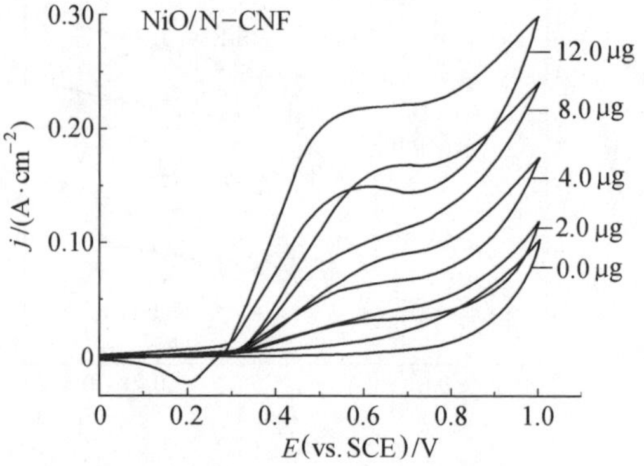

（c）不同 NiO 负载量催化剂在 1.0 mol·L^{-1} 甲醇溶液中的循环伏安图

续图 3.47

Thamer 等人利用静电纺丝技术，制备了氮掺杂的含有镍纳米粒子的碳纳米纤维，发现氮的掺杂能够大大增强甲醇的氧化活性，提高了对甲醇的催化性能，同时也增强了 Ni 基电催化剂的稳定性。掺杂 N 前后的纳米纤维微观形貌图如图 3.48 所示。

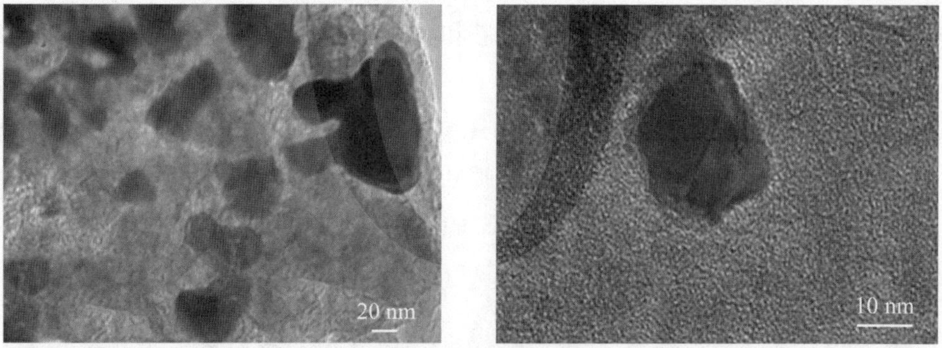

（a）Ni/CNFs 纳米纤维

图 3.48 含有镍纳米颗粒的氮掺杂碳纳米纤维（Ni/NCNFs）

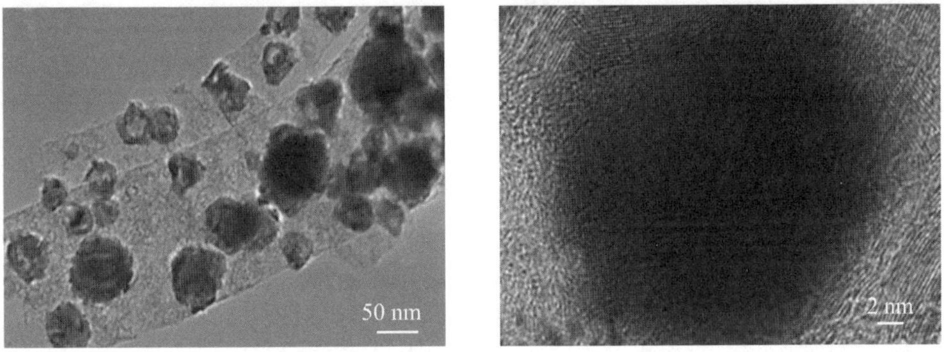

（b）掺杂氮的 Ni/NCNFs 纳米纤维的 TEM 图

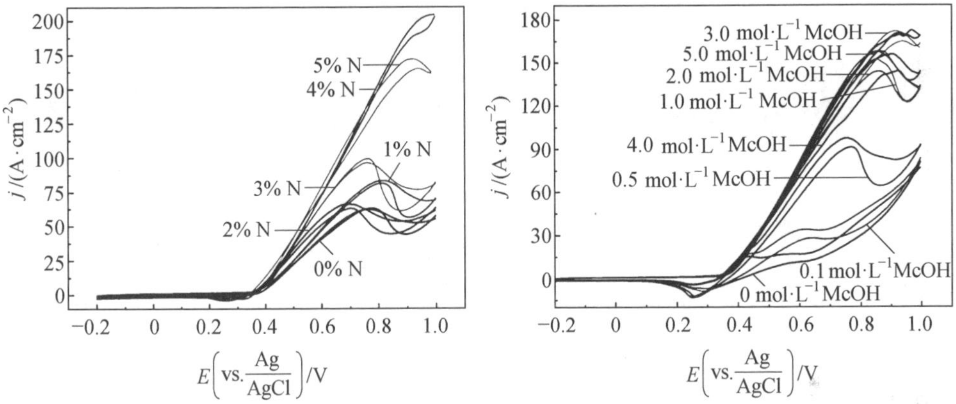

（c）不同氮量的纤维在甲醇的碱溶液中的 CV 图（d）纤维在不同甲醇浓度的碱溶液中的 CV 图

续图 3.48

金属钴可用作消除 Pt 中毒的助催化剂，已发现 Pt-Co 合金是优异的耐 CO 催化剂。Ayman 等人利用静电纺丝技术，制备了掺杂钴钛碳化物纳米颗粒（Co-TiC NPs）的碳纳米纤维，具有优异的甲醇氧化催化性能。Badr 等人通过静电纺丝技术，结合热处理工艺，制备了掺氮掺钴的碳纳米纤维，在甲醇氧化中表现出优异的电催化性能，如图 3.49 所示。结果表明，催化剂的活性随电纺丝溶液中尿素含量的增加而增加，最大电流密度为 100.84 mA·cm^{-2}，高于未掺杂的电流密度（63.56 mA·cm^{-2}）。

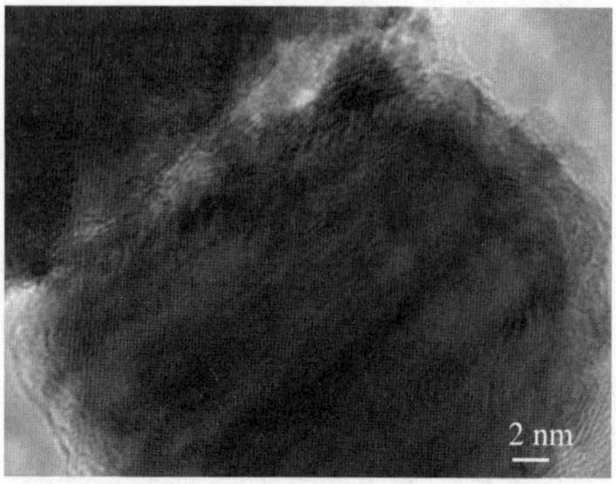

（a）煅烧后未掺杂的 Co/CNFs 的 TEM 图

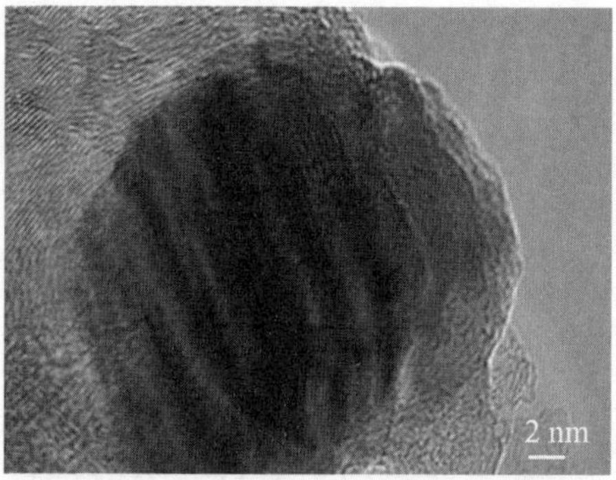

（b）氮掺杂的 Co/CNFs 纳米纤维的 TEM 图

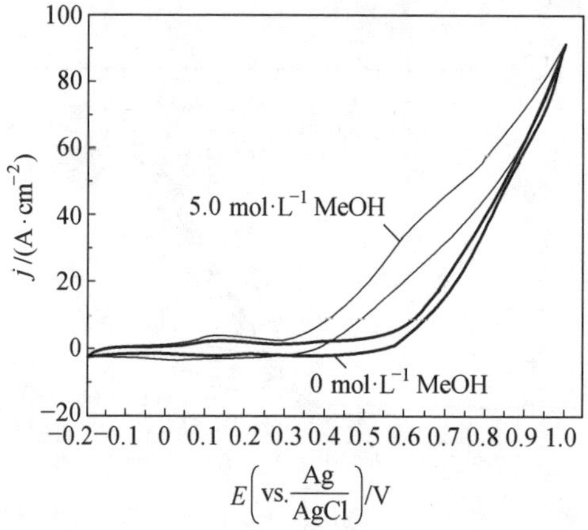

（c）Co/N-CNF（尿素为4%）在1 mol·L^{-1} KOH中的CV图

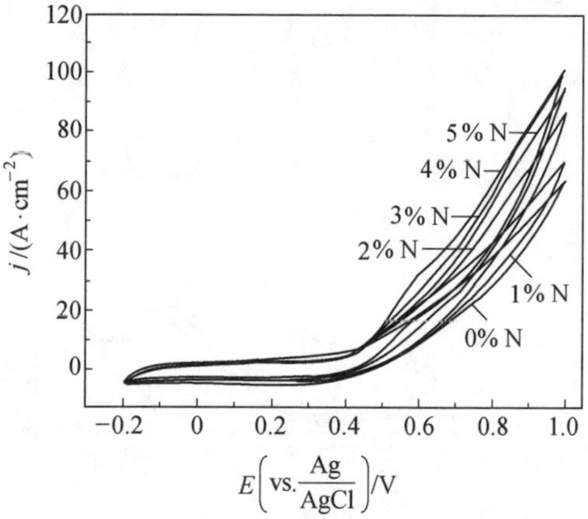

（d）氮掺杂浓度对甲醇催化的 CV 图

图 3.49　掺氮掺钴的碳纳米纤维

壳聚糖（CH）是一种可再生且可生物降解的生物聚合物，对金属具有很强的亲和力，它的成本较低，且在结构上具有 NH_2 和 OH 官能团，这使其成为良好的聚合物电解质和良好的催化剂载体材料。PVA 具有高亲水性，其结构中具有高密度的官能团，是催化剂的有效载体材料。Mehri-Saddat 等人通过静电纺丝技术制备了 CuO 和 Co_3O_4 纳米粒子掺杂的 Pt/聚乙烯醇-CuO-Co_3O_4/壳聚糖催化剂（Pt/PVA-CuO-Co_3O_4/CH）纳米纤维，对甲醇电氧化表现出优异的电催化性和稳定性，如图 3.50 所示，该纳米纤维是一种有应用前景的 DMFC 催化剂。

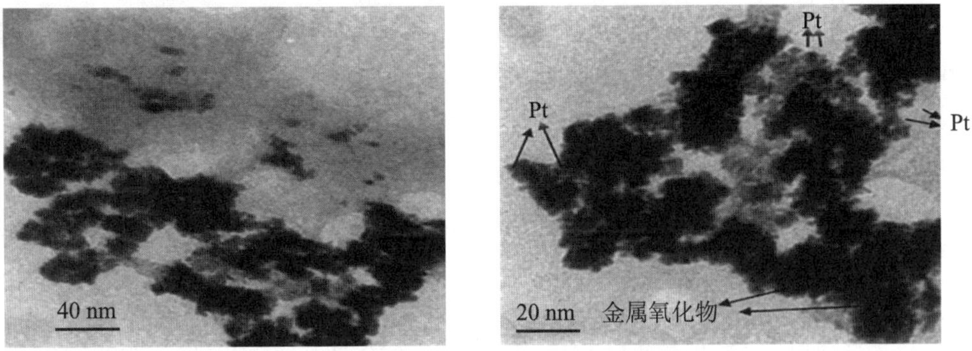

（a）$Pt/PVA-CuO-Co_3O_4/CH$ 的 TEM 图

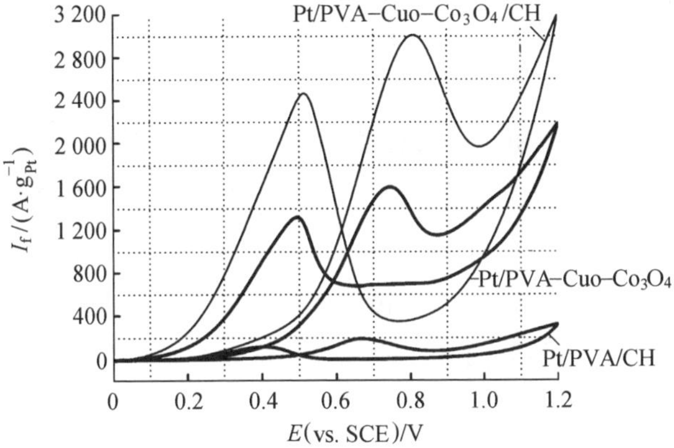

（b）1.83 $mol\cdot L^{-1}$ 甲醇和 0.5 $mol\cdot L^{-1}$ H_2SO_4 溶液中不同催化剂对甲醇的电催化作用对比图

图 3.50　Pt /聚乙烯醇–$CuO-Co_3O_4$/壳聚糖催化剂纳米纤维

3.4.4　甲烷转化催化剂

1. 甲烷氧化催化剂

甲烷（CH_4）是温室气体的第二大来源。它的温室效应比 CO_2 高约 20 倍。煤矿产业是向大气中排放甲烷的主要来源，不经过处理使甲烷随着煤矿通风气排出，造成了温室气体的大量排放，破坏生态环境。除此之外，甲烷是一种非常经济、清洁的燃料替代品，是一种新型能源。低浓度的甲烷在空气中不易燃烧，且在瓦斯开采过程中较高流速的气流会带走大量热量，造成温度的下降，进而降低甲烷的转化率。因此，寻找一种合适的甲烷燃烧的催化剂是至关重要的。

催化燃烧可以在适中的温度下高效、完全地燃烧甲烷，从而可以利用燃烧过程中产生的热量，同时减少空气污染物（例如未燃烧的碳氢化合物（UHC）、CO、NO_x和颗粒物等等）的形成，是一种有效的变废为宝的工艺。由于煤矿通风中的甲烷浓度较高，且空气浓度非常低（约 0.2%（体积分数）），催化剂必须具有极高的活性，才能催化其完全燃烧。催化剂还必须在长期运行中保持稳定。

甲烷氧化的化学方程式如下：

$$CH_4(g)+2O_2(g)=\!=CO_2(g)+2H_2O(l)\text{，}\quad \Delta H(25\,℃)=-192\ \text{kcal}\cdot\text{mol}^{-1}$$

贵金属材料是一种较为理想的催化剂，如 Pd 基催化剂的催化活性要高于其他贵金属催化剂。但由于价格较为昂贵，限制了其在商业上的大规模使用。因此，开发出一种价格低廉且催化活性好的新型催化剂是至关重要的。

钙钛矿材料（通式为 ABO_3）是甲烷和其他挥发性有机化合物的最重要的活性燃烧催化剂之一。Chen 等人通过静电纺丝技术结合热处理工艺制备了几种具有钙钛矿晶体结构的氧化物纳米纤维，包括 $LaMnO_3$、$LaFeO_3$、$La_{0.8}Sr_{0.2}CoO_3$ 和 $La_{0.9}Ce_{0.1}CoO_3$。其中，$La_{0.8}Sr_{0.2}CoO_3$ 钙钛矿纳米纤维具有大表面积和多孔结构，在 470 ℃的中等温度下可以达到 100%的甲烷燃烧转化率。四种纳米纤维的微观形貌图如图 3.51 所示。

（a）$LaMnO_3$ 纳米纤维的 SEM 和 TEM 图

图 3.51　钙钛矿晶体结构的氧化物纳米纤维

（b）$LaFeO_3$ 纳米纤维的 SEM 和 TEM 图

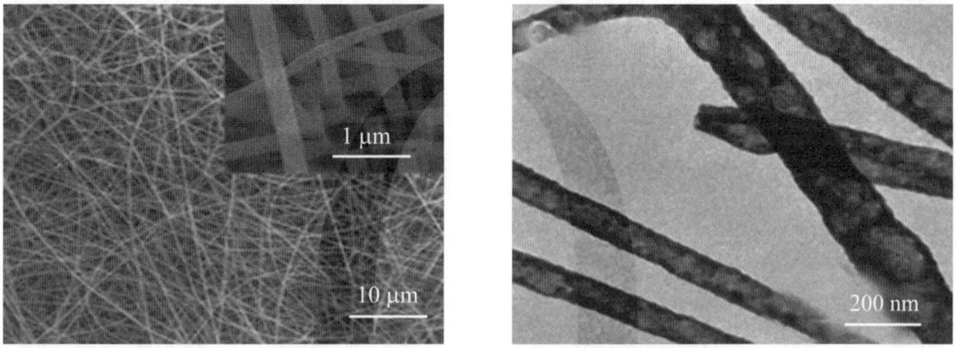

（c）$La_{0.8}Sr_{0.2}CoO_3$ 纳米纤维的 SEM 和 TEM 图

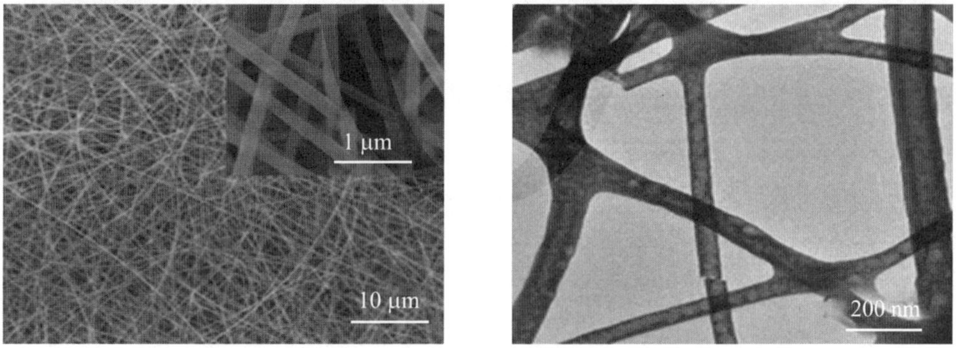

（d）$La_{0.9}Ce_{0.1}CoO_3$ 纳米纤维的 SEM 和 TEM 图

图 3.51　钙钛矿晶体结构的氧化物纳米纤维

Huang 等人通过静电纺丝技术合成了 $La_{0.75}Sr_{0.25}MnO_3$ 钙钛矿纳米纤维，与水热法合成的纳米颗粒进行比较，纳米纤维对 CH_4 燃烧显示出更高的催化活性，这可以归因于较高的表面积、较大的孔体积和较好的热稳定性。$La_{0.75}Sr_{0.25}MnO_3$ 纳米颗粒与纳米纤维微观形貌图和 CO 和 CH_4 转换率图如图 3.52 所示。

（a）LSMO 纳米纤维的 SEM 图　（b）LSMO 纳米纤维的 TEM 图

（c）LSMO 纳米颗粒的 SEM 图　（d）LSMO 纳米颗粒的 TEM 图

（e）CO 活性转换率图　（f）CH_4 活性转换率图

图 3.52　$La_{0.75}Sr_{0.25}MnO_3$（LSMO）钙钛矿纳米纤维

2. 甲烷干气重整催化剂

以甲烷与二氧化碳为原料，制备氢气和一氧化碳的反应称为甲烷干重整反应（DRM），其能够在减弱温室效应的同时，又能提供可利用的氢气资源，受到了越来越多的关注。其具有重要的研究意义和潜在的工业化应用价值。在不考虑副反应的情况下，甲烷干重整反应产物 H_2/CO 比例约为 1。甲烷干重整反应及其相关副反应的方程式如下所示：

$$CH_4 + CO_2 = 2CO + 2H_2,\ \Delta H_{298}^{\ominus} = +247\ kJ \cdot mol^{-1} \tag{3.24}$$

$$CO + H_2O = CO_2 + H_2,\ \Delta H_{298}^{\ominus} = -41\ kJ \cdot mol^{-1} \tag{3.25}$$

$$CH_4 = C + H_2,\ \Delta H_{298}^{\ominus} = +75\ kJ \cdot mol^{-1} \tag{3.26}$$

甲烷重装催化剂是阻碍工业化发展的主要瓶颈之一。重整催化剂通常分为两大类，即贵金属基催化剂和过渡金属基催化剂。其中，镍基催化剂因其出色的活性被认为是最有前途的候选物。但镍基催化剂也存在着一些问题限制了其在工业上的发展，如操作温度高和快速失活等。

Wang 等人通过静电纺丝法结合热处理工艺，制备了 Ni/Al_2O_3 纳米纤维催化剂，其具有较高的结构和热稳定性，以及较好的甲烷干重整催化活性。Liu 等人利用静电纺丝技术，制备了负载镍（Ni）催化剂的氧化铝（Al_2O_3）纳米纤维，如图 3.53 所示，发现金属 Ni 纳米粒子高度分散在 Al_2O_3 纳米纤维上，与传统工业制备的 Ni/Al_2O_3 纳米颗粒催化剂相比，具有更好金属分散性，且在甲烷的二氧化碳（CO_2）重整中具有更高的活性和优异的耐焦炭性。

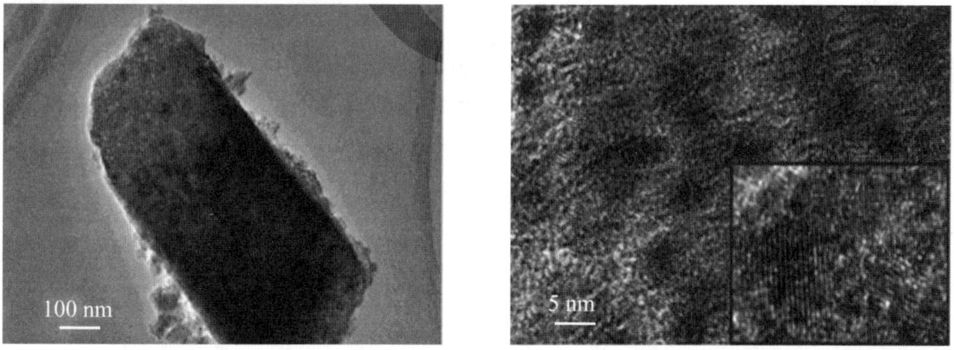

（a）Ni/Al_2O_3 纳米纤维 TEM 图　　（b）Ni/Al_2O_3 纳米纤维高倍 TEM 图

图 3.53　负载镍（Ni）的氧化铝（Al_2O_3）纳米纤维

（c）Ni/Al_2O_3 纳米颗粒的 TEM 图

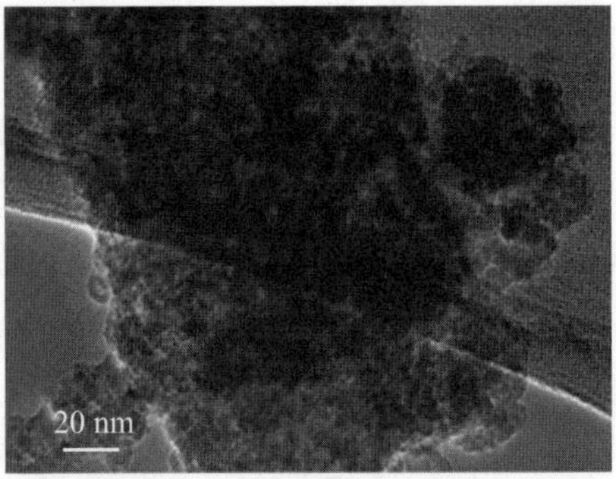

（d）催化后 Ni/Al_2O_3 纳米纤维的 TEM 图

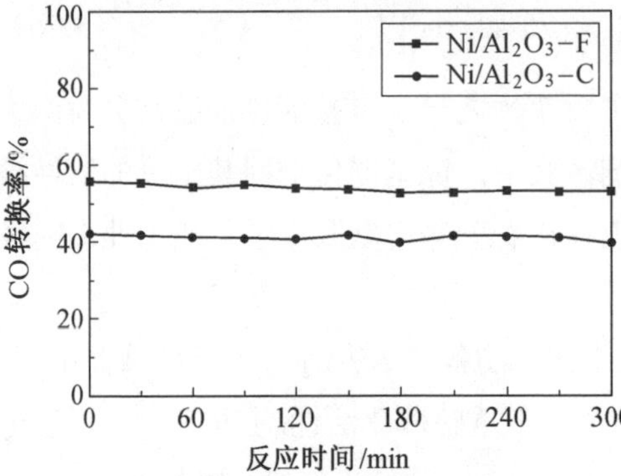

（e）CO 的转换率图

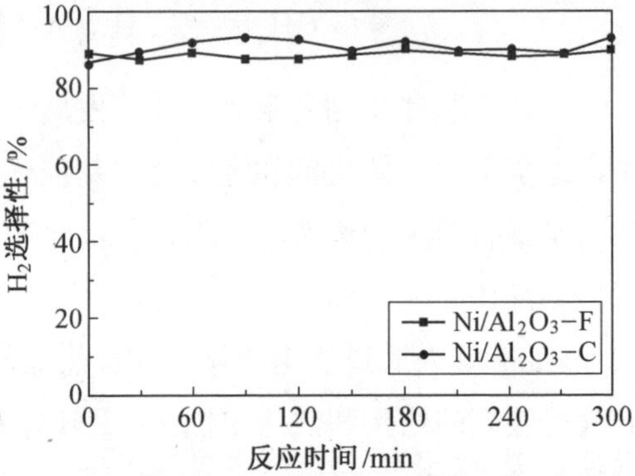

（f）H_2 的选择性随时间变化图

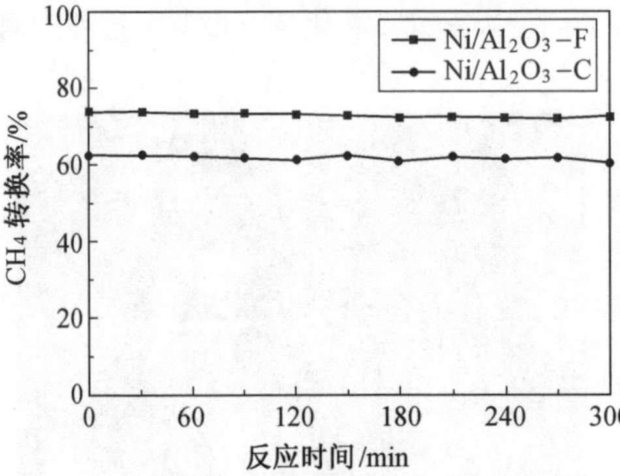

（g）CH_4 的转换率图

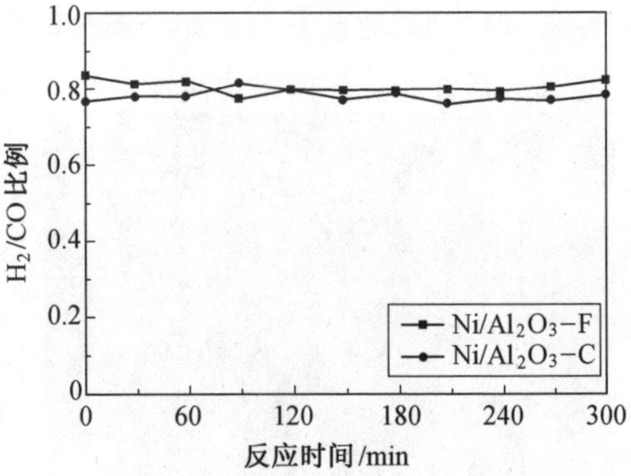

（h）H_2/CO 比例随时间变化图

续图 3.53

Wen 等人利用静电纺丝技术结合热处理工艺，制备了 Ni/SiO_2 纳米纤维，如图 3.54 所示，发现大多数 Ni 纳米粒子被限制在 SiO_2 纳米纤维内部，平均粒径为 8.1 nm。与传统的使用工业 SiO_2 粉末作为载体，通过初期浸渍方法制备的 Ni/SiO_2 催化剂相比，Ni/SiO_2 纳米纤维催化剂显示出更好的金属分散性和增强的金属-载体相互作用，从而显示出更高的甲烷重整的催化活性。

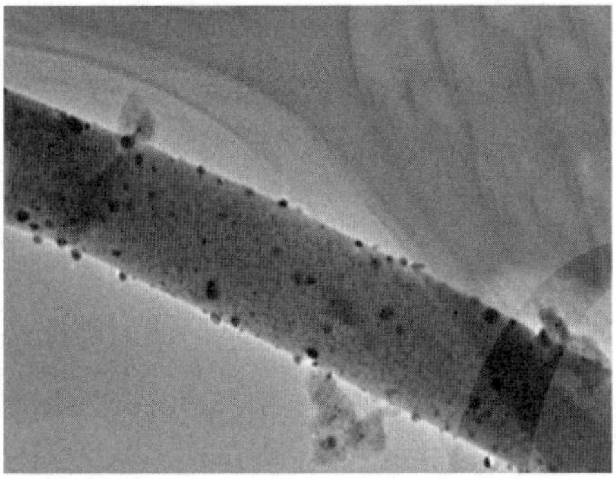

（a）Ni/SiO_2-F 催化剂

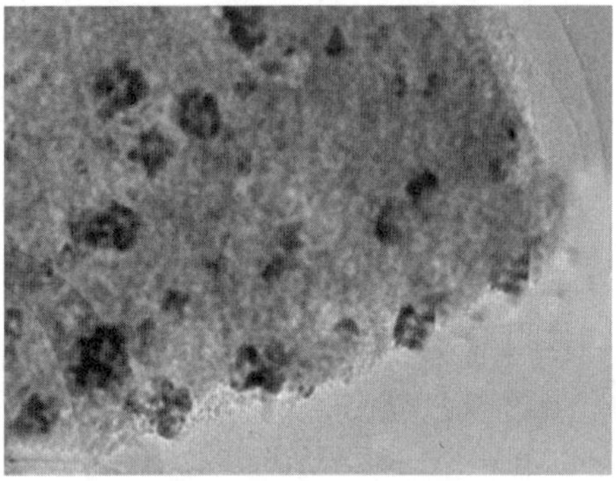

（b）Ni/SiO_2-C 催化剂

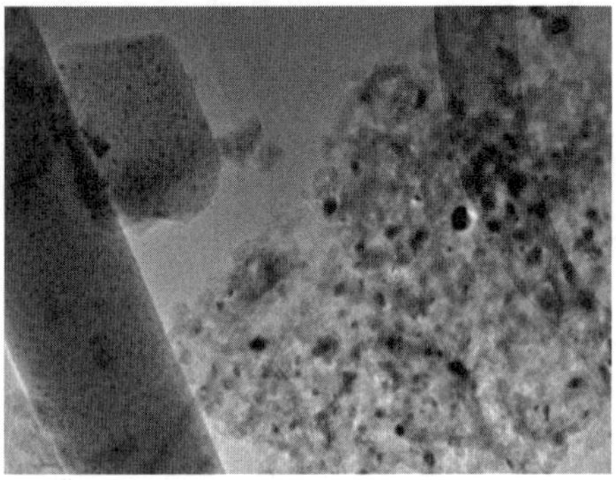

（c）360 min 反应后的 Ni/SiO_2-F 催化剂

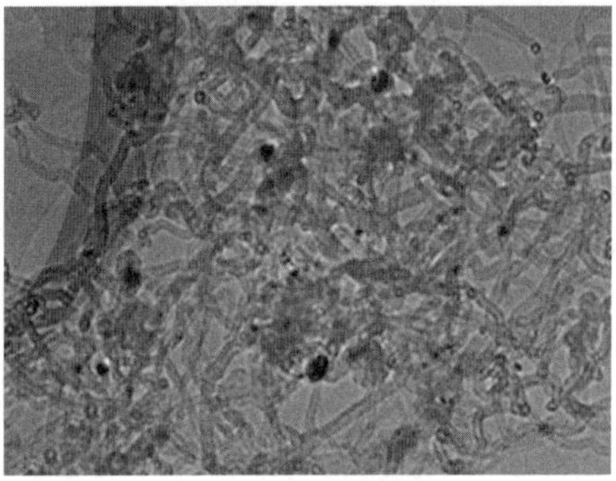

（d）360 min 反应后的 Ni/SiO_2-C 催化剂

图 3.54　Ni/SiO_2-F 和 Ni/SiO_2-C 纳米纤维

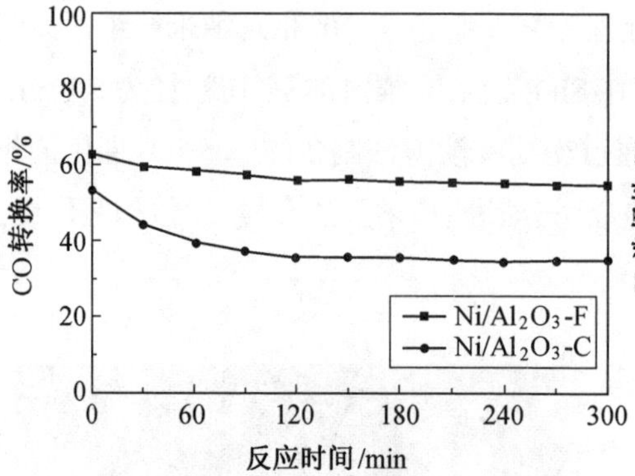

（e）CO 的转换率图

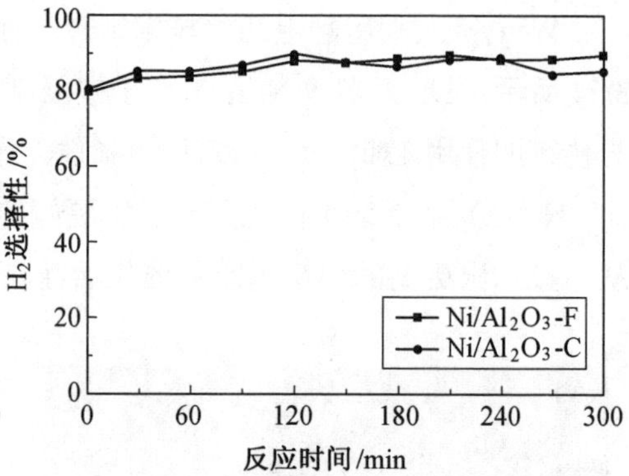

（f）H_2 的选择性随时间变化图

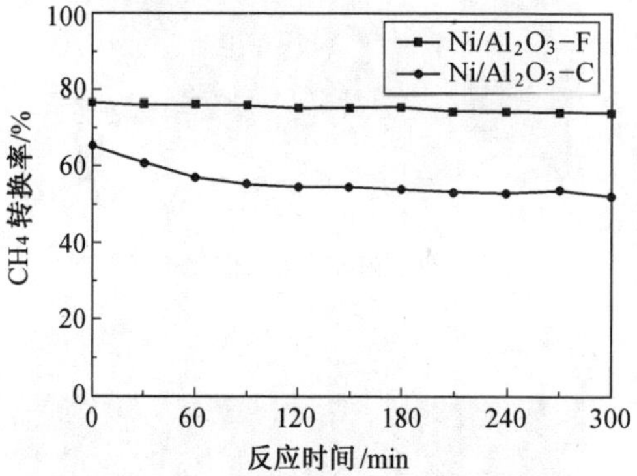

（g）CH_4 的转换率图

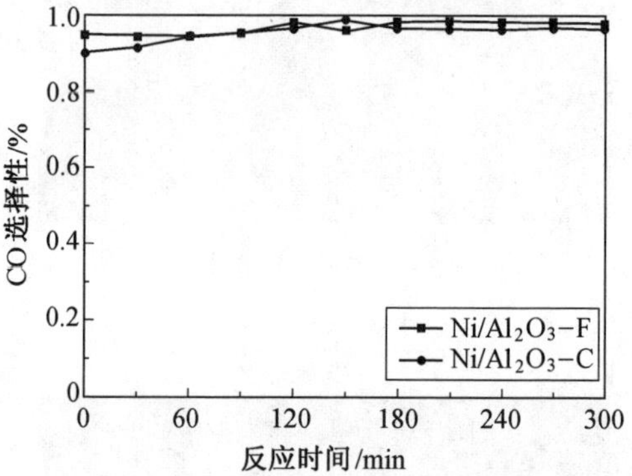

（h）CO 的选择性随时间变化图

续图 3.54

3.4.5 小结与展望

静电纺丝制备纳米纤维除了在上面提到的催化领域中发挥着重大的作用之外，还能应用于其他许多领域，如甲醇催化、尿素催化、甲醛催化以及甲烷催化等等。通过静电纺丝技术制得的一维纳米材料因其独特的结构，在催化领域有着巨大的应用前景。静电纺丝技术可以方便地进行结构设计、成分调控等，可以快速高效地制备催化活性好、稳定性能佳的催化材料。尽管在制备纳米纤维方面已经取得了很大的进展，但是仍需面对一些困难和挑战：

（1）首先，通过静电纺丝技术制得的一维纳米材料不仅可以提供更多的活性位点，同时能够改善电子和质量的传输，进而提高催化活性。制备中空纳米纤维通常采用同轴静电纺丝技术，然而目前该技术在对于内外径可控的大批量生产方面仍具有较大的挑战。

（2）其次，对于不同的催化反应，有许多催化机理尚不明确，仍需进一步研究和探索。

（3）最后，如何通过将静电纺丝与其他技术相结合，不同的物质对纳米纤维进行修饰，以进一步提高催化剂的催化活性和稳定性，也是未来的一大研究热点。

参考文献

[1] CHU S, MAJUMDAR A. Opportunities and challenges for a sustainable energy future [J]. Nature, 2012, 488: 294-303.

[2] COBO S, HEIDKAMP J, JACQUES P-A, et al. A Janus cobalt-basedcatalytic material for electrosplitting of water [J]. Nature Materials, 2012, 11: 802-807.

[3] HU X, BRUNSCHWIG B S, PETERS J C. Electrocatalytic hydrogen evolution at low overpotentials by cobalt macrocyclic glyoxime and tetraimine complexes[J]. Journal of the American Chemical Society, 2007, 129(29): 8988-8998.

[4] MORALES-GUIO C G, STERN L A, HU X. Nanostructured hydrotreating catalysts for electrochemical hydrogen evolution[J]. Chemical Society Reviews, 2014, 43(18): 6555-6569.

[5] SEH Z W, KIBSGAARD J, DICKENS C F, et al. Combining theory and experiment in electrocatalysis: insights into materials design[J]. Science, 2017, 355(6321): eaad4998.

[6] LI D, XIA Y. Electrospinning of nanofibers: reinventing the wheel?[J]. Advanced Materials, 2004, 16(14): 1151-1170.

[7] SUBBIAH T, BHAT G S, TOCK R W, et al. Electrospinning of nanofibers[J]. Journal of Applied Polymer Science, 2005, 96(2): 557-569.

[8] KAKORIA A, DEVI B, ANAND A, et al. Gallium oxide nanofibers for hydrogen evolution and oxygen reduction[J]. ACS Applied Nano Materials, 2018, 2: 64-74.

[9] ARICO A S, BRUCE P, SCROSATI B, et al. Nanostructured materials for advanced energy conversion and storage devices[J]. Materials for Sustainable Energy, 2011: 148-159.

[10] WANG Z, ZUO P, FAN L, et al. Facile electrospinning preparation of phosphorus and nitrogen dual-doped cobalt-based carbon nanofibers as bifunctional electrocatalyst[J]. Journal of Power Sources, 2016, 311: 68-80.

[11] LIN H, ZHANG W, SHI Z, et al. Electrospinning hetero-nanofibers of Fe_3C-Mo_2C/nitrogen-doped-carbon as efficient electrocatalysts for hydrogen evolution[J]. ChemSusChem, 2017, 10(12): 2597-2604.

[12] YANG P, ZHAO H, YANG Y, et al. Fabrication of N, P-codoped Mo_2C/carbon nanofibers via electrospinning as electrocatalyst for hydrogen evolution reaction[J]. ES Materials & Manufacturing, 2020, 7: 34-39.

[13] TONG J, LI Y, BO L, et al. CoP/N-doped carbon hollow spheres anchored on electrospinning core-shell N-doped carbon nanofibers as efficient electrocatalysts for water splitting[J]. ACS Sustainable Chemistry & Engineering, 2019, 7(20): 17432-17442.

[14] CHEN D, CHEN C, BAIYEE Z M, et al. Nonstoichiometric oxides as low-cost and highly-efficient oxygen reduction/evolution catalysts for low-temperature electrochemical devices[J]. Chemical Reviews, 2015, 115(18): 9869-9921.

[15] YAGI S, YAMADA I, TSUKASAKI H, et al. Covalency-reinforced oxygen evolution reaction catalyst[J]. Nature Communications, 2015, 6: 8249.

[16] ZHU Y, ZHOU W, ZHONG Y, et al. A perovskite nanorod as bifunctional electrocatalyst for overall water splitting[J]. Advanced Energy Materials, 2017, 7(8): 1602122.

[17] RHEEM Y, HAN Y, LEE K H, et al. Synthesis of hierarchical MoO_2/MoS_2 nanofibers for electrocatalytic hydrogen evolution[J]. Nanotechnology, 2017, 28(10): 105605.

[18] FANG Y, ZHANG T, WU Y, et al. Stacked Co_6W_6C nanocrystals anchored on N-doping carbon nanofibers with excellent electrocatalytic performance for HER in

wide-range pH[J]. International Journal of Hydrogen Energy, 2020, 45(3): 1901-1910.

[19] LI T, LUO G, LIU K, et al. Encapsulation of Ni_3Fe nanoparticles in N-doped carbon nanotube-grafted carbon nanofibers as high-efficiency hydrogen evolution electrocatalysts[J]. Advanced Functional Materials, 2018, 28(51): 1805828.

[20] LI M, ZHU Y, SONG N, et al. Fabrication of Pt nanoparticles on nitrogen-doped carbon/Ni nanofibers for improved hydrogen evolution activity[J]. Journal of colloid and interface science, 2018, 514: 199-207.

[21] LI Y, LI H, CAO K, et al. Electrospun three dimensional Co/CoP@ nitrogen-doped carbon nanofibers network for efficient hydrogen evolution[J]. Energy Storage Materials, 2018, 12: 44-53.

[22] ZHU Y, SONG L, SONG N, et al. Bifunctional and efficient CoS_2-C@ MoS_2 core-shell nanofiber electrocatalyst for water splitting[J]. ACS Sustainable Chemistry & Engineering, 2019, 7(3): 2899-2905.

[23] LU H, FAN W, HUANG Y, et al. Lotus root-like porous carbon nanofiber anchored with CoP nanoparticles as all-pH hydrogen evolution electrocatalysts[J]. Nano Research, 2018, 11(3): 1274-1284.

[24] ZHANG L, ZHU S, DONG S, et al. Co nanoparticles encapsulated in porous N-doped carbon nanofibers as an efficient electrocatalyst for hydrogen evolution reaction[J]. Journal of the Electrochemical Society, 2018, 165(15): J3271.

[25] CHEN N N, MO Q, HE L, et al. Heterostructured MoC-MoP/N-doped carbon nanofibers as efficient electrocatalysts for hydrogen evolution reaction[J]. Electrochimica Acta, 2019, 299: 708-716.

[26] GU H, FAN W, LIU T. Phosphorus-doped $NiCo_2S_4$ nanocrystals grown on electrospun carbon nanofibers as ultra-efficient electrocatalysts for the hydrogen evolution reaction[J]. Nanoscale Horizons, 2017, 2(5): 277-283.

[27] WEI L, LUO J, JIANG L, et al. $CoSe_2$ nanoparticles grown on carbon nanofibers derived from bacterial cellulose as an efficient electrocatalyst for hydrogen evolution reaction[J]. International Journal of Hydrogen Energy, 2018, 43(45): 20704-20711.

[28] CHINNAPPAN A, DONGXIAO J, JAYATHILAKA W, et al. Facile synthesis of electrospun C@ NiO/Ni nanofibers as an electrocatalyst for hydrogen evolution reaction[J]. International Journal of Hydrogen Energy, 2018, 43(32): 15217-15224.

[29] JI L, WANG J, TENG X, et al. N, P-doped molybdenum carbide nanofibers for efficient hydrogen production[J]. ACS Applied Materials & Interfaces, 2018, 10(17): 14632-14640.

[30] PATIL B, SATILMIS B, KHALILY M A, et al. Atomic layer deposition of $NiOOH/Ni(OH)_2$ on PIM-1-based N-doped carbon nanofibers for electrochemical water splitting in alkaline medium[J]. ChemSusChem, 2019, 12(7): 1469-1477.

[31] LI M, ZHU Y, WANG H, et al. Ni strongly coupled with Mo_2C encapsulated in nitrogen-doped carbon nanofibers as robust bifunctional catalyst for overall water splitting[J]. Advanced Energy Materials, 2019, 9(10): 1803185.

[32] WU W, ZHAO Y, LI S, et al. P doped MoS_2 nanoplates embedded in nitrogen doped carbon nanofibers as an efficient catalyst for hydrogen evolution reaction[J]. Journal of Colloid and Interface Science, 2019, 547: 291-298.

[33] RHEEM Y, PARK S H, YU S, et al. Electrospun hybrid MoS_2 nanofibers for high-efficiency electrocatalytic hydrogen evolution reaction[J]. Journal of The Electrochemical Society, 2020, 167(6): 066522.

[34] FABBRI E, HABEREDER A, WALTAR K, et al. Developments and perspectives of oxide-based catalysts for the oxygen evolution reaction[J]. Catalysis Science & Technology, 2014, 4(11): 3800-3821.

[35] JIN H, GUO C, LIU X, et al. Emerging two-dimensional nanomaterials for electrocatalysis[J]. Chemical Reviews, 2018, 118(13): 6337-6408.

[36] GUO Z, YE W, FANG X, et al. Amorphous cobalt-iron hydroxides as high-efficiency oxygen-evolution catalysts based on a facile electrospinning process[J]. Inorganic Chemistry Frontiers, 2019, 6(3): 687-693.

[37] WANG Y, ZHOU T, JIANG K, et al. Reduced mesoporous Co_3O_4 nanowires as efficient water oxidation electrocatalysts and supercapacitor electrodes[J]. Advanced Energy Materials, 2014, 4(16): 1400696.

[38] SANKAR S S, EDE S R, ANANTHARAJ S, et al. Electrospun cobalt-ZIF micro-fibers for efficient water oxidation under unique pH conditions[J]. Catalysis Science & Technology, 2019, 9(8): 1847-1856.

[39] GUO Z, YE W, FANG X, et al. Amorphous cobalt-iron hydroxides as high-efficiency oxygen-evolution catalysts based on a facile electrospinning process[J]. Inorganic Chemistry Frontiers, 2019, 6(3): 687-693.

[40] WANG X, LI Y, JIN T, et al. Electrospun thin-walled $CuCo_2O_4$@ C nanotubes as bifunctional oxygen electrocatalysts for rechargeable Zn-air batteries[J]. Nano Letters, 2017, 17(12): 7989-7994.

[41] YAN Y, XIA B Y, ZHAO B, et al. A review on noble-metal-free bifunctional heterogeneous catalysts for overall electrochemical water splitting[J]. Journal of Materials Chemistry A, 2016, 4(45): 17587-17603.

[42] PATIL B, SATILMIS B, KHALILY M A, et al. Atomic layer deposition of $NiOOH/Ni(OH)_2$ on PIM-1-based N-doped carbon nanofibers for electrochemical water splitting in alkaline medium[J]. ChemSusChem, 2019, 12(7): 1469-1477.

[43] CHEN H, HUANG X, ZHOU L J, et al. Electrospinning synthesis of bimetallic nickel-iron oxide/carbon composite nanofibers for efficient water oxidation electrocatalysis[J]. ChemCatChem, 2016, 8(5): 992-1000.

[44] CAO F, YANG X, SHEN C, et al. Electrospinning synthesis of transition metal alloy nanoparticles encapsulated in nitrogen-doped carbon layers as an advanced bifunctional oxygen electrode[J]. Journal of Materials Chemistry A, 2020, 8(15): 7245-7252.

[45] WEI P, SUN X, LIANG Q, et al. Enhanced oxygen evolution reaction activity by encapsulating nife alloy nanoparticles in nitrogen-doped carbon nanofibers[J]. ACS Applied Materials & Interfaces, 2020, 12(28): 31503-31513.

[46] ZHEN D, ZHAO B, SHIN H C, et al. Electrospun porous perovskite $La_{0.6}Sr_{0.4}Co_{1-x}Fe_xO_{3-\delta}$ nanofibers for efficient oxygen evolution reaction[J]. Advanced Materials Interfaces, 2017, 4(13): 1700146.

[47] LI T, LV Y, SU J, et al. Anchoring $CoFe_2O_4$ nanoparticles on N-doped carbon nanofibers for high-performance oxygen evolution reaction[J]. Advanced Science, 2017, 4(11): 1700226.

[48] SURENDRAN S, SHANMUGAPRIYA S, SIVANANTHAM A, et al. Electrospun carbon nanofibers encapsulated with NiCoP: a multifunctional electrode for supercapattery and oxygen reduction, oxygen evolution, and hydrogen evolution reactions[J]. Advanced Energy Materials, 2018, 8(20): 1800555.

[49] LI M, ZHU Y, WANG H, et al. Ni strongly coupled with Mo_2C encapsulated in nitrogen-doped carbon nanofibers as robust bifunctional catalyst for overall water splitting[J]. Advanced Energy Materials, 2019, 9(10): 1803185.

[50] ALJABOUR A. Long-lasting electrospun Co_3O_4 nanofibers for electrocatalytic oxygen evolution reaction[J]. ChemistrySelect, 2020, 5(25): 7482-7487.

[51] GUO D, WANG J, ZHANG L, et al. Strategic atomic layer deposition and electrospinning of cobalt sulfide/nitride composite as efficient bifunctional electrocatalysts for overall water splitting[J]. Small, 2020: 2002432.

[52] LI C, ZHANG Z, WU M, et al. FeCoNi ternary alloy embedded mesoporous carbon nanofiber: an efficient oxygen evolution catalyst for rechargeable zinc-air battery[J]. Materials Letters, 2019, 238: 138-142.

[53] LI G, HOU S, GUI L, et al. Carbon quantum dots decorated $Ba_{0.5}Sr_{0.5}Co_{0.8}Fe_{0.2}O_{3-\delta}$ perovskite nanofibers for boosting oxygen evolution reaction[J]. Applied Catalysis B: Environmental, 2019, 257: 117919.

[54] LI T, LI S, LIU Q, et al. Hollow Co_3O_4/CeO_2 heterostructures in situ embedded in N-doped carbon nanofibers enable outstanding oxygen evolution[J]. ACS Sustainable Chemistry & Engineering, 2019, 7(21): 17950-17957.

[55] LI W, LI M, WANG C, et al. Fe doped CoO/C nanofibers towards efficient oxygen evolution reaction[J]. Applied Surface Science, 2020, 506: 144680.

[56] TAO J, ZHANG Y, WANG S, et al. Activating three-dimensional networks of Fe@Ni nanofibers via fast surface modification for efficient overall water splitting[J]. ACS Applied Materials & Interfaces, 2019, 11(20): 18342-18348.

[57] SUN M, LIN D, WANG S, et al. Facile synthesis of $CoFe_2O_4$-$CoFe_x$/C nanofibers electrocatalyst for the oxygen evolution reaction[J]. Journal of the Electrochemical Society, 2019, 166(10): H412.

[58] ZHAO Y, ZHANG J, WU W, et al. Cobalt-doped MnO_2 ultrathin nanosheets with abundant oxygen vacancies supported on functionalized carbon nanofibers for efficient oxygen evolution[J]. Nano Energy, 2018, 54: 129-137.

[59] ZHANG Z, LIANG X, LI J, et al. Interfacial engineering of NiO/$NiCo_2O_4$ porous nanofibers as efficient bifunctional catalysts for rechargeable zinc-air batteries[J]. ACS Applied Materials & Interfaces, 2020, 12 (19): 21661-21669.

[60] ZHAO Y, ZHANG J, GUO X, et al. Fe_3C@ nitrogen doped CNT arrays aligned on nitrogen functionalized carbon nanofibers as highly efficient catalysts for the oxygen evolution reaction[J]. Journal of Materials Chemistry A, 2017, 5(37): 19672-19679.

[61] JI C, GUANG Y, ILANGO P R, et al. Molybdenum carbide-embedded multichannel hollow carbon nanofibers as bifunctional catalysts for water splitting[J]. Chemistry-an Asian Journal, 2020, 15 (13): 1957-1962.

[62] REN H, KOSHY P, CHEN W F, et al. Photocatalytic materials and technologies for air purification[J]. Journal of Hazardous Materials, 2017, 325: 340-366.

[63] ARSHAD M K, ADZHRI R, FATHIL M F, et al. Field-effect transistor-integration with TiO_2 nanoparticles for sensing of cardiac troponin I biomarker[J]. Journal of Nanoscience and Nanotechnology, 2018, 18(8): 5283-5291.

[64] SEO H O, WOO T G, PARK E J, et al. Enhanced photo-catalytic activity of TiO_2 films by removal of surface carbon impurities; the role of water vapor[J]. Applied Surface Science, 2017, 420: 808-816.

[65] PAL B, BAKR Z H, KRISHNAN S G, et al. Large scale synthesis of 3D nanoflowers of SnO_2/TiO_2 composite via electrospinning with synergisticproperties[J]. Mater. Lett, 2018, 225: 117-121.

[66] LEE C G, JAVED H, ZHANG D, et al. Porous electrospun fibers embedding TiO_2 for adsorption and photocatalytic degradation of water pollutants[J]. Environmental Science & Technology, 2018, 52(7): 4285-4293.

[67] ZHAO Y, LIU J, SHI L, et al. Surfactant-free synthesis uniform $Ti_{1-x}Sn_xO_2$ nanocrystal colloids and their photocatalytic performance[J]. Applied Catalysis B: Environmental, 2010, 100(1-2): 68-76.
[68] LIU Z, SUN D D, GUO P, et al. An efficient bicomponent TiO_2/SnO_2 nanofiber photocatalyst fabricated by electrospinning with a side-by-side dual spinneret method[J]. Nano letters, 2007, 7(4): 1081-1085.
[69] PENG X, SANTULLI A C, SUTTER E, et al. Fabrication and enhanced photocatalytic activity of inorganic core-shell nanofibers produced by coaxial electrospinning[J]. Chemical Science, 2012, 3(4): 1262-1272.
[70] LAVANYA T, SATHEESH K, DUTTA M, et al. Superior photocatalytic performance of reduced graphene oxide wrapped electrospun anatase mesoporous TiO_2 nanofibers[J]. Journal of Alloys and Compounds, 2014, 615: 643-650.
[71] NAINANI R K, THAKUR P. Facile synthesis of TiO_2-RGO composite with enhanced performance for the photocatalytic mineralization of organic pollutants[J]. Water Science and Technology, 2016, 73(8): 1927-1936.
[72] LIANG M, ZHI L. Graphene-based electrode materials for rechargeable lithium batteries[J]. Journal of Materials Chemistry, 2009, 19(33): 5871-5878.
[73] ZHANG L, ZHANG Q, XIE H, et al. Electrospun titania nanofibers segregated by graphene oxide for improved visible light photocatalysis[J]. Applied Catalysis B: Environmental, 2017, 201: 470-478.
[74] NASR M, BALME S, EID C, et al. Enhanced visible-light photocatalytic performance of electrospun rGO/TiO_2 composite nanofibers[J]. The Journal of Physical Chemistry C, 2017, 121(1): 261-269.
[75] WEN J, XIE J, CHEN X, et al. A review on g-C_3N_4-based photocatalysts[J]. Applied Surface Science, 2017, 391: 72-123.
[76] CHAI B, LIAO X, SONG F, et al. Fullerene modified C_3N_4 composites with enhanced photocatalytic activity under visible light irradiation[J]. Dalton Transactions, 2014, 43(3): 982-989.

[77] CHAI B, YAN J, WANG C, et al. Enhanced visible light photocatalytic degradation of rhodamine B over phosphorus doped graphitic carbon nitride[J]. Applied Surface Science, 2017, 391: 376-383.

[78] CAO S, LOW J, YU J, et al. Polymeric photocatalysts based on graphitic carbon nitride[J]. Advanced Materials, 2015, 27(13): 2150-2176.

[79] WANG C, HU L, CHAI B, et al. Enhanced photocatalytic activity of electrospun nanofibrous TiO_2/g-C_3N_4 heterojunction photocatalyst under simulated solar light[J]. Applied Surface Science, 2018, 430: 243-252.

[80] TANG Q, MENG X, WANG Z, et al. One-step electrospinning synthesis of TiO_2/g-C_3N_4 nanofibers with enhanced photocatalytic properties[J]. Applied Surface Science, 2018, 430: 253-262.

[81] ZNAD H, ANG M H, TADE M O. Ta/TiO_2-and Nb/TiO_2-mixed oxides as efficient solar photocatalysts: preparation, characterization, and photocatalytic activity[J]. International Journal of Photoenergy, 2012, 2012: 548158.

[82] UMEBAYASHI T, YAMAKI T, YAMAMOTO S, et al. Sulfur-doping of rutile-titanium dioxide by ion implantation: photocurrent spectroscopy and first-principles band calculation studies[J]. Journal of Applied Physics, 2003, 93(9): 5156-5160.

[83] MA D, XIN Y, GAO M, et al. Fabrication and photocatalytic properties of cationic and anionic S-doped TiO_2 nanofibers by electrospinning[J]. Applied Catalysis B: Environmental, 2014, 147: 49-57.

[84] SINGH N, PRAKASH J, MISRA M, et al. Dual functional Ta-doped electrospun TiO_2 nanofibers with enhanced photocatalysis and SERS detection for organic compounds[J]. ACS Applied Materials & Interfaces, 2017, 9(34): 28495-28507.

[85] RYU S Y, CHUNG J W, KWAK S Y. Dependence of photocatalytic and antimicrobial activity of electrospun polymeric nanofiber composites on the positioning of Ag-TiO_2 nanoparticles[J]. Composites Science and Technology, 2015, 117: 9-17.

[86] AWAZU K, FUJIMAKI M, ROCKSTUHL C, et al. A plasmonic photocatalyst consisting of silver nanoparticles embedded in titanium dioxide[J]. Journal of the American Chemical Society, 2008, 130(5): 1676-1680.

[87] HU C, LAN Y, QU J, et al. Ag/AgBr/TiO_2 visible light photocatalyst for destruction of azodyes and bacteria[J]. The Journal of Physical Chemistry B, 2006, 110(9): 4066-4072.

[88] NALBANDIAN M J, ZHANG M, SANCHEZ J, et al. Synthesis and optimization of Ag-TiO_2 composite nanofibers for photocatalytic treatment of impaired water sources[J]. Journal of Hazardous Materials, 2015, 299: 141-148.

[89] LIU Y, HOU C, JIAO T, et al. Self-assembled AgNP-containing nanocomposites constructed by electrospinning as efficient dye photocatalyst materials for wastewater treatment[J]. Nanomaterials, 2018, 8(1): 35.

[90] LIU L, LIU Z, BAI H, et al. Concurrent filtration and solar photocatalytic disinfection/degradation using high-performance Ag/TiO_2 nanofiber membrane[J]. Water Research, 2012, 46(4): 1101-1112.

[91] BIAN Z, TACHIKAWA T, ZHANG P, et al. Au/TiO_2 superstructure-based plasmonic photocatalysts exhibiting efficient charge separation and unprecedented activity[J]. Journal of the American Chemical Society, 2014, 136(1): 458-465.

[92] PARAMASIVAM I, MACAK J M, SCHMUKI P. Photocatalytic activity of TiO_2 nanotube layers loaded with Ag and Au nanoparticles[J]. Electrochemistry Communications, 2008, 10(1): 71-75.

[93] KUMAR L, SINGH S, HORECHYY A, et al. Hollow Au@ TiO_2 porous electrospun nanofibers for catalytic applications[J]. RSC Advances, 2020, 10(11): 6592-6602.

[94] BISCARAT J, BECHELANY M, POCHAT-BOHATIER C, et al. Graphene-like BN/gelatin nanobiocomposites for gas barrier applications[J]. Nanoscale, 2015, 7(2): 613-618.

[95] NASR M, VITER R, EID C, et al. Enhanced photocatalytic performance of novel electrospun BN/TiO_2 composite nanofibers[J]. New Journal of Chemistry, 2017, 41(1): 81-89.

[96] BOUVY C, MARINE W, SPORKEN R, et al. Photoluminescence properties and quantum size effect of ZnO nanoparticles confined inside a faujasite X zeolite matrix[J]. Chemical Physics Letters, 2006, 428(4-6): 312-316.

[97] KUMAR D P, SHANKAR M V, KUMARI M M, et al. Nano-size effects on CuO/TiO_2 catalysts for highly efficient H_2 production under solar light irradiation[J]. Chemical Communications, 2013, 49(82): 9443-9445.

[98] KAYACI F, VEMPATI S, OZGIT-AKGUN C, et al. Selective isolation of the electron or hole in photocatalysis: ZnO-TiO_2 and TiO_2-ZnO core-shell structured heterojunction nanofibers via electrospinning and atomic layer deposition[J]. Nanoscale, 2014, 6(11): 5735-5745.

[99] YAR A, HASPULAT B, ÜSTÜN T, et al. Electrospun TiO_2/ZnO/PAN hybrid nanofiber membranes with efficient photocatalytic activity[J]. RSC Advances, 2017, 7(47): 29806-29814.

[100] AN S, JOSHI B N, LEE M W, et al. Electrospun graphene-ZnO nanofiber mats for photocatalysis applications[J]. Applied surface science, 2014, 294: 24-28.

[101] DAS P P, ROY A, TATHAVADEKAR M, et al. Photovoltaic and photocatalytic performance of electrospun Zn_2SnO_4 hollow fibers[J]. Applied Catalysis B: Environmental, 2017, 203: 692-703.

[102] REN P, FAN H, WANG X. Electrospun nanofibers of $ZnO/BaTiO_3$ heterostructures with enhanced photocatalytic activity[J]. Catalysis Communications, 2012, 25: 32-35.

[103] LAINE D F, CHENG I F. The destruction of organic pollutants under mild reaction conditions: a review[J]. Microchemical Journal, 2007, 85(2): 183-193.

[104] LAKSHMI K, KADIRVELU K, MOHAN P S. Chemically modified electrospun nanofiber for high adsorption and effective photocatalytic decontamination of organophosphorus compounds[J]. Journal of Chemical Technology & Biotechnology, 2019, 94(10): 3190-3200.

[105] LV P, ZHENG M, WANG X, et al. Subsolidus phase relationships and photocatalytic properties in the ternary system TiO_2-Bi_2O_3-V_2O_5[J]. Journal of

Alloys and Compounds, 2014, 583: 285-290.

[106] ZHOU X, SHAO C, YANG S, et al. Heterojunction of g-C_3N_4/BiOI immobilized on flexible electrospun polyacrylonitrile nanofibers: facile preparation and enhanced visible photocatalytic activity for floating photocatalysis[J]. ACS Sustainable Chemistry & Engineering, 2018, 6(2): 2316-2323.

[107] WANG K, SHAO C, LI X, et al. Hierarchical heterostructures of p-type BiOCl nanosheets on electrospun n-type TiO_2 nanofibers with enhanced photocatalytic activity[J]. Catalysis Communications, 2015, 67: 6-10.

[108] 杨成武. g-C_3N_4基光催化剂材料的异质结构建与结构缺陷调控[D]. 秦皇岛：燕山大学, 2019.

[109] CHOI S K, KIM S, LIM S K, et al. Photocatalytic comparison of TiO_2 nanoparticles and electrospun TiO_2 nanofibers: effects of mesoporosity and interparticle charge transfer[J]. The Journal of Physical Chemistry C, 2010, 114(39): 16475-16480.

[110] HOU H, YUAN Y, SHANG M, et al. Significantly improved photocatalytic hydrogen production activity over ultrafine mesoporous TiO_2 nanofibers photocatalysts[J]. Chemistry Select, 2018, 3(36): 10126-10132.

[111] LIU K, ZHANG Z, LU N, et al. In situ generation of copper species nanocrystals in TiO_2 electrospun nanofibers: a multi-hetero-junction photocatalyst for highly efficient water reduction[J]. ACS Sustainable Chemistry & Engineering, 2018, 6(2): 1934-1940.

[112] NASIR M S, YANG G, AYUB I, et al. Tin diselinide a stable co-catalyst coupled with branched TiO_2 fiber and g-C_3N_4 quantum dots for photocatalytic hydrogen evolution[J]. Applied Catalysis B: Environmental, 2020: 118900.

[113] ONG C B, NG L Y, MOHAMMAD A W. A review of ZnO nanoparticles as solar photocatalysts: synthesis, mechanisms and applications[J]. Renewable and Sustainable Energy Reviews, 2018, 81: 536-551.

[114] ZHOU K, LIU M, YE X, et al. Electrospun highly crystalline ZnO nanofibers: super-efficient and stable photocatalytic hydrogen production activity[J]. Chemistry Select, 2020, 5(22): 6691-6696.

[115] SUN D, SHI J W, MA D, et al. CdS/ZnS/ZnO ternary heterostructure nanofibers fabricated by electrospinning for excellent photocatalytic hydrogen evolution without co-catalyst[J]. Chinese Journal of Catalysis, 2020, 41(9): 1421-1429.

[116] SHIMADA Y, WATANABE Y, SUGIHARA A, et al. Enzymatic alcoholysis for biodiesel fuel production and application of the reaction to oil processing[J]. Journal of Molecular Catalysis B: Enzymatic, 2002, 17(3-5): 133-142.

[117] SAGIROGLU A. Conversion of sunflower oil to biodiesel by alcoholysis using immobilized lipase[J]. Artificial Cells, Blood Substitutes, and Biotechnology, 2008, 36(2): 138-149.

[118] DE STEFANO G, PIACQUADIO P, SCIANCALEPORE V. Metal-chelate regenerable carriers in food processing[J]. Biotechnology Techniques, 1996, 10(11): 857-860.

[119] ODACI D, TIMUR S, PAZARLIOĞLU N, et al. Effects of mediators on the laccase biosensor response in paracetamol detection[J]. Biotechnology and Applied Biochemistry, 2006, 45(1): 23-28.

[120] ZHANG J, YE P, CHEN S, et al. Removal of pentachlorophenol by immobilized horseradish peroxidase[J]. International Biodeterioration & Biodegradation, 2007, 59(4): 307-314.

[121] TANG D, YUAN R, CHAI Y. Direct electrochemical immunoassay based on immobilization of protein-magnetic nanoparticle composites on to magnetic electrode surfaces by sterically enhanced magnetic field force[J]. Biotechnology Letters, 2006, 28(8): 559-565.

[122] MARTIN M T, PLOU F J, ALCALDE M, et al. Immobilization on eupergit C of cyclodextrin glucosyltransferase (CGTase) and properties of the immobilized biocatalyst[J]. Journal of Molecular Catalysis B: Enzymatic, 2003, 21(4-6): 299-308.

[123] NELSON J M, GRIFFIN E G. Adsorption of invertase[J]. Journal of the American Chemical Society, 1916, 38(5): 1109-1115.

[124] SHELDON R A, VAN PELT S. Enzyme immobilisation in biocatalysis: why, what

and how[J]. Chemical Society Reviews, 2013, 42(15): 6223-6235.

[125] KIM B C, NAIR S, KIM J, et al. Preparation of biocatalytic nanofibres with high activity and stability via enzyme aggregate coating on polymer nanofibres[J]. Nanotechnology, 2005, 16(7): S382.

[126] CAO L, VAN LANGEN L, SHELDON R A. Immobilised enzymes: carrier-bound or carrier-free?[J]. Current Opinion in Biotechnology, 2003, 14(4): 387-394.

[127] JIA H, ZHU G, VUGRINOVICH B, et al. Enzyme-carrying polymeric nanofibers prepared via electrospinning for use as unique biocatalysts[J]. Biotechnology Progress, 2002, 18(5): 1027-1032.

[128] KIM J, GRATE J W, WANG P. Nanostructures for enzyme stabilization[J]. Chemical Engineering Science, 2006, 61(3): 1017-1026.

[129] WU L, YUAN X, SHENG J. Immobilization of cellulase in nanofibrous PVA membranes by electrospinning[J]. Journal of Membrane Science, 2005, 250(1-2): 167-173.

[130] HUANG X J, CHEN P C, HUANG F, et al. Immobilization of Candida rugosa lipase on electrospun cellulose nanofiber membrane[J]. Journal of Molecular Catalysis B: Enzymatic, 2011, 70(3-4): 95-100.

[131] WANG Z G, WAN L S, LIU Z M, et al. Enzyme immobilization on electrospun polymer nanofibers: an overview[J]. Journal of Molecular Catalysis B: Enzymatic, 2009, 56(4): 189-195.

[132] MORADZADEGAN A, RANAEI-SIADAT S O, EBRAHIM-HABIBI A, et al. Immobilization of acetylcholinesterase in nanofibrous PVA/BSA membranes by electrospinning[J]. Engineering in Life Sciences, 2010, 10(1): 57-64.

[133] ZHANG L, LUO J, MENKHAUS T J, et al. Antimicrobial nano-fibrous membranes developed from electrospun polyacrylonitrile nanofibers[J]. Journal of Membrane Science, 2011, 369(1-2): 499-505.

[134] XU R, CHI C, LI F, et al. Laccase-polyacrylonitrile nanofibrous membrane: highly immobilized, stable, reusable, and efficacious for 2, 4, 6-trichlorophenol removal[J]. ACS Applied Materials & Interfaces, 2013, 5(23): 12554-12560.

[135] LI S F, CHEN J P, WU W T. Electrospun polyacrylonitrile nanofibrous membranes for lipase immobilization[J]. Journal of Molecular Catalysis B: Enzymatic, 2007, 47(3-4): 117-124.

[136] SAKAI S, LIU Y, YAMAGUCHI T, et al. Production of butyl-biodiesel using lipase physically-adsorbed onto electrospun polyacrylonitrile fibers[J]. Bioresource technology, 2010, 101(19): 7344-7349.

[137] LI Y, QUAN J, BRANFORD-WHITE C, et al. Electrospun polyacrylonitrile-glycopolymer nanofibrous membranes for enzyme immobilization[J]. Journal of Molecular Catalysis B: Enzymatic, 2012, 76: 15-22.

[138] WU D, FENG Q, XU T, et al. Electrospun blend nanofiber membrane consisting of polyurethane, amidoxime polyarcylonitrile, and β-cyclodextrin as high-performance carrier/support for efficient and reusable immobilization of laccase[J]. Chemical Engineering Journal, 2018, 331: 517-526.

[139] WU L, YUAN X, SHENG J. Immobilization of cellulase in nanofibrous PVA membranes by electrospinning[J]. J. Membr. Sci., 2005, 250:167-173.

[140] CHRONAKIS I S. Micro-and nano-fibers by electrospinning technology[M]. Boston: William Andrew Publishing, 2010: 264-286.

[141] PORTO M D A, DOS SANTOS J P, HACKBART H, et al. Immobilization of α-amylase in ultrafine polyvinyl alcohol (PVA) fibers via electrospinning and their stability on different substrates[J]. International Journal of Biological macromolecules, 2019, 126: 834-841.

[142] CHANDRAWATI R, OLESEN M T J, MARINI T C C, et al. Enzyme prodrug therapy engineered into electrospun fibers with embedded liposomes for controlled, localized synthesis of therapeutics[J]. Advanced Healthcare Materials, 2017, 6(17): 1700385.

[143] DOS SANTOS J P, DA ROSA ZAVAREZE E, DIAS A R G, et al. Immobilization of xylanase and xylanase–β-cyclodextrin complex in polyvinyl alcohol via electrospinning improves enzyme activity at a wide pH and temperature range[J]. International Journal of Biological Macromolecules, 2018, 118: 1676-1684.

[144] SAALLAH S, NAIM M N, LENGGORO I W, et al. Immobilisation of cyclodextrin glucanotransferase into polyvinyl alcohol (PVA) nanofibres via electrospinning[J]. Biotechnology Reports, 2016, 10: 44-48.

[145] JIA H, ZHU G, VUGRINOVICH B, et al. Enzyme-carrying polymeric nanofibers prepared via electrospinning for use as unique biocatalysts[J]. Biotechnology Progress, 2002, 18(5): 1027-1032.

[146] MARTROU G, LÉONETTI M, GIGMES D, et al. One-step preparation of surface modified electrospun microfibers as suitable supports for protein immobilization[J]. Polymer Chemistry, 2017, 8(11): 1790-1796.

[147] LIANG W, LV M, YANG X. The effect of humidity on formaldehyde emission parameters of a medium-density fiberboard: experimental observations and correlations[J]. Building and Environment, 2016, 101: 110-115.

[148] JIANG C, LI D, ZHANG P, et al. Formaldehyde and volatile organic compound (VOC) emissions from particleboard: Identification of odorous compounds and effects of heat treatment[J]. Building and Environment, 2017, 117: 118-126.

[149] RONG S, ZHANG P, YANG Y, et al. MnO_2 framework for instantaneous mineralization of carcinogenic airborne formaldehyde at room temperature[J]. Acs Catalysis, 2017, 7(2): 1057-1067.

[150] NAKAMURA H, KATO K, MASUDA Y, et al. Activity of formaldehyde dehydrogenase on titanium dioxide films with different crystallinities[J]. Applied Surface Science, 2015, 329: 262-268.

[151] ZHOU P, YU J, NIE L, et al. Dual-dehydrogenation-promoted catalytic oxidation of formaldehyde on alkali-treated Pt clusters at room temperature[J]. Journal of Materials Chemistry A, 2015, 3(19): 10432-10438.

[152] DUAN Y, SONG S, CHENG B, et al. Effects of hierarchical structure on the performance of tin oxide-supported platinum catalyst for room-temperature formaldehyde oxidation[J]. Chinese Journal of Catalysis, 2017, 38(2): 199-206.

[153] YAN Z, XU Z, YU J, et al. Effect of microstructure and surface hydroxyls on the catalytic activity of Au/AlOOH for formaldehyde removal at room temperature[J].

Journal of Colloid and Interface Science, 2017, 501: 164-174.

[154] ZHANG C, LI Y, WANG Y, et al. Sodium-promoted Pd/TiO_2 for catalytic oxidation of formaldehyde at ambient temperature[J]. Environmental Science & Technology, 2014, 48(10): 5816-5822.

[155] HUANG H, YE X, HUANG H, et al. Mechanistic study on formaldehyde removal over Pd/TiO_2 catalysts: oxygen transfer and role of water vapor[J]. Chemical Engineering Journal, 2013, 230: 73-79.

[156] XU F, LE Y, CHENG B, et al. Effect of calcination temperature on formaldehyde oxidation performance of Pt/TiO_2 nanofiber composite at room temperature[J]. Applied Surface Science, 2017, 426: 333-341.

[157] NIE L, YU J, FU J. Complete decomposition of formaldehyde at room temperature over a platinum-decorated hierarchically porous electrospun titania nanofiber mat[J]. ChemCatChem, 2014, 6(7): 1983-1989.

[158] BAI B, ARANDIYAN H, LI J. Comparison of the performance for oxidation of formaldehyde on nano-Co_3O_4, 2D-Co_3O_4, and 3D-Co_3O_4 catalysts[J]. Applied Catalysis B: Environmental, 2013, 142: 677-683.

[159] WANG Y, ZHU A, CHEN B, et al. Three-dimensional ordered mesoporous Co-Mn oxide: a highly active catalyst for "storage-oxidation" cycling for the removal of formaldehyde[J]. Catalysis Communications, 2013, 36: 52-57.

[160] WU Y, MA M, ZHANG B, et al. Controlled synthesis of porous Co_3O_4 nanofibers by spiral electrospinning and their application for formaldehyde oxidation[J]. RSC Advances, 2016, 6(104): 102127-102133.

[161] SIMKA W, PIOTROWSKI J, ROBAK A, et al. Electrochemical treatment of aqueous solutions containing urea[J]. Journal of Applied Electrochemistry, 2009, 39(7): 1137-1143.

[162] ROLLINSON A N, RICKETT G L, LEA-LANGTON A, et al. Hydrogen from urea-water and ammonia-water solutions[J]. Applied Catalysis B: Environmental, 2011, 106(3-4): 304-315.

[163] ONGLEY E D. Control of water pollution from agriculture[M]. Rome: Food and Agriculture Organization of the United Nations, 1996.

[164] MAHALIK K, SAHU J N, PATWARDHAN A V, et al. Kinetic studies on hydrolysis of urea in a semi-batch reactor at atmospheric pressure for safe use of ammonia in a power plant for flue gas conditioning[J]. Journal of Hazardous Materials, 2010, 175(1-3): 629-637.

[165] XUE C, WILSON L D. Kinetic study on urea uptake with chitosan based sorbent materials[J]. Carbohydrate Polymers, 2016, 135: 180-186.

[166] KRAJEWSKA B, UREASES I. Functional, catalytic and kinetic properties: a review[J]. Journal of Molecular Catalysis B: Enzymatic, 2009, 59(1-3): 9-21.

[167] VON AHNEN M, PEDERSEN L F, PEDERSEN P B, et al. Degradation of urea, ammonia and nitrite in moving bed biofilters operated at different feed loadings[J]. Aquacultural Engineering, 2015, 69: 50-59.

[168] BERNHARD A M, PEITZ D, ELSENER M, et al. Hydrolysis and thermolysis of urea and its decomposition byproducts biuret, cyanuric acid and melamine over anatase TiO_2[J]. Applied Catalysis B: Environmental, 2012, 115: 129-137.

[169] SHEN S, LI M, LI B, et al. Catalytic hydrolysis of urea from wastewater using different aluminas by a fixed bed reactor[J]. Environmental Science and Pollution Research, 2014, 21(21): 12563-12568.

[170] ZHU B, LIANG Z, ZOU R. Designing advanced catalysts for energy conversion based on urea oxidation reaction[J]. Small, 2020, 16(7): 1906133.

[171] LIU Q, XIE L, QU F, et al. A porous Ni_3N nanosheet array as a high-performance non-noble-metal catalyst for urea-assisted electrochemical hydrogen production[J]. Inorganic Chemistry Frontiers, 2017, 4(7): 1120-1124.

[172] WU M S, LIN G W, YANG R S. Hydrothermal growth of vertically-aligned ordered mesoporous nickel oxide nanosheets on three-dimensional nickel framework for electrocatalytic oxidation of urea in alkaline medium[J]. Journal of Power Sources, 2014, 272: 711-718.

[173] MILLER A T, HASSLER B L, BOTTE G G. Rhodium electrodeposition on nickel

electrodes used for urea electrolysis[J]. Journal of Applied Electrochemistry, 2012, 42(11): 925-934.

[174] BOGGS B K, KING R L, BOTTE G G. Urea electrolysis: direct hydrogen production from urine[J]. Chemical Communications, 2009 (32): 4859-4861.

[175] WANG L, REN L, WANG X, et al. Multivariate MOF-templated pomegranate-like Ni/C as efficient bifunctional electrocatalyst for hydrogen evolution and urea oxidation[J]. ACS Applied Materials & Interfaces, 2018, 10(5): 4750-4756.

[176] YU M, CHEN J, LIU J, et al. Mesoporous $NiCo_2O_4$ nanoneedles grown on 3D graphene-nickel foam for supercapacitor and methanol electro-oxidation[J]. Electrochimica Acta, 2015, 151: 99-108.

[177] ASGARI M, MARAGHEH M G, DAVARKHAH R, et al. Electrocatalytic oxidation of methanol on the nickel-cobalt modified glassy carbon electrode in alkaline medium[J]. Electrochimica Acta, 2012, 59: 284-289.

[178] ZHAO Y, YANG X, TIAN J, et al. Methanol electro-oxidation on Ni@ Pd core-shell nanoparticles supported on multi-walled carbon nanotubes in alkaline media[J]. International journal of hydrogen energy, 2010, 35(8): 3249-3257.

[179] BARAKAT N A M, ABDELKAREEM M A, EL-NEWEHY M, et al. Influence of the nanofibrous morphology on the catalytic activity of NiO nanostructures: an effective impact toward methanol electrooxidation[J]. Nanoscale Research Letters, 2013, 8(1): 1-6.

[180] BARAKAT N A M, YASSIN M A, YASIN A S, et al. Influence of nitrogen doping on the electrocatalytic activity of Ni-incorporated carbon nanofibers toward urea oxidation[J]. International Journal of Hydrogen Energy, 2017, 42(34): 21741-21750.

[181] LIU D, LI W, LI L, et al. Facile preparation of Ni nanowire embedded nitrogen and sulfur dual-doped carbon nanofibers and its superior catalytic activity toward urea oxidation[J]. Journal of Colloid and Interface Science, 2018, 529: 337-344.

[182] MUKAIBO H, MOMMA T, OSAKA T. Changes of electro-deposited Sn-Ni alloy thin film for lithium ion battery anodes during charge discharge cycling[J]. Journal

of Power Sources, 2005, 146(1-2): 457-463.

[183] JIANG D, MA X, FU Y. High-performance Sn-Ni alloy nanorod electrodes prepared by electrodeposition for lithium ion rechargeable batteries[J]. Journal of Applied Electrochemistry, 2012, 42(8): 555-559.

[184] BARAKAT N A M, AMEN M T, AL-MUBADDEL F S, et al. NiSn nanoparticle-incorporated carbon nanofibers as efficient electrocatalysts for urea oxidation and working anodes in direct urea fuel cells[J]. Journal of Advanced Research, 2019, 16: 43-53.

[185] BARAKAT N A M, EL-NEWEHY M H, YASIN A S, et al. Ni&Mn nanoparticles-decorated carbon nanofibers as effective electrocatalyst for urea oxidation[J]. Applied Catalysis A: General, 2016, 510: 180-188.

[186] UCHIDA H, MIZUNO Y, WATANABE M. Suppression of methanol crossover and distribution of ohmic resistance in Pt-dispersed PEMs under DMFC operation experimental analyses[J]. Journal of the Electrochemical Society, 2002, 149(6): A682.

[187] LIU H, SONG C, ZHANG L, et al. A review of anode catalysis in the direct methanol fuel cell[J]. Journal of Power Sources, 2006, 155(2): 95-110.

[188] WINTER M, BRODD R J. What are batteries, fuel cells, and supercapacitors[J]. Chemical Reviews, 2005, 105(3): 1021.

[189] LEE J K, CHOI J, KANG S J, et al. Influence of copper oxide modification of a platinum cathode on the activity of direct methanol fuel cell[J]. Electrochimica Acta, 2007, 52(6): 2272-2276.

[190] RALPH T R, HOGARTH M P. Catalysis for low temperature fuel cells[J]. Platinum Metals Review, 2002, 46(3): 117-135.

[191] LOVIĆ J. The kinetics and mechanism of methanol oxidation on Pt and PtRu catalysts in alkaline and acid media[J]. Journal of the Serbian Chemical Society, 2007, 72(7): 709-712.

[192] KIM J M, JOH H I, JO S M, et al. Preparation and characterization of Pt nanowire by electrospinning method for methanol oxidation[J]. Electrochimica Acta, 2010,

55(16): 4827-4835.

[193] PARK K W, HAN S B, LEE J M. Photo (UV)-enhanced performance of Pt-TiO_2 nanostructure electrode for methanol oxidation[J]. Electrochemistry Communications, 2007, 9(7): 1578-1581.

[194] MACAK J M, BARCZUK P J, TSUCHIYA H, et al. Self-organized nanotubular TiO_2 matrix as support for dispersed Pt/Ru nanoparticles: enhancement of the electrocatalytic oxidation of methanol[J]. Electrochemistry Communications, 2005, 7(12): 1417-1422.

[195] CHEN J M, SARMA L S, CHEN C H, et al. Multi-scale dispersion in fuel cell anode catalysts: role of TiO_2 towards achieving nanostructured materials[J]. Journal of Power Sources, 2006, 159(1): 29-33.

[196] SUBBAN C V, ZHOU Q, HU A, et al. Sol-gel synthesis, electrochemical characterization, and stability testing of $Ti_{0.7}W_{0.3}O_2$ nanoparticles for catalyst support applications in proton-exchange membrane fuel cells[J]. Journal of the American Chemical Society, 2010, 132(49): 17531-17536.

[197] FORMO E, PENG Z, LEE E, et al. Direct oxidation of methanol on Pt nanostructures supported on electrospun nanofibers of anatase[J]. The Journal of Physical Chemistry C, 2008, 112(27): 9970-9975.

[198] HEPEL M, DELA I, HEPEL T, et al. Novel dynamic effects in electrocatalysis of methanol oxidation on supported nanoporous TiO_2 bimetallic nanocatalysts[J]. Electrochimica Acta, 2007, 52(18): 5529-5547.

[199] MAILLARD F, PEYRELADE E, SOLDO-OLIVIER Y, et al. Is carbon-supported Pt-WO_x composite a CO-tolerant material?[J]. Electrochimica Acta, 2007, 52(5): 1958-1967.

[200] JAYARAMAN S, JARAMILLO T F, BAECK S H, et al. Synthesis and characterization of Pt-WO_3 as methanol oxidation catalysts for fuel cells[J]. The Journal of Physical Chemistry B, 2005, 109(48): 22958-22966.

[201] ZHAO Z G, YAO Z J, ZHANG J, et al. Rational design of galvanically replaced Pt-anchored electrospun WO_3 nanofibers as efficient electrode materials for

methanol oxidation[J]. Journal of Materials Chemistry, 2012, 22(32): 16514-16519.

[202] HAMPSON N A, WILLARS M J, MCNICOL B D. The methanol-air fuel cell: a selective review of methanol oxidation mechanisms at platinum electrodes in acid electrolytes[J]. Journal of Power Sources, 1979, 4(3): 191-201.

[203] LU Y, REDDY R G. Electrocatalytic properties of carbon supported cobalt phthalocyanine-platinum for methanol electro-oxidation[J]. International Journal of Hydrogen Energy, 2008, 33(14): 3930-3937.

[204] WANG C, WAJE M, WANG X, et al. Proton exchange membrane fuel cells with carbon nanotube based electrodes[J]. Nano Letters, 2004, 4(2): 345-348.

[205] DUAN Y, SUN Y, PAN S, et al. Self-stable WP/C support with excellent cocatalytic functionality for Pt: enhanced catalytic activity and durability for methanol electro-oxidation[J]. ACS Applied Materials & Interfaces, 2016, 8(49): 33572-33582.

[206] GAO M, ZENG L, NIE J, et al. Polymer-metal-organic framework core-shell framework nanofibers via electrospinning and their gas adsorption activities[J]. RSC Advances, 2016, 6(9): 7078-7085.

[207] LI M, ZHAO S, HAN G, et al. Electrospinning-derived carbon fibrous mats improving the performance of commercial Pt/C for methanol oxidation[J]. Journal of Power Sources, 2009, 191(2): 351-356.

[208] LI M, HAN G, YANG B. Fabrication of the catalytic electrodes for methanol oxidation on electrospinning-derived carbon fibrous mats[J]. Electrochemistry Communications, 2008, 10(6): 880-883.

[209] LIU X, LI M, HAN G, et al. The catalysts supported on metallized electrospun polyacrylonitrile fibrous mats for methanol oxidation[J]. Electrochimica Acta, 2010, 55(8): 2983-2990.

[210] WEI W, CHEN W. “Naked” Pd nanoparticles supported on carbon nanodots as efficient anode catalysts for methanol oxidation in alkaline fuel cells[J]. Journal of Power Sources, 2012, 204: 85-88.

[211] QIAO-HUI G U O, HUANG J S, TIAN-YAN Y O U. Electrospun palladium nanoparticle-loaded carbon nanofiber for methanol electro-oxidation[J]. Chinese Journal of Analytical Chemistry, 2013, 41(2): 210-214.

[212] BARAKAT N A M, MOTLAK M. Co_xNi_y-decorated graphene as novel, stable and super effective non-precious electro-catalyst for methanol oxidation[J]. Applied Catalysis B: Environmental, 2014, 154: 221-231.

[213] BARAKAT N A M, MOTLAK M, ELZATAHRY A A, et al. Ni_xCo_{1-x} alloy nanoparticle-doped carbon nanofibers as effective non-precious catalyst for ethanol oxidation[J]. International Journal of Hydrogen Energy, 2014, 39(1): 305-316.

[214] JIN G P, DING Y F, ZHENG P P. Electrodeposition of nickel nanoparticles on functional MWCNT surfaces for ethanol oxidation[J]. Journal of Power Sources, 2007, 166(1): 80-86.

[215] VINAYAN B P, SETHUPATHI K, RAMAPRABHU S. Facile synthesis of triangular shaped palladium nanoparticles decorated nitrogen doped graphene and their catalytic study for renewable energy applications[J]. International Journal of Hydrogen Energy, 2013, 38(5): 2240-2250.

[216] XU X, ZHOU Y, YUAN T, et al. Methanol electrocatalytic oxidation on Pt nanoparticles on nitrogen doped graphene prepared by the hydrothermal reaction of graphene oxide with urea[J]. Electrochimica Acta, 2013, 112: 587-595.

[217] AL-ENIZI A M, GHANEM M A, EL-ZATAHRY A A, et al. Nickel oxide/nitrogen doped carbon nanofibers catalyst for methanol oxidation in alkaline media[J]. Electrochimica Acta, 2014, 137: 774-780.

[218] THAMER B M, EL-NEWEHY M H, BARAKAT N A M, et al. Influence of nitrogen doping on the catalytic activity of Ni-incorporated carbon nanofibers for alkaline direct methanol fuel cells[J]. Electrochimica Acta, 2014, 142: 228-239.

[219] PAULUS U A, WOKAUN A, SCHERER G G, et al. Oxygen reduction on carbon-supported Pt-Ni and Pt-Co alloy catalysts[J]. The Journal of Physical Chemistry B, 2002, 106(16): 4181-4191.

[220] YOUSEF A, BROOKS R M, EL-NEWEHY M H, et al. Electrospun Co-TiC

nanoparticles embedded on carbon nanofibers: active and chemically stable counter electrode for methanol fuel cells and dye-sensitized solar cells[J]. International Journal of Hydrogen Energy, 2017, 42(15): 10407-10415.

[221] THAMER B M, EL-NEWEHY M H, AL-DEYAB S S, et al. Cobalt-incorporated, nitrogen-doped carbon nanofibers as effective non-precious catalyst for methanol electrooxidation in alkaline medium[J]. Applied Catalysis A: General, 2015, 498: 230-240.

[222] LI P C, LIAO G M, KUMAR S R, et al. Fabrication and characterization of chitosan nanoparticle-incorporated quaternized poly (vinyl alcohol) composite membranes as solid electrolytes for direct methanol alkaline fuel cells[J]. Electrochimica Acta, 2016, 187: 616-628.

[223] MOLLÁ S, COMPAÑ V. Polyvinyl alcohol nanofiber reinforced Nafion membranes for fuel cell applications[J]. Journal of Membrane Science, 2011, 372(1-2): 191-200.

[224] EKRAMI-KAKHKI M S, NAEIMI A, DONYAGARD F. Pt nanoparticles supported on a novel electrospun polyvinyl alcohol-$CuOCo_3O_4$/chitosan based on Sesbania sesban plant as an electrocatalyst for direct methanol fuel cells[J]. International Journal of Hydrogen Energy, 2019, 44(3): 1671-1685.

[225] SU S, AGNEW J. Catalytic combustion of coal mine ventilation air methane[J]. Fuel, 2006, 85(9): 1201-1210.

[226] SU S, CHEN H, TEAKLE P, et al. Characteristics of coal mine ventilation air flows[J]. Journal of Environmental Management, 2008, 86(1): 44-62.

[227] BIELACZYC P, WOODBURN J, SZCZOTKA A. An assessment of regulated emissions and CO_2 emissions from a European light-duty CNG-fueled vehicle in the context of Euro 6 emissions regulations[J]. Applied Energy, 2014, 117: 134-141.

[228] SETIAWAN A, FRIGGIERI J, KENNEDY E M, et al. Catalytic combustion of ventilation air methane (VAM)-long term catalyst stability in the presence of water vapour and mine dust[J]. Catalysis Science & Technology, 2014, 4(6): 1793-1802.

[229] BAE J S, SU S, YU X X. Enrichment of ventilation air methane (VAM) with carbon fiber composites[J]. Environmental Science & Technology, 2014, 48(10): 6043-6049.

[230] CIMINO S, LISI L, PIRONE R, et al. Methane combustion on perovskites-based structured catalysts[J]. Catalysis Today, 2000, 59(1-2): 19-31.

[231] WANG D, GONG J, LUO J, et al. Distinct reaction pathways of methane oxidation on different oxidation states over Pd-based three-way catalyst (TWC)[J]. Applied Catalysis A: General, 2019, 572: 44-50.

[232] MONAI M, MONTINI T, GORTE R J, et al. Catalytic oxidation of methane: Pd and beyond[J]. European Journal of Inorganic Chemistry, 2018, 2018(25): 2884-2893.

[233] VALDÉS-SOLÍS T, MARBÁN G, FUERTES A B. Preparation of nanosized perovskites and spinels through a silica xerogel template route[J]. Chemistry of materials, 2005, 17(8): 1919-22.

[234] CHEN C Q, LI W, CAO C Y, et al. Enhanced catalytic activity of perovskite oxide nanofibers for combustion of methane in coal mine ventilation air[J]. Journal of Materials Chemistry, 2010, 20(33): 6968-6974.

[235] HUANG K, CHU X, FENG W, et al. Catalytic behavior of electrospinning synthesized $La_{0.75}Sr_{0.25}MnO_3$ nanofibers in the oxidation of CO and CH_4[J]. Chemical Engineering Journal, 2014, 244: 27-32.

[236] WANG N, YU X, SHEN K, et al. Synthesis, characterization and catalytic performance of MgO-coated Ni/SBA-15 catalysts for methane dry reforming to syngas and hydrogen[J]. International Journal of Hydrogen Energy, 2013, 38(23): 9718-9731.

[237] ZHAO X, LI H, ZHANG J, et al. Design and synthesis of NiCe@ m-SiO_2 yolk-shell framework catalysts with improved coke-and sintering-resistance in dry reforming of methane[J]. International Journal of Hydrogen Energy, 2016, 41(4): 2447-2456.

[238] YAN Z F, DING R G, LIU X M, et al. Promotion effects of nickel catalysts of dry

reforming with methane[J]. Chinese Journal of Chemistry, 2001, 19(8): 738-744.

[239] SONG C. Global challenges and strategies for control, conversion and utilization of CO_2 for sustainable development involving energy, catalysis, adsorption and chemical processing[J]. Catalysis Today, 2006, 115(1-4): 2-32.

[240] LIU H, GUAN C, LI X, et al. The key points of highly stable catalysts for methane reforming with carbon dioxide[J]. ChemCatChem, 2013, 5(12): 3904-3909.

[241] XIE X, OTREMBA T, LITTLEWOOD P, et al. One-pot synthesis of supported, nanocrystalline nickel manganese oxide for dry reforming of methane[J]. ACS Catalysis, 2013, 3(2): 224-229.

[242] LIU C, YE J, JIANG J, et al. Progresses in the preparation of coke resistant Ni-based catalyst for steam and CO_2 reforming of methane[J]. ChemCatChem, 2011, 3(3): 529-541.

[243] WANG Z, HU X, DONG D, et al. Effects of calcination temperature of electrospun fibrous Ni/Al_2O_3 catalysts on the dry reforming of methane[J]. Fuel Processing Technology, 2017, 155: 246-251.

[244] LIU L, WANG S, GUO Y, et al. Synthesis of a highly dispersed Ni/Al_2O_3 catalyst with enhanced catalytic performance for CO_2 reforming of methane by an electrospinning method[J]. International Journal of Hydrogen Energy, 2016, 41(39): 17361-17369.

[245] WEN S, LIANG M, ZOU J, et al. Synthesis of a SiO_2 nanofibre confined Ni catalyst by electrospinning for the CO_2 reforming of methane[J]. Journal of Materials Chemistry A, 2015, 3(25): 13299-13307.

第4章　纳米纤维在环保领域的应用

除能源短缺外，环境污染已成为全球关注的另一个主要问题。工业、工艺的创新和进步产生了大量新污染物，其含量超出了环境的自净能力，对生态系统造成了严重的破坏。水和空气是最主要的两个环境系统，仍然存在着严重的环境污染问题。例如，重金属污染、有机废水、雾霾、一般挥发有机物（VOCs）和含油废水等环境问题严重威胁着人类健康和生态环境的安全。静电纺丝纳米纤维由于其高长宽比、高比表面积、优异柔韧性、结构多样和易于功能化等优点，目前已被广泛应用于环境领域，且取得了不错的成果。本章总结了电纺纳米纤维在水体净化、空气净化和油水分离等方面的应用和研究进展，并对其中尚存的问题进行了总结和展望，以提供对电纺纳米纤维的结构设计和环境应用的全面理解。

4.1　水体净化

4.1.1　引言

水是生命之源，因此确保充足的水资源对人类的生存和发展至关重要。然而，据世界卫生组织估计，由于人口增长以及各种气候和环境问题，超过11亿人口缺乏充足的饮用水。供水链中的主要挑战是各种有机和无机污染物对淡水资源的持续污染。在众多方法中，吸附和过滤由于其方便性、操作简便性、设计简单性和通用性的优点，被认为是去除污染物的通用和有前途的方法。电纺纳米纤维具有独特的特性，例如大表面积、可调的孔结构、高孔隙率和表面功能化的灵活性，因此可以用作去除污染物的高级吸附材料和高级过滤材料。在本节中，概述了用于去除水溶液中污染物（重金属离子和有机污染物）的电纺纳米纤维基吸附剂的研究和应用进展，另外对纳米纤维在水过滤处理以及海水淡化中的应用也进行了总结和展望。

4.1.2 重金属离子吸附

由于水中的重金属离子易于积累，无法被生物降解并会对生态系统和人类造成负面影响，因此去除重金属离子非常重要。吸附是去除这些重金属离子的最广泛应用的技术之一。如图 4.1 所示，吸附可以通过基于带正电的金属离子与带负电的基质之间的离子相互作用进行吸附，或通过螯合作用来配位金属离子与功能性基质之间的键，来除去重金属离子。然而，吸附效率、选择性、平衡时间、再生和稳定性通常取决于吸附剂的材料特性。因此，有必要开发新型的高效材料以满足从废水中去除重金属离子的要求。与传统的吸附剂（如活性炭和沸石等）相比，电纺纳米纤维具有比表面积大、可调的孔结构、良好的孔互连性以及在纳米级上整合活性化学或功能性的潜力，因此是重金属离子吸附的理想选择。目前，已有大量关于电纺丝纳米纤维在重金属离子去除应用的研究被报道。

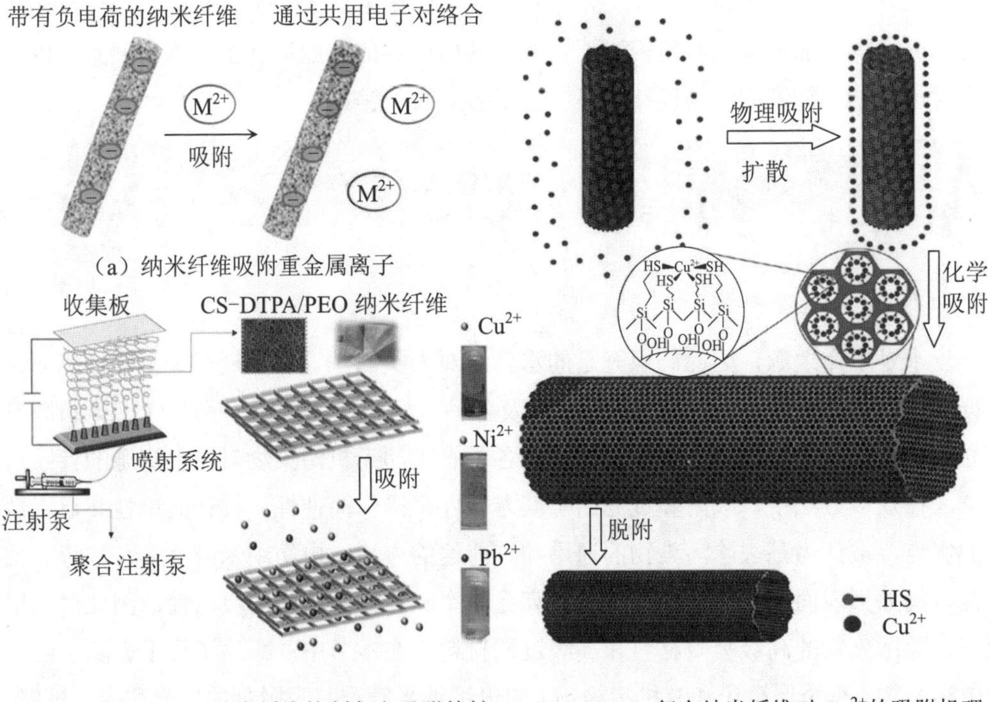

（b）CS-DTPA/PEO 纳米纤维的制备和吸附特性 （c）PVA/SiO_2 复合纳米纤维对 Cu^{2+} 的吸附机理

图 4.1 吸附重金属离子的纳米纤维（彩图见附录）

天然高分子纳米纤维由于其性质丰富、价格便宜而得到了深入研究。由亲水性氨基酸基团组成的羊毛角蛋白（WK）由于对金属阳离子的高亲和力而作为金属离子吸附剂受到越来越多的关注。Aluigi 等人通过静电纺丝制备了 WK/PA6（尼龙 6）纳米长丝毡，其对 Cu^{2+}的吸附能力达到 103.5 mg·g^{-1}。此外，经过多次循环处理后，制得的膜仍具有吸附能力。另一种有希望的重金属吸附聚合物是壳聚糖，它具有许多极性及可电离基团，可通过螯合吸附有毒的重金属离子。Haider 等人表明壳聚糖基电纺纤维膜可以有效去除有毒重金属离子，Cu^{2+}和 Pb^{2+}的平衡吸附容量分别为 485.44 mg·g^{-1} 和 263.15 mg·g^{-1}，且不会失去其生物相容性、亲水性、生物活性、非抗原性和无毒性。最近，Surgutskaia 等人通过静电纺丝制备了二乙烯三胺五乙酸改性的壳聚糖/聚环氧乙烷纳米纤维（CS-DTPA/PEO）以增强重金属离子的吸附（图 4.1（b））。结果表明，CS-DTPA/PEO 纳米纤维吸附金属离子的能力的顺序如下：Cu^{2+}> Pb^{2+}> Ni^{2+}，其对 Cu^{2+}、Pb^{2+}和 Ni^{2+}的最大吸附容量分别为 177 mg·g^{-1}、142 mg·g^{-1} 和 56 mg·g^{-1}。经过五次吸附-解吸测试，仍保持较好的稳定性。

除天然聚合物纳米纤维外，经官能团改性的多孔纳米纤维还因其特性而备受关注，这些特性包括：①高的体积比；②均匀的孔径分布；③回收利用的简便性；④对重金属离子的高平衡吸附能力。例如，PAN（聚丙烯腈）纳米纤维已被氨基肟基团修饰，以吸附金属离子。大多数吸附的金属离子可以在 1 mol·L^{-1} 的 HNO_3 溶液中在 1 h 内解吸，这表明功能性纳米纤维可以从废水中回收金属。Neghlani 等人用二亚乙基三胺修饰 PAN 纳米纤维以吸附铜离子。结果表明，改性纳米纤维的吸附效率是微纤维的三倍，饱和吸附量是其他报道材料的五倍。Bornillo 等人通过静电纺丝法制备了一种新型的电纺双响应（温度和 pH）聚醚砜-聚（二甲基氨基）甲基丙烯酸乙酯（PES-PDMAEMA）纳米纤维，用于从水溶液中去除 Cu^{2+}。该 PES-PDMAEMA 具有通过调节 pH 和温度来吸附和解吸 Cu^{2+}的能力，在 55 ℃下获得的最大吸附容量为 161.30 mg/g。

巯基中的硫原子也可用于去除金属离子，因为它可以与重金属离子形成螯合物。硫醇官能化的介孔聚乙烯醇/二氧化硅（PVA/SiO_2）复合纳米纤维的制备如图 4.1（c）所示。制备的膜 Cu^{2+}最大吸附容量为 489.12 mg·g^{-1}，且其吸附能力在六个吸附-解吸循环后仍可保持。此外，用巯基官能化的介孔结构的 PVA/SiO_2 复合纳米纤维显示出高的选择性和对 Hg^{2+}的吸附能力，以及良好的吸附和解吸循环特性。

综上所述，电纺纳米纤维在金属离子吸附方面具有巨大潜力，这主要是由于其异常高的表面体积比。未来的纳米纤维在吸附重金属离子方面的研究应侧重于更可靠、可扩展和环保的改性方法，并提高去除效率和循环寿命。

4.1.3 有机污染物吸附

水中的染料、持久性有机污染物（POPs）和其他有机污染物对环境和人体健康有害。由于表面积大、孔隙率高和表面修饰灵活性，基于多孔纳米纤维膜的吸附技术引起了人们的极大兴趣。

超大表面积和高孔体积使碳纳米纤维（CNF）具有强大的吸附量和快速的吸附速率，这使其成为从水中吸附染料的理想选择，并且近年来受到越来越多的关注。东华大学的 Si 等人通过电纺丝和原位聚合相结合，合成了基于聚苯并恶嗪（PBZ）前体的分层多孔磁性 Fe_3O_4/ACNF，用于从水溶液中去除染料（图 4.2（a））。所制造的纳米纤维的孔径分布主要集中在2.83 nm和4.92 nm处，具有典型的表面分形特征。得益于极大的比表面积（1 885 m^2/g）、高孔隙率（2.3 cm^3/g）和增强的多孔结构，所制备的 Fe_3O_4/ACNF 可以在 10 min 和 15 min 内完全吸附完水中的亚甲基蓝（MB）和若丹明（RhB）。更重要的是，吸附后的水悬浮液可通过外部磁体轻松分离，这对于实际应用而言非常重要。Qiao 等人使用酚醛树脂前体作为碳源，三嵌段聚合物 Pluronic F127 作为模板，通过溶胶-凝胶/电纺丝工艺制备了介孔 ACNF，以用作大染料分子（甲基亚硫酰氯、甲基橙、酸性红 1）的高效吸附剂。结果表明，除表面积外，不同染料的吸附还取决于适当的孔径大小分布。CNFC-3 具有最大的表面积和最大的中孔体积，比其他具有更大的吸附能力，例如，甲基亚硫酰氯的最大吸附量可以达到 567 mg/g。

其他电纺聚合物纳米纤维也已用于从水中吸附染料。Li 等人通过静电纺丝硫醇官能化的 PVA/SiO_2 复合纳米纤维膜合成了一种吸收剂，并研究了靛蓝胭脂红和酸性红的吸附性能。由于表面积大（140.1 m^2/g）、染料和膜之间的静电相互作用以及氢键相互作用（图 4.2（b）），靛蓝胭脂红和酸性红的最大吸附量分别为 266.77 mg/g 和 211.74 mg/g。Miao 等人通过电纺和水热反应相结合，制备了分级的 SiO_2/γ-AlOOH（勃姆石）芯/鞘纳米纤维，以从水中吸收刚果红。与勃姆石粉末相比，这样制备的自立式纤维膜具有更高的柔韧性、易于处理和可回收性（图 4.2（c）），且纳米纤维（21.3 mg/g）的吸附能力是勃姆石粉末（10.4 mg/g）的两倍以上（图 4.2（d））。

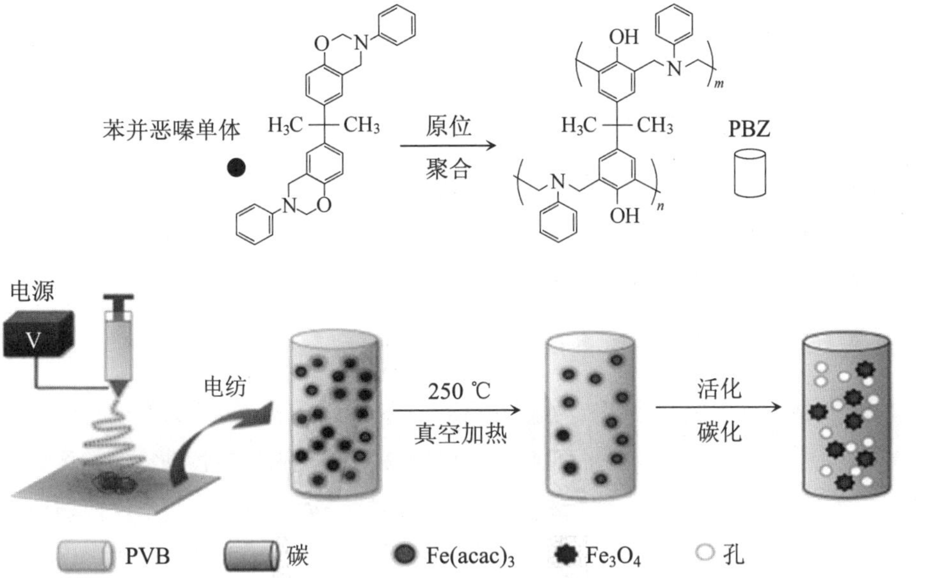

（a）通过静电纺丝和原位聚合相结合的方法来合成 Fe_3O_4/ACNF 的示意图

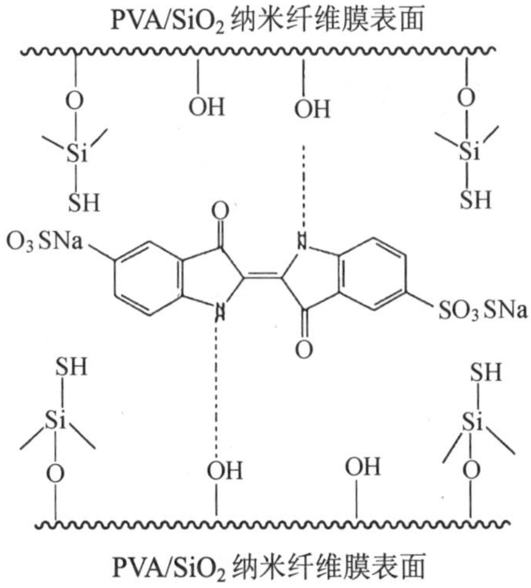

（b）氢键作用机理

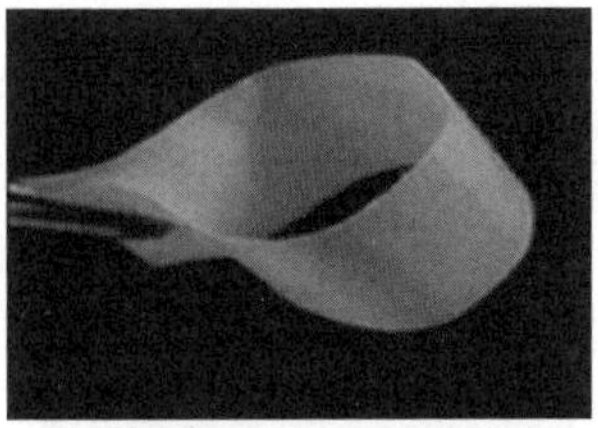

（c）自立式纤维膜的光学照片

（d）染料吸附性能

图 4.2　吸附有机污染物的纳米纤维

除了吸附废水中的染料外，电纺纳米纤维还被用于有效去除其他有机污染物，例如狄氏剂、雌激素和苯酚。Yue 等人制备了溴丙基官能化的二氧化硅纳米纤维膜。由于狄氏剂和溴丙基之间的强疏水性相互作用，溴丙基改性的二氧化硅纳米纤维被发现对去除低浓度的疏水性 POPs 有效，最大去除率达到为 91%，远高于商业颗粒活性炭（38.6%）和二氧化硅纳米纤维（20.9%）。Xu 等人使用静电纺丝制备了 PA-6 纳米纤维膜，来富集环境水中水样中的三种雌激素（雌二醇、乙炔基雌二醇和雌酮）。结果表明，由于高的比表面积（18.41 m^2/g）和纳米纤维与雌激素之间的增强的相互作用，雌激素可以定量地吸附在 PA-6 上，与市售 PA-6 微孔膜相比，纳米纤维膜在回收率、重现性和吸附量方面更好。

4.1.4 膜过滤净化水

过滤是一种膜分离技术，根据分离原理和过滤精度，可分为微滤（MF）、超滤（UF）、纳滤（NF）、反渗透（RO）。如图 4.3 所示，施加在膜进料侧的压力充当驱动力，将水分成两股流：渗透液和截留液。通常，渗透液是纯净水，而截留液是在排放前要处理或处理的浓缩溶液。根据技术类型的不同，可以从原水中保留悬浮颗粒、油乳液、细菌、细胞、胶状浑浊物、病毒、大分子、蛋白质、亚分子有机基团，离子等。

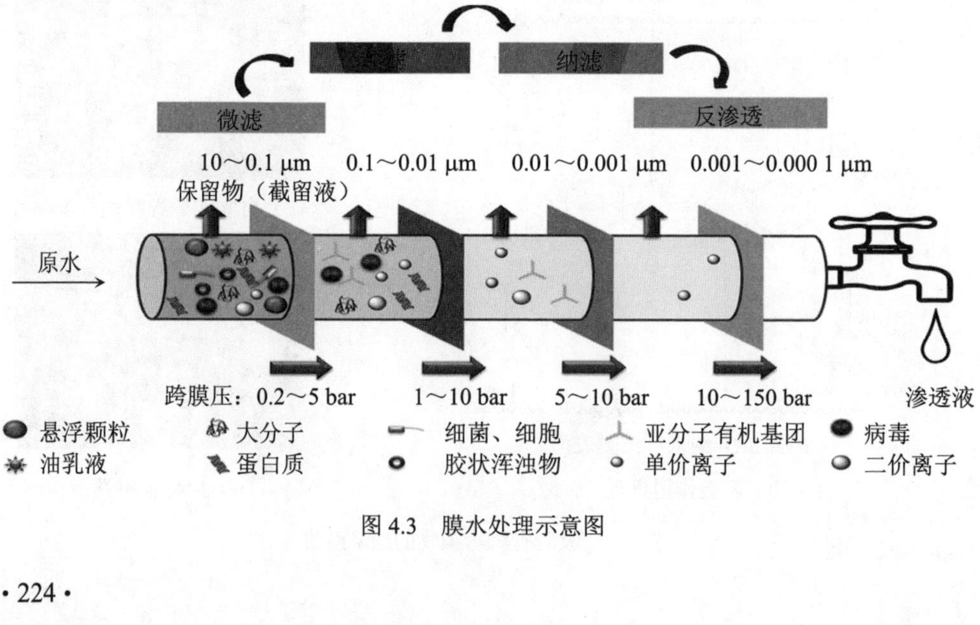

图 4.3　膜水处理示意图

MF 通常用于水的预处理，典型的 MF 过滤精度约为 0.1 μm 和 1.0 μm，用于预过滤以去除 1.0 μm 以上的颗粒。与其他方法制得的膜相比，静电纺丝制备的纳米纤维膜是 MF 滤膜良好的候选者，因为它们的孔隙率更高，并且孔径可以从亚微米水平定性优化至几微米。Gopal 等人首次对电纺纤维膜进行液体过滤的可行性进行探索，并通过分离 1 μm、5 μm 和 10 μm 的聚苯乙烯颗粒检查其过滤性能。如图 4.4 所示，超过 90%的微粒被成功过滤。用 PAN、PSU（聚砜）、PES（聚醚砜）和尼龙 6 制成了一系列纳米纤维膜，以进一步了解电纺纳米纤维结构对膜性能的影响。结果表明，电纺纳米纤维膜的纤维直径和厚度显著影响过滤性能。另外，化学修饰对于进一步优化电纺纳米纤维膜能是有效的。例如，Yoon 等人使用过硫酸铵（APS）进行的短时间氧化处理使电纺 PES 膜具有亲水性（接触角约 28°）。改性的 PES 电纺纳米纤维膜的纯净水通量得到提高，明显优于传统 MF 膜。然而，正如图 4.4（d）那样，颗粒沉积在膜表面的顶层上，产生结垢降低了渗透速率。

（a）过滤前

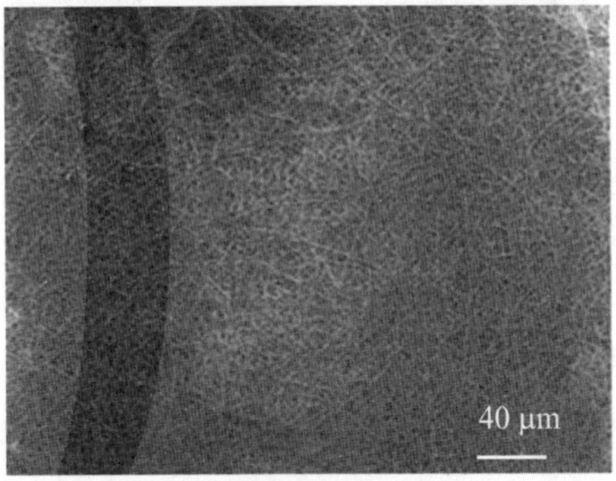

（b）过滤 10 μm 颗粒后

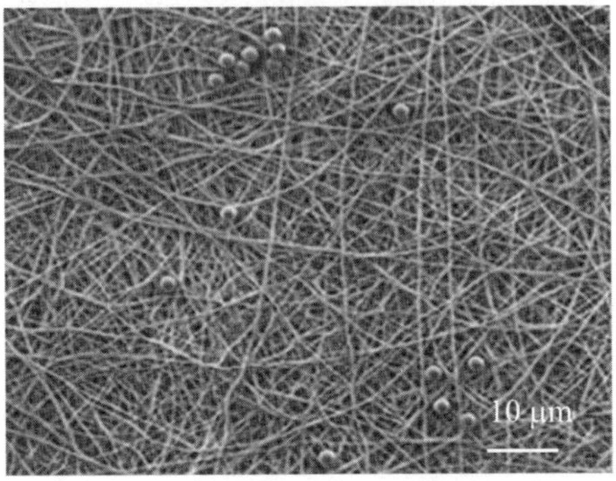

（c）过滤 5 μm 颗粒后

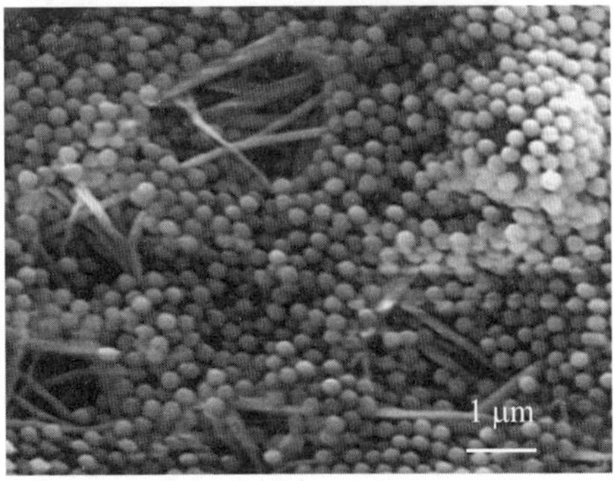

（d）过滤 1 μm 颗粒后

图 4.4　PVDF 膜的 FESEM 显微照片

因此，为了将电纺纳米纤维膜用于 MF，需要解决严重的膜污染问题。另外，由于纳米纤维可能在高通量和工作压力下被水流分离并冲走，因此纳米纤维膜可能会被破坏。因此，在将电纺纳米纤维膜大规模应用于水工业之前，对单个纳米纤维进行机械加固和对纳米纤维进行整合至关重要。单个纤维强度增强的方法包括掺入纳米粒子，如氧化锆；纤维间的整合包括热处理、溶剂诱导的纳米纤维间键合以及使用交联剂。

UF 是一种膜过滤过程，其孔径范围为 0.01～0.1 μm（10～100 nm），用于去除细菌、胶体和病毒，其应用包括饮用水处理、反渗透淡化、再生和膜生物反应器（MBR）之前的预处理。Halaui 等人首先使用通过共电纺丝技术生产的不对称微管开发了微型中空纤维超滤膜，包括平均内径为 4 μm 且厚度为 0.9 μm 的聚己内酯（PCL）微管和平均内径为 1.44 μm 且壁厚为 0.94 μm 的聚偏氟乙烯（PVDF）微管，微米级 UF 膜表现出较高的排斥值和相当大的通量。Yoon 等人构建了一种高通量和低污染的 UF 膜，如图 4.5 所示，膜由三层复合结构组成，即无孔亲水纳米复合材料顶层、静电纺丝聚丙烯腈纳米纤维基质中间层和常规无纺布超细纤维支撑层。该膜包含平均直径为 124～720 nm，孔隙率约为 70％的电纺 PAN 支架以及厚度约为 1 μm 的壳聚糖顶层，在操作 24 h 后的通量率仍比商用 NF 膜高一个数量级，同时对于含油废水过滤仍保持相同的截留效率（大于 99.9％）。Zhao 等人将壳聚糖（CTS）（由于其亲水性和高透水性）与电纺 PVDF 纳米纤维结合在一起，构成了一种新型 UF 膜，并经过戊二醛（GA）和对苯二甲酰氯（TPC）进行交联和改性。改性后的膜在 0.2 MPa，约 70.5 $L\cdot(m^2\cdot h)^{-1}$ 的牛血清白蛋白（BSA）过滤测试中保持了良好的通量率和截留效率（大于 98％），高于工业 UF 膜的 57.1 $L\cdot(m^2\cdot h)^{-1}$。

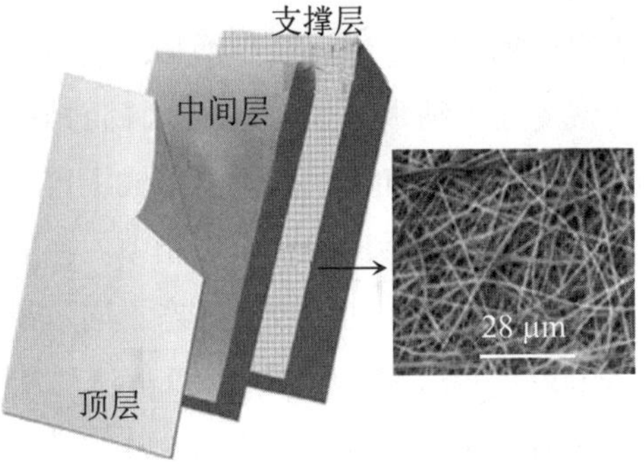

（a）高通量和低污染的 UF 膜

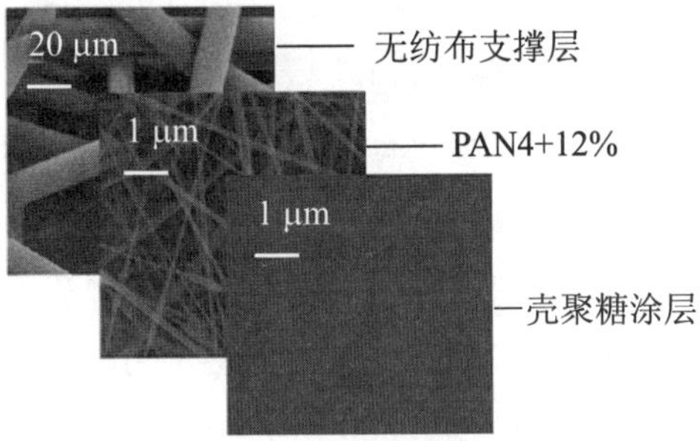

（b）UF 膜中各层材料的 SEM 图

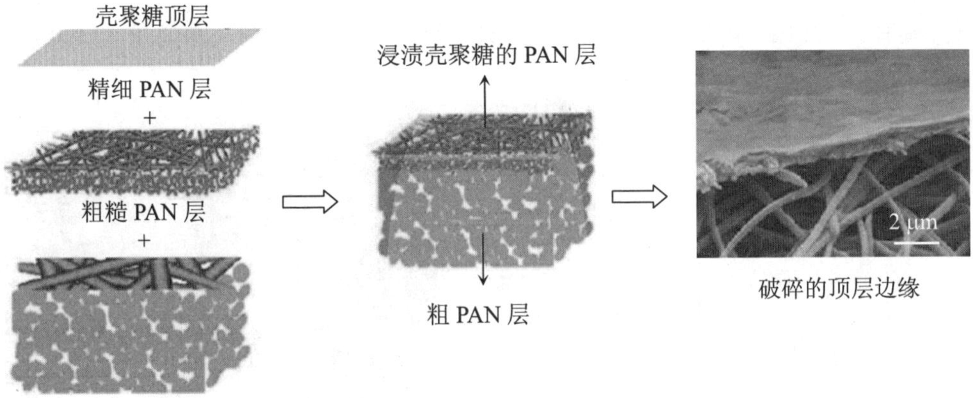

（c）壳聚糖/PAN 新型 UF 膜

图 4.5 超滤（UF）膜

NF 的典型分离尺寸范围为 100～1 000 Da，已被广泛用于水处理中，因为它能够软化和消毒水，并去除颜色、气味、一些微量有机污染物和二价离子。NF 中的分离涉及空间位阻效应和静电效应，并且是由高跨膜压驱动的。如图 4.6（a）所示，NF 与反渗透（RO）之间的区别在于 RO 能够排斥一价盐如氯化钠，而 NF 主要有效地保留二价离子和多价盐如硫酸钠。大多数 NF 膜是薄膜复合（TFC）膜，其具有通过界面聚合在多孔基材顶部上产生的超薄表层。PA 覆盖的 PAN-TFNC 膜的渗透通量是 TFC 膜的 2.4 倍以上，同时保持相同的排斥率（约 98%）。这说明纳米纤维薄膜在 NF 中具有独特的优势。TFNC 膜的典型表面和横截面形态如图 4.6（b）所示。影响 NF 性能的因素主要有以下三点：①纳米纤维结构。Kaur 等人表明，随着基底表面孔径和孔隙率的增加，TFNC 膜通量通过增加纳米纤维直径而增加。然而，存在纳米纤维直径和膜表面孔径的上限，超过该上限，纳米纤维膜不能支撑阻挡层。②界面聚合。Kaur 等人通过两种不同的界面聚合，在 ENM 的表面上形成了聚酰胺层。在方法 A 中，将电纺纳米纤维膜浸泡在水相中，然后浸泡在有机相中。在第二种方法 B 中，颠倒了这个顺序，这导致电纺纳米纤维膜表面的表面形态不同。最佳方法 B 在 70 psig（磅力/平方英寸）的压力下，2 000 mg·L^{-1} $MgSO_4$ 和 2 000 mg·L^{-1} NaCl 的分离率分别为 80.7%和 67.0%。获得的通量分别为 0.51 L·$(m^2$·h$)^{-1}$ 和 0.52 L·$(m^2$·h$)^{-1}$。③交联。Park 等人通过交联的超支化聚乙烯亚胺（PEI）和聚偏二氟烯（PVDF）构筑了纳米纤维复合材料（TNFC）膜，发现在交联的 7 bar（1 bar=10^5 Pa）跨膜压力下，具有交联 PEI / PVDF 网络的 TFNC 具有约 30 L·$(m^2$·h$)^{-1}$ 的高水通量，$MgCl_2$ 和 NaCl 的截留率分别为 88％和 65％。

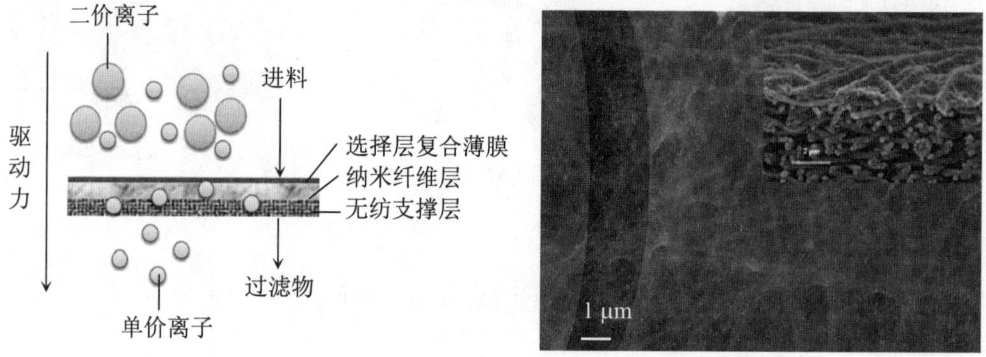

（a）薄膜纳米纤维复合材料（TFNC）膜的 NF 示意图　　（b）TFNC 典型的表面和横截面形貌

图 4.6　纳滤（NF）膜

4. 1. 5　海水淡化

由于人口急剧增长、全球变暖和淡水污染，寻找淡水替代资源已成为人类亟待解决的问题。海水淡化被认为是最可持续和最佳的水资源替代方案之一。膜蒸馏（MD）是用于水脱盐和净化的新兴技术，为热驱动过程，其中水蒸气通过微孔疏水膜传输。MD 具有四种模式（图 4.7），分别是直接接触膜蒸馏（DCMD）、气隙膜蒸馏（AGMD）、吹扫气膜蒸馏（SGMD）和真空膜蒸馏（VMD）。其中，DCMD 是研究最多的模式，在该过程中，挥发性分子在高温进料侧在液/气界面处蒸发，然后穿过膜，最后在冷渗透侧冷凝。与常规脱盐工艺相比，MD 工艺可以在更低的工作温度（相比于热脱盐）和静水压力（比反渗透更低）下进行，理论上可以 100%去除不挥发物。这些特征使 MD 比其他脱盐工艺更具吸引力。但是不论什么模式，膜都是其中最重要的部分，所以膜材料的性能及其设计是 MD 的关键。

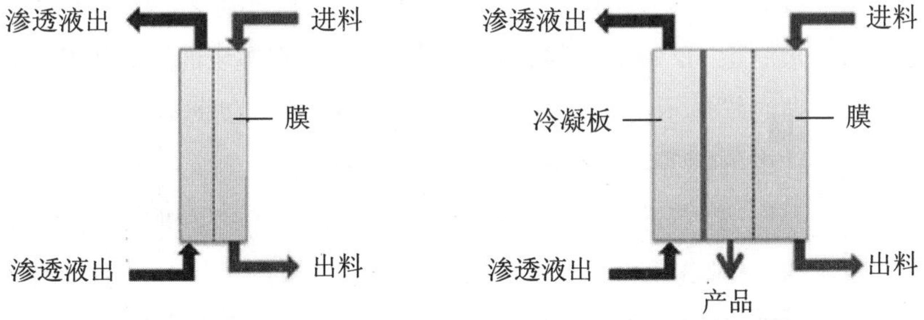

（a）DCMD　（b）AGMD

（c）SGMD　（d）VMD

图 4.7　膜蒸馏过程示意图

在 MD 工艺中，膜应表现出高渗透性、疏水性、在宽温度范围内的热稳定性和化学稳定性，并具有很强的机械强度。电纺纳米纤维膜的独特功能，包括高孔隙率、重叠的纳米纤维结构和相互连接的开孔，使其对 MD 应用具有独特的吸引力。Feng 等人首先制造了聚偏二氟乙烯（PVDF）纳米纤维膜，通过 AGMD 从盐水中产生饮用水。在进料温度为 323～358 K 的条件下，膜通量可与商用微滤膜（5～28 $kg\cdot(cm^2\cdot h)^{-1}$）获得的通量相媲美。Liao 等人表明，静电纺丝工艺参数和聚合物涂料溶液的性质是决定膜结构的主要因素。而且，热压后处理对于改善纳米纤维膜的完整性，增加渗透通量至关重要。

为了进一步提高 MD 的性能，开发超疏水纳米纤维成为目前的研究趋势。Liao 等人研究了整体改性和表面改性两种类型的超疏水改性。由于分层结构和表面修饰改变了膜表面的拓扑结构并使膜超疏水,当进料和渗透温度分别固定在 333 K 和 293 K 时，使用质量分数为 3.5%的 NaCl 水溶液作为进料，经过整体改性的 PVDF 膜实现了 31.6 $L\cdot(m^2\cdot h)^{-1}$ 的 MD 水通量。为了提高超疏水层的稳定性，Liao 等人分别在 PVDF 纳米纤维载体上形成的超疏水二氧化硅-PVDF 选择层，以及在无纺布载体上电纺超疏水层来制造 3D 超疏水膜。已经证明，超疏水性表面层在连续 DCMD 操作和经受超声处理时均显示出优异的耐久性，且具备一定的力学稳定性。

超疏水纳米纤维的制备还能解决 MD 时的结垢问题。Li 等人通过静电纺丝制聚砜（PSU）纳米纤维，接着进行聚二甲基硅氧烷（PDMS）涂层和冷压后处理，制备超疏水且自清洁的 DCMD 用聚砜-聚二甲基硅氧烷（PSU-PDMS）电纺纳米纤维膜。对于 30 $g\cdot L^{-1}$ 的进料 NaCl 溶液和 50 ℃的温差，其能够保持大约 21.5 $kg\cdot(m^2\cdot h)^{-1}$ 的稳定透过水通量，而不会检测到润湿纤维间的空间。同时，开发的超疏水性 PDMS/PVDF 杂化膜在 MD 处理废水中显示出优良防污性能。除了高分子材料改性外，掺入无机纳米材料制备纳米纤维复合材料是另一种改性方法。Tijing 等人通过静电纺丝制备了超疏水、坚固耐用的混合基质聚偏二氟乙烯-六氟丙烯（PcH）纳米纤维膜，其中掺入了不同质量分数（1%～5%）的碳纳米管（CNT）作为纳米填料，以赋予额外的机械和疏水性能。如图 4.8 所示，纯净的 PcH 纳米纤维的 CA 为 149°，高于商用 PVDF 膜的 131.1°，这主要归因于纳米纤维重叠层的较粗糙表面结构。当掺入 CNT 时，在膜表面上形成串珠，这进一步增加了膜的表面粗糙度，从而增加了接触角。测试结果表明，即使在最高的 NaCl 溶液为 70 $g\cdot L^{-1}$ 的情况下，纯 PcH 和 5CNT 纳米纤维膜也显示出比商用 PVDF 膜（18 $L\cdot(m^2\cdot h)^{-1}$）更高的通量（分别为 19.5 $L\cdot(m^2\cdot h)^{-1}$ 和 24 $L\cdot(m^2\cdot h)^{-1}$），且除盐效率达到 99.98%以上。此外，二氧化硅、TiO_2、Al_2O_3 纳米颗粒和石墨烯也已嵌入电纺纳米纤维中，以开发用于 MD 的分层超疏水或超两性膜。

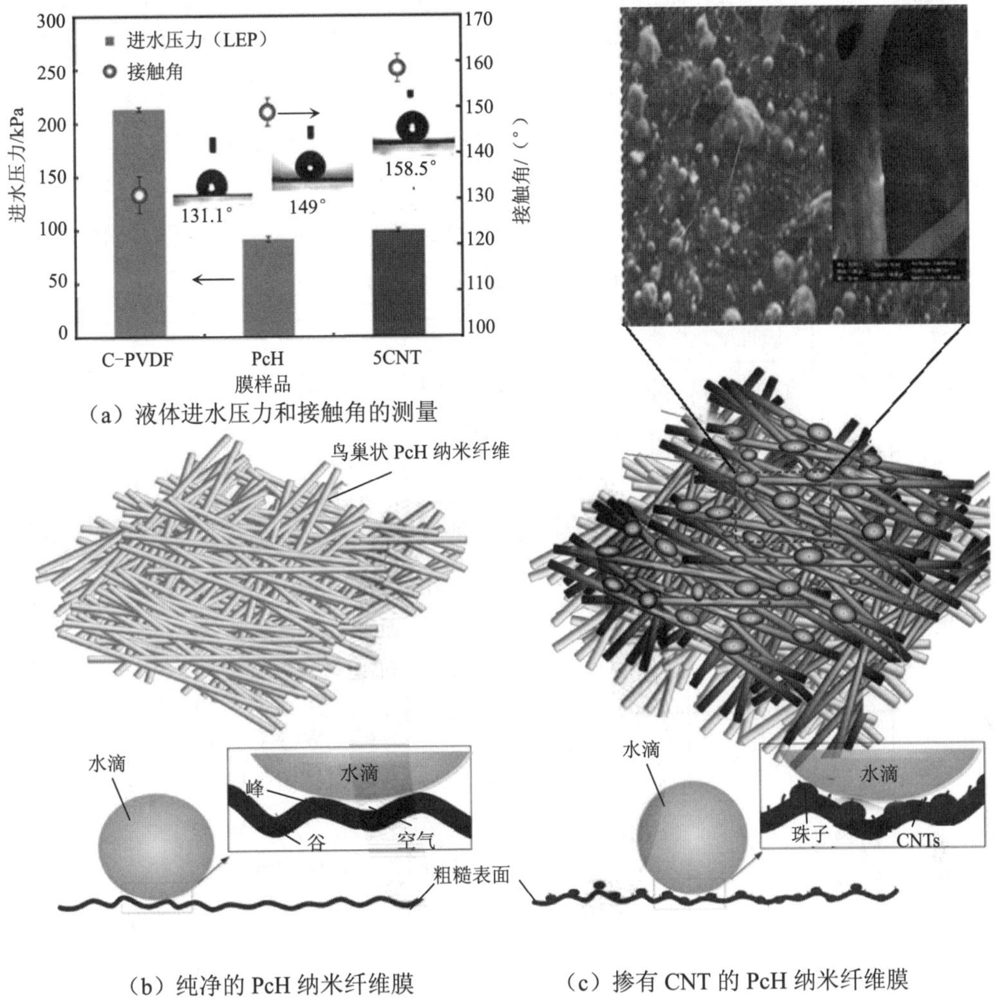

图 4.8　聚偏二氟乙烯-六氟丙烯（PcH）纳米纤维膜

最近发表的有关用于淡化和水处理应用中的 MF、UF、NF、FO 和 MD 的纳米纤维膜的材料、制备方法、分离特性、优点和缺点的研究见表 4.1。表中介绍了与制造工艺有关的主要挑战，例如成本问题、溶剂排放和回收问题、有限的组装配置等。此外，还讨论了每种技术的优点。可以看出，通用的静电纺丝技术可用于制造具有均匀孔分布、增强的孔互连性、高比表面积与孔体积比和更高表面积的高性能电纺纳米纤维膜，且特别适用于 MD、FO、UF、NF 和 MF 的水处理方面应用。显然，在静电纺丝过程中，选择合适的材料和合适的功能引入途径，对于开发满足特定需求的纳米纤维膜至关重要。在已有的研究中，电纺纳米纤维膜作为最有效的选择性/阻隔层，展现了非常优异的水净化能力。例如，Park 等人制备的含氟的热重排电纺膜具有通量为 114.8 $kg·(m^2·h)^{-1}$、除盐效率 99.99%的优异分离性能。然而，成本、溶剂排放和回收、有限的组装配置等是将电纺纤维膜大规模应用于水体净化的障碍。因此，研究人员需在提高电纺纳米纤维膜的耐用性和稳定性方面进行更大的努力。

表 4.1　不同材料制备的纳米纤维膜的特征

膜	方法	膜性能		分离特性			过程	优点	挑战
		孔径/μm	孔隙率/%	流量/[$kg·(m^2h)^{-1}$]	溶质排斥	稳定时间/h			
PU/PET 纳米纤维毡	静电纺丝/界面	0.25～0.64	85	—	—	—	MF	PU/PA 纳米纤维膜相通的孔结构和出色的孔隙率使其通量是市售膜的两到三倍	无针电纺装置无法精确控制纤维的形态和质量，限制了材料使用范围和纤维膜制备
PVA–MWNT/PAN	静电纺丝固溶处理	—	—	270.1	盐 −99.5%	—	UF	在 MWCNT 和聚乙烯醇链间生成的纳米级空腔可以增加最顶层阻隔层中的自由体积，从而增加阻隔层的水传输能力	溶剂（丙酮）排放到空气中，造成污染；溶剂回收的成本很高
PET/PA 复合膜	静电纺丝/反向界面聚合	0.001 8	85	34±2.3	盐 −78%	—	NF	巨大的开孔结构使通量增加；纳米纤维载体的较低水力阻力以及薄层的纳米复合形式增加了水的输送能力	使用两轴技术制备的纳米纤维具有有限的组装结构和功能

续表 4.1

膜	方法	膜性能		分离特性			过程	优点	挑战
		孔径/μm	孔隙率/%	流量/[kg·(m²h)⁻¹]	溶质排斥	稳定时间/h			
PES/PET/无纺布NF 膜	静电纺丝/层层组装/相转化	—	—	75	盐-80%	—	NF	当磺化聚醚酮含量较少时，增加的表面亲水性的影响将抵消减小的孔隙率，从而提高水通量	制造过程很麻烦。层层组装过程中存在材料浪费
PAN 支撑的 FO 膜	静电纺丝/界面聚合	—	79.8	55.05	盐-97%	—	FO	亲水性、结晶性、内部结构、粗糙度和表面形态等对分离和渗透性能具有协同作用	过程取决于几个变量
PI 支撑的微孔纳米纤维膜	静电纺丝/界面聚合	1.27	—	11.6±2	盐-49.8%	—	FO	正渗透膜可以帮助降低 ICP 并避免流失溶质	溶剂（乙醇）向大气中排放；增加了溶剂回收的费用
非晶聚丙烯/PVDF 复合膜	静电纺丝/热压热处理	0.46	79.7	135.3	盐-99.99%	50	DCMD	超疏水无定形聚丙烯表皮层以及纳米纤维支撑层的协同效应可在不严重影响孔隙率的情况下调整孔的大小	表层易出现缺陷。在长时间的 DCMD 工艺过程中，薄膜易损
含氟的热重排电纺膜	静电纺丝/热重排	0.8	81	114.8	盐-99.99%	250	DCMD	氟的引入加强了纤维的疏水性，复合膜表现出优异的通量和脱盐效率	制备过程复杂，且有溶剂污染的风险

续表 4.1

膜	方法	膜性能		分离特性			过程	优点	挑战
		孔径/μm	孔隙率/%	流量/(kg·(m²h)⁻¹)	溶质排斥	稳定时间/h			
CF4 等离子预处理的 PVDF 膜	静电纺丝/CF4 等离子体处理	0.8	86.4	15.2	盐-100%	—	AGMD	该电纺纤维膜具有处理有机污染物的可能性	优化加工参数以达到所需的纳米纤维特性和形态非常困难
碳纳米管包覆的超疏水电纺膜	静电纺丝/CNT 喷涂	0.19	70.2	28.4	盐-99.99%	26	VMD	该方法解决了纳米纤维中珠子发展和纳米颗粒嵌入的问题	在膜基质中包含纳米颗粒需要进行膜后处理，增加成本

4.1.6 小结与展望

静电纺丝已成为制造纳米纤维的全球公认的简便方法。纳米纤维的独有特征，包括高比表面积、高达 90%的高孔隙率、一维排列、功能性纳米材料的易于掺入和多样化的结构，使其在学术研究和工业应用中都具有很高的吸引力。本书对纳米纤维在吸附、水处理以及海水淡化等方面制造、改性和应用进行了探讨和总结。尽管目前取得了很大的进展，但在超细纳米纤维和优异孔隙率的纳米纤维膜的优化；在纳米纤维基底上开发坚固而超薄的选择性层；创建复杂的多功能纳米结构；大规模生产这些先进的纳米纤维膜等方面仍需进一步努力。具体如下：

（1）使用超细纳米纤维膜，可以提高较小颗粒的过滤效率，而不会显著牺牲由于纳米纤维基质的高孔隙率而引起的水渗透通量。同时，超细纳米纤维即使在升高的压力下也可以有效地支撑接枝或聚合的表层，并减轻其对基材的侵入，从而促进电纺纳米纤维膜在高压工艺中的应用。然而，目前大规模生产直径小于 100 nm 的纳米纤维仍然具有挑战性。

（2）应对基材材料、表面粗糙度、表面孔隙率和电荷的影响以及它们对选择性层和纳米纤维基材之间的协同作用进行更多研究，以增加两者间的黏合力，提高膜的力学性能。

（3）由于膜结垢而使通量减少到远低于理论容量，应该尝试更多的涂覆和改性方法，以防止污垢层在膜表面上生长。

（4）纳米纤维的可控和可靠生产仍然具有挑战性。稳定、连续批量生产的静电纺丝设备的设计和构造至关重要，值得进一步研究。

可以期待的是，上述挑战的解决能进一步推动电纺纳米纤维在水体净化中的发展，甚至工业化应用。

4.2　空气净化

空气污染（包括雾霾、沙尘暴和酸雨等）已成为最严重的环境问题之一，严重影响了公共卫生、生产效率乃至生态系统的安全。空气中的污染物包括颗粒物（PM）、挥发性有机气体（VOCs）、氮氧化物（NO_x）、硫氧化物（SO_x）、病原体等。据 2014 年世卫组织报告，2012 年 700 万人的死亡与空气污染有关。因此，开发新型、环保和高效的过滤介质，以解决日益严重的空气污染问题，成为目前的研究热点之一。静电纺丝纳米纤维膜具备大的比表面积、高孔隙率和相互连接的多孔结构。更重要的是，电纺纳米纤维易于功能化，这对解决空气中的有害气体和病原体有利。本章对电纺纳米纤维在颗粒物过滤、有害气体净化和病原体净化等领域的应用现状进行了总结，并对未来的发展提出了展望。

4.2.1　颗粒物的净化

随着快速工业化和城市化进程而排放的颗粒物（PM）严重威胁着生态系统和人类健康。危害最大的是空气动力学直径小于 2.5 μm 的颗粒（PM2.5），因为它们可以直接渗透并滞留在肺部深处。静电除尘器、离心收集器等是去除 PM 的常规方法。然而，这些方法具有能耗大、分离效率低和占用空间大的固有缺点。最近，膜空气过滤技术具有高能效、操作简便和环境友好的优点，已被研究用于从空气中捕获 PM。传统的空气过滤介质（微米级纤维）通常对亚微米级 PM 显示出较低的过滤效率，为了提高过滤效率，需制造更厚的介质。但是这样又会导致较高的压降和能源成本。具备直径小、比表面积大、良好的内部连通性和可调整的网络几何形状等优点的电纺纳米纤维膜，将改善高性能过滤膜中传统过滤材料的上述不足，近年来得到广泛的研究。

单一聚合物制备的空气滤膜通常不会表现出出色的过滤性能（例如过滤效率、压降和透气性）或缺少其他必要的膜性能（例如机械性能、超轻性、柔韧性等），故目前研究多集中在复合滤膜上。东华大学的 Wang 等人通过将电纺聚氯乙烯（PVC）/聚氨酯（PU）纤维沉积在常规滤纸载体上，制备出对空气中的颗粒显示优异过滤性能的两层复合过滤膜。其抗拉强度接近 9.9 MPa，且具有高耐磨性（134 个循环）、较高的透气性（154.1 $mm{\cdot}s^{-1}$）、高过滤效率（99.5%）和低压降（144 Pa）。如图 4.9（a）所示，添加 PU 后制备的复合膜在应力负荷下直至达到屈服点之前在第一区域中表现出线性弹性行为，这可能归因于聚合物共混膜中键合和非键合交叉结构的共存，从而显著提高抗拉强度。透气性和过滤效率高则归因于 3D 多孔结构中相互连通的通道（图 4.9（b））。另外 Wang 等人还制作了尼龙 6/PAN 复合膜，该膜具有超轻（2.94 $g{\cdot}m^{-2}$）的二元结构纳米纤维网，显示出 99.99%的颗粒物过滤效率。此外，最近通过将氟化聚氨酯（FPU）并入具有 PAN/PU 电纺复合膜的功能性纳米纤维膜（图 4.9（c））中，制造了具有超强防污性能的超两性纳米纤维膜。如图 4.9（d）所示，通过添加 FPU，创建了低表面能和粗糙的纳米级结构，赋予了膜疏水性和疏油性，水和油的接触角分别为 154° 和 151°。这些电纺纳米纤维膜的超疏水性和超疏油性可以通过调节 FPU 比例来调节。复合膜表现出相对较高的拉伸强度（12.28 MPa）、透气性（706.84 $mm{\cdot}s^{-1}$）和过滤效果（99.9%）。Liu 等人制造了电纺聚丙烯腈/聚丙烯酸（PAN/PAA）复合纳米纤维膜，且可以通过增加聚丙烯酸含量来增强其机械性能，从而使复合纤维毡的拉伸强度从 3.8 MPa 增加到 6.6 MPa，高于原始 PAN 纤维毡。该膜在 5.3 $cm{\cdot}s^{-1}$ 的气流速度下对 300～500 nm 氯化钠气溶胶颗粒显示出高过滤效率（99.994%）和最低的压降（160 Pa）。上述结果表明，单一聚合物制备的电纺空气过滤膜无法达到出色的过滤性能。通过引入另一种聚合物组分，不仅将优化空气过滤性能，还拓宽了过滤膜的应用范围。

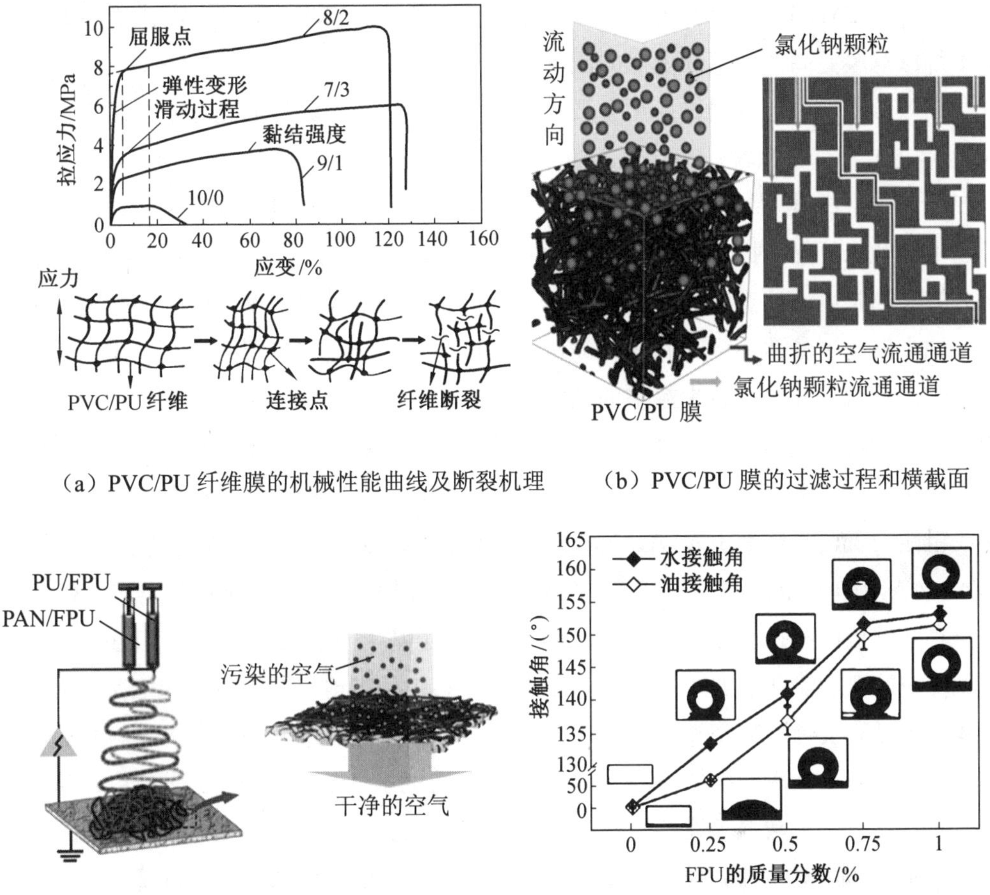

（a）PVC/PU 纤维膜的机械性能曲线及断裂机理　（b）PVC/PU 膜的过滤过程和横截面

（c）复合纳米纤维膜的电纺和过滤　（d）纳米纤维膜的水接触角和油接触角

图 4.9　电纺复合膜的功能性纳米纤维膜

电纺纳米纤维不仅可以通过多种聚合物复合制备，还能通过掺入无机纳米粒子获得。Wang 等人将 SiO_2 纳米颗粒负载到电纺聚丙烯腈纳米纤维膜上，SiO_2 纳米颗粒使纤维表面粗糙，具有非圆形的横截面。得益于逐层堆叠的结构和粗糙的纤维表面，与单层膜相比，多层 PAN/SiO_2 杂化膜具有 99.989%的高过滤效率（过滤 300～500 nm NaCl 气溶胶颗粒）和 117 Pa 的低空气阻力。Cho 等人制备了掺有 TiO_2 的 PAN 纳米纤维，他们认为随着 TiO_2 含量的增加，膜上离子和带电粒子的数量大大增加。

因此，杂化 PAN/TiO_2 膜对 100～500 nm 气溶胶颗粒的过滤性能显示出比原始 PAN 纳米纤维滤膜更高的效率和更低的压降。

高效颗粒空气过滤器的应用主要集中在高性能口罩上，以防止在户外活动中吸入大量 PM 和有害微粒。特别是，随着某病毒的传播，全球疫情的暴发，开发适用于口罩的纳米纤维材料变得越来越重要。与常规使用的口罩过滤介质相比，超薄直径的电纺纳米纤维可提供多种优异的性能，例如高的表面体积比、精细的多孔结构、相对较高的强度，这意味着更高的过滤效率而又不会牺牲渗透率。除了高过滤效率外，纳米纤维过滤器还应具有超低的压降，换句话说，这些口罩还应表现出出色的透气性。目前，开发了越来越多的功能性口罩以提高过滤性能并获得更好的用户体验。Zhu 等人采用静电纺丝法和热交联法制备了多功能的聚乙烯醇/聚丙烯酸（PVA-PAA）复合膜（图 4.10）。然后将超疏水性二氧化硅纳米粒子掺入到纤维中，形成粗糙的表面，然后引入 $AgNO_3$，从而通过 UV 还原形成 Ag 纳米粒子。发现 PVA-PAA-SiO_2-Ag 膜具有很高的空气过滤性能（对 PM2.5 的过滤效率大于 98%）以及有效的抗菌和抗病毒活性。Cheng 等人开发了基于电纺聚醚酰亚胺（PEI）驻极体无纺布的双功能智能面罩。由于 PEI 非织造布中稳定地保留着多余的电荷，因此智能面罩具有去除颗粒物和发电的双重功能。 此外，还展示了动态监控颗粒物去除能力和人类呼吸频率的独特应用。 除了提供独特的空气过滤器外，这项研究还将有望促进自供电可穿戴电子设备的发展。Yang 等人首先将热管理的概念引入口罩中，以提高用户的热舒适性。他们开发了一种在纳米多孔聚乙烯纳米纤维系统，其中具有强 PM 附着力的纳米纤维可确保低压降下的高 PM 捕获效率（对于 PM2.5 而言为 99.6%）和具有高红外（IR）透明性的纳米 PE 基材（92.1%，基于人体辐射的权重）可实现有效的辐射冷却。这些文献中报道的过滤性能和机械性能都非常好。然而，在绿色材料和溶剂的选择、制造方法、膜的改性以及这些膜的后处理等方面仍然有很多改进的空间。

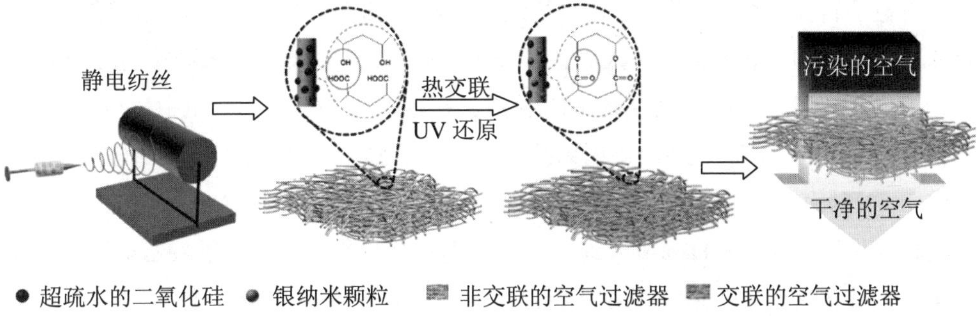

（a）电纺、UV 还原和热交联组合工艺制备 PVA/PAA/SiO_2-Ag 纳米纤维膜示意图

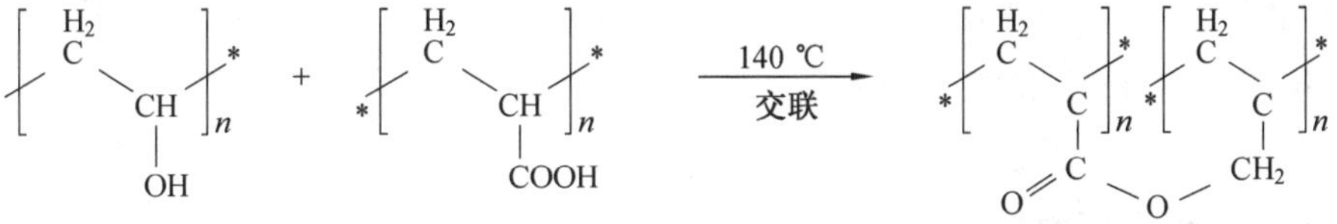

（b）PVA/PAA 间通过酯化反应的热交联机理

图 4.10　多功能聚乙烯醇/聚丙烯酸（PVA-PAA）复合膜（彩图见附录）

4.2.2　有害气体的净化

除了颗粒物外，空气中的有害气体也严重威胁人类健康和生态环境。例如甲醛等有机气体会导致白血病，SO_2 等气体会导致酸雨等。然而，相比于纳米纤维在颗粒物过滤上的广泛研究，纳米纤维在有害气体净化中的研究较少。Celebioglu 等人制备了羟丙基-β-环糊精（HP-β-CD）和羟丙基-γ-环糊精（HP-γ-CD）电纺丝纳米纤维，以截留苯和苯胺两种 VOC 气体。得益于纳米纤维多孔结构，因此 CD 纳米纤维比其粉末形式可从周围捕获更多的 VOC。Kim 等人报告了一种电纺粉煤灰/聚氨酯复合纤维膜，实验数据表明，在 5 种 VOC 中，无论 PU 纤维的组成如何，苯乙烯吸收率最高。粉煤灰质量分数为 30%的 PU 纤维显示出最高的 VOC 吸收能力，是原始 PU 纤维的 2.52～2.79 倍。Souzandeh 等人通过静电纺丝成功地制备了均匀的明胶纳米纤维垫。发现明胶纳米纤维不仅可以有效地去除 PM 颗粒，而且还具有出色的有毒化学物质吸收效率（例如，对于 CO 80%，HCHO 约 76%）。然而吸附/过滤的物理作用仅仅是截留了有害气体，并未使其净化。

得益于科技的发展，用光催化等高级氧化技术来使 VOC 等气体降解可能是一个新的选择。例如，Talukdar 等人制备了一种负载 CeO_2 和 Cu 纳米颗粒的 CNF/ACF 材料，以在室温下通过氧化控制 NO 排放。在富氧（20%）的气氛中，NO 浓度为 500 $mg \cdot L^{-1}$ 时，可获得约 80%的转化率。他们认为复合材料实现的相对较大的转化归因于 CNFs 和 Cu 纳米粒子的联合催化作用，以及二氧化铈与 Cu 纳米粒子之间的协同相互作用。最近，Dan 等人通过绿色静电纺丝和热交联制备了聚乙烯醇（PVA）和魔芋葡甘露聚糖（KGM）基纳米纤维膜，并在其中负载了 ZnO 纳米颗粒（图 4.11（a））。该纳米纤维膜不仅显示出高效的空气过滤性能（图 4.11（b）），而且显示出优异的光催化活性（图 4.11（c））和抗菌活性（图 4.11（d））。ZnO @ PVA/KGM 膜对超细颗粒（300 nm）的过滤效率高于 99.99%，优于商用过滤器。其凭借高的光催化活性，在 120 min 的阳光照射下，甲基橙被有效脱色，去除率高达 98%以上。但该领域目前仍处于探索阶段，还需要开展更多的研究。

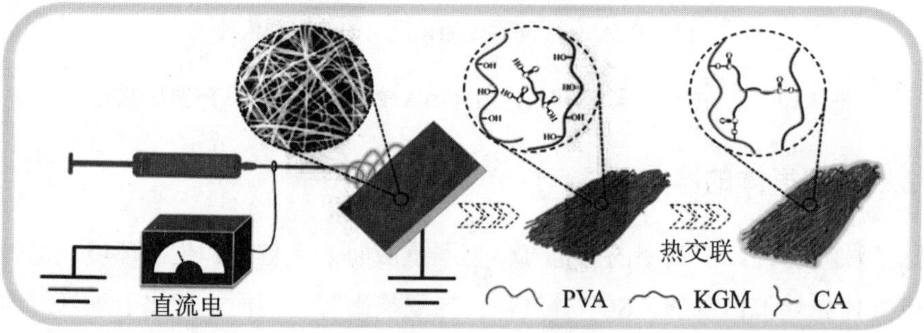

（a）ZnO @ PVA/KGM 膜的制备

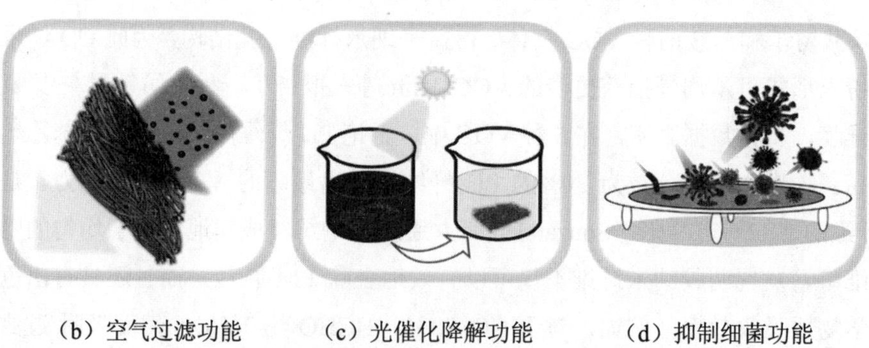

（b）空气过滤功能　（c）光催化降解功能　（d）抑制细菌功能

图 4.11　多功能 ZnO @ PVA/KGM 膜（彩图见附录）

4.2.3　细菌的净化

空气还包括各种细菌、病毒、真菌等，为了抑制病原体的传播和流行病的暴发，人们已经通过利用添加剂来制备抗菌膜，这些添加剂主要有无机（如 Ag 和 TiO_2 等）和天然植物提取物（如苦参提取物等）两种类型。Ag 纳米颗粒与细菌的相互作用如图 4.12（a）所示，带负电的细菌细胞壁和带正电的纳米粒子之间的静电相互作用导致细胞壁破裂，从而导致微生物死亡。Zhang 等人制备了具有抗菌活性的丝绸空气过滤介质（图 4.12（b）），其中包含通过甲酸还原 $AgNO_3$ 形成纳米级 Ag 颗粒（图 4.12（c））。该丝绸纳米纤维膜具有高效、低气流阻力、低质量和抗菌能力。Vanangamudi 等人通过电纺丝法制备了混合疏水复合 PVDF-Ag-Al_2O_3 纳米纤维空气滤膜，如图 4.12（d）所示。膜的抗菌活性证明，银的掺入提供了合适的消毒方式，所有膜的抗菌效率均高于 99.5%。TiO_2 是另一种常用的抗菌添加剂，Wang 等通过静电纺丝技术制备了具有优良的空气过滤性能和良好的抗菌活性的杂化聚乳酸/二氧化钛（PLA/TiO_2）纤维膜。PLA/TiO_2 纤维膜表现出较高的过滤效率（99.996%）和相对较低的压降（128.7 Pa），以及较高的抗菌性活性（99.5%）。

除无机纳米颗粒外，环保的植物提取物也可以添加到电纺纳米纤维膜中以实现抗菌过滤。例如，Choi 等人通过将苦参提取物掺入纤维中，制备了一种电纺抗菌膜，其对葡萄球菌的抗菌活性达到 99.98%，这种结合了无毒物质的静电纺丝空气过滤膜可能会成为未来个人防护的候选者。

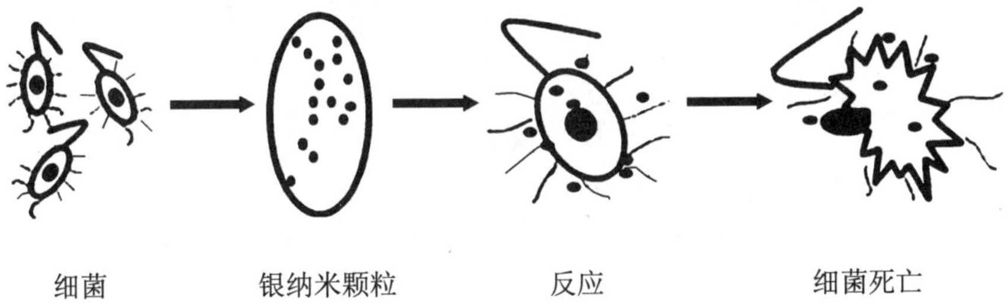

（a）Ag 纳米颗粒与细菌相互作用示意图

图 4.12　抗菌纤维膜

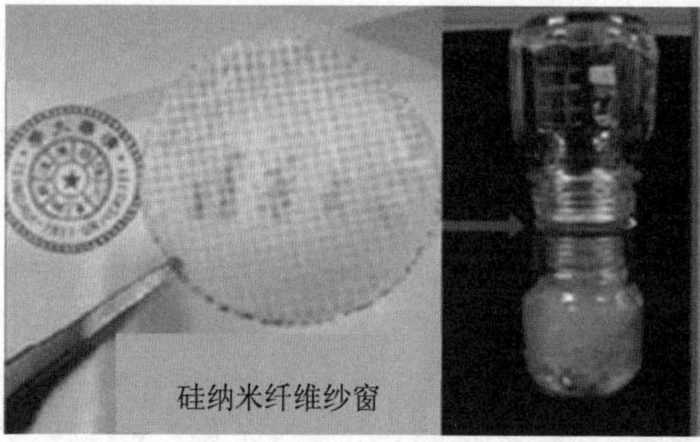

（b）硅纳米纤维纱窗及挡烟效果的光学照片

（c）Ag 掺杂的丝纳米纤维 TEM 图

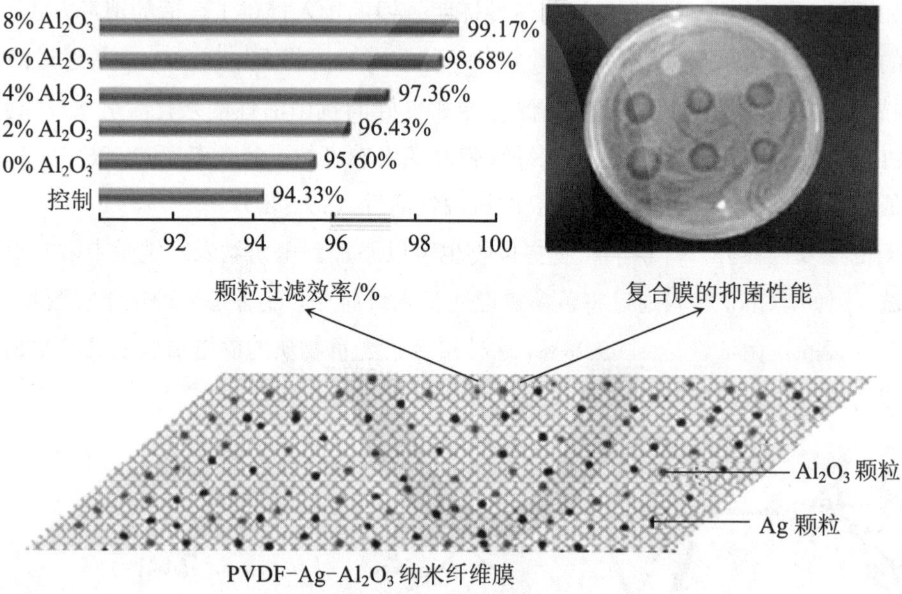

（d）纳米纤维微粒过滤效率和抗菌性能的示意图

续图 4.12

4.2.4　小结与展望

净化空气中的 PM、VOCs 和细菌等微生物，已成为确保人类健康和生态环境安全的重要问题。在过去的几十年中，电纺纳米纤维膜由于其直径小、比表面积大、孔隙率大等巨大优势，作为有效的空气过滤介质而受到了广泛关注和研究。目前，已经开发出一系列以聚合物、无机物为基体的电纺纳米纤维及其复合材料，且取得了显著进展，但仍然存在一些挑战尚未解决，这限制了它们的大规模应用。具体如下：

（1）需继续提高纳米纤维过滤材料的机械强度，以满足实际应用的需求。

（2）应进一步努力解决与制造有关的问题，例如生产效率、制造成本等。

（3）进一步开发绿色材料和水性溶剂或无溶剂静电纺丝方法来生产纳米纤维，以消除残留溶剂的负面影响。

（4）开发多功能纳米纤维复合材料，以应对目前在空气净化中的挑战。如通过结合光催化降解等技术，制备集吸附、过滤和净化功能于一体的纳米纤维复合材料，在高效过滤的同时，实现有害气体和微生物的高效净化。

4.3　油水分离

4.3.1　引言

工业含油废水及生活含油污水的大量排放使得环境污染问题日益严重，加上石油泄漏事故频繁发生，使得人类的健康受到了严重的威胁。因此，将体系复杂的油水混合物进行有效的分离成为目前急需解决的关键问题之一。处理含油污水的一个关键难点在于随着时间的推移，原油和成品油的理化性质发生了重大变化。这是由于水-油乳状液的蒸发、形成以及油在水中的分散。如图 4.13 所示，说明了油水乳化过程以及形成的稳定和不稳定的分散液和乳状液。传统的油水分离技术包含化学分离法、物理分离法和生物分离法三大类。化学分离法主要包括中和法、沉淀法、混凝法、氧化还原法等，化学分离法主要通过投加药剂产生的化学作用实现油水分离，根据含油废水的成分不同所需要添加的药剂及其投加量将不一样，即存在油水分离效率低、成本高，由于药剂的添加难以回收再利用且极易引发二次污染等问题。物理分离法又可以细分为聚结分离法、离心法、重力法等其他辅助分离方法。物理分离法是油水分离方法中应用领域最广的一种分离方法，它的优点在于分离过程中不会给分离物质带来新的杂质和污染物，但是它仍然具有分离时间过长、分离设备占地面积大、油水分离效果不够彻底等缺点。生物分离法主要包括生物膜反应器和强力的生物降解等方法，虽然生物法可以高效地使油水混合物得到有效的分离，但是微生物的量难以控制，灵活性较差。这些方法在操作上通常需要大量的资金注入，而且需要操作人员具备较高的工程专业知识，在操作过程中所用到的基础设施也比较昂贵。

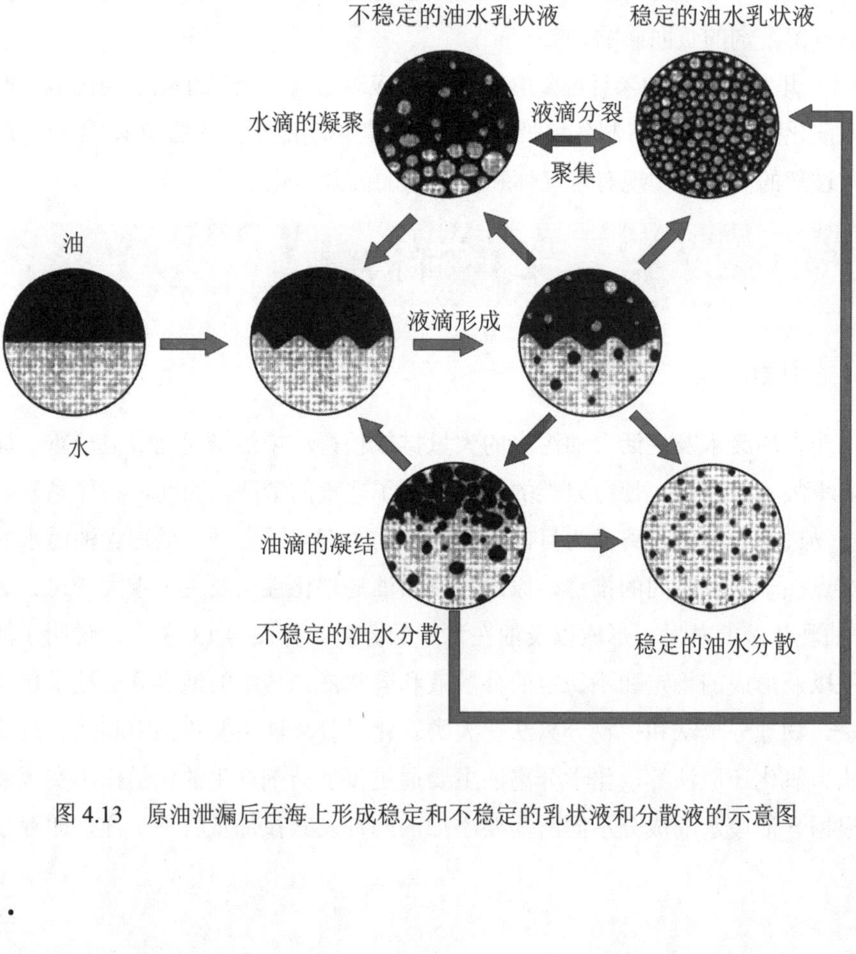

图 4.13　原油泄漏后在海上形成稳定和不稳定的乳状液和分散液的示意图

随着经济的高速发展，工业化进程越来越快，显然传统的油水分离技术无法满足人们的生活要求，更无法满足国家的环保标准，这迫使人们寻求更好的油水分离方法来减少环境污染。膜分离技术与传统的油水分离方法相比，具有更高的分离效率和更低的生产成本，且不会进一步加重环境压力，也不会因治疗本身而危害人体健康。而纳米纤维膜不仅具有良好的柔韧性和可调润湿性，而且具有连通的开孔结构，这些结构特点在油水分离、污水处理等领域具有很好的应用前景，是应对油泄漏、水污染的理想选择，并且在分离粒径小于 20 μm 的乳化油方面具有独特优势，它的出现为油水分离领域提供了一个可靠的解决思路。

4.3.2　亲油疏水型纳米纤维

有机合成纤维，如聚丙烯，由于其亲油疏水性、高油水选择性和低密度等特点，被广泛应用于油田溢油清理。近年来，由于聚苯乙烯具有非极性和低表面能、超疏水性和亲油性，PS 纳米纤维垫被广泛用于选择性吸油应用的研究。与商用 PP 超细纤维相比，PS 纳米纤维吸附剂具有非常低的密度（2～3 $mg\cdot cm^{-3}$）和非常高的孔隙率（99.7%）。结果证明了静电纺丝方法在制备高容量吸油剂中的有效性。Lin 等人利用静电纺丝技术制备了表面光滑、粗糙等不同形貌的 PS 纤维，并对其吸油性能进行了测试，以确定其在溢油清理中的潜在用途。表面光滑、纤维直径较小的多孔 PS 纳米纤维具有最高的吸油性能，其次是表面粗糙、纤维直径较大的多孔 PS 纤维。PS 纳米纤维的吸附容量是商业化聚丙烯纤维的 3～4 倍，这归因于吸附、毛细管作用或两者的结合。

直接静电纺丝制备疏水性亲油聚合物是制备滤油膜的一种简单有效的方法。U. P. Sarfaraz 等人以聚丙烯为原料，制备了疏水性纤维毡，用于去除柴油机中分散的水滴。用不同质量浓度的聚丙烯溶液进行静电纺丝，得到不同直径的亚微米级纤维。这些纤维膜对油和水具有良好的选择性润湿性，能有效地去除燃料中的水分，其去除效率达到 99%。

如图 4.14 所示，C.Shin 等人通过一步静电纺丝将 PS-NFs 直接沉积到基底（不锈钢网）上，成功制备了疏水超亲油膜。制备工艺简单，原料易得，价格低廉。由于 PS 的低自由能和膜的网状结构，因此膜的疏水性得到显著提高，同时保持了固有的亲油特性。因此，所获得的 PS 纤维膜可以直接有效地分离油/水混合物。其他疏水性亲油聚合物（如 PP、PVDF 和 PVB）也被直接用于通过一步静电纺丝法制备油水分离膜。

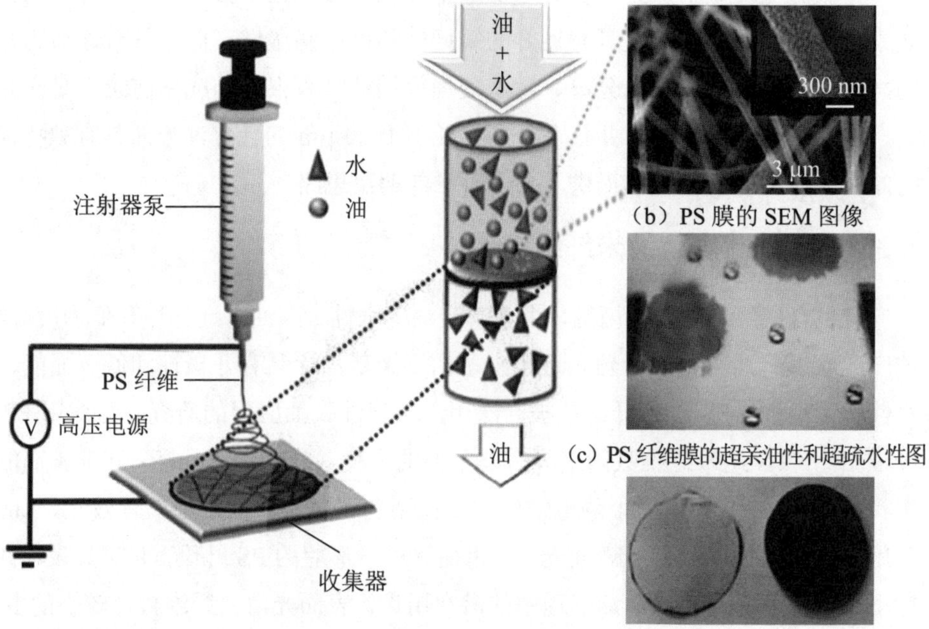

（b）PS 膜的 SEM 图像

（c）PS 纤维膜的超亲油性和超疏水性图

（a）用于油水分离的 PS 纤维膜的制备工艺示意图　（d）PS 纳米纤维膜与硬币大小对比图

图 4.14　PS 纳米纤维膜

Botao Song 等人利用静电纺丝技术，将电纺疏水亲油性聚乙烯醇缩丁醛（PVB）直接沉积在具有不同网孔数的不锈钢丝网上，得到具有不同孔径的纳米纤维膜，如图 4.15 所示。PVB 纳米纤维膜具有选择性润湿性和可调多孔结构的协同作用，具有分离不相容油水混合物和稳定的油包水乳液的能力。此外，通过控制孔径，分离膜的油通量大大提高，约为商用非织造布过滤膜的 10 倍。

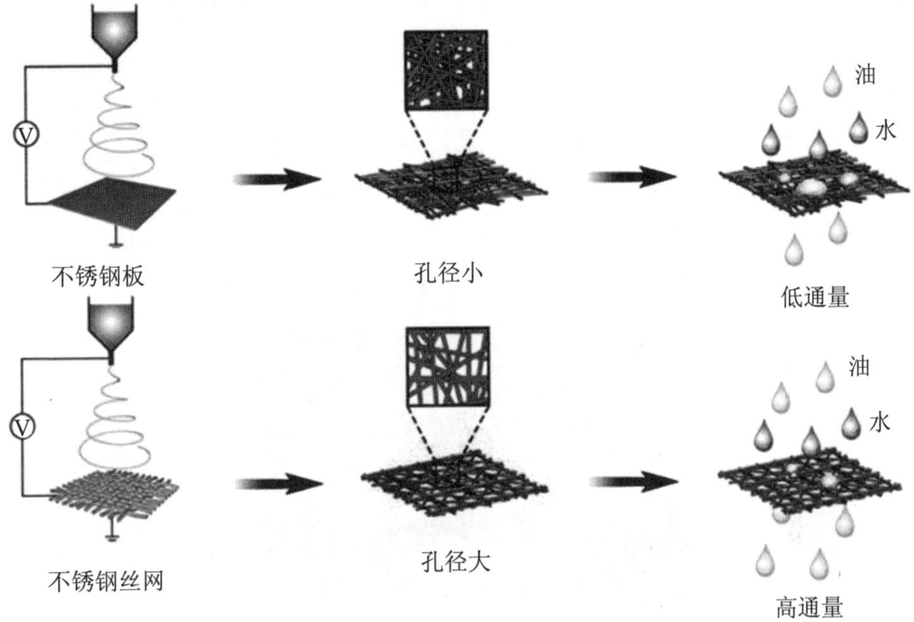

（a）设计孔径纳米纤维膜的制备工艺及油水分离机理示意图

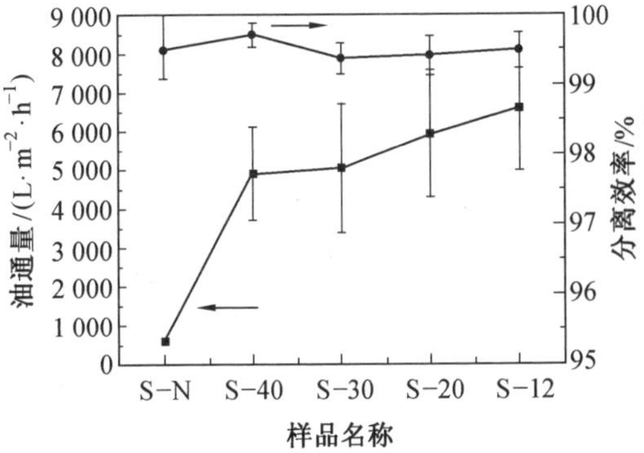

（b）相应纳米纤维膜的油通量和分离效率图

（c）分离前油包水乳液的图像和滤液的光学图

图 4.15　聚乙烯醇缩丁醛（PVB）纳米纤维膜

通过改变不同来源聚合物的分子量、溶剂组成、溶液浓度以及电纺 PS 纤维的微观和纳米结构，研究多孔 PS 纳米纤维膜的疏水性-亲油性及其作为吸油剂的用途（图 4.16）。多孔 PS 纳米纤维膜对机油和葵花籽油的吸附量分别为 84.41 $g \cdot g^{-1}$ 和 79.62 $g \cdot g^{-1}$，是商用 PP 无纺布的近 3 倍。通过静电纺丝技术控制 PS 溶液的纤维内孔隙率和纤维间孔隙率，并通过多喷嘴静电纺丝装置添加电纺聚氨酯（PU）纤维，以提高纤维毡的力学性能，特别是弹性和可恢复性。PU 含量较低的纤维毡对机油和葵花籽油的吸油量分别为 30.81 $g \cdot g^{-1}$ 和 24.36 $g \cdot g^{-1}$，具有较好的重复使用性。

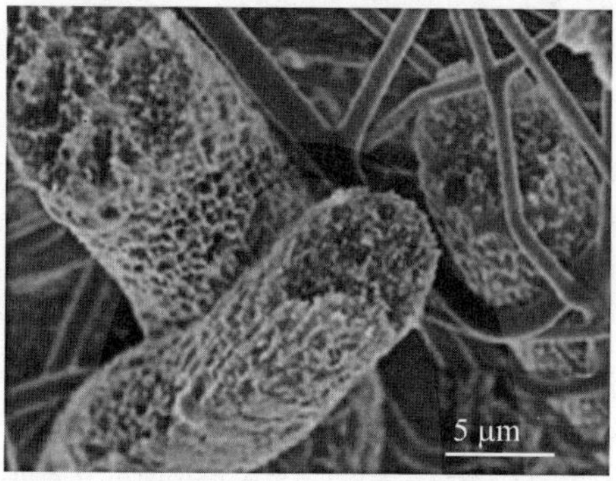

(a) 质量分数为 30%PS 溶液在 45%相对湿度、PS 与 PU 喷嘴比例为 4∶1 条件下制备的纳米纤维膜的 SEM 图

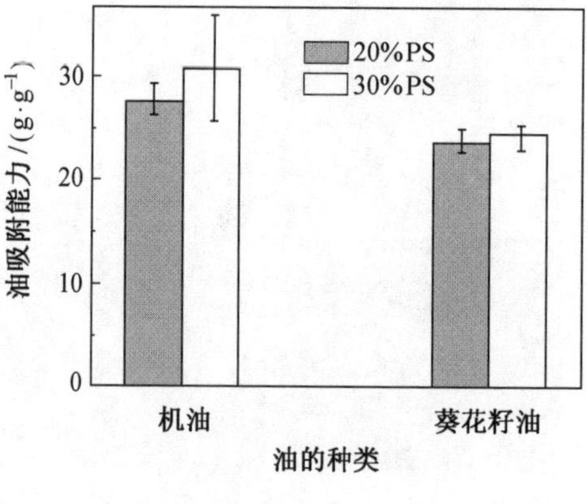

(b) 4/1 PS–PU 中质量分数为 20% PS 和 30 %PS 纤维的吸油性能对比图

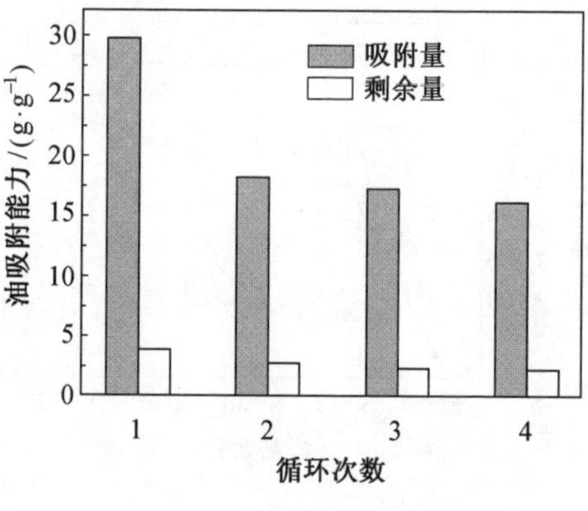

(c) 纳米纤维膜的可重复使用图

图 4.16　多孔 PS 基纳米纤维膜的 SEM 图和性能测试

纳米纤维吸油后的强度和弹性较差，导致其重复使用性差，是一个具有挑战性的问题。通过同轴静电纺丝，纤维的核-壳结构提供了从产品中获得独特性能的潜力。Lin 等人通过同轴静电纺丝制备了具有高比表面积的复合 PS-PU 纤维，用作吸油剂。复合纤维膜对机油（64.40 $g\cdot g^{-1}$）和葵花籽油（47.48 $g\cdot g^{-1}$）的吸油性能均显著提高，为常规聚丙烯纤维的 2～3 倍。即使经过五次吸附循环，纤维垫仍然保持较高的吸油能力。

4.3.3　超疏水型纳米纤维

超疏水型纳米纤维膜具有高的水接触角和低的接触角滞后特性，在许多潜在的应用领域，如自清洁和防护服、微流体动力学等方面有着巨大的应用，但它们作为油水分离膜的实用性是一个挑战，也是一个关注点。尚等人通过将原位聚合含氟聚苯并恶嗪（F-PBZ）和电纺的醋酸纤维素纳米纤维（CA）组合，制备出了独特的超亲油和超疏水纳米纤维膜，显示出强大的油水分离能力，该功能层包含二氧化硅纳米粒（SiO_2NPs）（图 4.17（a））。通过 F-PBZ/SiO_2 纳米粒子进行修饰，使原始的亲水 CA 纳米纤维膜具有 161° 水接触角的超疏水性和 3° 油接触角的超亲油性。该膜对油水混合物具有快速有效的分离效果，并且在广泛的 pH 条件下具有优异的稳定性，这将使其成为工业含油污水处理和溢油清理的良好候选物。Yang 等人将经静电纺丝制备的 PS 纳米纤维膜浸入多巴胺碱性缓冲液中 24 h，即可制得聚多巴胺涂层的 PS 纳米纤维。再分别用十一烷硫醇（UT）或 11-巯基十一烷酸（MUA）反应制备油水分离膜。固定在膜表面的长链 UT 明显增加了水的接触角，水被完全阻滞，而油则顺利通过膜（图 4.17（b）、（c））。

如图 4.18 所示，通过 PAN-NFs（APAN）胺化和用化学镀技术将 Ag 纳米团簇固定在纤维表面（APAN-Ag）的简单组合，可以在单一电纺 NF 的表面上构建均匀且分层的粗糙层。该 APAN-Ag 纳米纤维膜经烷基硫醇改性，改性纳米纤维膜具有超疏水和超亲油性，在高盐环境和广泛的 pH 条件下具有优异的油水分离能力。

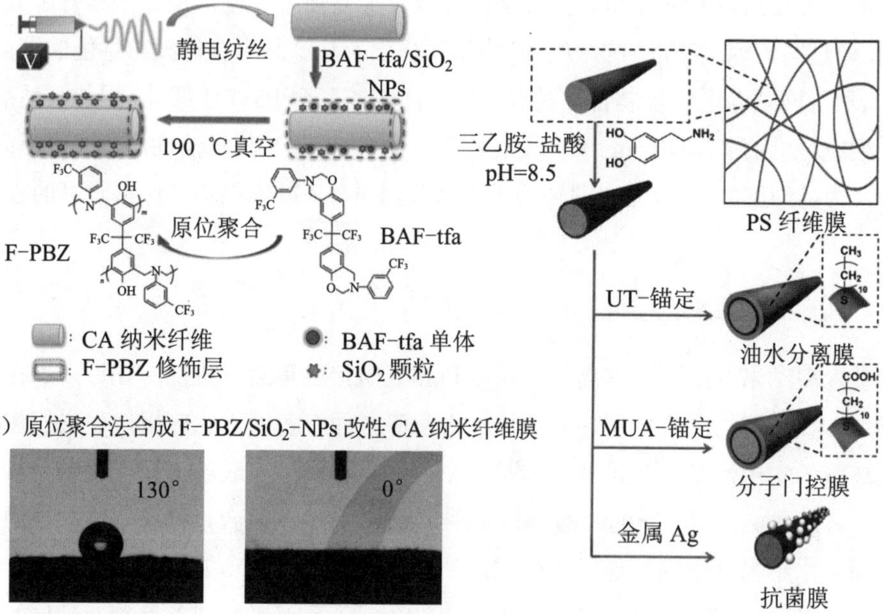

（a）原位聚合法合成 F-PBZ/SiO_2-NPs 改性 CA 纳米纤维膜

（b）聚多巴胺涂层 PS 纳米纤维膜的 WCA 光学图像（c）以 PS 纳米纤维膜为平台制备多孔功能膜

图 4.17　超亲油和超疏水纳米纤维膜

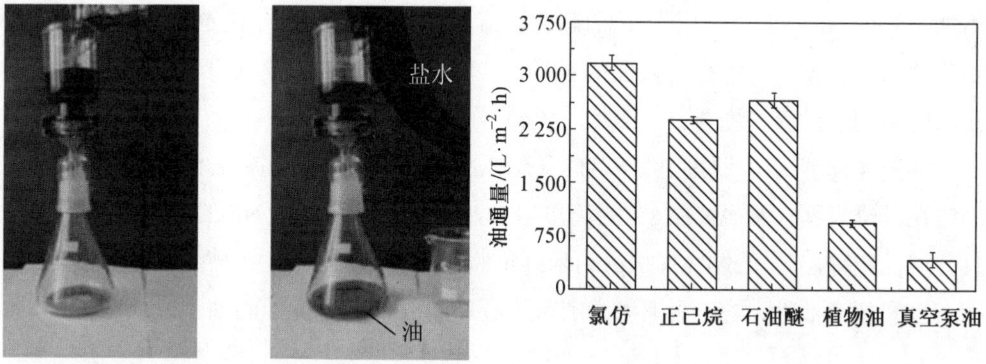

（a）APAN-Ag 膜的油水分离过程　　（b）不同油水混合物分离用纳米纤维膜的油通量

图 4.18　烷基硫醇改性的 APAN-Ag 膜

使用亲水膜的纳滤和超滤在诸如油-水乳液等水处理领域的应用变得越来越重要。传统的膜在去除水中的微乳液方面是有效的，但是由于渗透性和表面污染的限制，它们的通量通常很低。膜表面亲水性被广泛认为是控制膜污染发展的主要因素。与疏水膜相比，亲水膜表面通常具有更高的抗污染能力，并且为了改善污染问题，研究者已经做出了大量的尝试。

Wang 等人首次制备出高通量过滤介质，包括三层复合结构，即无孔亲水性聚醚双酚酰胺亲水性纳米复合涂层顶层、电纺聚乙烯醇（PVA）纳米纤维基材中层，以及用于油水乳化液分离的传统非织造微纤维支架。当在纳米纤维膜上沉积一层薄的亲水性、高透水性涂层时，由于涂层的水力阻力，复合膜的渗透通量通常会降低。Yoon 等人介绍了一种以聚丙烯腈（PAN）电纺纳米纤维为中间层支架、交联 PVA 为顶层涂层的膜体系。优化了膜厚和中间层厚度等关键参数，使油水混合物在实际压力范围（高达约 8.963×10^5 Pa）内长时间（测试时间长达 190 h）测试（水中的油微乳溶液浓度 500 $mg\cdot L^{-1}$）。该小组还研究了纳米纤维膜的油水分离特性，纳米纤维膜由电纺纳米纤维 PVA 中间层支架上的 UV 固化 PVA 水凝胶屏障层和聚对苯二甲酸乙二醇酯（PET）非织造基片组成（图 4.19）。经过长时间操作后，沉积在 PAN 纳米纤维支架上的离子液体中的纤维素，在乳化油水混合物时亦显示出了高渗透通量。而且直径为 5～10 nm 的超细多糖纳米纤维（即纤维素和甲壳素）也被当作屏障层，最后形成的薄膜纳米纤维复合膜的截留率超过 99.5%，且渗透通量提高了 10 倍，可以很好地应用在油水分离领域。

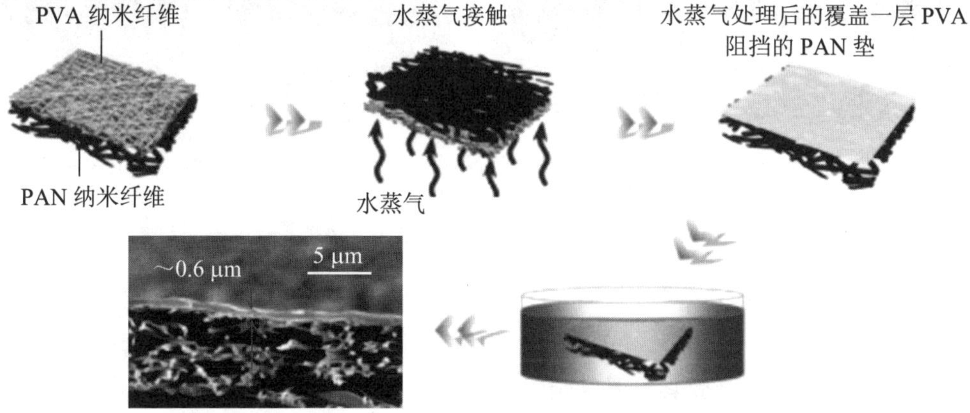

图 4.19　基于 PAN 电纺纳米纤维基底和交联 PVA 阻挡层的薄膜纳米纤维复合膜制备工艺示意图

4.3.4 具有可控润湿性的油水分离膜

除了具有单一疏水亲油或亲水亲油性能的纳米纤维外，具有可控表面润湿性的智能特种可湿膜因其在基础研究和工业应用中的良好性能而受到越来越多的关注。一般来说，在固体表面加载诸如 pH、温度、光，甚至气体等外部刺激物来控制其润湿性，因此相应的刺激响应活性材料是制备智能可湿表面的关键因素。因此，各种新型的可切换可湿性智能纳米纤维膜已被开发用于油水分离。如图 4.20 所示，Li 等人将聚甲基丙烯酸甲酯嵌段聚 4-乙烯基吡啶（PMMA-b-P4VP）的电纺 NFs 沉积在金属丝网上，制备了智能纳米纤维膜。利用 P4VP 的 pH 响应特性和 PMMA 的亲油/亲水特性，制备出的膜具有可切换的水/油润湿性，并能在重力作用下实现 pH 可控的油水分离。在分离过程中，油通过膜而水选择性地保留下来；当同一膜被酸性水预湿时，可以实现反分离过程。此外，分离效率高，经过多次分离循环后，膜能保持稳定的智能润湿性。

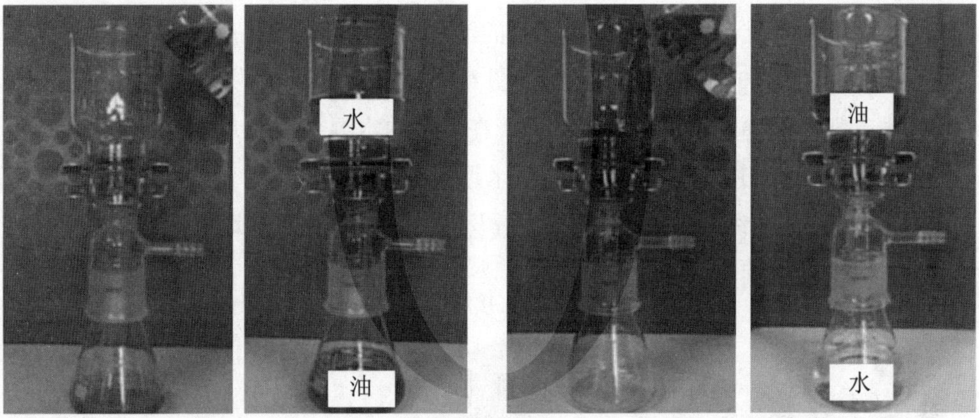

（a）油渗透到膜中，而水留在玻璃仪器上方　（b）水有选择地通过酸性水（pH=3）预湿膜

图 4.20　PMMA-b-P4VP 智能油水分离纳米纤维膜

Wang 等人采用表面引发原子转移自由基聚合法，利用热响应性聚合物（N-异丙基丙烯酰胺）（PNIPAAm）对电纺再生纤维素纳米纤维膜进行改性。PNIPAAm 接枝纤维素纳米纤维膜具有温度响应特性，具有可切换的超冻干/超冻干特性和良好的可控油水分离性能，在废水处理和油净化方面具有广阔的应用前景。在进一步的

研究中，制备了 TiO_2 掺杂的电纺 PVDF 纳米纤维膜，并在光（紫外线或阳光）照射和加热处理下显示出可切换的润湿性。智能膜通过选择性地允许水或油单独通过，可以有效地分离油/水混合物。更有趣的是，气体（如二氧化碳）可以作为控制膜润湿性的触发器，对膜的物理结构影响较小。Che 等人利用静电纺丝技术制备了 CO_2 响应共聚物 PMMA-co-poly（N，N-二乙氨基乙基甲基丙烯酸酯）（PMMA）的智能纳米纤维膜。由于水中的 CO_2 与叔胺基反应后可以形成一种延伸的亲水链，这种亲水链可以将纤维中的 PMMA 结构覆盖，不仅使材料具有良好的流动性，且具有亲水性。如果 N_2 鼓泡处理完全去除 CO_2，膜的亲水性表面润湿性可以恢复到最初的疏水性。利用二氧化碳诱导的选择性润湿行为和粗糙的多孔结构，这些智能可湿纳米纤维膜在水和油净化方面表现良好。

除这些刺激响应膜外，Janus 膜在水包油和水包油乳液的可切换分离方面也显示出良好的应用性能。Jiang 等人通过在电纺聚丙烯腈膜的单侧网络中涂覆一层疏水碳纳米管构建了 Janus 膜。获得的碳纳米管@聚丙烯腈膜的两侧均表现出不对称的润湿性（CNT 侧为水下亲油即为疏水性，PAN 侧为水下疏油即为亲水性）。因此，膜可在不同操作模式下显示出可切换的油水分离性能（即 CNT 侧高效油包水乳化分离和 PAN 侧为高效水包油乳化分离）。

4.3.5　用于油水分离的纳米纤维气凝胶

气凝胶作为一种新型的三维多孔大体积材料，已引起人们的广泛关注。气凝胶以其高孔隙率、低体积密度、高比表面积和曲折通道而闻名，这使其在油水分离方面具有巨大优势。因此，已经创造了几种用于油水分离的功能性气凝胶，例如二氧化硅胶体气凝胶、碳纳米管气凝胶、石墨烯气凝胶和聚合物气凝胶。制备用于油水分离的气凝胶的两个主要因素是表面润湿性（影响乳化液滴的截留和破乳）和细胞结构（决定一相（如油）通过气凝胶的渗透速率）。探索开发基于气凝胶的新型油水分离技术的相关技术是很有希望的。

Si 等人在纳米纤维气凝胶的制备方面取得了显著进展，如图 4.21 所示。他们展示了一种简单的方法来制备具有由黏结 NFs 组成的分层多孔结构的超弹性和超疏水气凝胶，并将其称为“纤维状、各向同性的复合弹性重建”（FIBER）气凝胶。如前所述，电纺 PAN/（苯并恶嗪（BA-a））NFs 和 SNFs 被选为主要构建块。制备过程从混合 NFs 使其均匀化并得到均匀的 NF 分散开始；制备的分散体被固化并冻干以产生无黏结纳米纤维气凝胶；最后一步是热交联无黏结纳米纤维气凝胶以获得纤维 NFA。得益于一种多用途且简单的组装工艺，纳米纤维气凝胶在各种应用中显示出很好的性能，并具有巨大的放大潜力。在此基础上，在气凝胶中加入适量的二氧化硅纳米颗粒以增强其层次多孔结构。采用低表面能的氟双功能苯并恶嗪（BAF-a）作为原位交联剂。所制备的功能性纳米纤维气凝胶具有超弹性、超疏水、超亲油性、高孔隙率等综合性能，能在重力作用下有效地分离油包水乳液。此外，其分离效率高，在相同分离条件下明显高于常规滤膜，表明在实际乳化液分离中具有稳健的性能。

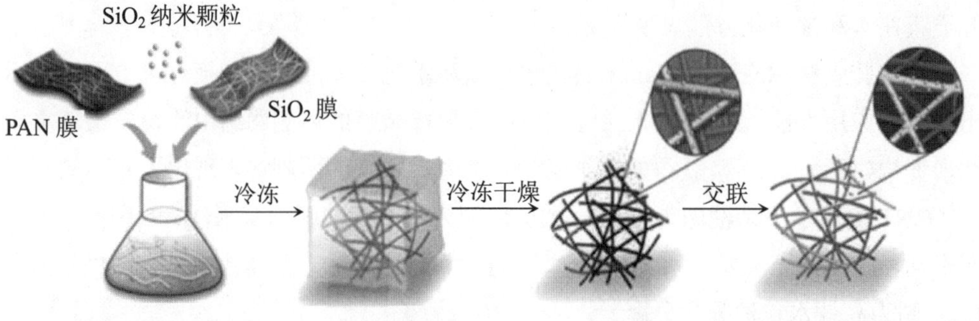

（a）均质纳米纤维分散　（b）冷冻的分散体　（c）游离的气凝胶　（d）结合纤维气凝胶

图 4.21　分层结构功能纤维气凝胶的制备工艺示意图

4.3.6　小结与展望

近年来，大量研究致力于工程复合纳米纤维膜，以改善纳米纤维膜的性能并扩大其潜在应用，电纺纳米纤维膜在环境应用（如油水分离）中显示出比传统介质更大的优势，但仍然存在一些难题需要解决，以实现纳米纤维膜在油水分离中的应用。尚需解决的问题如下：①如何制备具有高通量、高力学性能和高循环稳定性的超疏水/疏油纳米纤维膜材料，并可在超疏水和超亲水之间进行切换，以实现水（或油）的定向传递。②粗糙层在负载条件下容易损坏，这会使纳米纤维膜失去油/水选择性。此外，在过滤过程中纳米颗粒的逐渐脱离可能会产生二次污染。③油水分离过程非常复杂，其中涉及微流体力学、界面科学和工程学，但分离过程的理论模型并不完善，无法指导科学家合理设计和制造具有所需功能的纳米纤维膜。④缺乏实际油水混合物的分离效果，需进一步研究。⑤在连续式油水分离装置上的研究存在不足。解决方法：开发具有高通量、高力学性能和高循环稳定性的纳米纤维膜制备技术，建立油水分离理论模型，研究在纳米纤维上制备稳定的粗糙层的方法，设计合成超疏水/疏油可切换的纳米纤维材料，研究材料润湿性变化和定向传递能力的相互关联，阐明其构效机理。设计高效的连续式油水分离装置，开展超疏水/疏油可切换的纳米纤维膜在实际油水混合物中的分离效果和稳定性研究。在此基础上，进一步优化超疏水/疏油可切换的纳米纤维材料的结构和性能，实现纳米纤维膜分离材料在油水分离中的应用。

参考文献

[1] HALEEMA S, LEVENT T, ALI K, et al. Recent advances in nanofibrous membranes: production and applications in water treatment and desalination[J]. Desalination, 2020, 478: 114178.

[2] 世界卫生组织. 全球三之一的人无法获得安全饮用水[J]. 中国卫生政策研究, 2019, 12(7): 58.

[3] MUZAMMIL A, MIANDAD R, WAQAS M, et al. Remediation of wastewater using various nanomaterials[J]. Arabian Journal of Chemistry, 2019, 12: 4897-4919.

[4] LIAO Y, LOH C-H, TIAN M, et al. Progress in electrospun polymeric nanofibrous membranes for water treatment: fabrication, modification and applications[J]. Progress in Polymer Science, 2018, 77: 69-74.

[5] OMOMIYI P, CHRISh B-A, KATRI L, et al. Morphology, modification and characterisation of electrospun polymer nanofiber adsorbent material used in metal ion removal[J]. Journal of Polymers and the Environment, 2019, 27: 1843-1860.

[6] SAEED K, HAIDER S, OH T J, et al. Preparation of amidoxime-modified polyacrylonitrile (PAN-oxime) nanofibers and their applications to metal ions adsorption[J]. Journal of Membrane Science, 2008, 322: 400-405.

[7] MA H Y, HSIAO B S, CHU B. Electrospun nanofibrous membrane for heavy metal

ion adsorption[J]. Current Organic Chemistry, 2013, 17: 1361-1370.

[8] ALUIGI A, TONETTI C, VINEIS C, et al. Adsorption of copper (Ⅱ) ions by keratin/PA6 blend nanofibers[J]. European Polymer Journal, 2011, 47: 1756-1764.

[9] HAIDER S, PARK S Y. Preparation of the electrospun chitosan nanofibers and their applications to the adsorption of Cu(Ⅱ) and Pb(Ⅱ) ions from an aqueous solution[J]. Journal of Membrane Science, 2009, 328: 90-96.

[10] SURGUTSKAIA N S, MARTINO A D, ZEDNIK J, et al. Efficient Cu^{2+}, Pb^{2+} and Ni^{2+} ion removal from wastewater using electrospun DTPA-modified chitosan/ polyethylene oxide nanofibers[J]. Separation and Purification Technology, 2020, 247: 116914.

[11] SAEED K, HAIDER S, OH T J, et al. Preparation of amidoxime-modified polyacrylonitrile (PAN-oxime) nanofibers and their applications to metal ions adsorption[J]. Journal of Membrane Science, 2008, 322: 400-405.

[12] NEGHLANI P K, RAFIZADEH M, TAROMI F A. Preparation of aminated-polyacrylonitrile nanofiber membranes for the adsorption of metal ions: comparison with microfibers[J]. Journal of Hazardous Materials, 2011, 186:182-189.

[13] BORNILLO KAS, KIM S, CHOI H. Cu (Ⅱ) removal using electrospun dual-responsive polyethersulfone-poly (dimethyl amino) ethyl methacrylate (PES-PDMAEMA) blend nanofibers[J]. Chemosphere, 2020, 242: 125278.

[14] WU S, LI F, WANG H, et al. Effects of poly (vinyl alcohol) (PVA) content on preparation of novel thiol-functionalized mesoporous PVA/SiO_2 composite nanofiber membranes and their application for adsorption of heavy metal ions from aqueous solution[J]. Polymer, 2010, 51: 6203-6211.

[15] TENG M, WANG H, LI F, et al. Thioether-functionalized mesoporous fiber membranes: sol-gel combined electrospun fabrication and their applications for Hg^{2+} removal[J]. Journal of Colloid and Interface Science, 2011, 355: 23-28.

[16] SI Y, REN T, LI Y, et al. Fabrication of magnetic polybenzoxazine-based carbon nanofibers with Fe_3O_4 inclusions with a hierarchical porous structure for water treatment[J]. Carbon, 2012, 50: 5176-5185.

[17] QIAO J L, LI F T, BERA P K. Electrospun mesoporous carbon nanofibers produced from phenolic resin and their use inthe adsorption of large dye molecules[J]. Carbon, 2012, 50: 2877-2886.

[18] LI M, WANG H, WU S, et al. Adsorption of hazardous dyes indigo carmine and acid red on nanofiber membranes[J]. RSC Advance, 2012, 2: 900-907.

[19] MIAO Y E, WANG R, CHEN D, et al. Electrospun self-standing membrane of hierarchical SiO_2@gamma-AlOOH (boehmite) core/sheath fibers for water remediation[J]. ACS Applied Materials & Interfaces, 2012, 4: 5353-5359.

[20] YUE X, FENG S, LI S, et al. Bromopropyl functionalized silica nanofibers for effective removal of trace level dieldrin from water[J]. Colloids and Surfaces A, 2012, 406: 44-51.

[21] XU Q, WU S Y, WANG M, et al. Electrospun nylon6 nanofibrous membrane as spe adsorbent for the enrichment and determination of three estrogens in environmental water samples[J]. Chromatographia, 2010, 71: 487-492.

[22] FANE A G, WANG R, HU M X. Synthetic membranes for water purification: status and future[J]. Angewandte Chemie International Edition, 2015, 54: 3368-3386.

[23] YOON K, HSIAO B S, CHU B. Functional nanofibers for environmental applications[J]. Journal of Materials Chemistry, 2008, 18: 5326-5334.

[24] GOPAL R, KAUR S, MA Z, et al. Electrospun nanofibrous filtration membrane[J]. Journal of Membrane Science, 2006, 281: 581-586.

[25] WANG R, LIU Y, LI B, et al. Electrospun nanofibrous membranes for high flux microfiltration[J]. Journal of Membrane Science, 2012, 392-393: 167-174.

[26] BARHATE R S, LOONG C K, RAMAKRISHNA S. Preparation and characterization of nanofibrous filtering media[J]. Journal of Membrane Science, 2006, 283: 209-218.

[27] GOPAL R, KAUR S, FENG C Y, et al. Electrospun nanofibrous polysulfone membranes as pre-filters: particulate removal[J]. Journal of Membrane Science, 2007, 289: 210-219.

[28] AUSSAWASATHIEN D, TEERAWATTANANON C, VONGACHARIYA A.

Separation of micron to sub-micron particles from water: electrospun nylon-6 nanofibrous membranes as pre-filters[J]. Journal of Membrane Science, 2008, 315: 11-19.

[29] HOMAEIGOHAR S S, BUHR K, EBERT K. Polyethersulfone electrospun nanofibrous composite membrane for liquid filtration[J]. Journal of Membrane Science, 2010, 365: 68-77.

[30] YOON K, HSIAO B S, CHU B. Formation of functional polyethersulfone electrospun membrane for water purification by mixed solvent and oxidation processes[J]. Polymer, 2009, 50: 2893-2899.

[31] HOMAEIGOHAR S, KOLL J, LILLEODDEN E T, et al. The solvent induced interfiber adhesion and its influence on the mechanical and filtration properties of polyethersulfone electrospun nanofibrous microfiltration membranes[J]. Separation and Purification Technology, 2012, 98: 456-463.

[32] HALAUI R, MOLDAVSKY A, COHEN Y, et al. Development of micro-scale hollow fiber ultrafiltration membranes[J]. Journal of Membrane Science, 2011, 379:370-377.

[33] YOON K, KIM K, WANG X, et al. High flux ultrafiltration membranes based on electrospun nanofibrous PAN scaffolds and chitosan coating[J]. Polymer, 2006, 47: 2434-2441.

[34] ZHAO Z, ZHENG J, WANG M, et al. High performance ultrafiltration membrane based on modified chitosan coating and electrospun nanofibrous PVDF scaffolds[J]. Journal of Membrane Science, 2012, 394-395: 209-217.

[35] WANG X L, ZHANG C, OUYANG P. The possibility of separating saccharides from a NaCl solution by using nanofiltration in diafiltration mode[J]. Journal of Membrane Science, 2002, 204: 271-281.

[36] YOON K, HSIAO B S, CHU B. High flux nanofiltration membranes based on interfacially polymerized polyamide barrier layer on polyacrylonitrile nanofibrous scaffolds[J]. Journal of Membrane Science, 2009, 326: 484-492.

[37] KAUR S, SUNDARRAJAN S, RANA D, et al. Influence of electrospun fiber size on the separation efficiency of thin film nanofiltration composite membrane[J]. Journal

of Membrane Science, 2012, 392-393: 101-111.

[38] KAUR S, SUNDARRAJAN S, GOPAL R, et al. Formation and characterization of polyamide composite electrospun nanofibrous membranes for salt separation[J]. Journal of Applied Polymer Science, 2012, 124: 205-215.

[39] PARK S J, CHEEDRALA R, DIALLO M, et al. Nanofiltration membranes based on polyvinylidene fluoride nanofibrous scaffolds and crosslinked polyethyleneimine networks[J]. Journal of Nanoparticle Research, 2012, 14: 1-14.

[40] YALCINKAYA F. A review on advanced nanofiber technology for membrane distillation[J]. Journal of Engineered Fibers and Fabrics, 2018, 14: 1-12.

[41] YANG X, WANG R, SHI L, et al. Performance improvement of PVDF hollow fiber-based membrane distillation process[J]. Journal of Membrane Science, 2011, 369: 437-447.

[42] TIJING L D, WOO Y C, CHOI J S, et al. Fouling and its control in membrane distillation—a review[J]. Journal of Membrane Science, 2015, 475: 215-244.

[43] FENG C, KHULBE K C, MATSUURA T, et al. Production of drinking water from saline water by air-gap membrane distillation using polyvinylidene fluoride nanofiber membrane[J]. Journal of Membrane Science, 2008, 311: 1-6.

[44] LIAO Y, WANG R, TIAN M, et al. Fabrication of polyvinylidene fluoride (PVDF) nanofiber membranes by electro-spinning for direct contact membrane distillation[J]. Journal of Membrane Science, 2013, 425-426: 30-39.

[45] LIAO Y, WANG R, FANE A G. Engineering superhydrophobic surface on poly (vinylidene fluoride) nanofiber membranes for direct contact membrane distillation[J]. Journal of Membrane Science, 2013, 440: 77-87.

[46] LIAO Y, WANG R, FANE A G. Fabrication of bioinspired composite nanofiber membranes with robust superhydrophobicity for direct contact membrane distillation[J]. Environmental Science & Technology, 2014, 48: 6335-6341.

[47] LIAO Y, LOH C H, WANG R, et al. Electrospun superhydrophobic membranes with unique structures for membrane distillation[J]. ACS Applied Materials & Interfaces, 2014, 6: 16035-16048.

[48] LI X, GARCíA-PAYO M C, KHAYET M, et al. Superhydrophobic polysulfone/polydimethylsiloxane electrospun nanofibrous membranes for water desalination by direct contact membrane distillation[J]. Journal of Membrane Science, 2017, 542: 308-319.

[49] AN A K, GUO J, LEE E J, et al. PDMS/PVDF hybrid electrospun membrane with superhydrophobic property and drop impact dynamics for dyeing wastewater treatment using membrane distillation [J]. Journal of Membrane Science, 2017, 525: 57-67.

[50] TIJING L D, WOO Y C, SHIM W G, et al. Superhydrophobic nanofiber membrane containing carbon nanotubes for high-performance direct contact membrane distillation[J]. Journal of Membrane Science, 2016, 502: 158-170.

[51] SU C, CHANG J, TANG K, et al. Novel three-dimensional superhydrophobic and strength-enhanced electrospun membranes for long-term membrane distillation[J]. Separation and Purification Technology, 2017, 178: 279-287.

[52] REN L F, XIA F, CHEN V, et al. TiO_2-FTCS modified superhydrophobic PVDF electrospun nanofibrous membrane for desalination by direct contact membrane distillation[J]. Desalination, 2017, 423: 1-11.

[53] LEE E J, AN A K, HADI P, et al. Advanced multi-nozzle electrospun functionalized titanium dioxide/polyvinylidene fluoride-co-hexafluoropropylene (TiO_2/PVDF-HFP) composite membranes for direct contact membrane distillation[J]. Journal of Membrane Science, 2017, 524: 712-720.

[54] HOU D, LIN D, DING C, et al. Fabrication and characterization of electrospun superhydrophobic PVDF-HFP/SiNPs hybrid membrane for membrane distillation[J]. Separation and Purification Technology, 2017, 189: 82-89.

[55] MOSLEHI M, MAHDAVI H. Controlled pore size nanofibrous microfiltration membrane via multi-step interfacial polymerization: preparation and characterization[J]. Separation and Purification Technology, 2019, 223: 96-106.

[56] MAHDAVI H, MOSLEHI M. A new thin film composite nanofiltration membrane based on PET nanofiber support and polyamide top layer: preparation and

characterization[J]. Journal of Polymer Research, 2016, 23: 257.
[57] XU G R, LIU X-Y, XU J-M, et al. High flux nanofiltration membranes based on layer-by-layer assembly modified electrospun nanofibrous substrate[J]. Applied Surface Science, 2018, 434: 573-581.
[58] SHI J, KANG H, LI N, ET AL. Chitosan sub-layer binding and bridging for nanofiber-based composite forward osmosis membrane[J]. Applied Surface Science, 2019, 478: 38-48.
[59] CHI X-Y, ZHANG P-Y, GUO X-J, et al. A novel TFC forward osmosis (FO) membrane supported by polyimide (PI) microporous nanofiber membrane[J]. Applied Surface Science, 2018, 427: 1-9.
[60] DENG L, LI P, LIU K, et al. Robust superhydrophobic dual layer nanofibrous composite membranes with a hierarchically structured amorphous polypropylene skin for membrane distillation[J]. Journal of Materials Chemistry A, 2019, 7: 11282-11297.
[61] PARK S H, KIM J H, MOON S J, et al. Enhanced, hydrophobic, fluorine-containing, thermally rearranged (TR) nanofiber membranes for desalination via membrane distillation[J]. Journal of Membrane Science, 2018, 550: 545-553.
[62] WOO Y C, CHEN Y, TIJING L D, et al. CF4 plasma-modified omniphobic electrospun nanofiber membrane for produced water brine treatment by membrane distillation[J]. Journal of Membrane Science, 2017, 529: 234-242.
[63] YAN K-K, JIAO L, LIN S, et al. Superhydrophobic electrospun nanofiber membrane coated by carbon nanotubes network for membrane distillation[J]. Desalination, 2018, 437: 26-33.
[64] WANG S-X, YAP C C, HE J, et al. Electrospinning: a facile technique for fabricating functional nanofibers for environmental applications[J]. Nanotechnology Reviews, 2016, 5: 51-73.
[65] NELIN T D, JOSEPH A M, GORR M W, et al. Direct and indirect effects of particulate matter on the cardiovascular system[J]. Toxicology Letters, 2012, 208: 293-299.

[66] KOO W T, JANG J S, QIAO S, et al. Hierarchical metal-organic framework-assembled membrane filter for efficient removal of particulate matter [J]. ACS Applied Materials & Interfaces, 2018, 10: 19957-19963.

[67] ZHU M, HAN J, WANG F, et al. Electrospun nanofibers membranes for effective air filtration, macromol[J]. Macromolecular Materials & Engineering, 2017, 302: 1600353.

[68] WANG N, RAZA A, SI Y, et al. Tortuously structured polyvinyl chloride/polyurethane fibrous membranes for high-efficiency fine particulate filtration[J]. Journal of Colloid and Interface Science, 2013, 398: 240-246.

[69] WANG N, YANG Y, AL-DEYAB S S, et al. Ultra-light 3D nanofibre-nets binary structured nylon 6-polyacrylonitrile membranes for efficient filtration of fine particulate matter[J]. Journal of Materials Chemistry, 2015, 3: 23946-23954.

[70] WANG N, ZHU Z, SHENG J, et al. Superamphiphobic nanofibrous membranes for effective filtration of fine particles [J]. Journal of Colloid and Interface Science, 2014, 428: 41-48.

[71] LIU Y, PARK M, DING B, et al. Facile electrospun polyacrylonitrile/poly (acrylic acid) nanofibrous membranes for high efficiency particulate air filtration [J]. Fibers and Polymers, 2015, 16: 629-633.

[72] WANG N, SI Y, WANG N, et al. Multilevel structured polyacrylonitrile/silica nanofibrous membranes for high-performance air filtration[J]. Separation and Purification Technology, 2014, 126: 44-51.

[73] CHO D, NAYDICH A, FREY M W, et al. Further improvement of air filtration efficiency of cellulose filters coated with nanofibers via inclusion of electrostatically active nanoparticles[J]. Polymer, 2013, 54: 2364-2372.

[74] ZHU M, HUA D, PAN H, et al. Green electrospun and crosslinked poly(vinyl alcohol)/poly(acrylic acid) composite membranes for antibacterial effective air filtration[J]. Journal of Colloid and Interface Science, 2018, 511: 411-423.

[75] CHENG Y, WANG C, ZHONG J, et al. Electrospun polyetherimide electret nonwoven for bi-functional smart face mask[J]. Nano Energy, 2017, 34: 562-569.

[76] YANG A, CAI L, ZHANG R, et al. Thermal management in nanofiber-based face mask[J]. Nano Letter, 2017, 17: 3506.

[77] CELEBIOGLU A, SEN H S, DURGUN E, et al. Molecular entrapment of volatile organic compounds (VOCs) by electrospun cyclodextrin nanofibers [J]. Chemosphere, 2016, 144: 736-744.

[78] KIM H J, PANT H R, CHOI N J, et al. Composite electrospun fly ash/polyurethane fibers for absorption of volatile organic compounds from air[J]. Chemical Engineering Journal, 2013, 230: 244-250.

[79] SOUZANDEH H, WANG Y, ZHONG W H. "Green" nano-filters: fine nanofibers of natural protein for high efficiency filtration of particulate pollutants and toxic gases[J]. RSC Advance, 2016, 6: 105948-105956.

[80] PRIYANKAR T, BHASKAR B, NISHITH V. Catalytic oxidation of NO over CNF/ACF-supported CeO_2 and Cu nanoparticles at room temperature [J]. Industrial & Engineering Chemistry Research, 2014, 53: 12537-12547.

[81] DAN L, WANG R X, TANG G M Z, et al. Ecofriendly electrospun membranes loaded with visible-light responding nanoparticles for multifunctional usages: highly efficient air filtration, dye scavenging, and bactericidal activity[J]. ACS Applied Materials & Interfaces, 2019, 11: 12880-12889.

[82] SELVAM A K, NALLATHAMBI G. Polyacrylonitrile/silver nanoparticle electrospun nanocomposite matrix for bacterial filtration[J]. Fibers & Polymers, 2015, 16: 1327-1335.

[83] WANG C, WU S, JIAN M, et al. Silk nanofibers as high efficient and lightweight air filter[J]. Nano Research, 2016, 9: 2590-2597.

[84] VANANGAMUDI A, HAMZAH S, SINGH G. Synthesis of hybrid hydrophobic composite air filtration membranes for antibacterial activity and chemical detoxification with high particulate filtration efficiency (PFE)[J]. Chemical Engineering Journal, 2015, 260: 801-808.

[85] WANG Z, PAN Z, WANG J, et al. A novel hierarchical structured poly(lactic acid)/titania fibrous membrane with excellent antibacterial activity and air filtration

performance[J]. Journal of Nanomaterials, 2016, 2016: 1-17.

[86] CHOI J, YANG B J, BAE G N, et al. Herbal extract incorporated nanofiber fabricated by an electrospinning technique and its application to antimicrobial air filtration[J]. ACS Applied Materials & Interfaces, 2015, 7: 25313-25320.

[87] KOTA A K, KWON G, CHOI W, et al. Hygro-responsive membranes for effective oil-water separation[J]. Nature Communications, 2012, 3(1): 1-8.

[88] NORDVIK A B, SIMMONS J L, BITTING K R, et al. Oil and water separation in marine oil spill clean-up operations[J]. Spill Science & Technology Bulletin, 1996, 3(3): 107-122.

[89] MOATMED S M, KHEDR M H, EL-DEK S I, et al. Highly efficient and reusable superhydrophobic/superoleophilic polystyrene@Fe_3O_4 nanofiber membrane for high-performance oil/water separation[J]. Journal of Environmental Chemical Engineering, 2019, 7(6): 103508.

[90] ZHU H, QIU S, JIANG W, et al. Evaluation of electrospun polyvinyl chloride/polystyrene fibers as sorbent materials for oil spill cleanup[J]. Environmental Science & Technology, 2011, 45(10): 4527-4531.

[91] LIN J, SHANG Y, DING B, et al. Nanoporous polystyrene fibers for oil spill cleanup[J]. Marine Pollution Bulletin, 2012, 64(2): 347-352.

[92] PATEL S U, CHASE G G. Separation of water droplets from water-in-diesel dispersion using superhydrophobic polypropylene fibrous membranes[J]. Separation and Purification Technology, 2014, 126: 62-68.

[93] SHIN C. Filtration application from recycled expanded polystyrene[J]. Journal of Colloid and Interface Science, 2006, 302(1): 267-271.

[94] SONG B, XU Q. Highly hydrophobic and superoleophilic nanofibrous mats with controllable pore sizes for efficient oil/water separation[J]. Langmuir, 2016, 32(39): 9960-9966.

[95] LIN J, DING B, YANG J, et al. Subtle regulation of the micro-and nanostructures of electrospun polystyrene fibers and their application in oil absorption[J]. Nanoscale, 2012, 4(1): 176-182.

[96] LIN J, TIAN F, SHANG Y, et al. Facile control of intra-fiber porosity and inter-fiber voids in electrospun fibers for selective adsorption[J]. Nanoscale, 2012, 4(17): 5316-5320.

[97] LIN J, TIAN F, SHANG Y, et al. Co-axial electrospun polystyrene/polyurethane fibres for oil collection from water surface[J]. Nanoscale, 2013, 5(7): 2745-2755.

[98] WANG H, HUANG X, LI B, et al. Facile preparation of super-hydrophobic nanofibrous membrane for oil/water separation in a harsh environment[J]. Journal of Materials Science, 2018, 53(14): 10111-10121.

[99] LI K, HOU D, FU C, et al. Fabrication of PVDF nanofibrous hydrophobic composite membranes reinforced with fabric substrates via electrospinning for membrane distillation desalination[J]. Journal of Environmental Sciences, 2019, 75: 277-288.

[100] HE S, ZHAN Y, BAI Y, et al. Gravity-driven and high flux super-hydrophobic/super-oleophilic poly (arylene ether nitrile) nanofibrous composite membranes for efficient water-in-oil emulsions separation in harsh environments[J]. Composites Part B: Engineering, 2019, 177: 107439.

[101] JAYARAMULU K, GEYER F, PETR M, et al. Shape controlled hierarchical porous hydrophobic/oleophilic metal-organic nanofibrous gel composites for oil adsorption[J]. Advanced Materials, 2017, 29(12): 1605307.

[102] SHANG Y, SI Y, RAZA A, et al. An in situ polymerization approach for the synthesis of superhydrophobic and superoleophilic nanofibrous membranes for oil-water separation[J]. Nanoscale, 2012, 4(24): 7847-7854.

[103] YANG H, LAN Y, ZHU W, et al. Polydopamine-coated nanofibrous mats as a versatile platform for producing porous functional membranes[J]. Journal of Materials Chemistry, 2012, 22(33): 16994-17001.

[104] LI X, WANG M, WANG C, et al. Facile immobilization of ag nanocluster on nanofibrous membrane for oil/water separation [J]. ACS Appl Mater Interfaces, 2014, 6(17): 15272-15282.

[105] GUI H, ZHANG T, GUO Q. Nanofibrous, emulsion-templated syndiotactic polystyrenes with superhydrophobicity for oil spill cleanup [J]. ACS Applied

Materials & Interfaces, 2019, 11(39): 36063-36072.

[106] YANG T, LIU F, XIONG H, et al. Fouling process and anti-fouling mechanisms of dynamic membrane assisted by photocatalytic oxidation under sub-critical fluxes [J]. Chinese Journal of Chemical Engineering, 2019, 27(8): 1798-1806.

[107] LIU L, HUANG L, SHI M, et al. Amphiphilic PVDF-g-PDMAPMA ultrafiltration membrane with enhanced hydrophilicity and antifouling properties [J]. Journal of Applied Polymer Science, 2019, 136(42): 48049.

[108] WANG X, CHEN X, YOON K, et al. High flux filtration medium based on nanofibrous substrate with hydrophilic nanocomposite coating [J]. Environmental Science & Technology, 2005, 39(19): 7684-7691.

[109] WANG X, FANG D, YOON K, et al. High performance ultrafiltration composite membranes based on poly (vinyl alcohol) hydrogel coating on crosslinked nanofibrous poly (vinyl alcohol) scaffold [J]. Journal of Membrane Science, 2006, 278(1-2): 261-268.

[110] YOU H, YANG Y, LI X, et al. Low pressure high flux thin film nanofibrous composite membranes prepared by electrospraying technique combined with solution treatment [J]. Journal of Membrane Science, 2012: 394-395.

[111] YOON K, HSIAO B S, CHU B. High flux ultrafiltration nanofibrous membranes based on polyacrylonitrile electrospun scaffolds and crosslinked polyvinyl alcohol coating [J]. Journal of Membrane Science, 2009, 338(1): 145-152.

[112] TANG Z, WEI J, YUNG L, et al. UV-cured poly(vinyl alcohol) ultrafiltration nanofibrous membrane based on electrospun nanofiber scaffolds [J]. Journal of Membrane Science, 2009, 328(1-2): 1-5.

[113] WANG X, ZHANG K, YANG Y, et al. Development of hydrophilic barrier layer on nanofibrous substrate as composite membrane via a facile route [J]. Journal of Membrane Science, 2010, 356(1): 110-116.

[114] MA H Y, YOON K, RONG L X, et al. High-flux thin-film nanofibrous composite ultrafiltration membranes containing cellulose barrier layer [J]. Journal of Materials Chemistry, 2010, 20(22): 4692-4704.

[115] MA H, BURGER C, HSIAO B S, et al. Ultrafine polysaccharide nanofibrous membranes for water purification [J]. Biomacromolecules, 2011, 12(4): 970-976.

[116] LI J J, ZHOU Y N, LUO Z H. Smart fiber membrane for pH-induced oil/water separation [J]. ACS Applied Materials & Interfaces, 2015, 7(35): 19643-19650.

[117] SI Y, FU Q, WANG X, et al. Superelastic and superhydrophobic nanofiber-assembled cellular aerogels for effective separation of oil/water emulsions[J]. ACS Nano, 2015, 9(4): 3791-3799.

[118] SI Y, YAN C, HONG F, et al. A general strategy for fabricating flexible magnetic silica nanofibrous membranes with multifunctionality[J]. Chemical Communications, 2015, 51(63): 12521-12524.

[119] CHE H, HUO M, PENG L, et al. CO_2-responsive nanofibrous membranes with switchable oil/water wettability [J]. Angewandte Chemie International Edition, 2015, 127(31): 9062-9066.

[120] JIANG Y, HOU J, XU J, et al. Switchable oil/water separation with efficient and robust Janus nanofiber membranes [J]. Carbon, 2017, 115(0): 477-485.

[121] ZANG L, MA J, LV D, et al. A core-shell fiber-constructed pH-responsive nanofibrous hydrogel membrane for efficient oil/water separation [J]. Journal of Materials Chemistry A, 2017, 5(36): 19398-19405.

[122] YONG L, MINGJI S, YANBO F, et al. Corrosive environments tolerant, ductile and self-healing hydrogel for highly efficient oil/water separation [J]. Chemical Engineering Journal, 2018, 354: 1185-1196.

[123] ADEBAJO M O, FROST R L, KLOPROGGE J T, et al. Porous materials for oil spill cleanup: a review of synthesis and absorbing properties [J]. Journal of Porous Materials, 2003, 10(3): 159-170.

[124] HU H, ZHAO Z, GOGOTSI Y, et al. Compressible carbon nanotube-graphene hybrid aerogels with superhydrophobicity and superoleophilicity for oil sorption [J]. Environtechnollett, 2014, 1(3): 214-220.

[125] LI J, LI J, MENG H, et al. Ultra-light, compressible and fire-resistant graphene aerogel as a highly efficient and recyclable absorbent for organic liquids [J].

Journal of Materials Chemistry A, 2014, 2(9): 2934-2941.

[126] JIANG F, HSIEH Y L. Amphiphilic superabsorbent cellulose nanofibril aerogels [J]. Journal of Materials Chemistry A, 2014, 2(18): 6337-6342.

[127] GE J, YE Y D, YAO H B, et al. Pumping through porous hydrophobic/oleophilic materials: an alternative technology for oil spill remediation [J]. Angewandte Chemie, 2014, 126(14): 3685-3690.

[128] SI Y, YU J, TANG X, et al. Ultralight nanofibre-assembled cellular aerogels with superelasticity and multifunctionality [J]. Nature Communications, 2014, 5: 5802.

第 5 章　纳米纤维在生物领域的应用

人类对生命科学的认识逐步深入，对现代医学的研究尤其是再生医学和癌症医学提出了越来越高的需求，例如在体外完全模拟体内的细胞环境，并需要将外部药物精确地传递到相应部位。静电纺织纳米纤维具有孔隙率高、比表面积大、纤维精细程度与均一度高、长径比大等优点，并且与天然细胞外基质十分相似，与生物分子附着力强，因此近些年在组织工程、药物传递和癌症治疗领域得到了广泛的关注和研究。通过调节参数和制备工艺，包括直径、孔隙率、排列、堆叠、图案化、表面官能团、机械特性和生物降解性，设计并制造了组织工程支架来控制细胞迁移和干细胞分化增强修复或再生各种类型的组织（例如神经、皮肤、心脏、血管和肌肉骨骼系统），通过引入生物活性分子作为原料制备了多种需要定向传递并控制释放的药物的载体，并综合以上技术实现癌症诊断和构建三维癌症支架。本章将从上述的细胞迁移、组织工程、创伤修复、药物控释、癌症研究五个方面阐述纳米纤维在生物领域的巨大作用。

5.1　细胞迁移调控

5.1.1　引言

细胞迁移在各种生物现象中起着中心作用，影响胚胎形成、伤口愈合、组织再生和癌症扩散等诸多生理过程。例如，皮肤再生涉及真皮的重建，可以通过促进成纤维细胞向受伤区域的迁移来加速这个过程。因此，充分理解控制细胞迁移的机理和因素对设计具有理想效果的生物工程支架至关重要。静电纺丝这种可控性强、使用要求低的材料制备技术在细胞迁移方面也有很大的应用前景。基于细胞迁移的机理，研究者发现可以通过调整电纺过程的参数和纤维材料的性能以达到对细胞迁移过程的调控。

5.1.2 纳米纤维调节细胞迁移

细胞迁移是一个复杂的过程，包括多个连续的步骤，如黏附、极化和向前移动。单个细胞的迁移过程包括细胞骨架的极化排列、膜运输的极化组织和信号级联的极化。在细胞的前部，细胞骨架重建导致膜突的形成，如丝状伪足、片状伪足，提供运动的驱动力。细胞后部也通过肌动球蛋白收缩积极参与细胞置换。对于细胞的集体迁移，其分子机制与单细胞迁移类似。细胞集体迁移时，先导细胞在与细胞外基质和生长因子及趋化因子相互作用的刺激下明显极化，从而启动细胞迁移。之后，跟随的细胞和先导细胞之间的细胞间通信也将调节和促进细胞的集体运动。

细胞迁移可以使用各种类型的信号来引导，包括形貌、化学和机械信号。当制备出的纳米纤维具有合适的直径、取向、表面化学和机械性能时，纳米纤维就可以作为有效的基质或支架来控制细胞的迁移。研究者通过调节纳米纤维的参数和性质等因素，来研究正常细胞或癌细胞在体外和体外的迁移机理以及组织修复、体内肿瘤消除的方法。

通过施加在电池上的接触引导，电纺纳米纤维可以排列成不同的图案来控制细胞的迁移方向和速度。在随机排列的纳米纤维上，细胞倾向于无方向性地迁移，这样会导致较短的迁移距离。而在同一方向排列的纳米纤维上，细胞被引导沿纳米纤维迁移，这极大地提高了迁移速度。以星形胶质细胞在随机和单轴排列的 PLA 纤维支架上的迁移为例（图 5.1（a）和图 5.1（b）），在单轴排列的 PLA 纤维上培养的细胞在 2 d 内就能部分弥合 2.25 mm 宽的间隙，而在随机排列的 PLA 纤维上培养的细胞即使在 5 d 后仍保持相对静止。纳米纤维排列方向对细胞迁移的调控也适用于干细胞。对于直径相似的纳米纤维，人类神经前体细胞和骨髓间充质干细胞（MSCs）在单向排列的纳米纤维上的迁移速度比随机纳米纤维的迁移速度更快，如图 5.1（c）和图 5.1（d）所示。

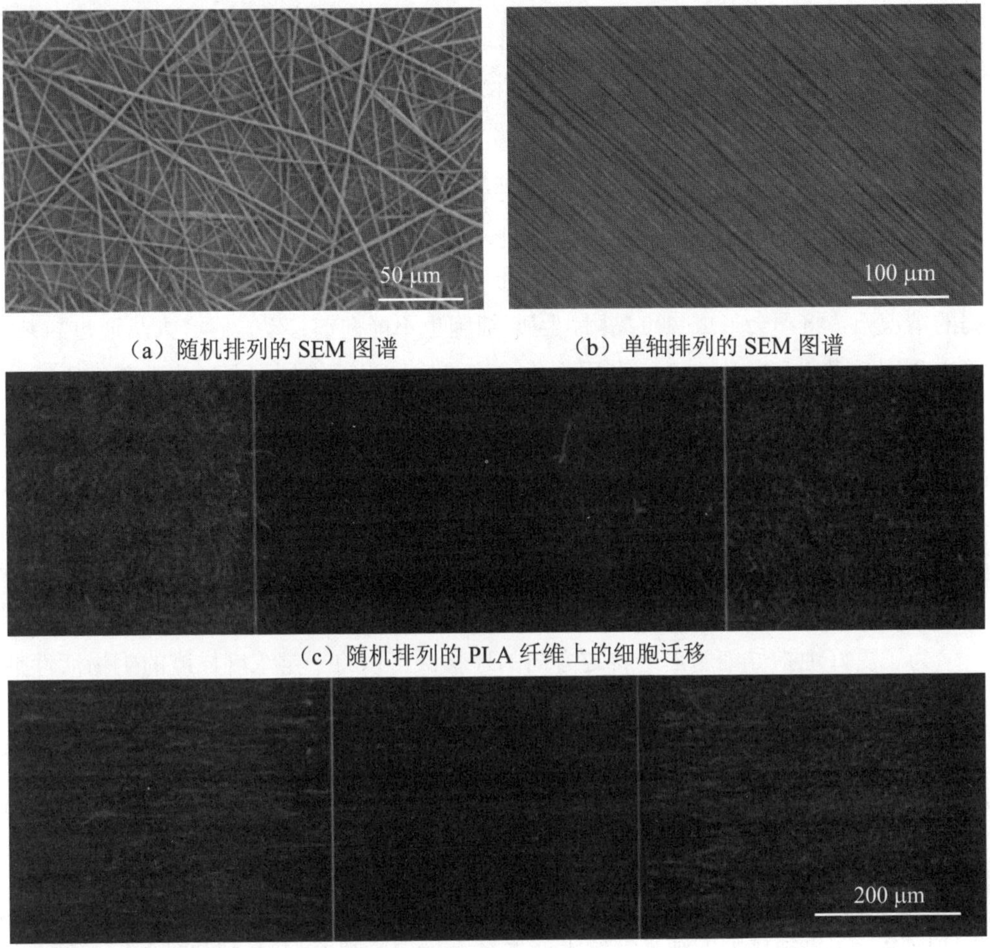

（a）随机排列的 SEM 图谱

（b）单轴排列的 SEM 图谱

（c）随机排列的 PLA 纤维上的细胞迁移

（d）单轴排列的 PLA 纤维上的细胞迁移

图 5.1　随机和单轴排列的 PLA 纤维支架上的细胞迁移（彩图见附录）

除了排列方式，纳米纤维的直径是影响细胞迁移的另一个重要参数。Wang 等人设计了雪旺细胞在三种不同直径的单轴排列的 PLA 纤维上的迁移：粗纤维（(1 325±383) nm）、中纤维（(759±179) nm）和细纤维（(293±65) nm）。在三组纳米纤维中，细胞在直径最大的纤维上迁移距离最远。作者认为，直径较大的纤维堆积得更密集，从而阻碍了雪旺细胞进入附近的纤维。而纤维直径对细胞迁移的影响还取决于聚合物的组成和细胞类型。例如，当单轴排列的蚕丝蛋白纤维作为 MSCs 生长和迁移的支架时，对比不同直径（400 nm、800 nm、1 200 nm）的纤维发现，400 nm 的纳米纤维因为直径小展现出最佳的性能，能够最大限度地促进 MSCs 迁移。因此，根据所调控的细胞的类型，纳米纤维的直径应该根据具体情况细致地调整。

纳米纤维表面的官能团也可以通过调节细胞和纳米纤维之间的相互作用来影响细胞迁移。由于细胞迁移是一个涉及细胞黏附和随后向前运动的过程，细胞黏附到纳米纤维上的强度在控制细胞迁移方面非常重要。当纳米纤维的黏附强度太强时，细胞将因为受到纤维对细胞表面的牵制而失去流动性。而在纳米纤维黏附强度非常弱的情况下，将不会形成焦点黏附，因此细胞也不能向前移动。研究人员通过封装、静电吸引、吸附和共价结合，利用生化线索修饰纳米纤维的表面，为改善细胞迁移做出了许多努力。最简单的方法是用生物活性剂（例如蛋白质和生长因子）覆盖纳米纤维的表面，以调节细胞与下部纤维之间的相互作用。Xie 等人比较了原始硬脑膜成纤维细胞在带有或不带有纤连蛋白涂层的径向排列纳米纤维制成的支架上的迁移。他们发现纤连蛋白涂层可大大增强细胞黏附力，改善细胞分布的均匀性并提高细胞迁移速度。

为了更好地指导细胞沿着特定方向的迁移，可以将一定浓度梯度的生物活性剂施加在纳米纤维上。在这种情况下，细胞倾向于沿着生物活性剂的浓度梯度向上迁移。目前较为常见的方法有通过修改物理吸附、化学共轭和电沉积的方案以及通过使用微流体混合来产生梯度；通常采用沿特定方向逐渐改变纳米纤维垫在生物活性剂溶液中的浸泡时间来产生梯度。随着生物活性剂溶液的逐渐加入，沿轴向纳米纤维垫表面沉积的药剂量逐渐增加，结果会形成生物活性剂的梯度。这种方法已经成功地应用于在单轴排列的纳米纤维上产生纤连蛋白的梯度，并且观察到 NIH-3T3（小鼠胚胎成纤维细胞系）细胞的数量和蛋白质含量有很强的关联。这种方法的主要缺点是梯度的形成需要相当大量的溶液和昂贵的生物活性剂，实验成本较高。Michael 等人提出另一种新颖的实验方案，他们在纳米纤维表面上生成牛血清白蛋白（BSA）梯度，然后在裸露区域填充生物活性剂，生成与 BSA 梯度相反的梯度。这种方法典型的过程中，通过铜线将纳米纤维支架的中央部分抬高以呈锥形，然后将 BSA 溶液滴加到该容器中，沿着每个纳米纤维径向方向都产生了 BSA 的浓度梯度（图 5.2(a)）。与在径向排列的 PCL 纳米纤维表面均匀覆盖层粘连蛋白或表皮生长因子的情况相比，该梯度显著促进了成纤维细胞或角质形成细胞从外围向中心的迁移（图 5.2(e)）。这一方法也可用于产生活性蛋白沿径向排列的纳米纤维的环形梯度。

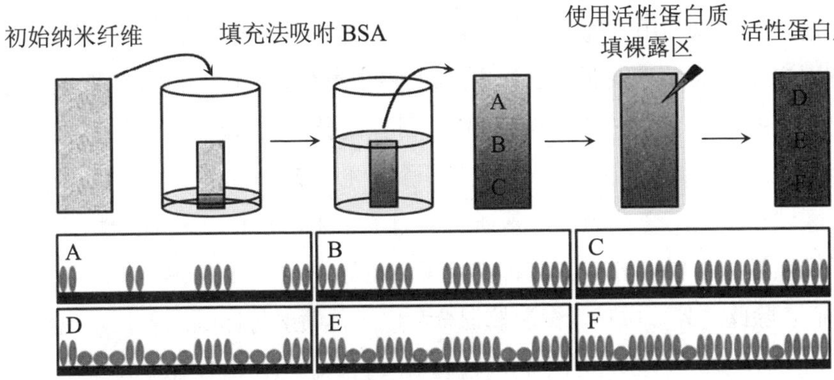

（a）纳米纤维支架表面产生生物活性蛋白梯度反应示意图

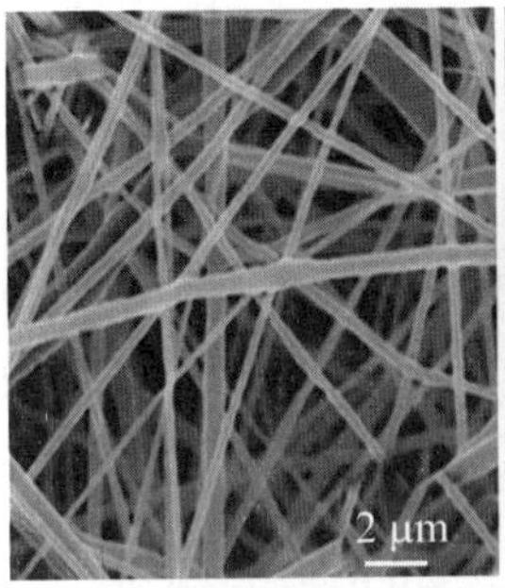

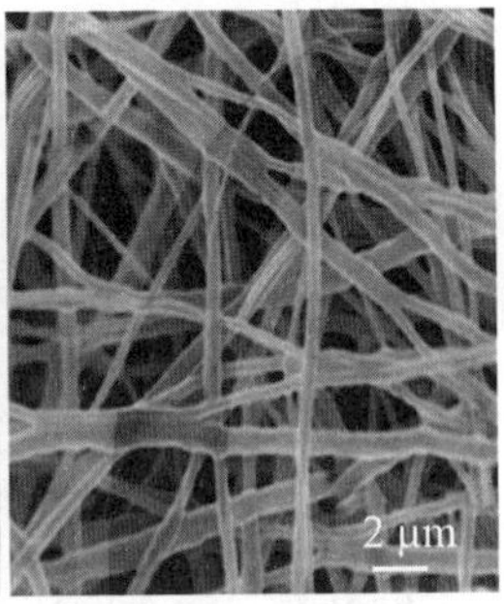

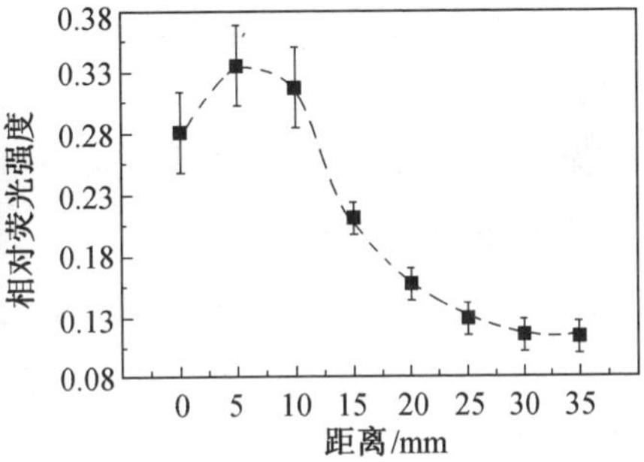

（b）纤维 SEM 图　（c）纤维吸附 BSA 的 SEM 图（d）吸附 BSA 后不同位置的相对荧光强度

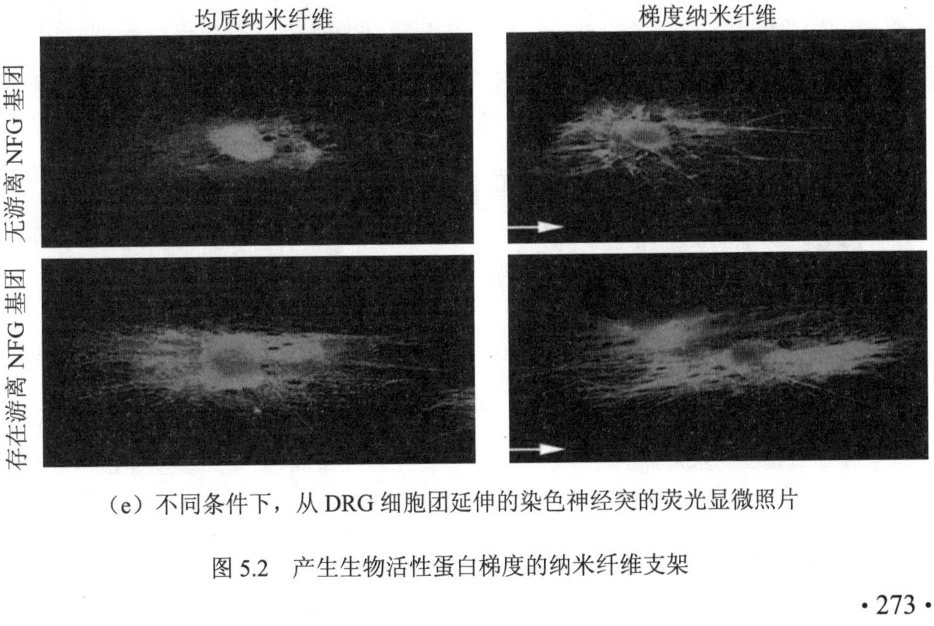

（e）不同条件下，从 DRG 细胞团延伸的染色神经突的荧光显微照片

图 5.2　产生生物活性蛋白梯度的纳米纤维支架

除此之外还可以利用生物活性剂与纳米纤维之间的静电吸引或共价结合，将生物活性剂固定在纳米纤维表面。这种方法可控制纳米纤维上官能团的密度，是一种产生生化梯度的有效途径。在一项研究中，研究人员以放射状的方式排列 PCL 和胶原的混合物制成的纳米纤维，以产生从外围到中心逐渐增加的纤维密度。当纳米纤维的胶原区域与胶原结合域融合的基质衍生因子 alpha（SDF-1α）结合时，生物活性剂沿纳米纤维呈圆形梯度分布。这种梯度可以诱导神经干细胞从纳米纤维垫的外围迁移到中心。除此之外，电喷、微流控混合和电流体喷射打印等制备技术也已与电纺纳米纤维相结合，以产生梯度的生物活性物质。例如，当微流控梯度发生器覆盖在由透明质酸制成的纳米纤维上时，会产生沿纤维排列方向的血管内皮生长因子梯度。

细胞迁移对纳米纤维的机械性能也很敏感。在不同模量的纳米纤维上，细胞的迁移速度会有很大的差异。在 Rao 等人对癌细胞在核心–鞘纳米纤维上迁移的研究中，PCL 作为“鞘”保留纳米纤维的表面化学特性，而不同类型的聚合物，包括明胶、聚醚砜和 PDMS 被用作“核心”，以调节纳米纤维的机械性能，如图 5.3 所示。在中等模量（约 8 MPa）的纯 PCL 纳米纤维上培养的单个多形性胶质母细胞瘤细胞的迁移速度最快（约 11 $\mu m \cdot h^{-1}$）。相比之下，模量较低和较高的纳米纤维的迁移速度都较慢，在约 2 MPa 的明胶/PCl 纳米纤维中的迁移速度约为 3.5 $\mu m \cdot h^{-1}$，在约 30 MPa PDMS/PCl 纳米纤维中的迁移速度约为 6.3 $\mu m \cdot h^{-1}$，在约 30 MPa（聚醚砜/PCl）中的迁移速度约为 5.8 $\mu m \cdot h^{-1}$（图 5.3（i））。以上结果表明通过改变力学性能，电纺纳米纤维可以用来检测癌细胞的迁移行为，从而获得一种独特的高通量的体外培养基质，以开发抑制癌细胞迁移疗法用于癌症的治疗。

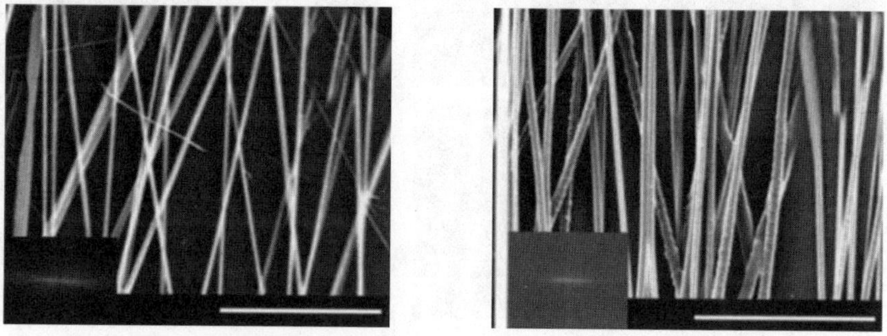

（a）凝胶–PCL 的 SEM 图　　（b）PCL 的 SEM 图

图 5.3　癌细胞在核心–鞘纳米纤维上迁移（比例尺为 20 μm）

（c）PDMS-PCL 的 SEM 图

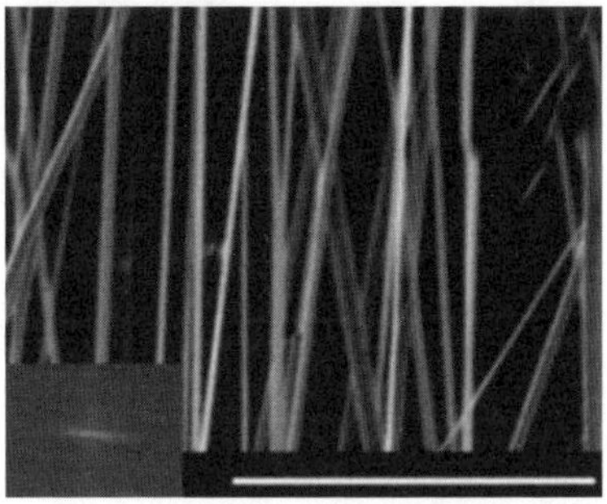

（d）PES-PCL 的 SEM 图

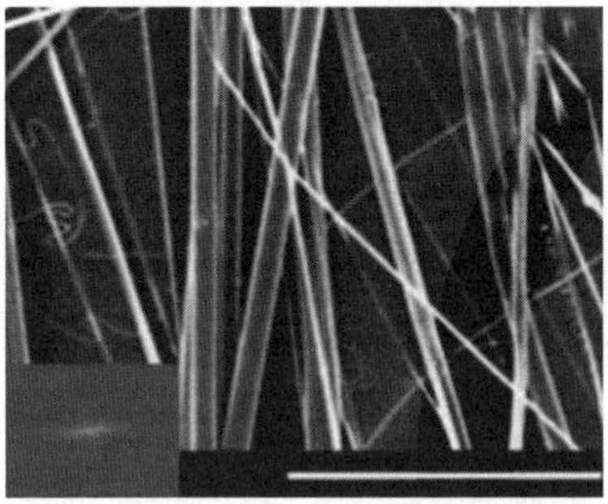

（e）PCL-胶原蛋白的 SEM 图

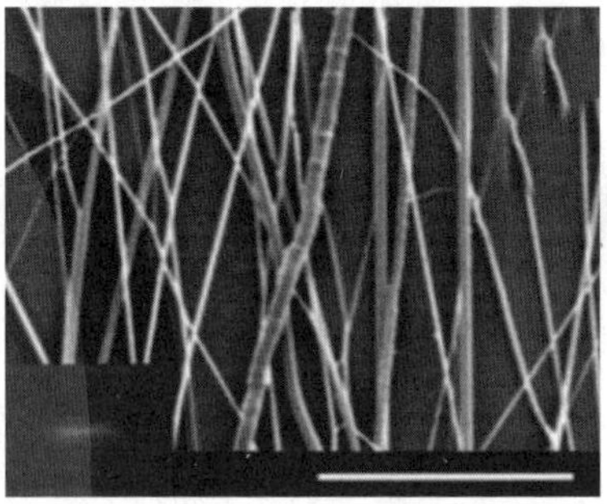

（f）PCL-HA 的 SEM 图

（g）PCL-基质胶的 SEM 图

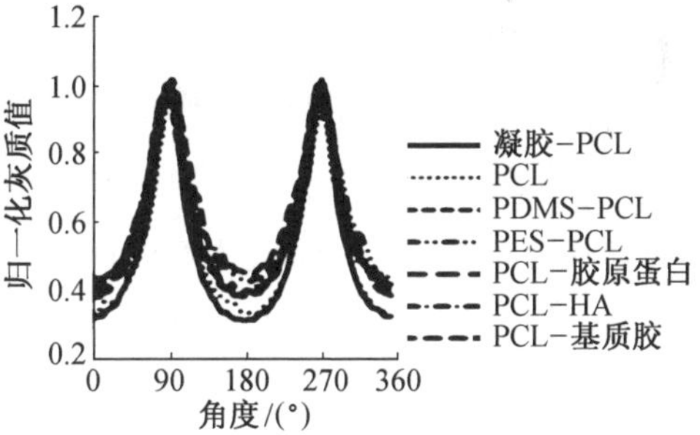

（h）样品的 FFT 分析结果

续图 5.3

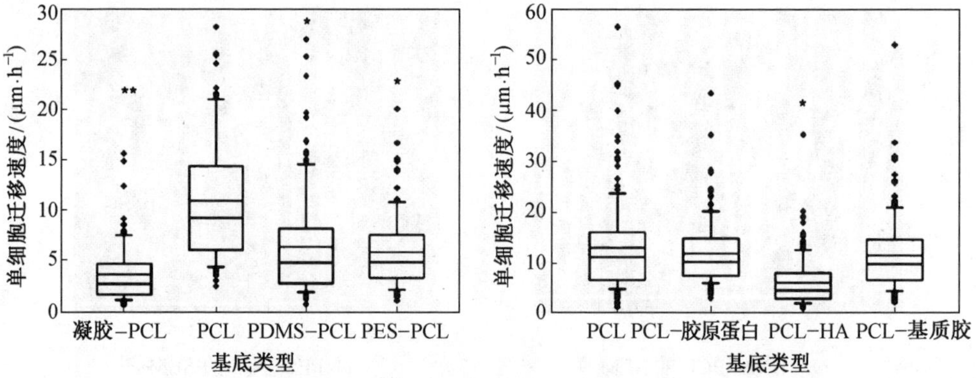

(i) 纳米纤维表面细胞迁移速度与力学和表面化学态的函数

续图 5.3

5.1.3 总结和展望

不论在生命起源还是生命活动的过程中，细胞迁移都起着非常重要的作用，是进一步研究人类疾病机理的基础。近年来，随着细胞生物学与生物物理、生物化学、遗传学、分子生物学、生物信息学等其他学科的紧密交叉，细胞运动的机理逐渐被揭开，体内体外实验也证明细胞迁移可以在一定程度上受到人为的调控。尤其是，研究人员采用纳米纤维材料作为细胞迁移基底时发现，通过改变纳米纤维的直径、取向、负载的生物活性剂、表面化学和机械性能时，细胞会沿着设定的方向移动，证明了纳米纤维在生物医学领域的巨大潜力。

然而要满足纳米纤维自由调控细胞的移动，仍有待解决的问题：

(1) 在分子水平上，推动力产生的机理目前仍然停留在假说水平，因此纳米纤维负载的生物活性分子还要继续研究。

(2) 目前的研究集中在体外调控细胞迁移，体内成功实现定向迁移的案例很少。

(3) 要实现体内调控细胞移动，纳米纤维与体内生物环境的生物亲和性还需进一步研究。

5.2　组织工程

5.2.1　引言

以移植细胞、组织器官植入、药物诱导为主要手段的再生医学是目前世界医学研究的前沿，代表着人类迄今为止最好的器官组织损伤修复水平，实现了许多令人惊叹的成就，例如让卵巢早衰的女性重新获得生育能力；促进急性脊髓损伤患者损伤修复等。电纺纳米纤维在这一领域已经引起了许多研究者的重视，目前已有多种生物材料通过静电纺丝方法制备成纳米纤维，例如天然聚合物材料、合成聚合物材料以及复合材料，其中天然聚合物包括胶原蛋白、明胶、甲壳素及其衍生物、纤维蛋白等。

电纺纳米纤维支架具有较高的表面积比、孔隙率和增强细胞黏附性的潜力；与天然细胞外基质相似的三维结构为细胞生长和实现其生物功能提供了微观环境。因此，纳米纤维结构的支架已经广泛用于组织工程和生物医学应用，受到了人们的强烈追捧。目前，采用静电纺丝制备的纳米纤维已经广泛应用在血管、软骨、骨、神经等组织工程修复等方面，本节将详细展开介绍。

5.2.2　组织工程支架

生物医药领域组织工程的原理是从机体获取少量活组织，在体外环境下培养扩增，然后将扩增得到的细胞或组织与生物相容性良好、可吸收的生物材料按比例混合，细胞或组织与生物材料形成细胞组织材料复合体。将该复合体植入机体的组织或器官受损部位，植入的细胞或组织在体内不断增殖，最终达到修复重建的目的。组织工程可以由三个基本组成部分来表示：细胞、细胞外基质生物模拟支架和生物信号。其中支架材料是组织工程的最重要的应用领域，包括支架制备方法和材料研究。制备方法决定了支架的结构和力学性能；材料决定了支架的生物相容性、生物降解性等。

不可生物降解的组装支架无法被人体吸收，在人体内长期使用会造成不良影响，需要再次手术才能去除。因此，制备可生物降解支架是当前组织工程支架的发展趋势。而在生物降解过程中，支架材料会降解为小分子物质，为了保证安全性，还需要研究这种小分子物质的生物相容性。支架的相容性和生物降解性主要取决于纳米纤维的材料。目前，在组织工程中应用最为广泛的可降解合成材料主要有聚乳酸、聚乙醇酸以及它们的复合纤维材料等。天然材料（如胶原和蚕丝）具有足够的细胞亲和力和组织取向性，这类材料缓慢的生物降解速率可以保证载荷从支架逐渐转移到愈合组织，也被应用于纳米纤维支架的制备。其他天然高分子材料，如壳聚糖（CS）和海藻酸钠胶体，可以保持支架的正常结构，具有良好的组织相容性。事实上，人工合成的生物材料可以提供足够的机械支持，但细胞识别能力差是人工合成材料在生物学上的应用障碍；天然材料具有更好的生物相容性，但也有力学性能较差的缺点。因此，将合成的材料与天然材料混合，这样可以融合两者的优点并且弥补各自的缺点。

5.2.3 血管组织工程支架

如图 5.4 所示，血管有三层结构：内膜、中膜和外膜。就细胞结构而言，内膜由连续的单层内皮细胞构成，以保证管腔通畅；中膜在弹性组织中含有密集的周向排列的平滑肌细胞，以确保机械强度；外膜主要由成纤维细胞和嵌入胶原细胞外基质的血管周围神经细胞组成。结构上的差异决定了组织工程支架在血管中的多样性。

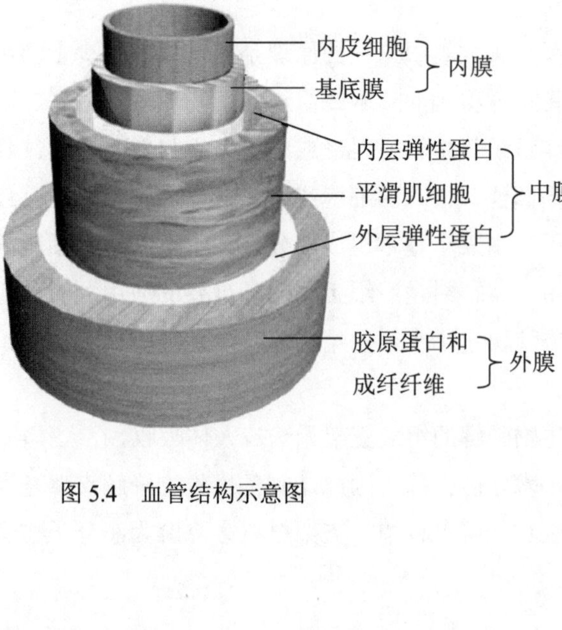

图 5.4　血管结构示意图

与已经有实际应用的非组织工程移植物的大直径动脉的情况相比，用于临床再生小尺寸（通常小于 6 mm）血管的仿生支架迫切需要开发和应用。目前所使用的移植物由于血液相容性差、植入移植物与逐渐再生组织之间的生物降解性不匹配、机械强度低而不能支持血流动力学而存在血液流通不理想和血栓的问题。从功能上看，内膜在防止血栓形成和加速内皮化进程方面起着最重要的作用，而中膜则提供力学支持，以避免吻合口的不良重塑。因此，该领域的研究主要集中在构建血管支架时内、中层的功能化。在植入活体血管内部之前，必须对支架进行测试，以确保管状结构能够支持管腔内的血液流动而不会渗漏，并且支架材料具有防止闭塞的抗凝能力，这对管腔的通畅性至关重要。为此，通常采用平面结构来研究血管内膜的重建和平滑肌细胞层的再生。然后，将支架折叠成管状结构，用于原位组织再生的研究，接下来将相关方向的研究分成几个小的方向说明。

1. 促进管腔内的抗凝和快速内皮化的研究

血管支架内层的关键作用包括抗凝和快速内皮化，管腔表面融合的内皮细胞单层对于防止血液凝结至关重要。当前研究人员已经探索了几种加速内皮形成的策略。血管内皮生长因子和基质细胞衍生因子-1α 是两种常用的生物因子，它们可以通过嫁接到纳米纤维表面或包裹到纳米纤维核心，促进内皮母细胞和内皮细胞的补充、迁移和增殖。另一种有效的方法是使用靶向递送系统局部递送 mRNA，分别下调和上调相关基因的表达来调节内皮细胞的增殖，通过静电纺超细纤维局部传递 miRNA 模拟物或抑制剂来调控 miRNAs 的表达已显示出其在组织再生中的应用前景，例如 Zhou 等人将纤维表面的 Arg-Glu-Asp-Val（REDV）肽修饰成双功能电纺膜，以增强血管内皮细胞在电纺纤维中对 miRNA-126 复合物的黏附和包裹，促进血管内皮细胞的增殖，如图 5.5 所示。首先采用聚乙烯醇（PELCL）和端基聚己内酯（质量比 1：1）的乳液静电纺丝法制备了新型静电纺丝膜，其中 miRNA-126 经 REDV 肽修饰的三甲基壳聚糖-g-聚乙二醇包覆。通过引入分子量较低的 PCL，得到的电纺纤维可以在其表面用 REDV 肽进行改性（图 5.5（a）～（f）），并能获得相对较快的 miRNA-126 复合物释放曲线（图 5.5（g）），有利于血管内皮细胞的增殖。在电纺膜上直接接种管内皮细胞的结果表明，血管内皮细胞的黏附和增殖得到了增强（图 5.5（i））。因此，将 REDV 肽修饰电纺纤维膜和可控 miRNA 释放结合起来，提供了一种由表面引导和生化信号的协同作用来调节血管内皮细胞、促进血管组织再生的方法。

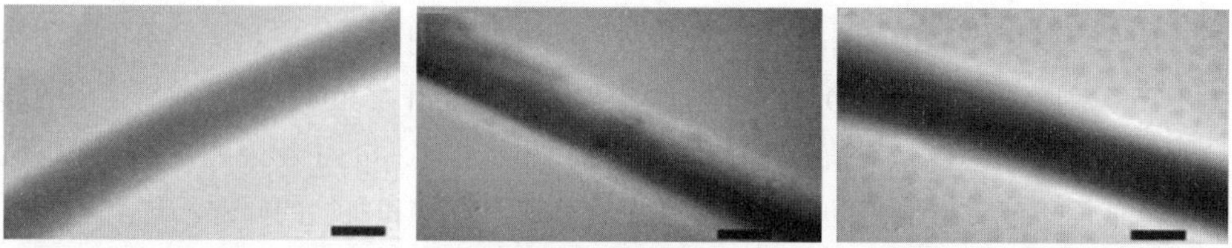

（a）PELCL 的 TEM 图　（b）PELC/PR 的 TEM 图　（c）PELCL/PPR 的 TEM 图

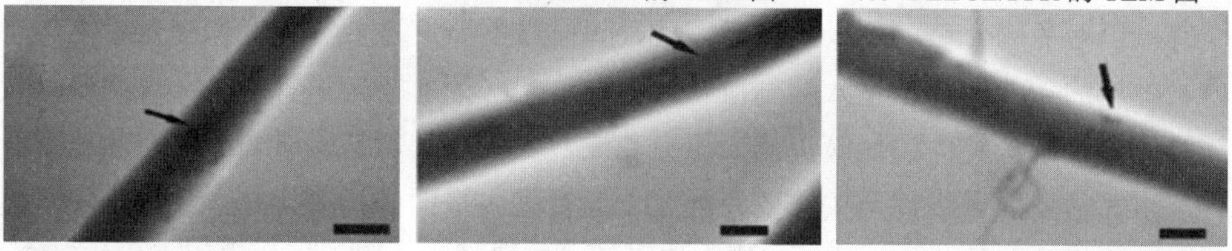

（d）PELCL(miR)的 TEM 图　（e）PELCL/PR(miR)的 TEM 图　（f）PELCL/PPR(miR)的 TEM 图

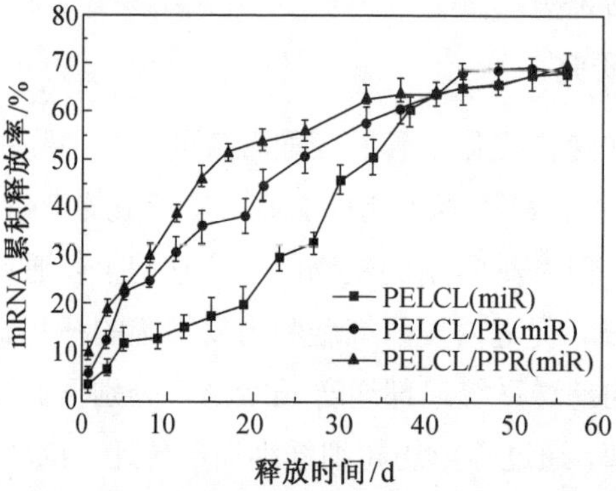

（g）miRNA 的体外释放曲线

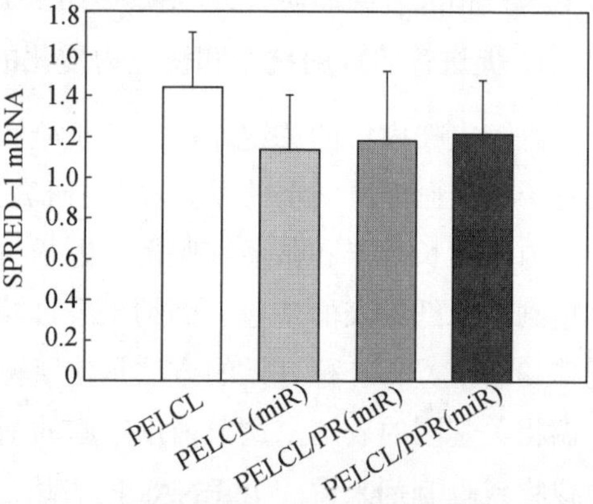

（h）SPRED-1 mRNA 的表达水平

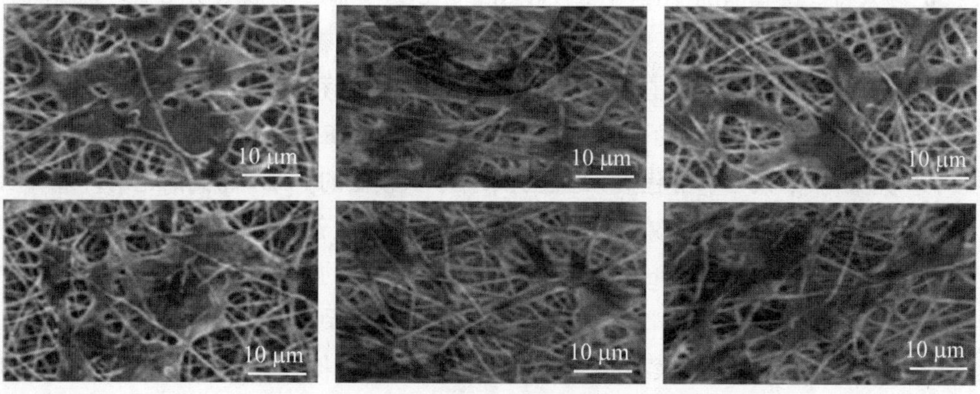

（i）第 6 天 VEC 在电纺膜上散布的 SEM 显微照片

图 5.5　采用 Arg-Glu-Asp-Val（REDV）肽修饰成的双功能电纺膜

2. 原位血管组织工程

由多层纳米纤维制成并负载各种生长因子（如血管内皮生长因子和血小板衍生生长因子）的管状支架已被用于血管组织的原位再生。Pan 等人通过疏水作用，对小口径（2.0 mm）电纺聚己内酯（PCL）血管移植物进行了融合蛋白 VEGF-HGFI 的修饰，该融合蛋白由 I 类疏水蛋白（HGFI）和血管内皮生长因子（VEGF）组成。抗 VEGF 抗体免疫荧光染色显示，VEGF-HGFI 在移植物纤维表面形成一层蛋白层。无论是改性的 PCL 席还是裸露的 PCL 垫，溶血率都是相同的。人脐静脉内皮细胞（HUVECs）体外培养表明，VEGF-HGFI 修饰能显著促进内皮细胞产生一氧化氮（NO）、前列环素 2（PGI 2）释放和摄取乙酰化低密度脂蛋白（AC-LDL）。免疫荧光染色显示大鼠腹主动脉置换术后 1 个月内，VEGF-HGFI 修饰的 PCL 移植物内皮化、血管化和平滑肌细胞（SMC）再生明显改善。这些结果表明，融合蛋白 VEGF-HGFI 修饰是提高人工血管移植再生能力的有效方法。

5.2.4　肌肉骨骼组织工程

肌肉骨骼系统由骨骼、肌肉、软骨、肌腱、韧带、关节和结缔组织组成，主要功能是为身体提供支持、稳定、保护和运动能力。肌肉骨骼损伤是指由创伤、先天性缺陷或肿瘤消融引起的肌肉或骨骼组织损伤。通常，基于纳米纤维的非织造布被用作贴片，以诱导干细胞向成骨、软骨或肌腱组织的特异性分化。另外，纳米纤维也可制作三维支架，为细胞浸润和组织重塑提供空间。

1. 骨组织再生

骨骼作为一种硬组织，主要由 I 型胶原纤维和羟基磷灰石纳米粒子组成。由于具有很强的机械强度和较高的矿物含量，骨再生支架通常采用生物相容性无机相（如羟基磷灰石、生物活性玻璃、二氧化硅和氧化锌纳米颗粒）增强可生物降解聚合物。一个理想的三维支架不仅要在结构上模拟大块组织，还要在机械上支持骨愈合过程，并提供生物化学线索诱导成骨，如图 5.6 所示。

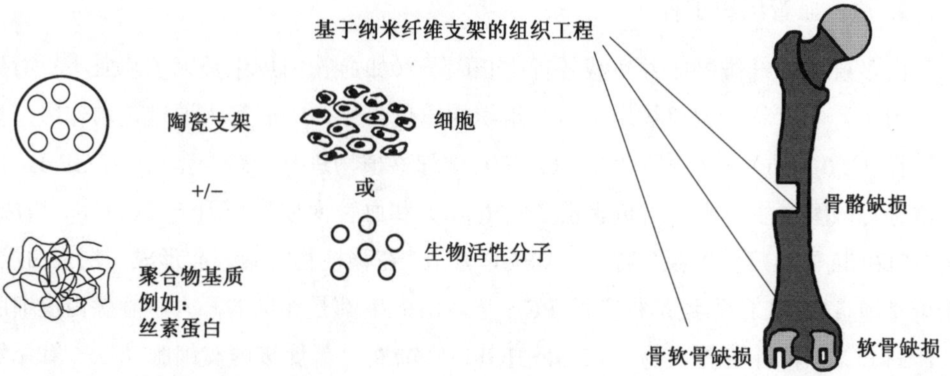

图 5.6　用于骨再生的可以植入生物分子的纳米复合支架

Chamundeswari 团队制备了一种由聚乳酸（PLA）、聚己内酯（PCL）和聚氧化乙烯（PEO）混合而成的海绵状 3D 纳米纤维支架，具有高度相互连接的孔隙、交织的亚微米纤维形态、大的孔径并能控制类固醇的释放。该支架可指导人骨髓间充质干细胞的成骨分化，而无须使用任何分化培养基，此外，还可以将矿物质和生物活性物质引入支架，以进一步提高其性能。由形状记忆聚合物制成的纳米纤维支架也显示出支持骨组织再生的潜力。聚（D，L-丙交酯-共三氯乙烯碳酸酯）的二维平面或三维圆柱形支架具有良好的形状记忆性能、高形状恢复率（大于 94 %）和高恢复速率（39 ℃下，恢复只需要约 10 s）。这种支架可以制成任何形状以促进成骨、碱性磷酸酶表达和矿物质沉积，具有修复各种类型骨缺损的潜力，包括用作引导骨再生的屏障膜或用作愈合骨螺钉洞的植入物。

2. 软骨组织再生

软骨是一种弹性并且光滑的组织，覆盖和保护关节处长骨的末端。高度专业化的软骨细胞负责外基质中胶原和蛋白多糖的产生。从浅区到软骨深区，II 型胶原含量降低，蛋白多糖含量增加。对于角质层缺陷的修复，电纺纳米纤维已显示出作为软骨形成或软骨分化的基础的前景。然而，当使用 2D 纳米纤维垫时，使细胞从表面到最深处完全浸润缺陷部位是一个存在已久的问题。因此，需要研制一种具有分层结构和相互连接的孔隙、具有抗压强度和促进软骨形成的能力的三维纳米纤维支架，从而实现理想的较厚软骨组织的浸润。一种透明质酸修饰的三维纳米纤维支架，由明胶和聚乳酸混合而成，有效地修复了兔躯体中的软骨缺陷。该支架具有多孔纤维结构、高抗压强度、高吸水性和形状恢复性能。此外，作为软骨中细胞外基质的天然成分，透明质酸的加入增强了支架的亲水性，引入了更多的细胞识别位点。因此，纳米纤维支架促进软骨细胞的生长，实现了大块软骨组织的再生，并在实验兔手术 12 周后填补了缺损部位。

3. 肌腱、配体组织再生

肌腱是一种致密的纤维结缔组织，其作用是将肌肉连接到骨头上，并且能够承受张力。同样，韧带是连接骨和另一根骨的纤维结缔组织。组织学上，肌腱和韧带都是由致密的规则结缔组织束平行排列组成的，也就是说，紧密填充的胶原纤维和胶原蛋白束被包裹在由致密不规则结缔组织组成的鞘中。由单轴排列的纳米纤维组成的各种二维垫被用于肌腱和韧带组织工程，因为它们可以模拟天然组织的排列。然而，健康的肌腱或韧带组织通常体积较大，结构和力学性能具有高度各向异性。为此，仿生三维支架需要提供结构、物理和生化特征来模拟天然组织。例如，一种由聚氨酯纤维制成的可逆膨胀管，其粗糙的内表面与血小板衍生的生长因子结合，可以应用于治疗肌腱破裂部位。此外，纳米纤维纱线和编织、机织或针织纱线网络由于其高机械强度、各向异性结构和细胞浸润的大孔隙率，对肌腱和韧带组织再生具有吸引力。例如，两根索纱（长度为 20 mm）通过一根长丝相互连接，通过用缝合线将所得到的索纱固定在肌腱上进行肌腱修复（图 5.7（b））。在另一项研究中，以聚己内酯纳米纤维纱线为纬纱，多聚乳酸长丝为经纱，采用平织工艺制备了纺织纳米纤维支架，呈三维排列结构，随着其孔径增大，拉伸力学性能增强，促进肌腱再生（图 5.7（e））。在机械刺激条件下，将纤维支架与脂肪来源的细胞外基质、人的肌腱细胞和人的脐静脉内皮细胞共培养时，胶原分泌和肌腱分化效果均提高。另一种策略是应用含有纤维和水凝胶的复合材料进行韧带再生。例如，通过将壳聚糖和透明质酸的混合物制成的水凝胶涂覆在单向排列的聚己内酯纤维上制成的双层支架可以促进韧带再生。

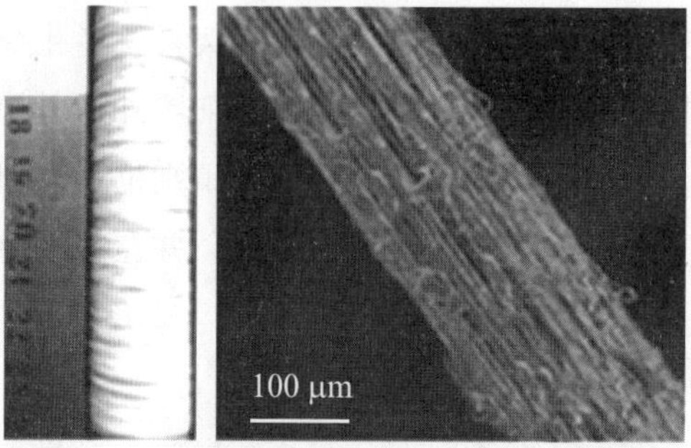

（a）纤维织物的光学照片和电镜图

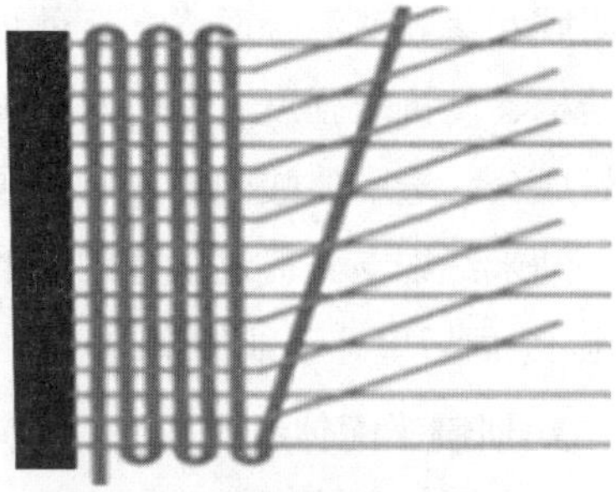

（b）纤维织物示意图

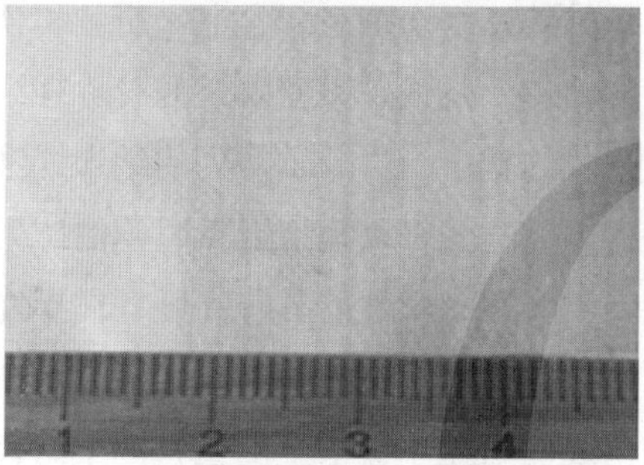

（c）纤维织物的光学照片

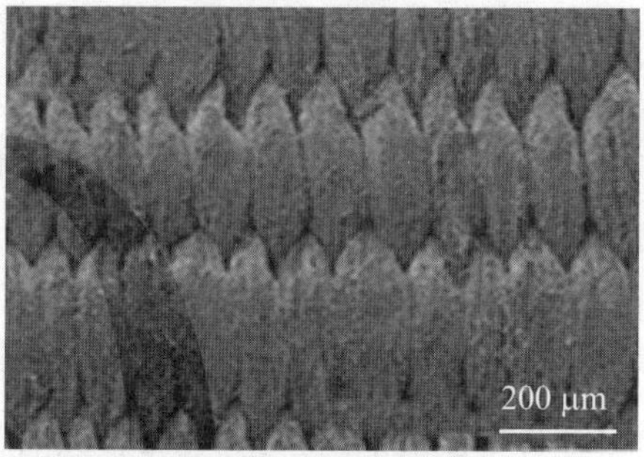

（d）纤维织物的 SEM 图

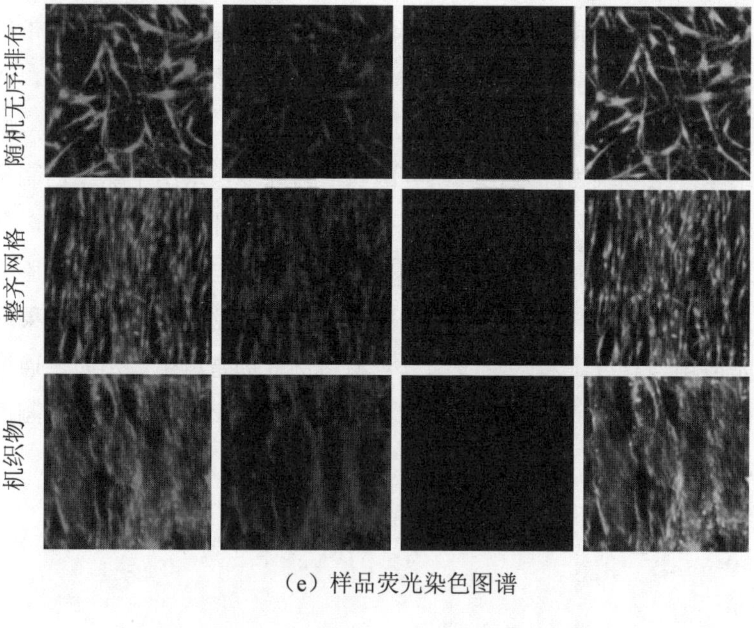

（e）样品荧光染色图谱

图 5.7 促进肌腱组织再生的纳米纤维

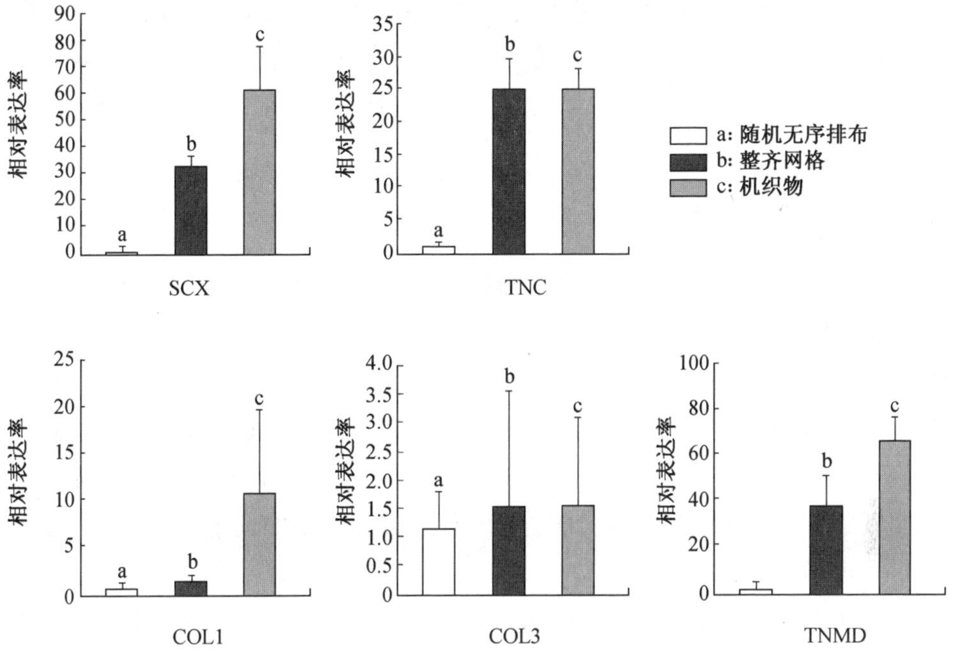

（f）样品的 qPCR 表征结果

续图 5.7

4. 骨骼肌再生

骨骼肌由多核肌肉细胞中高度定向、密集排列的肌纤维束组成，通过肌腱与骨骼相连。如果肌纤维排列不正确，就不可能在功能性肌纤维再生过程中进行有效的力传递和力收缩。因此，引导肌肉细胞排列和按照肌肉形貌、化学或生物信息形成肌管的支架被认为是骨骼肌形成的关键。在最近的研究中，各向异性纳米纤维材料的形态和功能与天然骨骼肌组织非常相似，已被研究和应用于制作促进骨骼肌再生的支架。例如，2016 年 Jana 课题组研制出一种具有单轴纤维排列的三维多孔支架，有利于肌细胞的预排列，从而促进早期肌源性分化并且可以加快长而厚的肌管的形成。

在各种设计中，纤维束结构决定了骨骼肌再生的可行性。用丝素蛋白制成的电纺纤维束模仿了骨骼肌单轴排列的纤维的分层组装，经聚吡咯（PPy）进一步修饰后，电活性纤维束在生物相关电解质溶液中，在低电位作用下进行机电驱动，从而模拟肌肉的收缩功能。此外，纳米纤维还可以与水凝胶等其他材料相结合，构建三维复合支架。以纳米纤维纱为芯材，以光固化水凝胶为鞘材，在合适的三维环境下诱导成肌细胞三维排列和伸长。

5.2.5 周围神经组织工程

周围神经可被视为一个封闭的、电缆状的神经纤维束或轴突。周围神经在中枢神经系统（CNS）和身体其他部分之间传递运动、感觉和自主神经信息，如果周围神经被切断，例如面部或四肢受伤，这些功能就会丧失。值得庆幸的是，与脊髓损伤后大部分不能再生的过程不同，周围神经系统（PNS）的轴突能够再生，同时周围神经损伤的长期影响结果取决于损伤的严重程度。周围神经系统的神经元胞体位于脊髓腹角（运动神经元）、背根神经节（感觉神经元）或靠近脊柱的交感神经链神经节。

哺乳动物外周神经纤维的损伤可以激活轴突切除后的神经元细胞体中的生长程序，使其轴突再生，但条件是找到了允许生长的基质。在简单的横断伤后，通过重新连接单个的近端和远端神经束，可以手术修复周围神经。然而，当被破坏的神经桩之间的间隙太大时，外科医生需要移植从病人其他地方取出的自体神经的一部分，例如腓肠神经或腓肠神经经常被用于这种手术，很容易导致失去这部分的感觉功能。另外尽管感觉神经通常被移植来修复运动纤维，但实验数据表明，感觉神经移植没有运动神经效果好。由于现有自体移植供体材料的局限性和供体部位发病的额外风险，需要在外体诱导神经组织生长从而替代自体神经移植。

周围神经系统的再生过程如图 5.8 所示：（a）周围神经系统的神经元位于脊髓腹角、背根神经节或交感神经链神经节，轴突细胞有髓鞘。（b）神经损伤后，髓鞘和轴突在损伤部位的远端退化；雪旺细胞增生，巨噬细胞清除退化纤维碎片；在损伤部位，神经元形成轴突生长锥。（c）轴突能够沿着胶质细胞和细胞外基质的纵向带再生；随后，雪旺细胞重新髓化新轴突。（d）在完成再生过程后，近端和远端神经收回。通常，当轴突生长无法穿过损伤部位时就会形成神经瘤。神经和肌肉纤维的远端不再受神经支配而变得萎缩。

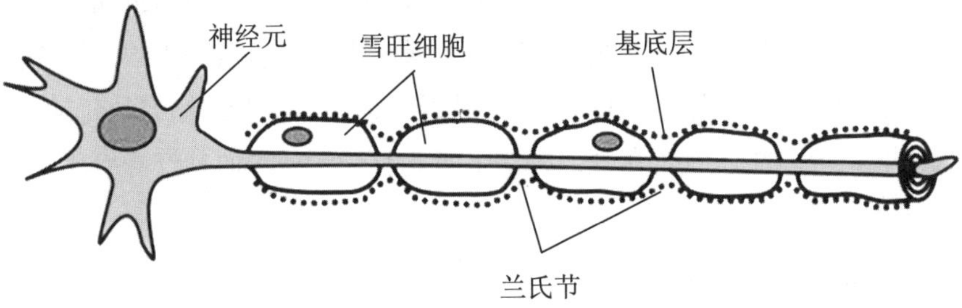

（a）

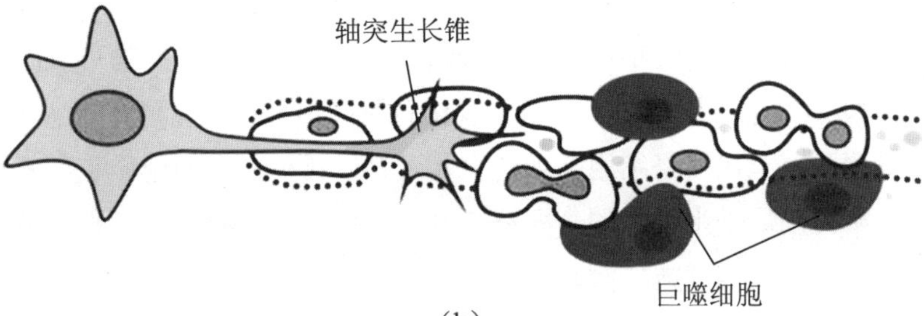

（b）

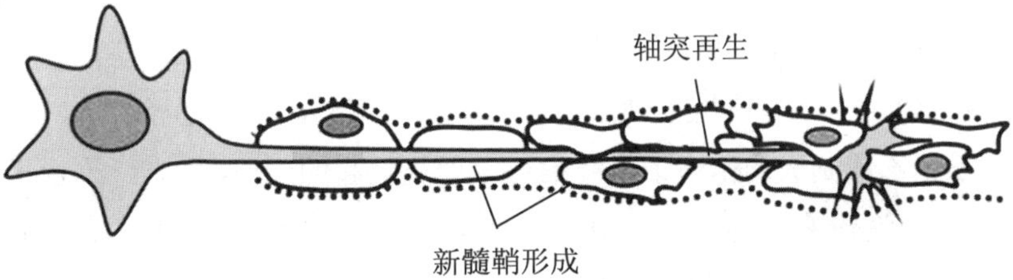

（c）

（d）

图 5.8　周围神经系统的再生过程

当前周围神经系统修复的研究重点是控制形貌因素，以及在某些情况下与电化学因素和生化因素的整合。在体外研究中，通常使用平面支架，以大鼠嗜铬细胞瘤（PC12 细胞）和背根神经节（DRG）体为神经元模型，研究神经突的延伸，以及雪旺细胞的生长情况。在体内研究中，通常采用具有管状结构的神经导管（NGCs）的支架。

1. 体外促进轴突延长和雪旺细胞生长的研究

单轴排列的电纺纳米纤维在神经组织工程中得到了广泛的应用。特别是单轴排列的纳米纤维可以提供类细胞外基质的微环境来指导细胞排列和轴突延伸。纳米纤维可以进一步修饰孔隙、沟槽或其他二级结构，还可以作为额外的形貌因素，促进突起的延伸和/或雪旺细胞的生长。当用氧化石墨烯纳米片包覆聚乳酸纳米纤维以增加表面粗糙度时，大鼠嗜铬细胞瘤细胞突起沿纤维排列方向的延伸和雪旺细胞的增殖均得到显著提高。纳米纤维的表面粗糙度也可以通过电喷雾粒子来调节。为此，将可生物降解微粒的连续或离散密度梯度电喷涂在通过面罩的玻璃载玻片上，以控制不同位置的沉积持续时间，从而控制背根神经节体神经突起的延伸生产。

除了纳米纤维改变的物理因素外，电活化产生的电化学因素也影响突起的延伸或雪旺细胞的生长。电活性材料，包括压电和导电聚合物，可以在神经修复中直接将电信号、电化学信号或机电刺激传递给细胞。例如，目前已经研制好的采用聚偏氟乙烯-三氟乙烯制成的压电纳米纤维可以促进突起的延伸，纳米纤维由于自身的压电属性，可以在没有外部电源的情况下将压电性电刺激传递给细胞。然而，由于外在机械应变的限制，目前对这种压电性电刺激的控制是有限的。与压电材料不同的是，导电聚合物允许通过施加电位下降来很好地控制电刺激。利用高导电性，用导电聚合物制成的纳米纤维被用来促进突起的延伸和施旺细胞的生长。在 Liu 等人的研究中，将聚吡咯包覆在聚己内酯纳米纤维表面，通过静电纺丝与水聚合相结合，形成内导电核-外保护套结构的纳米纤维（图 5.9（a）、（b））。在电刺激下，无论纳米纤维的取向如何，都促进了背根神经节的突起延伸（图 5.9（f））。聚吡咯还被涂覆在由聚 L-lactidecoε-已内酯（PLCL）和丝素蛋白混合而成的纳米纤维上，以显示大鼠嗜铬细胞瘤细胞的突起延伸增强，以及在电刺激下雪旺细胞的增殖增强。

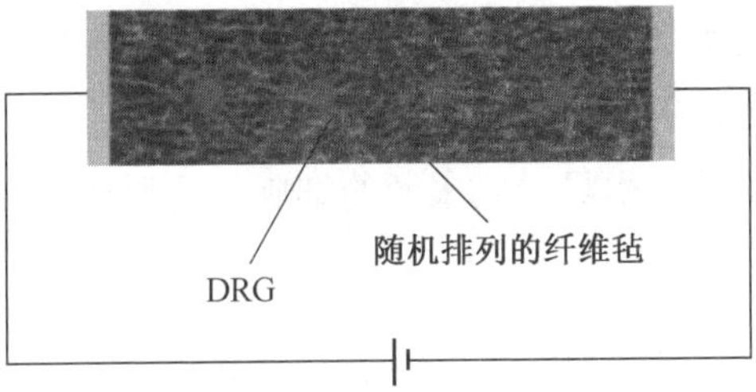

（a）DRG 神经突向在随机排列纤维生长图

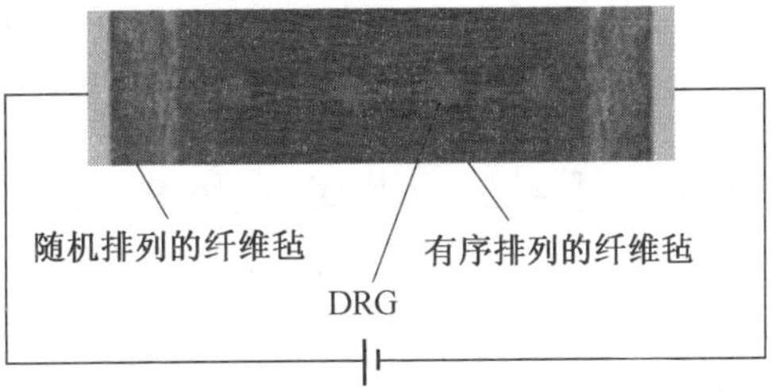

（b）DRG 神经突向在有序排列纤维生长图

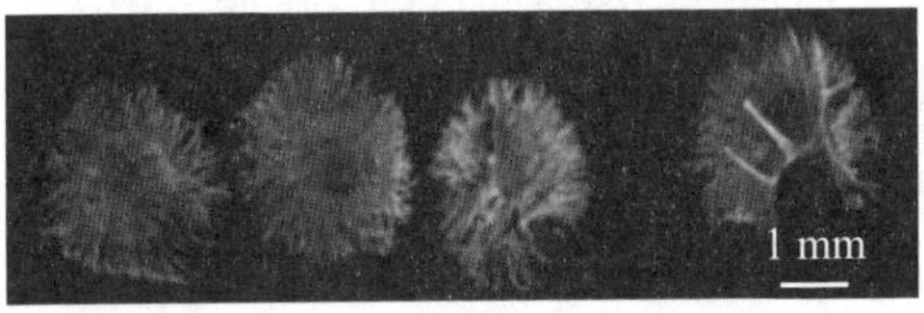

（c）随机纤维的 DRG 神经突向荧光显微镜图

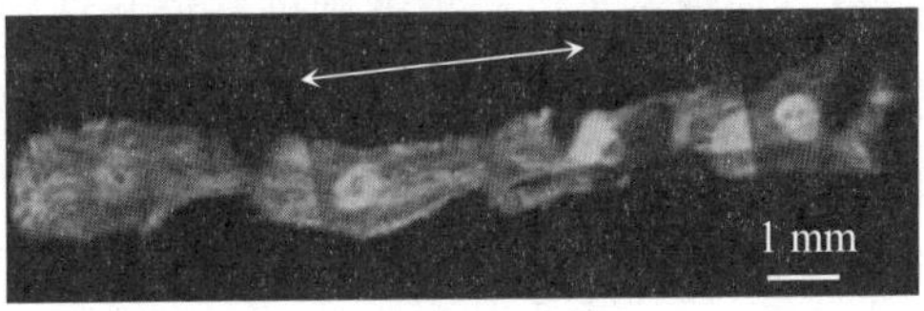

（d）单向纤维的 DRG 神经突向荧光显微镜图

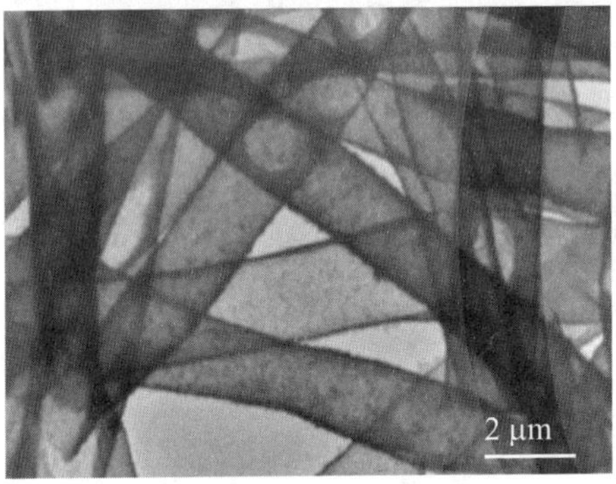

（e）纳米管的 TEM 图谱

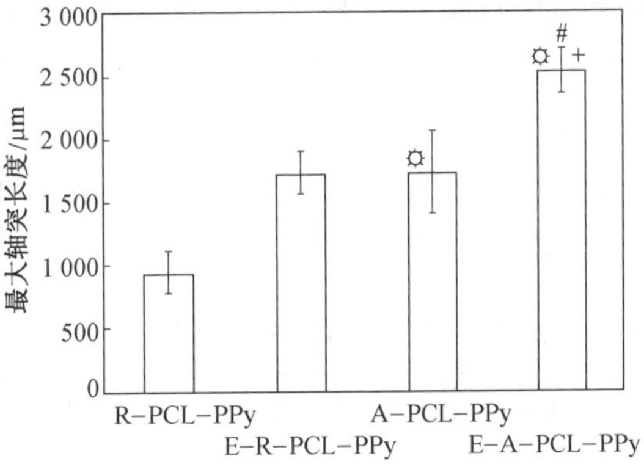

（f）四组样品中神经突起的最大长度

图 5.9　聚己内酯复合纳米纤维调节 DRG 神经突向的生长

生物因素通常与纳米纤维提供的形貌因素和电化学引导相结合，以促进突起的延伸和雪旺细胞的增殖。例如，通过将神经生长因子纳入由聚苯胺（PANI）、PLCL 和丝素蛋白混合而成的纳米纤维的核心，电刺激和神经生长因子的协同作用促进了大鼠嗜铬细胞瘤细胞和雪旺细胞的突起延伸。此外，用生物活性剂（例如层粘连蛋白和神经生长因子）功能化的单轴排列纳米纤维也能够有效增强沿纤维排列的突起延伸。特别是通过用神经生长因子梯度修饰单轴排列的纳米纤维，背根神经节细胞在增加梯度延伸方向上的突起明显长于相对于梯度延伸的突起，表明纳米纤维提供的地形线索和神经生长因子梯度提供的趋化性的协同作用。此外，单轴取向的聚己内酯纳米纤维已被用作底物，在体外化学诱导下引导骨髓来源的细胞外基质向雪旺细胞分化。

2. 用于体内评估的神经导管研究

研究表明，神经导管最初是由多孔泡沫棒或单个空心管制成的，但是使用单一的空心导管不足以模拟细胞外基质和细胞在本地神经中的空间排列。因此，用于桥接周围神经缺陷的支架已经从单一的空心管发展到更复杂的多管导管。这些导管可以进一步结合物理、化学和生物信息，对雪旺细胞的迁移和增殖以及轴突的伸长都很重要。管腔内通道可加入神经导管以构建多通道管道，例如由 PLCL 和丝素蛋白共混而成的具有丰富大孔和高孔隙率的纳米纤维海绵可以作为由相同材料制成的纤维管内的填料。另外还有以电纺聚乳酸（PLA）或层粘连蛋白包覆的聚乳酸-聚乙醇酸纳米纤维为填料，在神经导管内腔中制备了一系列具有纵向取向的可生物降解填料。总的来说，与单管神经导管相比，多通道神经导管或以生物材料为基础的神经导管能更好地模拟神经束的结构，从而减少再生轴突在神经导管管腔内的分散，大大提高功能的恢复。

5.2.6 其他组织工程支架

由电纺纳米纤维制成的支架也可以用于其他组织的修复或再生，如胆管、输尿管、膀胱和气管。然而，由于其复杂的结构，这些组织的修复或再生远未得到很好的研究。例如，胆管可以被认为是一个长的管状结构，连接总肝和胆囊管。胆道手术后，病人经常会出现各种并发症，如胆道损伤、胆管狭窄和胆漏。临床上，人们探索使用金属或塑料支架来修复受损的胆管。但是此类支架的生物和组织相容性较差，容易导致闭塞和感染。对此，可以在金属支架表面覆盖特定的纳米纤维垫优化其生物相容性，PCL 纳米纤维为基底，同时在每根纳米纤维的核心负载乙二胺四乙酸和胆酸钠。其中 PLCL 纳米纤维具有良好的生物相容性，药物的持续释放达到了溶解胆结石的效果，从而成功抑制胆管的阻塞。此外，由于胆管的主要功能是输送胆汁，模仿胆总管的假体应该有相当大的顺应性来抵抗这种动力学。到目前为止，除了对在不同电纺条件下制成的一套聚氨酯纳米纤维制成的管子进行测试外，对于潜在用于胆总管修复的纳米纤维基支架的顺应性方面的研究还很有限。

膀胱是一个中空的肌肉器官，在排尿前从肾脏收集和储存尿液。许多临床情况都会导致膀胱顺应性差、容量减少和大小便失禁，需要进行膀胱扩张或使用再生技术和支架。理想的膀胱组织工程支架应具有特殊的中空结构、光滑的尿路上皮表面，以及多孔的平滑肌细胞外层隔间。此外，由于膀胱需要扩张和收缩以存储和排出尿液，因此支架的拉伸力学性能和顺应性也应予以考虑。Ajalloueia 等人将一层静电纺丝的 PLGA 纳米纤维夹在两层塑料压缩的胶原纳米纤维之间，形成 3D 支架（图 5.10（a））。然后，通过将切碎的组织分布在支架顶部或同时分布在支架上和支架内，支架被用于扩张自体碎块的尿路上皮。在一项关于膀胱组织扩张的研究中，将一层静电纺丝的 PLGA 纳米纤维夹在两层塑料压缩的胶原纳米纤维之间，形成 3D 支架。结果表明，组织培养 2 周后，组织块中的细胞迁移重组为支架顶端的融合细胞层，4 周后形成多层尿路上皮，并具有典型的尿路上皮细胞形态和表型（图 5.10（d））。另一种尝试中，将可降解的聚酯聚氨酯电纺以产生纤维，并收集在湿膀胱脱细胞基质的外侧，形成双层杂化膀胱支架。由于聚酯氨酯纤维的生物相容性、力学性能和生物可降解性，与 PLGA 纤维和膀胱无细胞基质杂化支架相比，重建膀胱壁的平滑肌和尿路上皮再生能力明显增强，炎症反应减少。

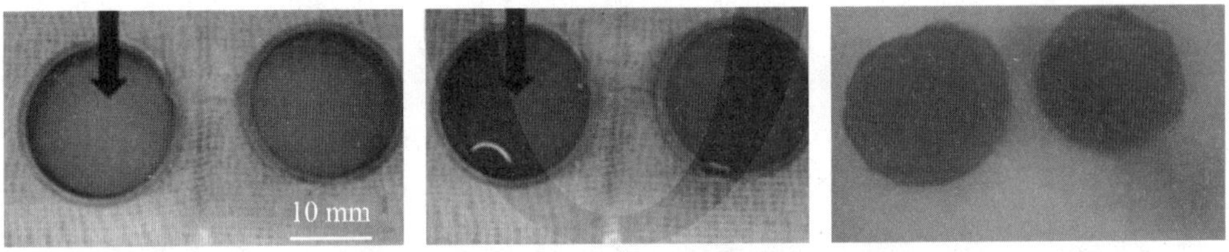

（a）PLGA-胶原蛋白复合结构的制备过程

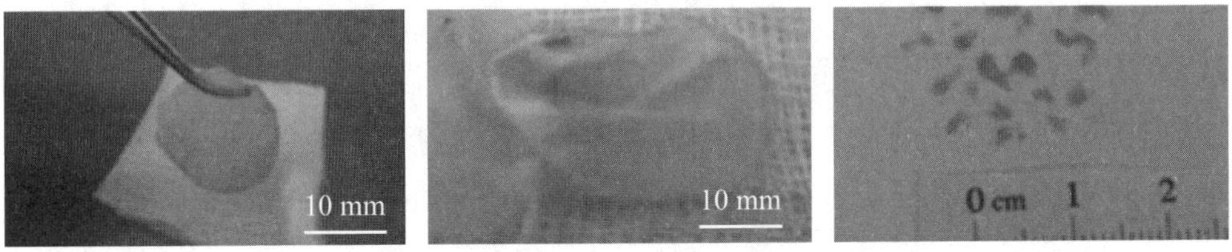

（b）PLGA-胶原蛋白的 SEM 图谱

图 5.10　电纺 PLGA 纤维与胶原纤维复合形成的 3D 支架

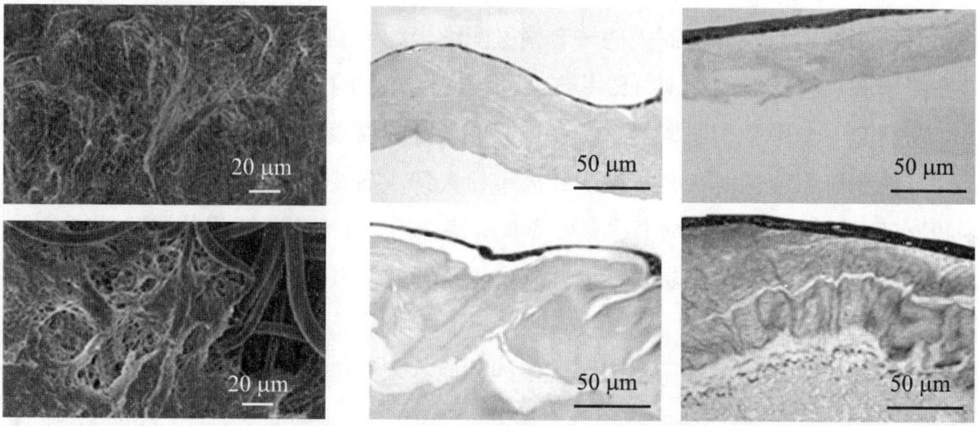

（c）从单层到多层上皮的组织学外观（HTX-Eosin）

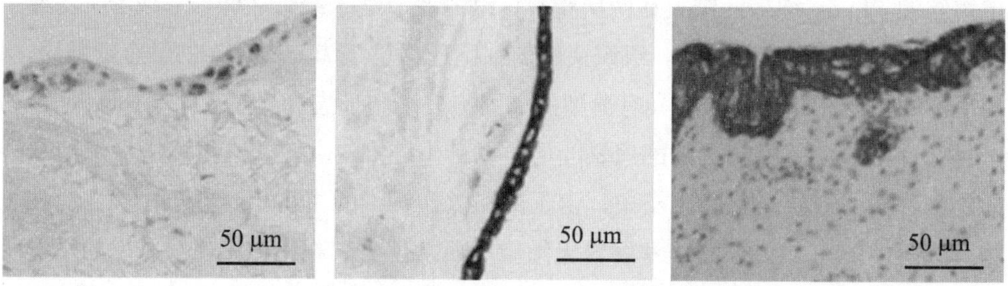

（d）切碎的组织与原生膀胱（右一）对比图

续图 5.10

5.2.7 小结与展望

为使制备的纳米纤维具有更好的生物相容性和机械性能，目前大部分纳米纤维组织工程支架材料都是将天然高分子和合成高分子通过接枝、共混等方法进行复合。纳米纤维直径通常在几十到几百纳米之间，因此其纤维支架能最大限度地仿生体内细胞外基质。同时，纳米纤维支架具有高孔隙率和连续的孔结构，利于细胞黏附、生长、分化，细胞外基质的沉积、氧气和营养物质进入、代谢产物的排出。因此，研究人员采用纳米纤维支架进行了多种人体器官组织的修复，是理想的组织工程支架，具有广阔的应用前景。但是组织工程支架修复实际损伤案例的效果还并不理想，在机理和制备方面仍有问题亟待解决。

（1）纳米纤维组织支架的机械强度、组织的再生速率和降解速度之间的矛盾还无法解决。

（2）纳米纤维组织支架在体内仍有一定的抗原性，针对不同的组织需要采用不同的天然生物原料制备纳米纤维，其机理仍需研究。

（3）纳米纤维组织支架在人体内的可能会致畸、致瘤，需进一步研究来确定。

（4）细胞与组织工程支架的附着性和亲和性需进一步提高。

5.3　伤口愈合

5.3.1　引言

皮肤是人体最大的器官，具有阻挡异物和病原体侵入，防止体液流失等功能，既是人体重要的保护屏障，又是机体免疫系统的重要组成部分。哺乳动物皮肤是一种多层结构，主要由表皮和真皮组成，它们位于富含脂肪的皮下组织之上。表皮和真皮分别包含角质形成细胞和成纤维细胞，嵌入细胞外基质中。在皮肤不断更新的过程中，角质形成细胞聚集在表皮的底部，然后逐渐向表皮的顶层迁移。到达最外层后，角质形成细胞经历一种特殊形式的程序性细胞死亡，以产生所谓的角质层。

然而由于暴露在外界环境的独特位置，皮肤组织特别容易受伤，从而造成进一步的体内损伤。皮肤创伤通常是创伤引起的缺损，它涉及不同类型细胞的许多反应，包括免疫细胞和可修复细胞。这种伤口的愈合通常经历四个连续的阶段：止血、发炎、细胞迁移增殖和重塑。轻微伤口可以通过身体的内在修复过程愈合，而大面积或全层伤口的愈合，如大面积烧伤或糖尿病患者的慢性伤口，则很需要外部支架材料的帮助来促进可修复细胞的迁移和浸润。

电纺纳米纤维作为伤口愈合的敷料材料已被广泛研究。通过设计纳米纤维来提供形貌和生物线索，可修复细胞的迁移和渗透可以如预期的那样得到加强。一旦基于纳米纤维的支架在体外被优化以促进细胞迁移或生物分子的传递，它们将接受使用小鼠、大鼠或兔模型在体内愈合的评估，接下来将从实际应用中的纳米纤维疗愈伤口作用的角度具体介绍研究进展。

5.3.2 抑制感染和发炎

发展创面敷料并配合局部用药以抑制感染和消炎，是治愈大面积创面和促进有效修复的一条途径。为此，常用的抗生素，如阿莫西林、四环素、环丙沙星、左氧氟沙星、莫西沙星、头孢唑林和呋西地酸，都被集成到纳米纤维中，以提供抗菌能力。Yang 等人使用一种小分子（6-氨基青霉烷酸，APA）包覆金纳米粒子来抑制 MDR 细菌，将金纳米粒子掺入到聚己内酯（PCL）/明胶的电纺纤维中，以获得抑制 MDR 细菌感染的材料，APA 修饰的金纳米颗粒即使在多药耐药菌的作用下也表现出显著的抗菌活性。另外，在体细菌感染的伤口愈合实验也表明，它有优良的治疗 MDR 细菌伤口感染的能力，如图 5.11 所示。此外，负载抗炎药物的纳米纤维已经被用于促进伤口的愈合，典型的药物包括地塞米松、泼尼松、对乙酰氨基酚、布洛芬、萘普生和酮洛芬。

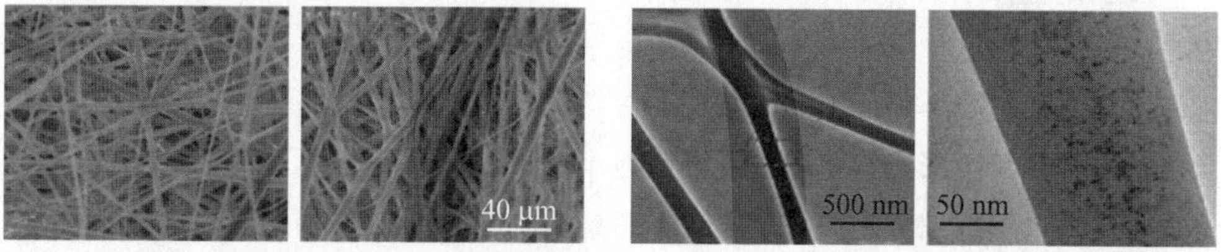

（a）APA-Au 和 PCL/明胶的 SEM 图　　（b）APA-Au 的电纺纤维 TEM 图

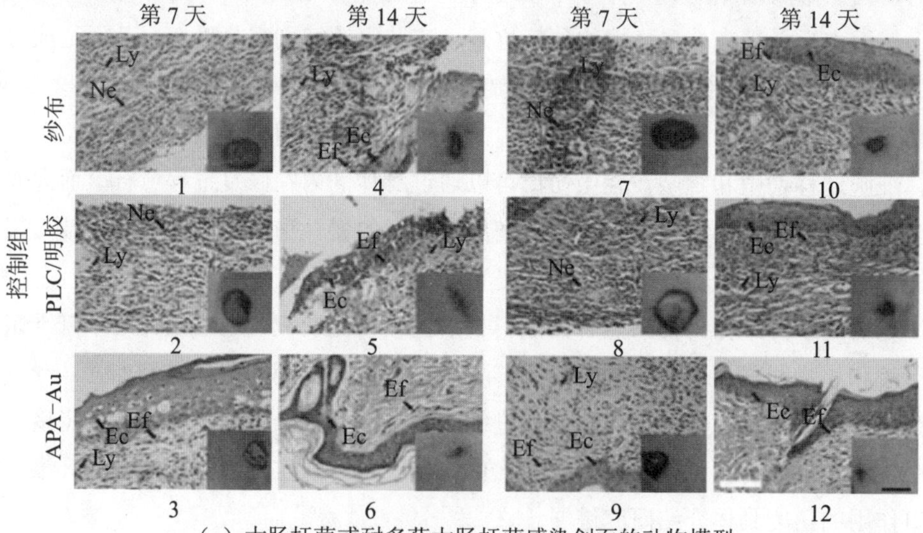

（c）大肠杆菌或耐多药大肠杆菌感染创面的动物模型

图 5.11　抑制感染和发炎的纳米纤维

5.3.3　预防伤口愈合和皮肤癌

手术切除虽然可以完全切除皮肤肿瘤组织，但同时也会导致皮肤缺损。创面敷料作为一种新的发展趋势，不仅可以增强皮肤的抗肿瘤功能，而且可以促进皮肤缺损的愈合。为此，具有局部皮肤肿瘤治疗和皮肤组织再生能力的双功能支架显示出很有希望能够修复肿瘤引起的伤口并且避免肿瘤复发。例如，Wang 等人将 Cu_2S 纳米花加入到由 PLA 和 PCL 的混合物制成的微图案化纳米纤维支架中，使皮肤肿瘤细胞的死亡率达到 90%以上，如图 5.12 所示，展现出优良的伤口愈合和抗癌效果。

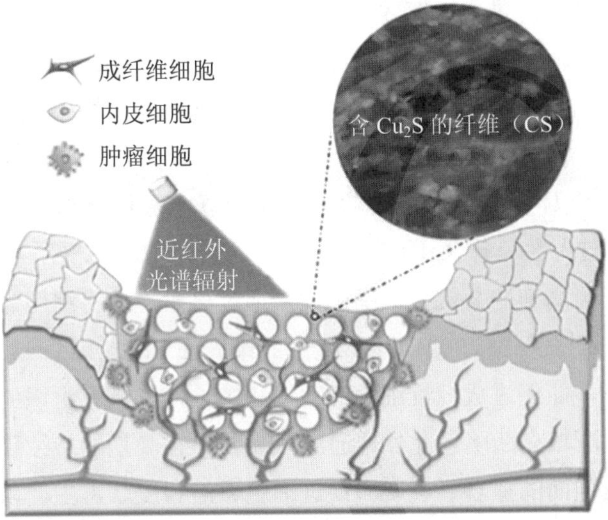

（a）局部治疗皮肤肿瘤和伤口愈合的示意图

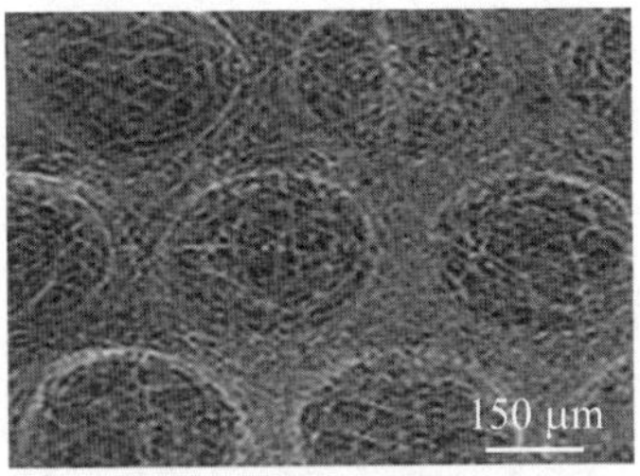

（b）微图案化支架形态的 SEM 图

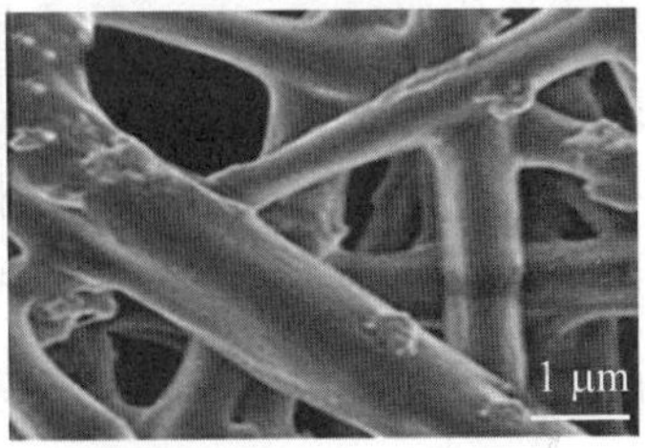

（c）掺有 Cu_2S 的纳米纤维 SEM 图

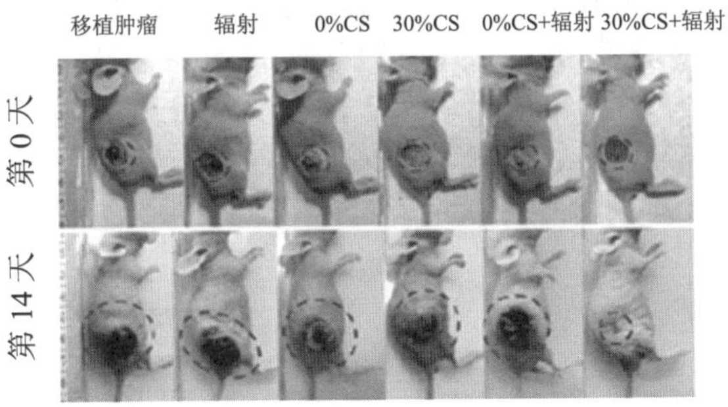

（d）小鼠不同处理手段得到第 0 天和第 14 天的愈合对比图

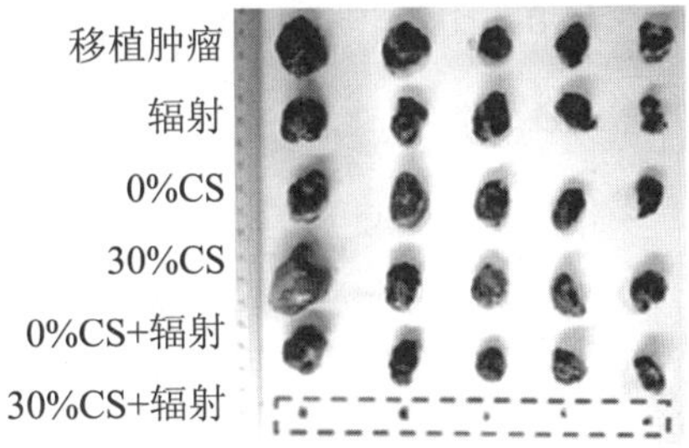

（e）各种治疗后肿瘤的照片

图 5.12　伤口愈合和抗癌的纳米纤维创面敷料

5.3.4 抑制瘢痕形成

在伤口愈合过程后期抑制瘢痕形成是临床上长期存在的难题。瘢痕的形成通常是由于成纤维细胞的异常增殖和胶原沉积。增生性瘢痕是一种皮肤疾病，通常发生在深度烧伤后伤口愈合过程中。有几种细胞信号分子或途径可用于抑制瘢痕形成，包括基质金属蛋白酶、碱性成纤维细胞生长因子、mitsugumin 53 蛋白、转化生长因子 β1 和 20（R）-人参皂苷 Rg_3。例如，Cheng 等人发现抑制转化生长因子-β1 信号能有效地防止瘢痕的形成。在兔耳全层创面愈合过程中，将转化生长因子-β1 抑制剂加入到由聚磷脂和明胶组成的纳米纤维中，有效地抑制了成纤维细胞的增殖，并防止了增生性瘢痕的形成。同时，他们又将 PLGA 纤维负载人参皂苷-Rg_3，然后用碱性成纤维细胞生长因子表面固定，以促进兔耳创面早期愈合，抑制晚期增生性瘢痕形成。

Xu 等人在 ECM 的物理化学环境的启发下，将电纺光交联水凝胶纤维支架 γ-PGA-Nor/γ-PGA-SH（PNS）与人参皂苷 Rg_3（GS-Rg_3）结合，形成 PNS-G，并用于组织修复和伤口治疗。为了提高细胞与支架的黏附性，通过化学方法将细胞黏附肽（精氨酸-甘氨酸-天冬氨酸-半胱氨酸，RGDC）锚定在纤维表面（图 5.13（a））。仿生纤维支架在附着肽的修饰下，能促进成纤维细胞的萌发和生长，形成有组织的空间填充基底，在早期封闭前逐渐填充凹陷。此外，通过 GS-Rg_3 在晚期的持续释放，纤维支架在体内促进无瘢痕创面的愈合，表现为促进细胞通信和皮肤再生，以及随后减少血管生成和胶原积累（图 5.13（c），Ⅰ为 PNS，Ⅱ为 PNS-G，Ⅲ为 PNS-RGDC，Ⅳ为 PNS-G-RGDC，Ⅴ为对照组（无支架和药物）以及各组创面愈合率与 0 d 比较（n=4，*表示 p<0.05））。

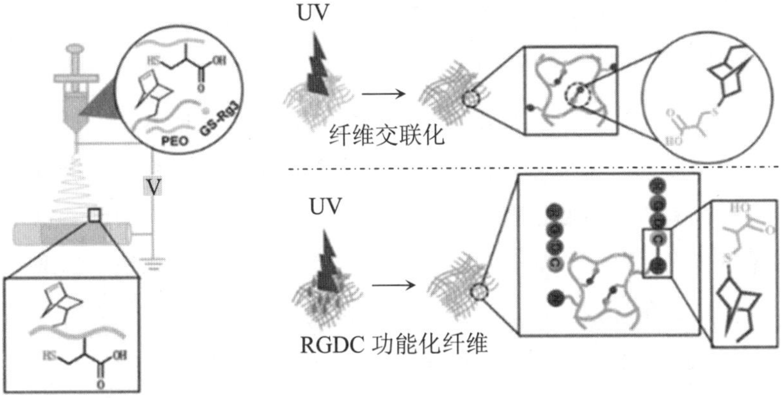

（a）仿生 ECM 电纺丝-PGA 水凝胶支架的制备

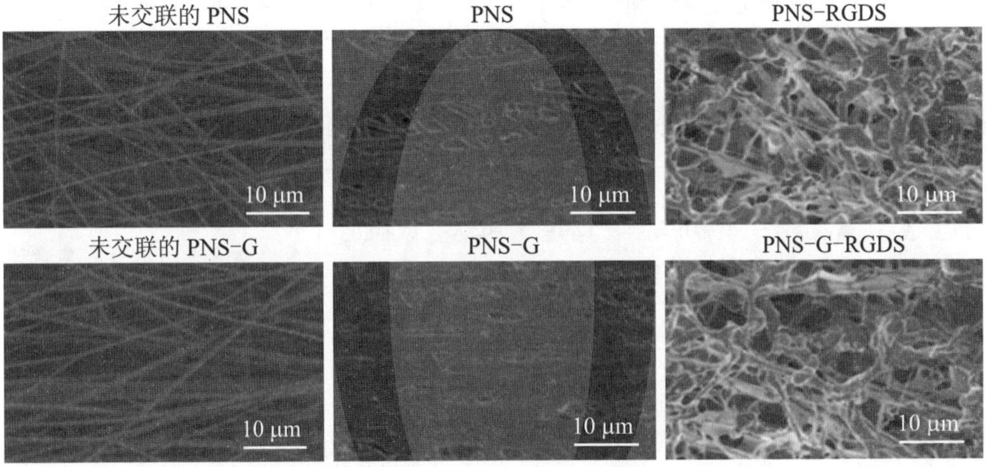

（b）纳米纤维支架负载 $GS-Rg_3$ 前后的微观结构 SEM 图

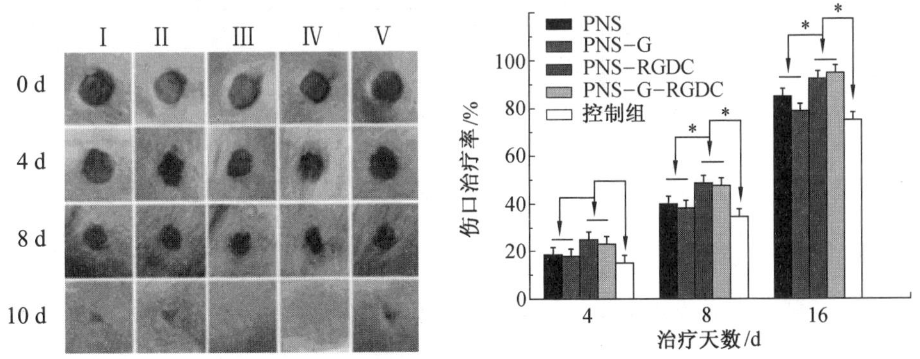

（c）大鼠背部皮肤瘢痕的宏观评价及隆起指数

图 5.13 抑制瘢痕的纳米纤维

5.3.5 载药材料

在静电纺丝纤维中，电纺纤维原料决定了纺丝膜体系的基本性质和作用。一般来说，形成基底膜的纤维原料是聚合纤维，主要分为生物活性聚合纤维和化学合成聚合纤维两类。其中，生物活性聚合纤维通常有胶原蛋白、壳聚糖、透明质酸以及藻酸盐多糖纤维、纤维素、聚甲基丙烯酸磺基甜菜碱，各种动植物提取蛋白如大豆蛋白、玉米蛋白、角蛋白、丝蛋白等。化学合成聚合纤维通常有聚己内酯（PCL）、聚乙烯醇（PVA）、聚乳酸-乙醇酸（PLGA）、聚乙烯吡咯烷酮（PVP）、聚氨酯等（表 5.1）。

表 5.1 用于伤口愈合的纳米纤维膜的载药材料、制备方法、优点和缺点的研究

载药材料	负载药物	制备方法	拉伸强度/MPa	纤维直径/nm	水润湿角/(°)	性能	特点
PCL/胶原蛋白	负载胰岛素的壳聚糖纳米颗粒	共混电纺+浸泡涂敷	2.69±0.05	294.5±21.92	57.03±1.02	2 周后伤口愈合 96.90%±1.11%	刚性结构易附着、愈合速度快
壳聚糖/PVA/氧化石墨烯	—	共混电纺+真空干燥	—	83±10	51～80（不同比例）	对大肠杆菌和金黄色葡萄球菌有较好的抗菌性	抗菌性好
透明质酸/明胶	—	共混电纺+真空干燥+戊二醛蒸气交联	—	47.4～70.6（不同成分比例）	—	2 周后伤口愈合约 80%	提高表皮形成速度，降低炎症细胞数量
PCL	他扎罗汀（AT）	共混电纺+真空干燥	—	785.7±154.6	78.42±0.27	2 周后经 PCL/AT 处理的伤口部位显示出密集和明显的胶原蛋白重建	与常规的 PCL 膜相比，促进新血管形成和再上皮形成效果更好

续表 5.1

载药材料	负载药物	制备方法	拉伸强度/MPa	纤维直径/nm	水润湿角/(°)	性能	特点
PVA/壳聚糖/淀粉	—	共混电纺+真空干燥	2.8～5.8（不同壳聚糖含量）	300～400	—	水蒸气透过率在 2 340～3 318 g·$(m^2 \cdot 24\,h)^{-1}$ 的良好透过率范围内	高孔隙率、吸水率、水蒸气透过率；适当的机械强度和柔韧性；有一定抗菌性
PVA	负载双氯芬酸（DLF）的玉米醇溶蛋白(Zein)纳米颗粒	共混电纺+GA/HCl 蒸气交联+真空干燥	6.12±0.97	324.42±72.80	(55.128±1.858)～(68.848±0.938)（不同 Zein 含量）	纤维中的纳米颗粒可以将释放时间延长至 5 d	将药物封装到纳米颗粒中，可控制释放；有一定生物相容性
PLGA/牛血清白蛋白	水溶性维生素、脂溶性维生素、类固醇激素氢化可的松等	共混电纺+真空干燥	2.8±0.29	210±62	—	融合多种因素共同促进角质形成细胞和成纤维细胞的增殖	实现在纤维上负载多种活性生物分子
PVP	马鞭草精（EP）	共混电纺+真空干燥	—	约 375	—	不仅在体外和体内均保持 MRSA 抗性，而且有更好的药物保留性	良好的药物保留性（活性和剂量）

续表 5.1

载药材料	负载药物	制备方法	拉伸强度/MPa	纤维直径/nm	水润湿角/(°)	性能	特点
PU/醋酸纤维素	—	同轴电纺+真空干燥	12.5±0.5	380±170	—	使植物嫁接成活率从使用塑料膜时的30%提高到近63%	高机械强度、良好的水相容性；促进植物嫁接时伤口愈合，提高嫁接成活率
壳聚糖/PEO	—	共混电纺+真空干燥	4.5±1.8	619±227	0	4周后伤口愈合（95.5%±1.5%）	在细胞黏附和增殖方面表现不佳，但有效阻断了伤口的收缩，并增强了新真皮的产生和伤口的再上皮化
明胶	—	共混电纺+真空干燥+戊二醛蒸气交联	2.2±0.6	858±96	—	4周后伤口愈合（94.9%±2.3%）	体外促进细胞增殖效果很好，提高了愈合速度；在体内效果不明显
PCL	—	共混电纺+真空干燥	2.3±0.1	1 833±369	124±3	4周后伤口愈合（84%±5%）	细胞黏附和增殖表现不佳，延缓伤口收缩；可引起大量多核巨噬细胞的异物反应

5.3.6　小结与展望

利用静电纺丝技术制备的纳米纤维具有高比表面积和高孔隙率，并且能模拟天然细胞外基质的结构和生物功能，促进细胞黏附、迁移和增殖。同时以天然聚合物为原料的电纺纳米纤维具有较好的生物相容性、生物可降解性、高水汽透过性以及合成聚合物可控的机械强度和物理特性，目前广泛应用于伤口敷料领域。如本章所述，利用不同载药原料和负载药物可开发出具有一定抗菌性、提高伤口上皮化进程的纳米纤维，可用于抑制感染、愈合伤口、抑制瘢痕和预防皮肤癌。但当前研究出的用于伤口愈合的纳米纤维仍存在着不足：

（1）无负载生物活性分子作为药物的纳米纤维伤口敷料的伤口恢复速度仍不够理想。

（2）作为伤口敷料的纳米纤维负载的药物和小分子物质的缓释速度的可控性不好。

（3）组织工程皮肤成分复杂，含有多种细胞、生长因子和支架结构。细胞之间、细胞与基质材料及细胞与生长因子之间的相互作用仍是盲点。

（4）支架材料的功能与演变规律、细胞分化的调控机制尚未阐明；尤其是干细胞应用的定向诱导分化以及皮肤附属结构再生的相关机制尚需要更多的探索。

5.4　药物传递与释放控制

5.4.1　引言

进入人体内的药物如果浓度过高可导致中毒，浓度过低则无治疗效果；同时，当药物到达病灶部位时才能发挥疗效，到达其他部位不起治疗作用甚至产生毒性和副作用。因此想达到理想的治疗效果不仅需要有效的药物还需要能够精准传递和控制药物释放的药物载体。

在药物传递方面，静电纺织纳米纤维与其他纳米结构相比具有独特的优势，一是与细胞外基质形态结构相近，便于细胞黏附、扩增及迁移；二是比表面积大，可增大疏水性成分的分散和溶出，进而提高其生物利用度；三是同轴或多轴静电纺织形成的壳-核结构便于制备长效控释系统，尤其可实现水溶性成分的长效释药；四是可包裹敏感性的生物大分子，避免与有机溶剂的直接接触。此外，还具有操作简单、载药量和包封率高、成本低、适应性强等特点。鉴于静电纺织纳米纤维的诸多优点，静电纺丝技术为载药系统的构建开辟了新方向，静电纺织纳米纤维支架可以作为各种药物、基因和生长因子的载体，通过调节支架的形貌、孔隙率和组成可以很好地控制药物释放的过程。因此，电纺支架作为药物载体在生物医学领域的应用前景广阔，特别是在预防手术后组织粘连和感染、术后局部化疗等方面。

5.4.2 抗生素类药物传递

作为功能半导体纳米材料，改性二氧化钛纤维膜具有理想的异质结构、良好的力学性能和优异的可见光驱动光催化活性，在环境修复方面具有广阔的应用前景，但构建这种奇特的纤维膜仍然是一个巨大的挑战。Song 等人通过简单的静电纺丝和随后的热聚合工艺制备了柔软的异质结构 g-C_3N_4@Co-TiO_2(CNCT)纳米纤维膜，如图 5.14 所示。超薄 g-C_3N_4 纳米壳由原位合成，均匀包裹在 Co-TiO_2 纳米纤维上，形成核壳量子异质结，通过简单调节前驱体三聚氰胺的含量，可以精确控制 g-C_3N_4 纳米壳的厚度和负载量。通过构建三维多孔网络、增强可见光响应和均匀致密的异质结诱导电荷转移，所合成的 CNCT 膜在 60 min 内对盐酸四环素的光降解率高达 90.8 %，且具有优异的抗菌性能，可见光照射 90 min 后对大肠杆菌的灭活达到 6 log 杀菌率。此外，稳定的核壳结构和坚固的机械强度也提供了良好的可逆性和易回收性。

作用范围小、能精确控制时间的给药系统，对于获得高局部生物活性和低副作用的治疗牙周感染和骨感染的抗生素至关重要。在这个研究领域，Kai 研究团队开发了一种 3D 多孔组织工程支架，具有控制释放抗生素的能力，能够长期抑制细菌生长。实验中，研究者使用改良的油包水乳化法成功地将高可溶性抗生素药物植入 PLGA 纳米球中，然后将 PLGA 纳米球结合到预制的、具有良好大孔结构的 PLLA 纳米纤维支架中。结果表明，所研制的含药纳米纤维支架能够有效地以控制的方式传递多西环素，延长了药物作用的持续时间。这些可生物降解的 PLLA 支架具有良好的大孔和纳米纤维结构，加入含有多西环素的 PLGA 纳米球可抑制普通细菌生长 6 周以上。多西环素在三维支架中的成功结合及其在支架中的控制释放，使纳米纤维支架的应用从大分子（如生长因子）的输送扩展到小分子亲水性药物的输送，从而使其具有更广泛和更复杂的组织工程应用。

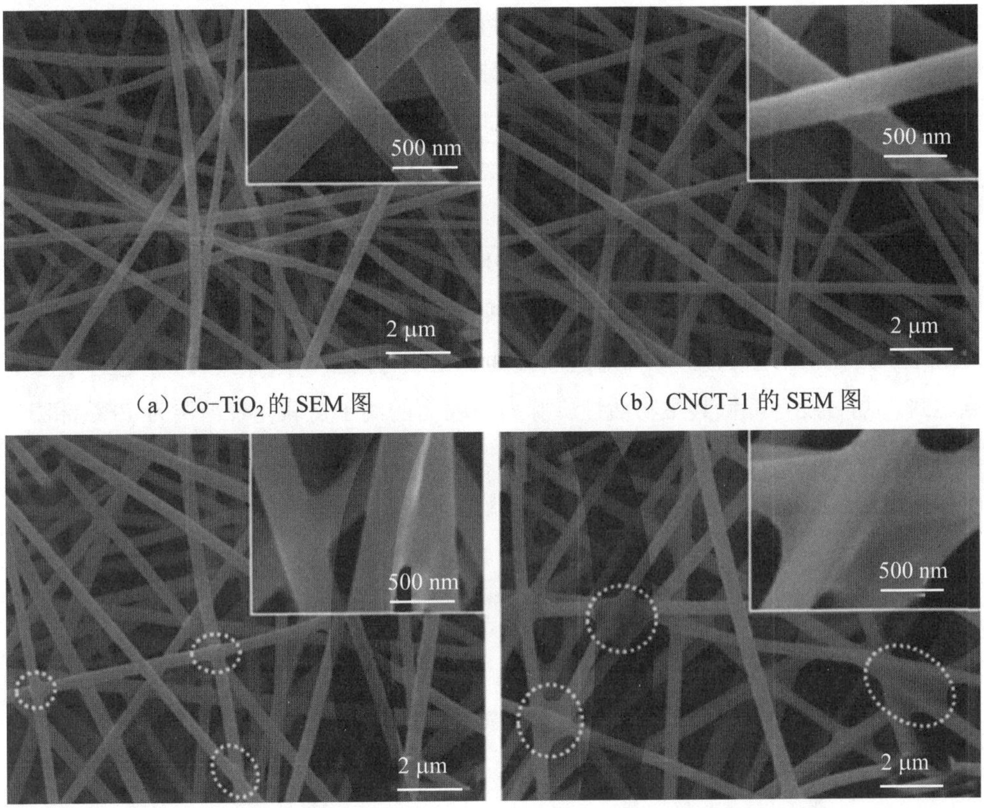

（a）$Co-TiO_2$ 的 SEM 图　　（b）CNCT-1 的 SEM 图

CNCT-3 的 SEM 图　　（d）CNCT-5 的 SEM 图

图 5.14　异质结构 $g-C_3N_4@Co-TiO_2$(CNCT)纳米纤维膜

5.4.3　抗癌药物传递

近年来，随着静电纺织技术的发展，静电纺织纤维作为药物载体，具有治疗效果好、毒性小、操作方便等优点，是一种很有前途的抗癌药物递送方法，特别是在术后局部化疗中有很好的应用潜力。由自组装肽 V6K3 形成的球形纳米颗粒可作为难溶性抗肿瘤药物的载体，有效地将药物送入肿瘤细胞。Gong 等人实现了 V6K3 在等离子体胺氧化酶（PAO）诱导下实现纳米颗粒到纳米纤维的形状转变。PAO 功能上与赖氨酰氧化酶（LO）相似，后者广泛存在于血清中。添加胎牛血清（FBS）或

PAO 后，肽的二级结构发生了变化，球形纳米粒子被拉伸并转化为纳米纤维。自组装形态的转变揭示了这种两亲性肽对细微化学修饰的敏感性，以及通过酶诱导来控制自组装结构的可行性。酶自组装 V6K3 在添加 PAO 后具有杀菌性能，其杀菌性能明显优于添加 PAO 前的杀菌性能，使该肽可用于预防感染。两亲性肽 V6K3 在哺乳动物细胞中表现出抗肿瘤特性和低毒性，具有良好的生物相容性和杀菌性能，可防止细菌污染。这些优点表明酶自组装 V6K3 在肿瘤治疗中具有巨大的生物医学应用潜力。

为了实现有效的肿瘤基因治疗，系统性传递肿瘤靶向性腺病毒复合物对原发性和转移性肿瘤的治疗都是至关重要的。Park 等人采用静电纺丝法制备了壳聚糖-PEG-FA 共轭物（图 5.15），并通过原位离子交联与三聚磷酸盐（TPP）进行了腺病毒包封，同时没有降低其生物活性。腺病毒表面的离子交联壳聚糖层为 PEG 和 FA 提供了化学结合位点，作为异官能团 PEG 末端的靶基。电子显微镜照片表明分散性好，单个腺病毒纳米复合物没有聚集或降解。腺病毒/壳聚糖-聚乙二醇-脂肪酸纳米复合物的转导依赖于脂肪酸受体的表达，表明脂肪酸受体是靶向的病毒转导途径。这些结果清楚地表明，腺病毒/壳聚糖-PEG-FA 纳米复合物的癌细胞靶向病毒转导可有效地用于转移性肿瘤治疗。

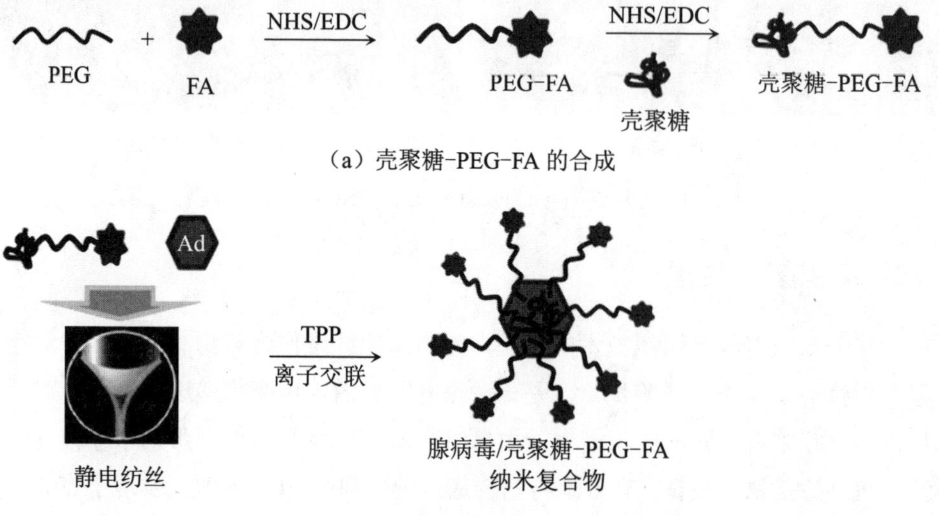

（a）壳聚糖-PEG-FA 的合成

（b）静电纺丝制备腺病毒壳聚糖-PEG-FA 纳米复合物

图 5.15　作为病毒靶向的纳米纤维复合物

5.4.4　基因传递

以支架为载体的基因传递或反向传递目前已被用于传递 DNA 分子，到目前为止，所使用的传递载体通常是水凝胶和由气体形成或颗粒分离技术形成的多孔支架材料，可以确保遗传物质对细胞的长期可用性。

目前组织工程中的病毒基因治疗肿瘤或癌症的方法包括在支架或水凝胶的情况下或者无支架和水凝胶条件下植入体外转基因细胞。虽然病毒基因转移在组织工程中实现转基因表达是有效的，但由于病毒会引起免疫反应、病毒传播、毒性和瞬时基因表达，还未能实现广泛应用。因此，许多组织工程研究选择在体外对细胞进行基因工程，然后再导入体内。例如 Xie 等人在研究中，利用同轴静电纺丝技术制备了具有生物活性的 PEI（聚乙烯亚胺)/pBMP2-（骨形态发生蛋白-2 质粒）PLGA 核壳支架，用于控制基因传递给 hPDLSCs（人牙周韧带干细胞)。结果表明，pBMP2 在最初几天释放水平较高，并在接下来的 28 d 持续释放，同时 PEI/pBMP2 转染效率高。此外，qRT-PCR 和 Western Blot 检测结果显示，与单轴静电纺织支架相比，核壳静电纺织支架 BMP2 表达时间更长（超过 28 d)。此外，Kazantseva 等人采用混合三维纳米纤维支架对乳腺癌（MDA-MB 231）、大肠癌（$CaCO_2$）、黑色素瘤（WM239A）和神经母细胞瘤（Kelly）进行选择性基因表达调节，不需要任何其他刺激，如图 5.16 所示。丝裂霉素 C 处理 MDA-MB 231 后，其基因表达也发生了变化，为抗癌药物的检测提供了一个合适的平台，实现了有选择地、“干净”地指导人们深入了解癌细胞的生长和肿瘤形成的机制，而不需要使用特定的标记物进行操作，从而避免副作用。

纤维制备工艺也会影响负载的基因表达效果。Saraf 等人通过分数因子设计，制备了包含质粒 DNA（pDNA）、非病毒基因传递载体聚（乙烯胺）-透明质酸（PEI-HA）的纤维网支架，并在两个层面研究了四个工艺参数的影响。研究了 PCL 聚合物浓度、PEG 聚合物分子量和浓度以及 pDNA 浓度对 PEI-HA 平均纤维直径、释放动力学和传递效率的影响。结果表明，增加每个参数的值均会导致纤维的平均直径增加。此外，在含有 PEI-HA 和 pDNA 的同轴纤维网支架上直接种植的成纤维细胞样细胞在 60 d 内也合成了绿色荧光蛋白，并且合成量明显高于单独使用 pDNA 支架观察到的绿色荧光蛋白。这些结果证实了具有可变和持续转染特性的静电纺织同轴纤维网支架可用于组织工程和其他涉及基因治疗的基因传递应用。

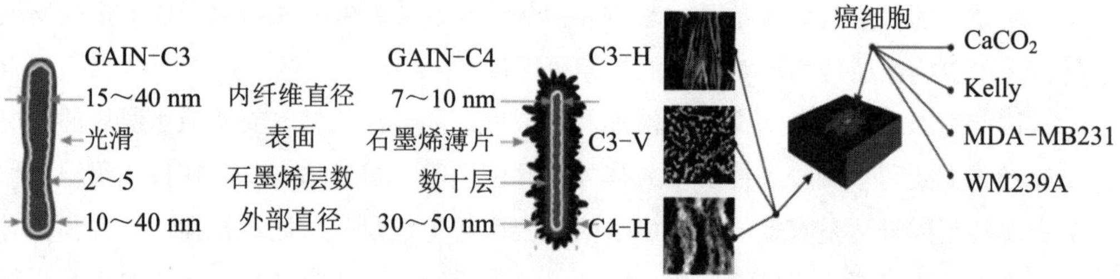

（a）C3 和 C4 类型的支架及其差异示意图以及对应的 SEM 图谱

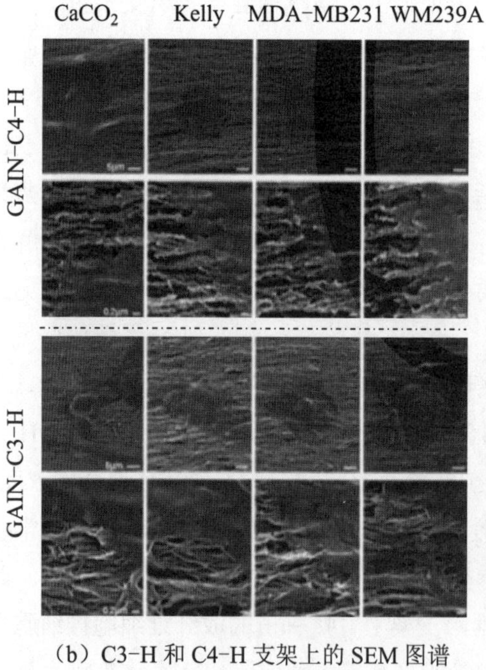

（b）C3-H 和 C4-H 支架上的 SEM 图谱

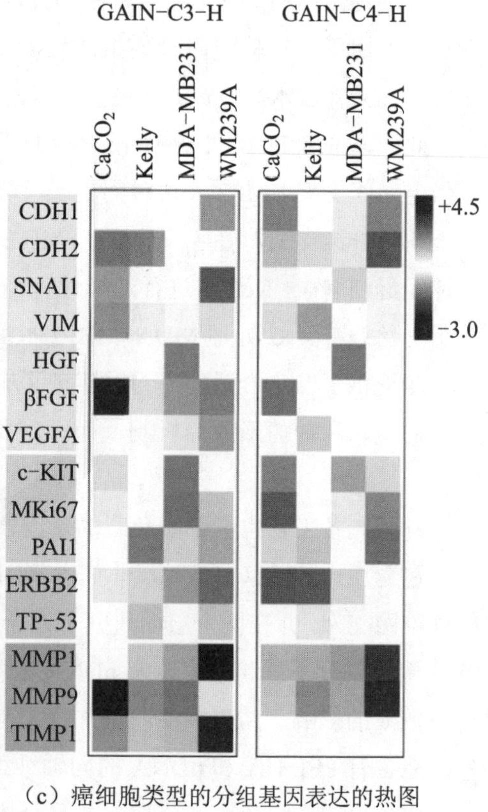

（c）癌细胞类型的分组基因表达的热图

图 5.16 调节细胞基因表达的三维纳米纤维支架（彩图见附录）

5.4.5 小结与展望

静电纺丝是一种简单、高效地制备微米及纳米尺度纤维的技术。由于具有较高的表面积-体积比、多孔性和可调的多孔性等特点静电纺丝制备的纳米纤维在药物缓释、组织工程支架、创伤敷料等领域具有良好的应用前景。黏度、分子量、载体的浓度、应用的电压、收集的距离和导电性等溶液和工艺操作参数等会影响纤维的形态，通过调节这些参数以满足具体应用的需要。如本章所述，利用静电纺织制备的载药纳米纤维具有较好的抗菌性、靶向性、降解性，可用于传递和控释抗生素药物、抗癌药物和基因分子等物质，在治疗牙周炎感染、抑制肿瘤细胞扩散等方面有良好的效果。但是目前为止的递药纳米纤维仍有不理想的方面，需研究人员进一步改良：

（1）静电纺丝缺乏明确剂型，这导致静电纺织膜在缓控释方面的应用存在局限性。

（2）药物在静电纺织膜中的含量均一性可能随着工艺参数及实验设备的改变发生变化，对质量评价、工业化生产及实际应用造成负面影响。

（3）静电纺织纳米纤维给药仍处于研究阶段，关于控释和给药的生物模型还需进一步研究。

（4）静电纺丝技术提高了成品的吸水性，这也提高了它的储存标准。

（5）现有静电纺丝技术生产效率低，生产时间长，仍然难以实现大规模生产。

5.5 癌症研究

5.5.1 引言

癌症是严重威胁人类生命的疾病之一，近十年来全球范围内癌症危害呈持续增长态势。在我国，肿瘤死亡占所有死因的 1/4。癌症不仅给患者家庭和社会带来沉重负担，也对公共安全和社会稳定有一定的影响。纳米纤维由于尺寸易调控同时细胞相容性较强，在癌症诊断、传递抗癌药物、构建肿瘤模型等方面都有很好的应用，在当前已有的抑制癌细胞的药物和治疗手段中也存在。

相对于表面光滑的固体基质，静电纺织纳米纤维提供了更高的表面积-体积比，因此与分析物和癌细胞有更大的接触面积，可以很容易地负载药物。因此，静电纺织纳米纤维包括那些由无机和有机材料制成的纤维，以及无机和有机杂化材料，在癌症研究方面的应用有癌症诊断、构建3D肿瘤模型和癌症治疗，本节将依次介绍。

5.5.2 癌症诊断

研究表明，癌细胞通常会表达特定的蛋白质生物标志物，这为早期诊断提供了机会。静电纺织纳米纤维已被用于在癌症发展的早期阶段检测血液或体液中的肿瘤标志物或循环中的肿瘤细胞。例如在Paul等人对检测肿瘤标志物的研究中，含有多壁碳纳米管的ZnO纳米纤维可以与抗癌抗原-125抗体共价偶联。这种纳米纤维对卵巢癌的特异性标志物癌症抗原-125的检测范围很广。此外，纳米纤维还可以与微流控技术相结合，以提高对特定癌细胞的灵敏度和选择性。例如，一种由多孔石墨烯泡沫制成并经碳掺杂纳米纤维修饰的微流控免疫生物芯片被用来检测乳腺癌生物标记物。通过作为免疫电极生物传感器，该生物芯片可以检测很大的浓度范围的目标抗原，并且在毫摩尔检测水平上具有很高的灵敏度。除了特定的生物标志物，氧也是癌细胞生物学中的一个重要指标。Xue设计了一系列由静电纺织纳米纤维制成的氧传感器。在他的另一项研究中，将氧敏发光探针Pt（Ⅱ）或Pd（Ⅱ）引入到聚醚砜@PCL或聚砜@PCL芯-鞘纳米纤维的芯中。纳米纤维表现出快速的对氧的响应时间。

循环肿瘤细胞（CTC）为早期癌症诊断提供了另一条途径。循环肿瘤细胞在癌症组织中脱落会随着循环系统转移到身体的别的部位，因此通过针对CTC的捕获会减慢癌症的转移。静电纺织的纳米纤维通过修饰具有特定功能基的纳米纤维表面以靶向肿瘤细胞来捕获CTC。Xiao等人制备了PEI/PVA短纳米纤维，DNA适配体功能化后得到磁性短纳米纤维（适配体MSNFs），捕获和释放CTCs的效率分别达到了87%和91%，展现出极强的应用前景。此外，纳米纤维与微流体技术结合使用也可以提高捕获效率。Wang等人制备了透明质酸（HA）功能化的静电纺织壳聚糖纳米纤维（CNF），嵌入到微流控芯片中后，纳米纤维捕获癌细胞的效率达91%（A549，人肺癌细胞系）。在另一项研究中，Xiao等人将取向的PEI / PVA纳米纤维功能化后，与可用于动态捕获和释放癌细胞的集成的微流体平台相结合，形成PMPC功能化纳米纤维。该纳米纤维具有良好的血液相容性，并且能够高效地捕获高纯度叶酸（FA）受体过表达的癌细胞（例如，HeLa 细胞）。动态微流体条件下，30 min 内捕获效率可高达 92.7%。

提高稀有癌细胞的纯度和丰富其种群也是癌症研究的迫切需要。为此，Yu 等人设计了一种同时实现 CTC 动态捕获和释放的生物电子器件。实验中使用聚-L-赖氨酸-g-PEG 生物素、链霉亲和素和抗上皮细胞黏附分子的顺序的生物活性剂对 PEO-PEDOT/PSS 芯-鞘结构的纳米纤维改性。然后将纳米纤维沉积在由氧化铟锡（ITO）制成，并与包含微流体通道的 PDMS 室集成在一起的可单独寻址电极的平行阵列上。该设备可以用作生物电子接口，通过每个电极的电激活来分离、检测、收集和富集极少量的 CTC。实验结果显示，使用收集的纳米纤维 10 min，该设备捕获了 90%以上的靶向癌细胞。此外，由于 PEDOT 的电化学掺杂和去掺杂特性，超过 87%的捕获的癌细胞可以通过解吸聚-L-赖氨酸-g-聚乙二醇-生物素而被电触发从电极释放。这种类型的装置可以用于进一步探索检测和捕获，并释放多种类型的癌细胞。

5.5.3　构建体外三维肿瘤模型

体外三维肿瘤模型工程是一种可行的识别致癌物、测试新药、筛选药物以及研究肿瘤生长和转移机制并且经济有效的方法。主要目标是设计一个类似于肿瘤细胞外环境的特定 3D 矩阵，支持细胞间和细胞外基质间的通信，并在肿瘤转移过程中提供生化信号。静电纺织纳米纤维可以模拟天然肿瘤细胞外基质的组成和结构，通过表面修饰或药物包封实现生物化学功能化，为构建体外三维肿瘤模型提供了一个有吸引力的平台。为此，Janan 等人使用胶原包覆的明胶纳米纤维构建了体外 3D 模型，以研究乳腺癌的转移。他们的肿瘤模型模拟了结缔组织的细胞外成分、表面复杂性和机械特性。除此之外，研究人员还发现细胞微环境特别是细胞外基质的成分在调节癌细胞转移中起重要作用。例如在 Ma 等人的研究中将层粘连蛋白亚型阵列涂覆在聚苯乙烯纳米纤维上，并在之后创建了用于 U251 胶质母细胞瘤细胞培养的纳米纤维 3D 模型，并发现 3D 结构和层粘连蛋白修饰的组合促进了神经胶质瘤干细胞的生长。相关的 3D 模型纳米支架如图 5.17 所示。

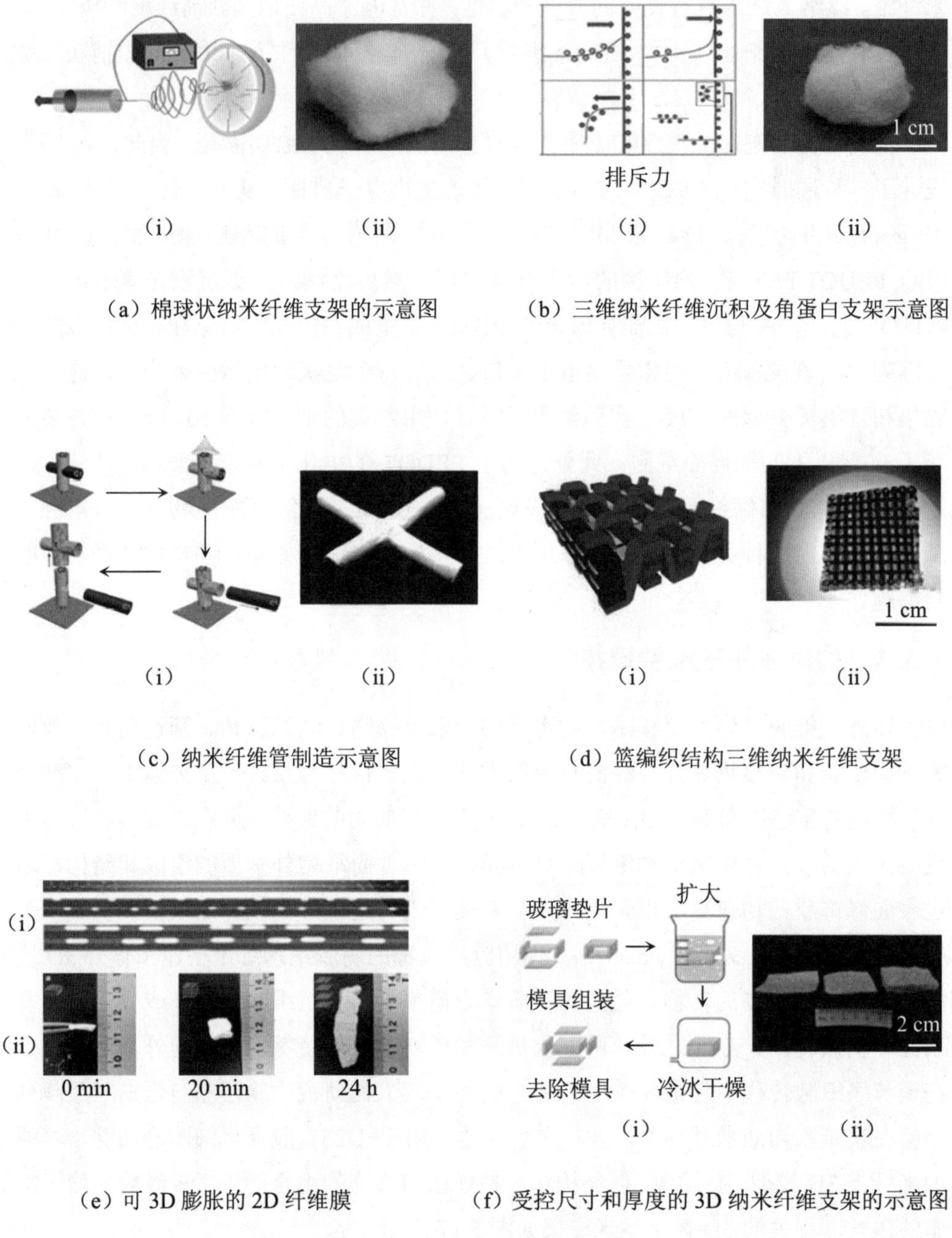

（a）棉球状纳米纤维支架的示意图

（b）三维纳米纤维沉积及角蛋白支架示意图

（c）纳米纤维管制造示意图

（d）篮编织结构三维纳米纤维支架

（e）可 3D 膨胀的 2D 纤维膜

（f）受控尺寸和厚度的 3D 纳米纤维支架的示意图

图 5.17　多种结构的 3D 模型纳米纤维支架

5.5.4　癌症治疗

静电纺织纳米纤维可以控制癌细胞的迁移、控制抗肿瘤药物的释放和提供同时抑制癌细胞增殖、血管生成和侵袭的协同平台，因此也被应用于癌症治疗。

1. 控制癌细胞的迁移

通过控制形态和排列，静电纺织纳米纤维可以通过调节癌细胞的迁移和诱导细胞凋亡，为抑制原发性肿瘤的生长提供新的策略。Jain 等人将单轴排列的 PCL 纳米纤维的非织造垫插入导管中，以引导胶质母细胞瘤肿瘤细胞从初始肿瘤部位迁移到皮质外细胞毒性水凝胶，在该水凝胶中癌细胞经凋亡而消除。体内实验结果表明，与含有光滑 PCL 膜的导管和空白对照组相比，含有纳米纤维的导管治疗后，动物脑肿瘤的总体积显著减小。

2. 通过释放抗肿瘤药物进行化疗

静电纺织纳米纤维也可用于癌症化疗，利用它们的能力以可控的方式包封和释放抗肿瘤药物。功能化后的纳米纤维可以实现局部给药从而精准定位被治疗部位。例如 Wang 等人制备了阿霉素和吲哚菁绿共载介孔二氧化硅纳米颗粒（DIMSN）。然后将纳米粒子与壳聚糖/聚乙烯醇（CS/PVA）静电纺织法制备了多功能复合纳米纤维（DIMSN/F）。在模拟阴道分泌物侵蚀的情况下，DIMSN/F 出现了位点特异性药物释放。与系统注射 DIMSN 相比，阴道植入 DIMSN/F 可以使药物在小鼠阴道内积累最大化，能大幅度提升宫颈癌和阴道癌的肿瘤抑制率（TIR），使其高达 72.5%，显示出在宫颈癌治疗方面的巨大潜力。

静电纺织纳米纤维通常与多功能纳米颗粒结合使用，以提供可长期局部释放的药物载体。例如 Chen 课题组制备了一种瘤内注射复合物，以实现在肿瘤微环境快速响应而从静电纺织纤维片段中持续释放载药胶束。实验具体流程是将喜树碱与透明质酸偶联，然后将偶联物嫁接到 PLA 纳米纤维上，制成可注射的纤维片段。实验结果证明瘤内注射纤维碎片可确保胶束在肿瘤组织内累积超过 3 周。此外，相对于瘤内注射游离胶束，胶束从纤维碎片中持续释放有助于显著提高抗肿瘤功效，为靶向癌症化疗提供了新途径。在另一项研究中，Zhao 等人开发了一种使用复合材料抑制癌症复发的涉及肿瘤触发的阿霉素释放的智能系统，由静电纺织 PLA 纤维和 $CaCO_3$

封端的介孔 SiO_2 纳米颗粒构成。实验结果证明该系统通过利用 $CaCO_3$ 与酸之间的反应，能够在癌症组织的酸性环境中持续释放阿霉素以杀死癌细胞。相反，抗肿瘤药仅在保持在生理 pH 的正常组织中少量释放。这种智能的递送系统表现出 pH 触发的抗肿瘤功效，可以持续 40 d。控制释放抗肿瘤药物的纳米纤维如图 5.18 所示。

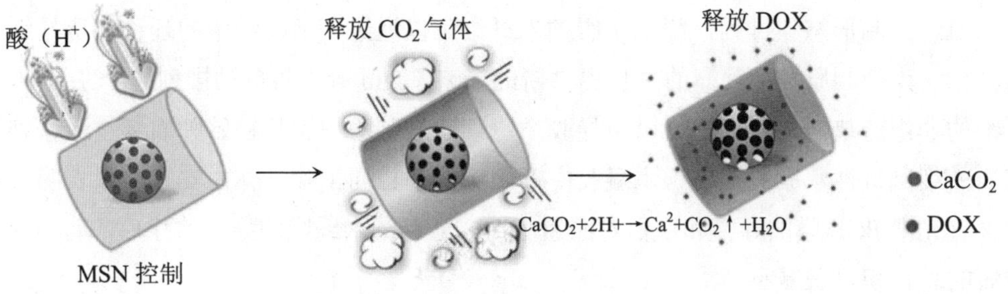

（a）pH 响应控制释放系统释放机理示意图

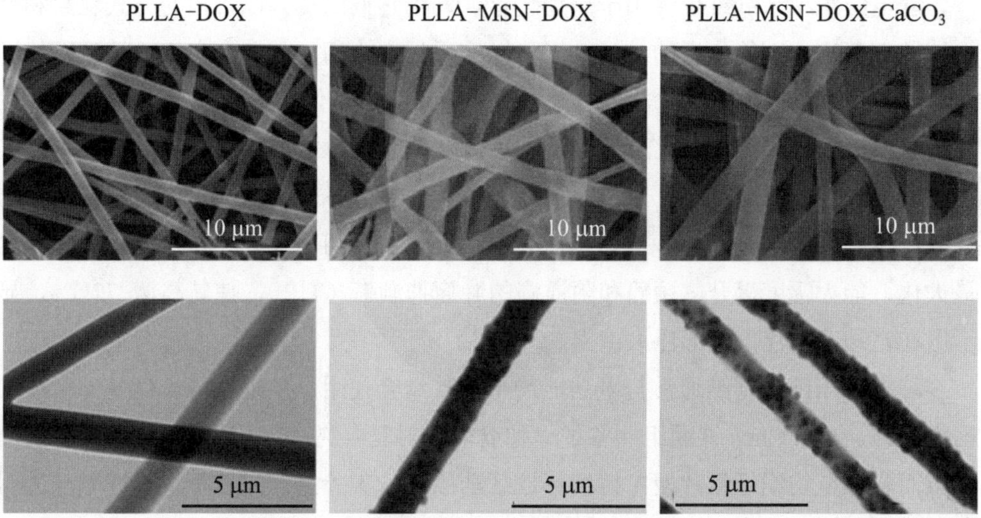

（b）DOX、PLLA-MSN-DOX 和 PLLA-MSN-DOC-$CaCO_3$ 三种纳米纤维的微观结构

图 5.18　控制释放抗肿瘤药物的纳米纤维

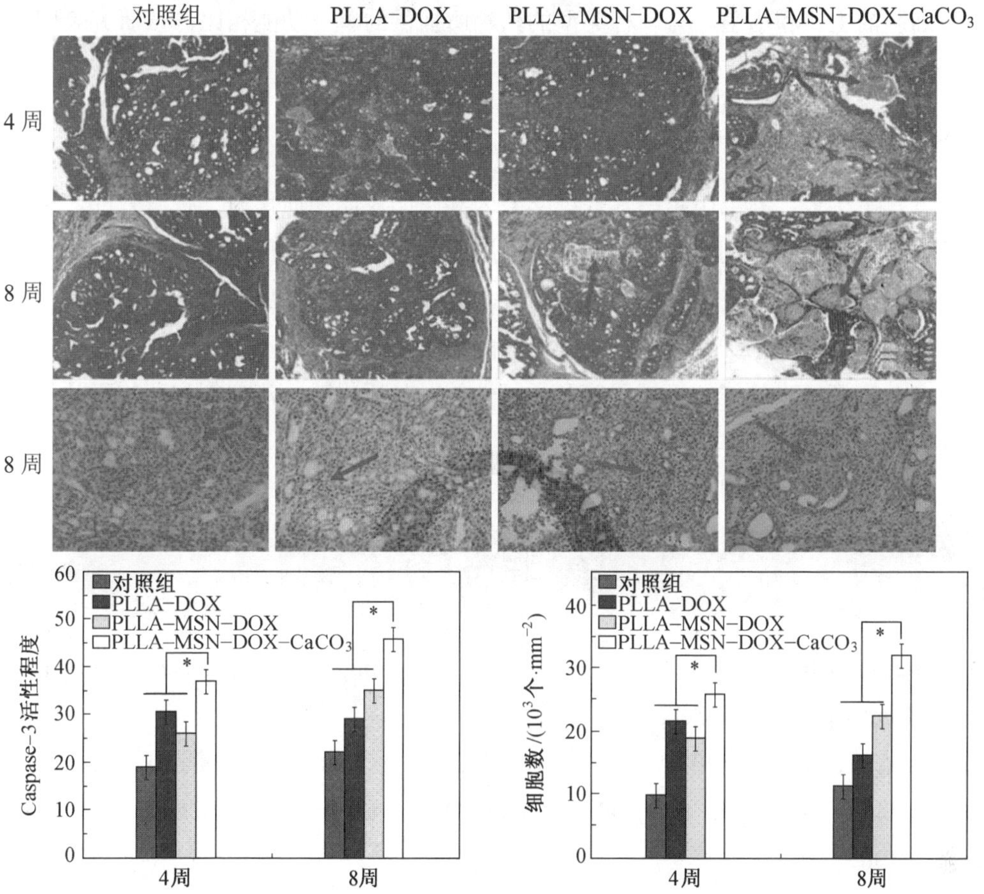

（c）不同静电纺织纤维对肿瘤细胞的反应

续图 5.18

3. 协同癌症治疗

协同疗法可以同时抑制癌细胞的增殖、血管生成和侵袭。例如，Lei 开发了一种双重递送系统，将 RNA 干扰物（RNAi）和化学疗法相结合，用于脑肿瘤治疗。实验中，将设计的 RNAi 质粒复合物与基因载体聚乙烯亚胺复合，来特异性抑制神经胶质瘤信号通路中基质金属蛋白酶 2 的表达。然后将 RNAi 质粒与紫杉醇（一种细胞毒性药物）一起封装在 PLGA 纤维中，以实现两种药物的持续释放。抑制的基质金属蛋白酶 2 表达抑制了脑肿瘤的侵袭和血管生成，同时细胞毒性紫杉醇也能够阻止癌细胞的生长和增殖。与一种类型的药物所提供的单一功能相比，该研究表明纳米纤维参与了协同治疗，疗效显著提高。

5.5.5 小结与展望

近几十年来，静电纺织纳米纤维支架因其高比表面积而作为一种有效的、可植入的多功能传递和控制释放药物的平台应用于局部和术后癌症治疗而备受关注。然而，要在工业生产和医疗应用中全面实施，仍有一些挑战需要解决：

（1）聚合物纳米纤维静电纺丝中溶剂的相对毒性。

（2）陶瓷纳米纤维的机械强度和柔韧性差，基质材料、治疗剂和药物释放优化之间的矛盾。

（3）实际病例应用时是否有副作用还需进一步研究来明确。

参 考 文 献

[1] LAUFFENBURGER D A, HORWITZ A F. Cell migration: a physically integrated molecular process[J]. Cell, 1996, 84(3): 359-369.

[2] TRACY L E, MINASIAN R A, CATERSON E J. Extracellular matrix and dermal fibroblast function in the healing wound[J]. Advanced Wound Care, 2016, 5(3): 119-136.

[3] MAYOR R, ETIENNE-MANNEVILLE S. The front and rear of collective cell migration[J]. Nature Reviews Molecular Cell Biology, 2016, 17(2): 97-109.

[4] SUNDARARAGHAVAN H G, SAUNDERS R L, HAMMER D A, et al. Fiber alignment directs cell motility over chemotactic gradients[J]. Biotechnology and Bioengineering, 2013, 110(4): 1249-1254.

[5] LIU Y, FRANCO A, HUANG L, et al. Control of cell migration in two and three dimensions using substrate morphology[J]. Experimental Cell Research, 2009, 315(15): 2544-2557.

[6] MI H Y, SALICK M R, JING X, et al. Electrospinning of unidirectionally and orthogonally aligned thermoplastic polyurethane nanofibers: fiber orientation and cell

migration[J]. Journal of Biomedical Materials Research Part A, 2015, 103(2): 593-603.

[7] OTTOSSON M, JAKOBSSON A, JOHANSSON F. Accelerated wound closure-differently organized nanofibers affect cell migration and hence the closure of artificial wounds in a cell based in vitro model[J]. Public Library of Science One, 2017, 12(1): 0169419.

[8] HURTADO A, CREGG J M, WANG H, et al. Robust CNS regeneration after complete spinal cord transection using aligned poly-l-lactic acid microfibers[J]. Biomaterials, 2011, 32(26): 6068-6079.

[9] QU J, ZHOU D, ZHANG F, et al. Optimization of electrospun TSF nanofiber alignment and diameter to promote growth and migration of mesenchymal stem cells[J]. Applied Surface Science, 2012, 261: 320-326.

[10] ENGLUND-JOHANSSON U, NETANYAH E, JOHANSSON F. Tailor made electrospun culture scaffolds control human neural progenitor cell behavior-studies on cellular migration and phenotypic differentiation[J]. Biomaterial Nanobiotechnology, 2017, 8: 72746.

[11] WANG H B, MULLINS M E, CREGG J M, et al. Varying the diameter of aligned electrospun fibers alters neurite outgrowth and schwann cell migration[J]. Acta Biomaterialia, 2010, 6(8): 2970-2978.

[12] XIE J, MACEWAN M R, RAY W Z, et al. Radially aligned, electrospun nanofibers as dural substitutes for wound closure and tissue regeneration applications[J]. ACS Nano, 2010, 4(9): 5027-5036.

[13] SHI J, WANG L, ZHANG F, et al. Incorporating protein gradient into electrospun nanofibers as scaffolds for tissue engineering[J]. ACS Applied Materials & Interfaces 2010, 2(4): 1025-1030.

[14] TANES M L, XUE J, XIA Y. General strategy for generating gradients of bioactive proteins on electrospun nanofiber mats by masking with bovine serum albumin[J]. Journal of Materials Chemistry B, 2017, 5(28): 5580-5587.

[15] LI X, LI M, SUN J, et al. Radially aligned electrospun fibers with continuous

gradient of SDF1α for the guidance of neural stem cells[J]. Small, 2016, 12: 5009-5018.

[16] SUNDARARAGHAVAN H G, SAUNDERS R L, HAMMER D A, et al. Fiber alignment directs cell motility over chemotactic gradients[J]. Biotechnology and Bioengineering, 2013, 110(4): 1249-1254.

[17] RAO S S, NELSON M T, XUE R, et al. Mimicking white matter tract topography using core-shell electrospun nanofibers to examine migration of malignant brain tumors[J]. Biomaterials, 2013, 34(21): 5181-5190.

[18] 何创龙, 黄争鸣, 张彦中, 等. 静电纺丝法制备组织工程/纳微米纤维支架[J]. 自然科学进展, 2005, 15: 1175-1182.

[19] 陈宗刚. 静电纺丝制备胶原蛋白壳聚糖纳米纤维仿生细胞外基质[D]. 上海: 东华大学, 2007, 1: 1-3.

[20] 曹谊林. 组织工程学的建立与发展[J]. 组织工程与重建外科杂志, 2005, 1: 5-8

[21] ZHANG Y, LI X, GUEX A G, et al. A compliant and biomimetic three-layered vascular graft for small blood vessels[J]. Biofabrication, 2017, 9(2): 025010.

[22] LEE K W, JOHN N R, GAO Y, et al. Human progenitor cell recruitment via SDF-1alpha coacervate-laden PGS vascular grafts [J]. Biomaterials, 2013, 34: 9877-9885.

[23] RUJITANAROJ P O, AID-LAUNAIS R, CHEW S Y, et al. Polysaccharide electrospun fibers with sulfated poly(fucose) promote endothelial cell migration and vegf-mediated angiogenesis[J]. Biomaterials Science, 2014, 2(6): 843-852.

[24] ZHOU F, WEN M, ZHOU P, et al. Electrospun membranes of PELCL/PCL-REDV loading with miRNA-126 for enhancement of vascular endothelial cell adhesion and proliferation[J]. Materials Science and Engineering: C, 2018, 85: 37-46.

[25] WANG K, ZHANG Q, ZHAO L, et al. Functional modification of electrospun poly(ε-caprolactone) vascular grafts with the fusion protein VEGF–HGFI enhanced vascular regeneration[J]. ACS Applied Materials & Interfaces, 2017, 9(13): 11415-11427.

[26] HAJIALI F, TAJBAKHSH S, SHOJAEI A. Fabrication and properties of

polycaprolactone composites containing calcium phosphate-based ceramics and bioactive glasses in bone tissue engineering: a review[J]. Polymer Reviews, 2018, 58(1): 164-207.

[27] LUO J, ZHANG H, ZHU J, et al. 3-D mineralized silk fibroin/polycaprolactone composite scaffold modified with polyglutamate conjugated with BMP-2 peptide for bone tissue engineering[J]. Colloids and Surfaces B: Biointerfaces, 2018, 163: 369-378.

[28] CHAMUNDESWARI V N, SIANG L Y, CHUAH Y J, et al. Sustained releasing sponge-like 3D scaffolds for bone tissue engineering applications[J]. Biomedical Materials, 2018, 13(1): 015019.

[29] BAO M, LOU X, ZHOU Q, et al. Electrospun biomimetic fibrous scaffold from shape memory polymer of PDLLA-co-TMC for bone tissue engineering[J]. ACS Applied Materials & Interfaces, 2014, 6(4): 2611-2621.

[30] GARRIGUES N W, LITTLE D, SANCHEZ-ADAMS J, et al. Electrospun cartilage-derived matrix scaffolds for cartilage tissue engineering[J]. Journal of Biomedical Materials Research Part A, 2014, 102(11): 3998-4008.

[31] EVROVA O, BURGISSER G M, EBNOTHER C, et al. Bioactive, elastic, and biodegradable emulsion electrospun degrapol tube delivering PDGF-BB for tendon rupture repair[J]. Macromolecular Bioscience, 2016, 16(7): 1048-1063.

[32] MOUTHUY P A, ZARGAR N, HAKIMI O, et al. Fabrication of continuous electrospun filaments with potential for use as medical fibres[J]. Biofabrication, 2015, 7(2): 025006.

[33] WU S, STREUBEL P N, DUAN B, et al. Living nanofiber yarn-based woven biotextiles for tendon tissue engineering using cell tri-culture and mechanical stimulation[J]. Acta Biomaterialia, 2017, 62: 102-115.

[34] DEEPTHI S, JEEVITHA K, SUNDARARAM M N, et al. Chitosan-hyaluronic acid hydrogel coated poly(caprolactone) multiscale bilayer scaffold for ligament regeneration[J]. Chemical Engineering Journal, 2015, 260: 478-485.

[35] JANA S, LEVENGOOD S K, ZHANG M. Anisotropic materials for skeletal-

muscle-tissue engineering[J]. Advanced Materials, 2016, 28(48): 10588- 10612.

[36] ZHANG K, ZHENG H, LIANG S, et al. Aligned PLLA nanofibrous scaffolds coated with graphene oxide for promoting neural cell growth[J]. Acta Biomaterialia, 2016, 37: 131-142.

[37] LEE Y S, COLLINS G, ARINZEH T L, et al. Extension of primary neurons on electrospun piezoelectric scaffolds[J]. Acta Biomaterialia, 2011, 7(11):3877-3886.

[38] LIU W, THOMOPOULOS S, XIA Y. Electrospun nanofibers for regenerative medicine[J]. Advanced Healthcare Materials, 2012, 1: 10-25.

[39] SUN B, WU T, WANG J, et al. Polypyrrole-coated poly(l-lactic acid-co-ε-caprolactone)/silk fibroin nanofibrous membranes promoting neural cell proliferation and differentiation with electrical stimulation[J]. Journal of Materials Chemistry B, 2016, 4:670-6679.

[40] XUE J, YANG J, O'CONNOR D M, et al. Differentiation of bone marrow stem cells into schwann cells for the promotion of neurite outgrowth on electrospun fibers[J]. ACS Applied Materials & Interfaces, 2017, 9(14): 2299-12310.

[41] SUN B, ZHOU Z, WU T, et al. Development of nanofiber sponges-containing nerve guidance conduit for peripheral nerve regeneration in vivo[J]. ACS Applied Materials & Interfaces, 2017, 9(32): 26684-26696.

[42] HUANG L, ZHU L, SHI X, et al. A compound scaffold with uniform longitudinally oriented guidance cues and a porous sheath promotes peripheral nerve regeneration in vivo[J]. Acta Biomaterialia, 2018, 68: 223-236.

[43] ZONG C, WANG M, YANG F, et al. A novel therapy strategy for bile duct repair using tissue engineering technique: PCL/PLGA bilayered scaffold with hMSCs[J]. Tissue Engineering and Regenerative Medicine, 2017, 11(4): 966-976.

[44] MOAZENI N, VADOOD M, SEMNANI D, et al. Modeling the compliance of polyurethane nanofiber tubes for artificial common bile duct[J]. Materials Research Express, 2018, 5(2): 025004.

[45] AJALLOUEIAN F, LEMON G, HILBORN J, et al. Bladder biomechanics and the use of scaffolds for regenerative medicine in the urinary bladder[J]. Nature Reviews

Urology, 2018, 15(3): 155-174.

[46] AJALLOUEIAN F, ZEIAI S, FOSSUM M, et al. Constructs of electrospun plga, compressed collagen and minced urothelium for minimally manipulated autologous bladder tissue expansion[J]. Biomaterials, 2014, 35(22): 5741-5748.

[47] DIAS J R, GRANJA P L, BARTOLO P J. Advances in electrospun skin substitutes[J]. Progress in Materials Science. 2016, 84: 314-334.

[48] SUNDARAMURTHI D, KRISHNAN U M, SETHURAMAN S. Electrospun nanofibers as scaffolds for skin tissue engineering[J]. Polymer Reviews, 2014, 54(2): 348-376.

[49] DHAND C, VENKATESH M, BARATHI V A, et al. Bio-inspired crosslinking and matrix-drug interactions for advanced wound dressings with long-term antimicrobial activity[J]. Biomaterials, 2017, 138: 153-168.

[50] YANG X, YANG J, WANG L, et al. Pharmaceutical intermediate-modified gold nanoparticles: against multidrug-resistant bacteria and wound healing application via an electrospun scaffold[J]. ACS Nano, 2017, 11(6): 5737-5745.

[51] WANG X, LV F, LI T, et al. Electrospun micropatterned nanocomposites incorporated with Cu_2S nanoflowers for skin tumor[J]. ACS Nano, 2017, 11(11): 11337-11349.

[52] LEE E J, HUH B K, KIM S N, et al. Application of materials as medical devices with localized drug delivery capabilities for enhanced wound repair[J]. Progress in Materials Science, 2017, 89: 392-410.

[53] WANG L, YANG J, YANG X, et al. Small molecular TGF-β1-inhibitor-loaded electrospun fibrous scaffolds for preventing hypertrophic scars[J]. ACS Applied Materials & Interfaces, 2017, 9: 32545-32553.

[54] CHENG L, SUN X, ZHAO X, et al. Surface biofunctional drug-loaded electrospun fibrous scaffolds for comprehensive repairing hypertrophic scars[J]. Biomaterials, 2016, 83: 169-181.

[55] XU T, YANG R, MA X, et al. Bionic poly(gamma-glutamic acid) electrospun fibrous scaffolds for preventing hypertrophic scars[J]. Advanced Healthcare Materials, 2019,

8(13): 1900123.

[56] GOMES S R, RODRIGUES G, Martins G G, t al. In vitro and in vivo evaluation of electrospun nanofibers of PCL, chitosan and gelatin: a comparative study[J]. Materials Science & Engineering C-Materials for Biological Applications, 2015, 46: 348-358.

[57] GHOSAL K, MANAKHOV A, MANAKHOV A, et al. Structural and surface compatibility study of modified electrospun poly(epsilon-caprolactone) (PCL) composites for skin tissue engineering[J]. AAPS PharmSciTech, 2017, 18(1): 72-81.

[58] LAW J X, LIAU L L, SAIM A, et al. Electrospun collagen nanofibers and their applications in skin tissue engineering[J]. Journal of Tissue Engineering and Regenerative Medicine, 2017, 14(6): 699-718.

[59] LUO X, GUO Z, HE P, et al. Study on structure, mechanical property and cell cytocompatibility of electrospun collagen nanofibers crosslinked by common agents[J]. International Journal of Biological Macromolecules, 2018, 113: 476-486.

[60] EHTERAMI A, SALEHI M, FARZAMFAR S, et al. In vitro and in vivo study of PCL/COLL wound dressing loaded with insulin-chitosan nanoparticles on cutaneous wound healing in rats model[J]. International Journal of Biological Macromolecules, 2018, 117: 601-609.

[61] YOUNES I, RINAUDO M. Chitin and chitosan preparation from marine sources. Structure, properties and applications[J]. Marine Drugs, 2015, 13(3): 1133-1174.

[62] YANG S, LEI P, SHAN Y, et al. Preparation and characterization of antibacterial electrospun chitosan/poly (vinyl alcohol)/graphene oxide composite nanofibrous membrane[J]. Applied Surface Science, 2018, 435: 832-840.

[63] ARDESHIRZADEH B, ANARAKI N A, IRANI M, et al. Controlled release of doxorubicin from electrospun PEO/chitosan/graphene oxide nanocomposite nanofibrous scaffolds[J]. Materials Science & Engineering C-Materials for Biological Applications, 2015, 48: 384-390.

[64] CHAND A, ADHIKARI J, GHOSH A, et al. Electrospun chitosan/ polycaprolactone-hyaluronic acid bilayered scaffold for potential wound healing applications[J].

International Journal of Biological Macromolecules, 2018, 116: 774-785.

[65] FIGUEIRA D R, MIGUEL S P, DE SA K D, et al. Production and characterization of polycaprolactone-hyaluronic acid/chitosan-zein electrospun bilayer nanofibrous membrane for tissue regeneration[J]. International Journal of Biological Macromolecules, 2016, 93: 1100-1110.

[66] 蒋丹, 黄建文, 邵惠丽, 等. 丝素蛋白/膀胱脱细胞基质/透明质酸复合纤维支架的制备及生物学性能研究[J]. 功能材料, 2017, 48(6): 6124-6128.

[67] EBRAHIMI-HOSSEINZADEH B, PEDRAM M, HATAMIAN-ZARMI A, et al. In vivo evaluation of gelatin/hyaluronic acid nanofiber as burn-wound healing and its comparison with chitoheal gel[J]. Fiber Polymer, 2016, 17(6): 820-826.

[68] ADUBA D C, YANG H. Polysaccharide fabrication platforms and biocompatibility assessment as candidate wound dressing materials[J]. Bioengineering (Basel), 2017, 4(4): 1.

[69] SAMADIAN H, SALEHI M, FARZAMFAR S, et al. In vitro and in vivo evaluation of electrospun cellulose acetate/gelatin/hydroxyapatite nanocomposite mats for wound dressing applications[J]. Artificial Cells Nanomedicine and Biotechnology, 2018, 46: S964- S974.

[70] LALANI R, LIU L. Electrospun zwitterionic poly(sulfobetaine methacrylate) for nonadherent, superabsorbent, and antimicrobial wound dressing applications[J]. Biomacromolecules, 2012, 13(6): 1853-1863.

[71] KUMAR N S, SANTHOSH C, SUDAKARAN S V, et al. Electrospun polyurethane and soy protein nanofibres for wound dressing applications[J]. Iet Nanobiotechnology, 2018, 12(2): 94-98.

[72] ALHUSEIN N, BLAGBROUGH I S, BEETON M L, et al. Electrospun zein/pcl fibrous matrices release tetracycline in a controlled manner, killing staphylococcus aureus both in biofilms and ex vivo on pig skin, and are compatible with human skin cells[J]. Pharmaceutical Research, 2016, 33(1): 237-246.

[73] WANG Y, LI P, XIANG P, et al. Electrospun polyurethane/keratin/AgNP biocomposite mats for biocompatible and antibacterial wound dressings[J]. Journal

of Materials Chemistry B, 2016, 4(4): 635-648.
[74] SUN H F, MEI L, SONG C, et al. The in vivo degradation, absorption and excretion of PCL-based implant[J]. Biomaterials, 2006, 27(9): 1735-1740.
[75] ZHU Z, LIU Y, XUE Y, CHENG X, et al. Tazarotene released from aligned electrospun membrane facilitates cutaneous wound healing by promoting angiogenesis[J]. ACS Applied Materials & Interfaces, 2019, 11(39): 36141-36153.
[76] MALIKMAMMADOV E, TANIR T E, KIZILTAY A, et al. PCL and PCL-based materials in biomedical applications[J]. Journal of Biomaterials Science, Polymer Edition, 2018, 29(7-9): 863-893.
[77] BAKER S R, BANERJEE S, BONIN K, et al. Determining the mechanical properties of electrospun poly-epsilon-caprolactone (PCL) nanofibers using AFM and a novel fiber anchoring technique[J]. Materials Science & Engineering C-Materials for Biological Applications, 2016, 59: 203-212.
[78] RATHER H A, THAKORE R, SINGH R, et al. Antioxidative study of cerium oxide nanoparticle functionalised PCL-Gelatin electrospun fibers for wound healing application[J]. Bioactive Materials, 2018, 3(2): 201-211.
[79] PELIPENKO J, KOCBEK P, GOVEDARICA B, et al. The topography of electrospun nanofibers and its impact on the growth and mobility of keratinocytes[J]. European Journal of Pharmaceutics and Biopharmaceutics, 2013, 84(2): 401-411.
[80] ADELI H, KHORASANI M T, PARVAZINIA M. Wound dressing based on electrospun PVA/chitosan/starch nanofibrous mats: fabrication, antibacterial and cytocompatibility evaluation and in vitro healing assay[J]. International Journal of Biological Macromolecules, 2019, 122: 238-254.
[81] GHALEI S, ASADI H, GHALEI B. Zein nanoparticle-embedded electrospun PVA nanofibers as wound dressing for topical delivery of anti-inflammatory diclofenac[J]. Journal of Applied Polymer Science, 2018, 135(33): 46643.
[82] RAFIQ M, HUSSAIN T, ABID S, et al. Development of sodium alginate/PVA antibacterial nanofibers by the incorporation of essential oils[J]. Materials Research Express, 2018, 5(3): 035007.

[83] PEH P, LIM N S J, BLOCKI A, et al. Simultaneous delivery of highly diverse bioactive compounds from blend electrospun fibers for skin wound healing[J]. Bioconjugate Chemistry, 2015, 26(7): 1348-1358.

[84] LUO M, MING Y, WANG L, et al. Local delivery of deep marine fungus-derived equisetin from polyvinylpyrrolidone (PVP) nanofibers for anti-MRSA activity[J]. Chemical Engineering Journal, 2018, 350: 157-163.

[85] GUO Z, TANG G, ZHOU Y, et al. Fabrication of sustained-release CA-PU coaxial electrospun fiber membranes for plant grafting application[J]. Carbohydrate Polymers, 2017, 169: 198-205.

[86] GOMES S R, RODRIGUES G, MARTINS G G, et al. In vitro and in vivo evaluation of electrospun nanofibers of PCL, chitosan and gelatin: a comparative study[J]. Materials Science and Engineering: C, 2015, 46: 348-358.

[87] SU L, ZHENG J, WANG Y, et al. Emerging progress on the mechanism and technology in wound repair[J]. Biomed Pharmacother, 2019, 117: 109191.

[88] STRONG A L, NEUMEISTER M W, LEVI B. Stem cells and tissue engineering: regeneration of the skin and its contents[J]. Clinics in Plastic Surgery, 2017, 44(3): 635-650.

[89] HUR W, LEE H Y, MIN H S, et al. Regeneration of full-thickness skin defects by differentiated adipose-derived stem cells into fibroblast-like cells by fibroblast-conditioned medium[J]. Stem Cell Research & Therapy, 2017, 8(1): 92.

[90] WON C H, PARK G H, WU X, et al. The basic mechanism of hair growth stimulation by adipose-derived stem cells and their secretory factors[J]. Current Stem Cell Research & Therapy, 2017, 12(7): 535-543.

[91] BAINBRIDGE A, WALKER S, SMITH J, et al. IKBKE activity enhances AR levels in advanced prostate cancer via modulation of the Hippo pathway[J]. Nucleic Acids Research, 2020, 48: 5366-5382.

[92] AYTAC Z, UYA R T. Core-shell nanofibers of curcumin/cyclo-dextrin inclusion complex and polylactic acid: enhanced water solubility and slow release of curcumin[J]. International Journal of Pharmaceutics, 2017, 518(1-2): 177-184.

[93] LU Y, HUANG J, YU G, et al. Coaxial electrospun fibers: applications in drug delivery and tissue engineering[J]. Wires Nanomed Nanobiotechnol, 2016, 8(5): 654-677.
[94] SEDGHI R, SHAABNI A. Electrospun biocompatible core/shellpolymer-free core structure nanofibers with superior antimicrobial potency against multi drug resistance organisms[J]. Polymer, 2016, 101(9): 151-157.
[95] ZHANG Q, LI Y, LIN Z, et al. Electrospun polymeric micro/nanofibrous scaffolds for long-term drug release and their bio-medical applications[J]. Drug Discovery Today, 2017, 22 (9): 1351-1366.
[96] SONG J, ZHANG M, LIU C, et al. Highly flexible, core-shell hetero structured, and visible-light-driven TiO_2 based nanofibrous membranes for antibiotic removal and E. coil inactivation[J]. Chemical Engineering Journal, 2020, 379: 122269.
[97] KAI F, SUN H, BRADLEY M, et al. Novel antibacterial nanofibrous PLLA scaffolds[J]. Control Release, 2010, 146(3): 363-369.
[98] GONG Z, SHI Y, TAN H, et al. Plasma amine oxidase-induced nanoparticle-to-nanofiber geometric transformation of an amphiphilic peptide for drug encapsulation and enhanced bactericidal activity[J]. ACS Applied Materials & Interfaces, 2020, 12(4): 4323-4332.
[99] PARK Y, KANG E, KWON O, et al. Ionically crosslinked Ad/chitosan nanocomplexes processed by electrospinning for targeted cancer gene therapy[J]. Control Release, 2010, 148: 175.
[100] XIE Q, JIA L, XU H, et al. Fabrication of core-shell PEI/pBMP2-PLGA electrospun scaffold for gene delivery to periodontal ligament stem cells[J]. Stem Cells International, 2016: 5385137.
[101] KAZANTSEVA J, IVANOV R, GASIK M, et al. Graphene-augmented nanofiber scaffolds trigger gene expression switching of four cancer cell types[J]. ACS Biomaterials Science & Engineering, 2018, 4(5): 1622-1629.
[102] SARAF A, BAGGETT L S, RAPHAEL R M, et al. Regulated non-viral gene delivery from coaxial electrospun fiber mesh scaffolds[J]. Control Release, 2010,

143(1): 95-103.

[103] YANAN Z, GUO S, HUANG H, et al. Silicon nanodot-based aptasensor for fluorescence turn-on detection of mucin 1 and targeted cancer cell imaging[J]. Analytica Chimica Acta, 2018, 1035: 154-160.

[104] PAUL K B, SINGH V, VANJARI S R K, et al. One step biofunctionalized electrospun multiwalled carbon nanotubes embedded zinc oxide nanowire interface for highly sensitive detection of carcinoma antigen-125[J]. Biosensors and Bioelectronics, 2017, 88: 144-152.

[105] XUE R, BEHERA P, VIAPIANO M, et al. Rapid response oxygen-sensing nanofibers[J]. Materials Science and Engineering: C, 2013, 33(6): 3450-3457.

[106] XUE R, GE C, RICHARDSON K, et al. Microscale sensing of oxygen via encapsulated porphyrin nanofibers: effect of indicator and polymer "core" permeability[J]. ACS Applied Materials & Interfaces, 2015, 7(16): 8606-8614.

[107] CHEN S, BODA S K, BATRA S K, et al. Emerging roles of electrospun nanofibers in cancer research[J]. Advanced Healthcare Materials, 2018, 7(6): 1701024.

[108] XIAO Y, LIN L, SHEN M, et al. Design of DNA aptamer-functionalized magnetic short nanofibers for efficient capture and release of circulating tumor cells[J]. Bioconjugate Chemistry, 2020, 31(1): 130-138.

[109] XU G, TAN Y, XU T, et al. Hyaluronic acid-functionalized electrospun PLGA nanofibers embedded in a microfluidic chip for cancer cell capture and culture[J]. Biomaterials Science,2017, 5(4): 752-761.

[110] WANG M, XIAO Y, LIN L, et al. A microfluidic chip integrated with hyaluronic acid-functionalized electrospun chitosan nanofibers for specific capture and nondestructive release of CD44-overexpressing circulating tumor cells[J]. Bioconjugate Chemistry, 2018, 29: 1081-1090.

[111] XIAO Y, WANG M, LIN L, et al. Integration of aligned polymer nanofibers within a microfluidic chip for efficient capture and rapid release of circulating tumor cells [J]. Materials Chemistry Frontiers, 2018, 2: 891-900.

[112] YU C C, HO B, JUANG R, et al. Poly (3,4-ethylenedioxythiophene)-based

nanofiber mats as an organic bioelectronic platform for programming multiple capture/release cycles of circulating tumor cells[J]. ACS Applied Materials & Interfaces, 2017, 9(36): 30329-30342.

[113] CHEN S, LI R Q, LI X R, et al. Electrospinning: an enabling nanotechnology platform for drug delivery and regenerative medicine[J]. Advanced Drug Delivery Reviews, 2018, 132: 188-213.

[114] JANANI G, PILLAI M M, SELVAKUMAR R, et al. An in vitro 3D model using collagen coated gelatin nanofibers for studying breast cancer metastasis[J]. Biofabrication, 2017, 9(1): 015016.

[115] MA N K, LIM J K, LEONG M F, et al. Collaboration of 3D context and extracellular matrix in the development of glioma stemness in a 3D model[J]. Biomaterials, 2016, 78: 62-73.

[116] BLAKENEY B, TAMBRALLI A, ANDERSON J M, et al. Cell infiltration and growth in a low density, uncompressed three-dimensional electrospun nanofibrous scaffold[J]. Biomaterials, 2011, 32(6): 1583-1590.

[117] XU H, CAI S B, XU L, et al. Water-stable three-dimensional ultrafine fibrous scaffolds from keratin for cartilage tissue engineering[J]. Langmuir, 2014, 30(28): 8461-8470.

[118] ZHANG D, CHANG J. Electrospinning of three-dimensional nanofibrous tubes with controllable architectures[J]. Nano letters, 2008, 8(10): 3283-3287.

[119] XIE J, MA B, MICHAEL P L. Fabrication of novel 3D nanofiber scaffolds with anisotropic property and regular pores and their potential applications[J]. Advanced Healthcare Materials, 2012, 1(5): 1674-1678.

[120] JIANG J, CARLSON M A, XIE J W. Expanding two-dimensional electrospun nanofiber membranes in the third dimension by a modified gas-foaming technique[J]. ACS Biomaterials Science & Engineering, 2015, 1(10):991-1001.

[121] JIANG J, LI Z R, WANG H, et al. Expanded 3D nanofiber scaffolds: cell penetration, neovascularization, and host response[J]. Advanced Healthcare Materials, 2016, 5(23): 2993-3003.

[122] DOMURA R, SASAKI R, OKAMOTO M, et al. Comprehensive study on cellular morphologies, proliferation, motility, and epithelial-mesenchymal transition of breast cancer cells incubated on electrospun polymeric fiber substrates[J]. Journal of Materials Chemistry B, 2017, 5(14): 2588-2600.

[123] JAIN A, BETANCUR M, PATEL G D, et al. Guiding Intracortical brain tumour cells to an extracortical cytotoxic hydrogel using aligned polymeric nanofibres[J]. Nature Materials, 2014, 13(3): 308-316.

[124] WANG X, WANG L, ZONG S. et al. Use of multifunctional composite nanofibers for photothermal chemotherapy to treat cervical cancer in mice[J]. Biomaterials Science, 2019, 7(9): 3846-3854.

[125] CHEN Z, LIU W, ZHAO L, et al. Acid-labile degradation of injectable fiber fragments to release bioreducible micelles for targeted cancer therapy[J]. Biomacromolecules, 2018, 19(4): 1100-1110.

[126] ZHAO X, YUAN Z, YILDIRIMER L, et al. Tumor-triggered controlled drug release from electrospun fibers using inorganic caps for inhibiting cancer relapse[J]. Small, 2015, 11(34): 4284-4291.

[127] CHEN S, BODA S K, BATRA S K, et al. Emerging roles of electrospun nanofibers in cancer research[J]. Advanced Healthcare Materials, 2018, 7(6): 1701024.

第 6 章　纳米纤维在传感器领域的应用

静电纺织纳米纤维由于接触面积大、响应速度快、尺寸小以及功耗小，在传感器领域的应用一直是研究热点。传感器是一种可以感知、测量和转换为可用输出信号的设备。传感响应材料是传感元件的核心，也是决定传感器精确性的元件。

二十多年来，研究人员将纳米纤维引入传感响应材料的结构设计中，并获得了很好的效果。值得注意的是，传感性能一方面取决于传感元件的组成部分，另一方面取决于传感元件的材料结构和几何尺寸。当材料的尺度从宏观向纳米转变时，小尺度效应、表面和界面效应以及量子尺寸效应开始显著地影响传感性能，传感器表现出在宏观尺度下所不具备的一些特性。此外，比表面积的增加提供了大量增强行列式和纳米纤维之间相互作用的区域和通道，从而进一步提高了灵敏度。同时，使用纳米纤维可以降低传感器的能耗和总体尺寸，从而扩大其应用范围。本章将依次介绍电化学传感器、电阻传感器、光学传感器和生物传感器的研究现状。

6.1　电化学传感器

6.1.1　引言

电化学传感器是基于待测物的电化学性质并将待测物化学量转化为电学量进行传感检测的一种传感器，通过与待测气体发生反应并产生与气体浓度成正比的电信号来工作。典型的电化学传感器由电极和薄电解层组成，采用比表面积高、透气性好的纳米纤维薄电解层，具有灵敏度高、响应快等特点，本节将依次介绍基于金属、导电聚合物的纳米纤维电化学传感器。

6.1.2　金属电化学传感器

通常，在电化学传感器（图 6.1）中，电催化反应产生可量化电流（安培传感）、可测量电位或电荷积累（电位传感）和可检测的电极间介质的导电性（电导传感）。典型的电化学传感系统通常包括参比电极（大多数情况下由 Ag/AgCl 或玻璃碳材料制成）、对电极以及工作电极或氧化还原电极（传感电极）。工作电极在反应中传递信息，而对电极和工作电极满足良好的导电性和化学稳定性的要求。在这三种类型的电极中，工作电极在传感系统中起着至关重要的作用。由于能够根据应用采用各种功能材料制备纳米纤维，同时可以实现相对较高的比表面积，以及它们的结构特性的可调性，纳米纤维可以用于工作电极的改性。

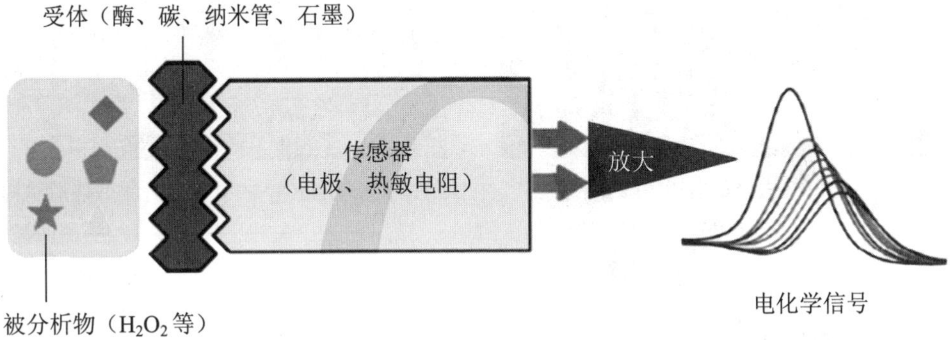

图 6.1　电化学传感器原理图

大多数金属纳米纤维是由前驱体的烧结产生的。一般来说，制备过程分为三步：首先通过溶胶-凝胶方法制备含有聚合物和金属盐或陶瓷源的前驱体溶液，然后以适当的工艺参数对前驱体溶液进行静电纺丝以获得前驱体纳米纤维，最后在气体气氛中对前驱体纳米纤维进行热处理，其中温度和升温速率具有决定性的作用。所获得的静电纺丝金属纳米纤维通常由纳米颗粒或晶体制成。颗粒的大小会影响传感器的灵敏度，从而影响传感器的性能。

研究者不断尝试使用金属纳米纤维直接构建电极，如铂等。铂是一种贵金属，在较宽的 pH 范围内（从碱性到中性）具有良好的催化活性，之前的研究者已经证明了它对葡萄糖氧化的显著电催化作用，并且它也是 C_1 位碳分子脱氢反应中最好的催化剂，但是铂也有低灵敏度和选择性差的缺点。针对这一点，Liu 等人通过静电纺丝设计了基于 Pt-Au 纳米珊瑚结构的纳米纤维。实验中将聚乙烯吡咯烷酮（PVP）溶于二甲基甲酰胺与水混合溶剂中，再与铂盐（H_2PtCl_6）和金盐（$HAuCl_4$）混合，制备出前驱体聚合物溶液。之后烧结以除去聚乙烯吡咯烷酮组分，形成 Pt-Au 纳米珊瑚结构（图 6.2（c））。与单金属材料相比，双金属材料或合金通常表现出更好的催化性能，这表明二元组分界面有望产生协同效应，以获得更好的电催化活性和对葡萄糖的选择性。此外，有必要指出，由于中间体的化学吸附和氯离子的吸附，金属制备的电化学传感器稳定性可能会降低。

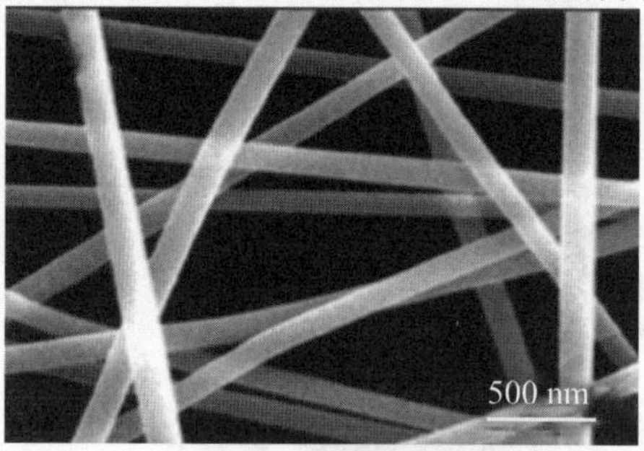

（a）H_2PtCl_6-$HAuCl_4$-PVP 纳米纤维

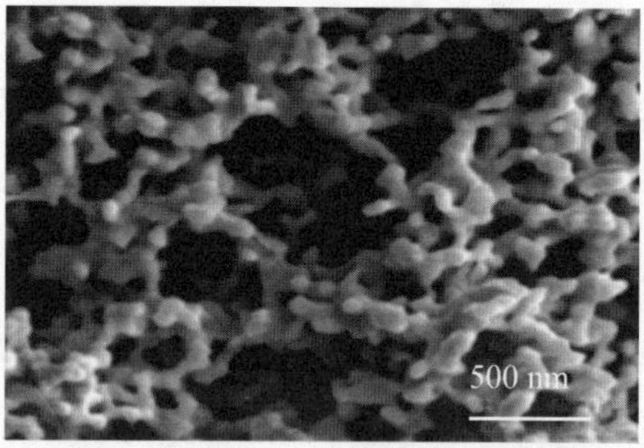

（b）H_2PtCl_6-$HAuCl_4$-PVP 纳米纤维煅烧后结构

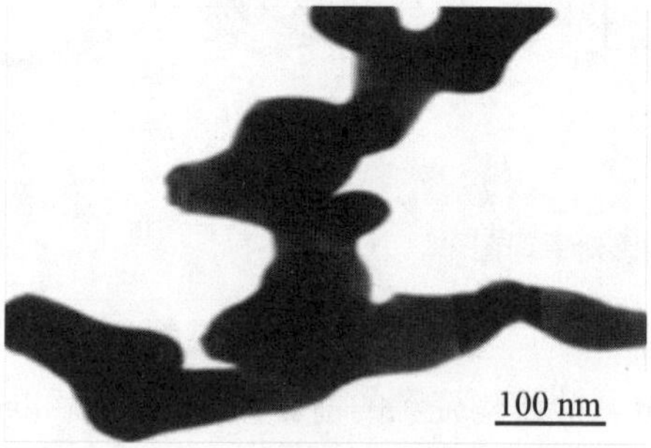

（c）Pt-Au 纳米珊瑚结构的 TEM 图像

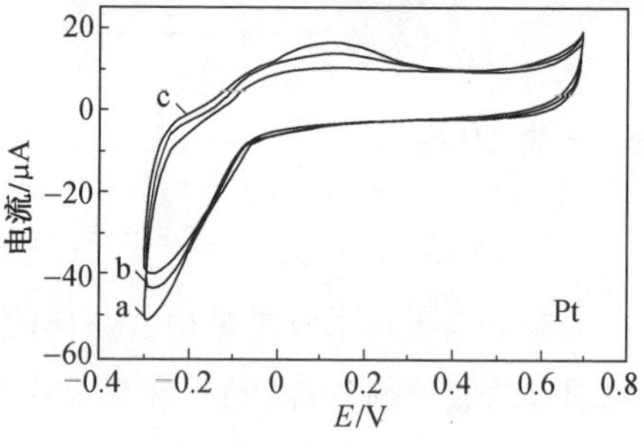

（d）Pt-NFs/GCE 的 CV 图

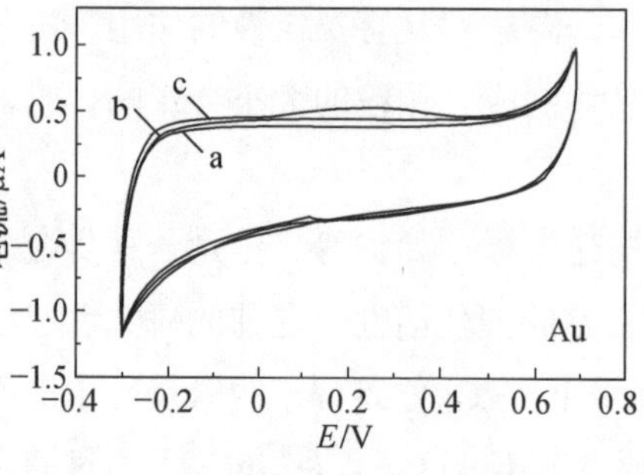

（e）Au-MPs/GCE 的 CV 图

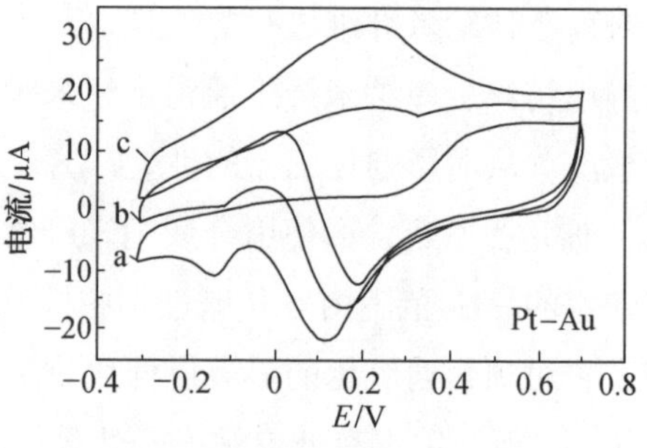

（f）Pt-Au NCs/GCE 的 CV 图

图 6.2 金属纳米纤维 SEM 图及电化学测试

a—不存在葡萄糖；b—存在 4 $mmol·L^{-1}$ 葡萄糖；c—存在 8 $mmol·L^{-1}$ 葡萄糖

由于成本低廉、特异性高和高灵敏度，金属氧化物、陶瓷和复合纳米电化学传感器有望在环境监测和生物医学诊断中发挥关键作用。Cabrita 等人用纳米晶半导体硫化铋（Bi_2S_3）纳米粒子对钛矿型纳米材料进行改性，制备了一种新型的钛矿型纳米抗坏血酸电极。硫化铋是 V～VI族的重要组成部分，是一种直接的带隙材料（E_g=1.2 eV）。用该复合材料对钛矿型纳米材料进行改性时，可以实现较高的电化学储氢能力和电化学相互作用。在最佳 pH 条件（pH=7）下，实现了 38 $\mu A \cdot mmol^{-1} \cdot cm^{-2}$ 的灵敏度，以及较高的稳定性和 1～10 $mmol \cdot L^{-1}$ 的线性检测范围。Zhang 等人采用静电纺丝法合成多孔双金属（Mo-W）氧化物纳米纤维，由于材料中异质结的存在，比纯 WO_3 在传感器中展现出更佳的丙酮传感性能。在电纺丝过程中可以通过调节 Mo 和 W 的摩尔比，从而合理地调节双金属 Mo-W 氧化物纳米纤维的组成、表面积、孔隙率和氧簇吸收能力，实现不同的传感性能。当 Mo / W 摩尔比为 6%时，Mo-W 氧化物纳米纤维在 375 ℃时对 100 $mg \cdot L^{-1}$ 丙酮表现出最高响应 26.5，最低检测限为 19.2 $\mu g \cdot L^{-1}$，分别是纯 WO_3 纳米纤维的 2.6 和 4.2 倍，同时也展现出极佳的稳定性以及较强的选择性。纳米结构的金属氧化物和陶瓷纳米材料也为固定化生物元素（酶、抗体、受体蛋白、细胞、核酸等）开辟了可能性，这些生物元件在传感器材料中起传感元件的作用，并传输信号。Mehrabi 以二氧化锡（SnO_2）和聚环氧乙烯（PEO）为基材，制备了一种新型四氢大麻酚（THC）传感器，如图 6.3 所示。首先对 PEO 和 SnO_2 进行静电纺丝，并采用溅射法对复合纳米纤维进行金掺杂，并在高温烘箱中进行煅烧。结果表明，在 350 ℃下煅烧 4 h 后，纳米纤维的表面积与体积比较大，这使传感器对 THC 和甲醇的敏感度显著提高（图 6.3（d））。另外，金掺杂影响了敏感层电子吸收率，提高了传感器的灵敏度。将该传感器的性能与商用传感器的性能进行了比较，结果表明其性能优于商用传感器。

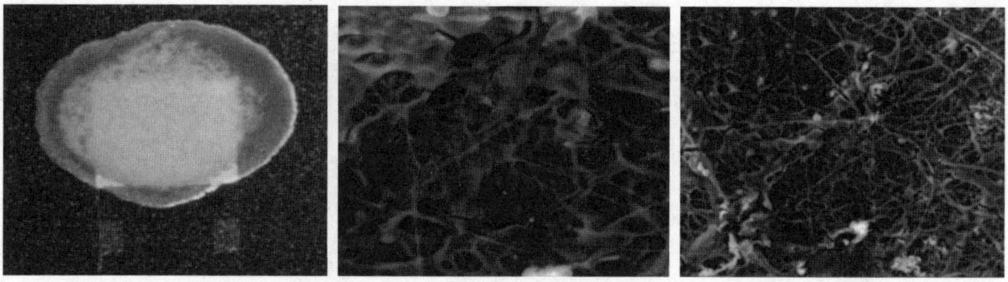

（a）基于 SnO_2 复合纳米纤维制备 THC 传感器　（b）350 ℃、煅烧 4 h 后感应层 SEM 图　（c）400 ℃、煅烧 4 h 后感应层 SEM 图

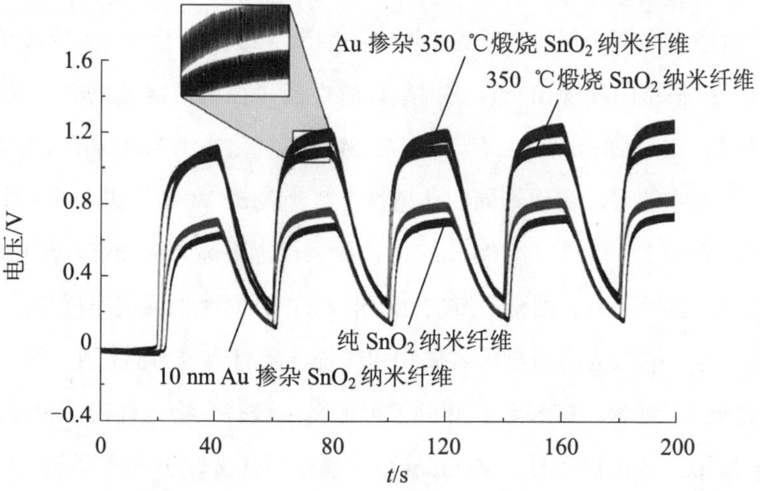

（d）不同材料的传感器对 1 000 $mg \cdot L^{-1}$ 甲醇的响应效果

图 6.3　新型四氢大麻酚（THC）及甲醇传感器

6.1.3　导电聚合物电化学传感器

聚（3,4-亚乙基二氧噻吩）（PEDOT）是一种典型的导电聚合物，在不同的环境中具有不同的结构和反应机理。Ali 等人利用氧化石墨烯纳米片和 PEDOT 纳米纤维构建了微流控硝酸盐阻抗传感器，如图 6.4 所示。此外，Ali 等人还将界面与硝酸还原酶分子接枝，实现硝酸盐的传感。该传感器已证明能够准确地检测和量化从土壤中提取的实际样品中的硝酸盐离子，加入氧化石墨烯纳米片后进一步增强了其电极的电荷转移电阻。结果显示，该传感器的灵敏度为 61.15 $\Omega \cdot (mg \cdot L^{-1}) \cdot cm^{-2}$，浓度范围较宽（0.44～442 $mg \cdot L^{-1}$），检测极限为 0.135 $mg \cdot L^{-1}$。

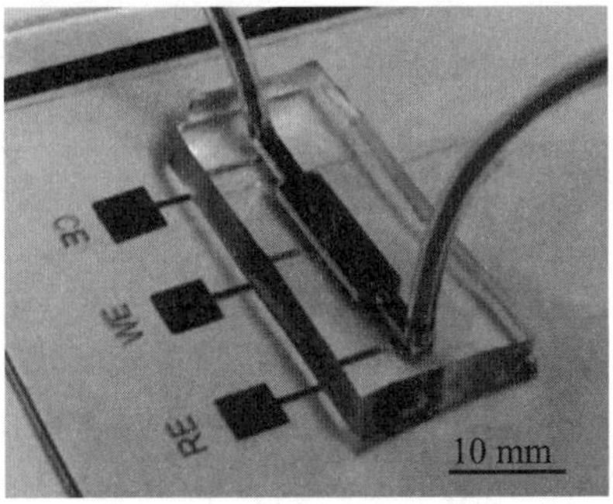

（a）传感器实物图

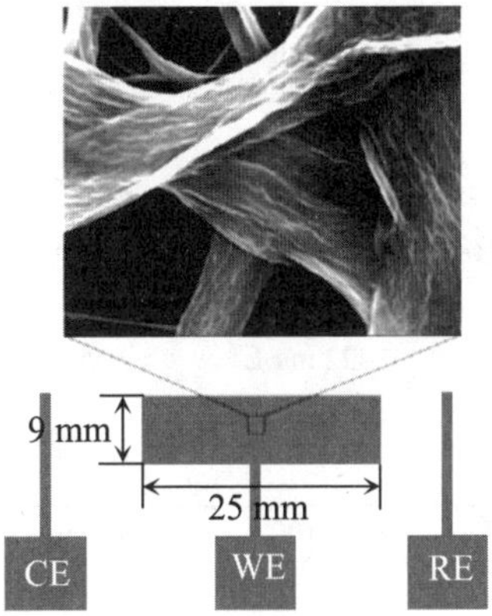

（b）传感器示意图及纤维 SEM 图

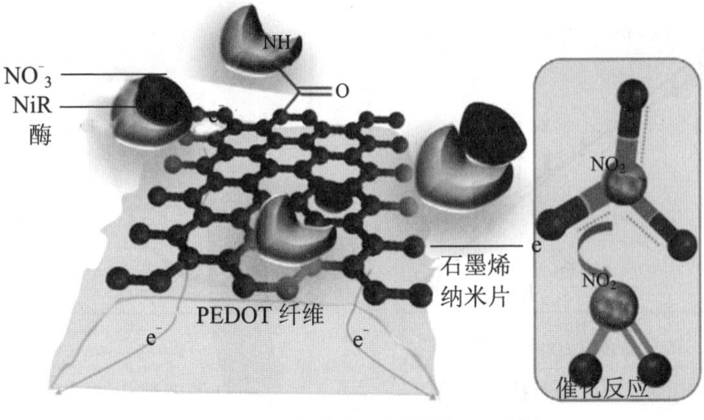

（c）电化学检测硝酸盐的原理图

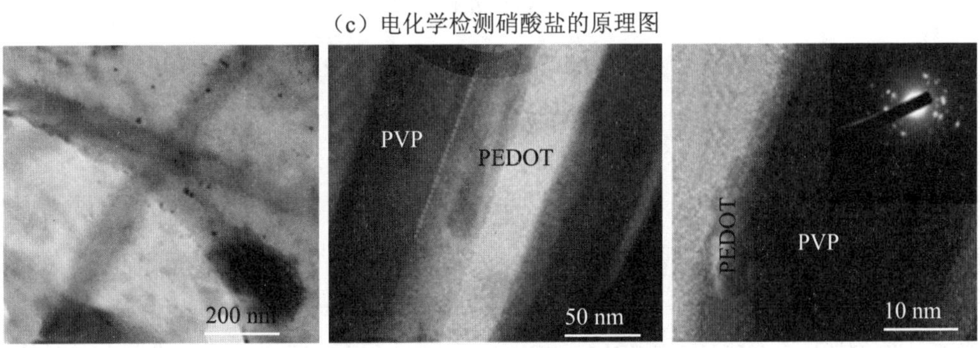

（d）石墨烯纳米片修饰 PDEOT 微观结构 TEM 图

图 6.4　微流控硝酸盐阻抗传感器

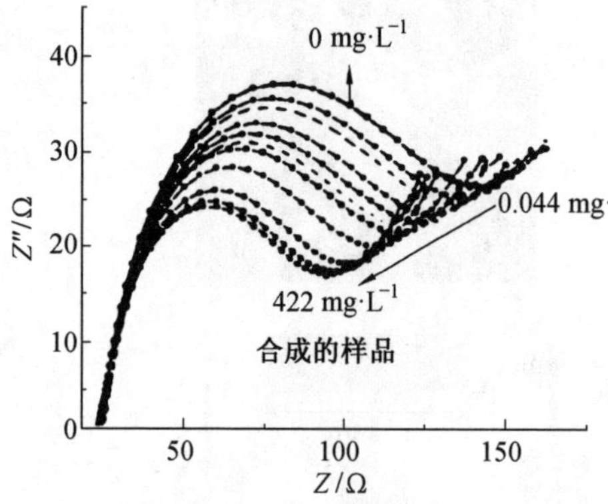

（e）合成样品的阻抗测试

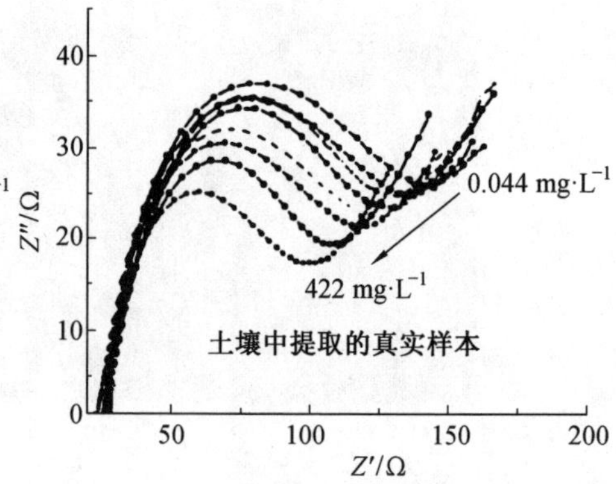

（f）真实样品的阻抗测试

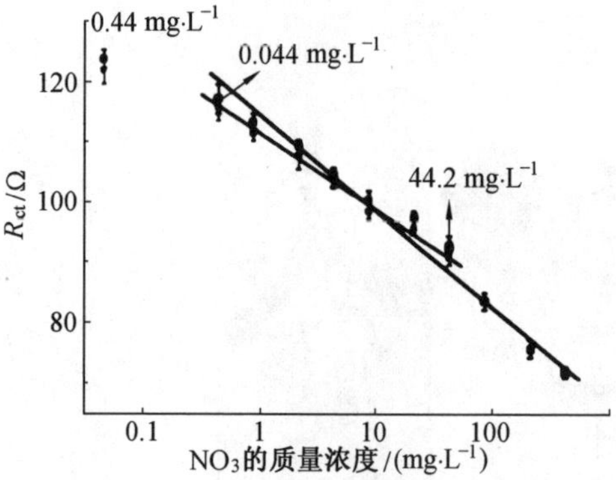

（g）硝酸盐浓度的阻抗测试

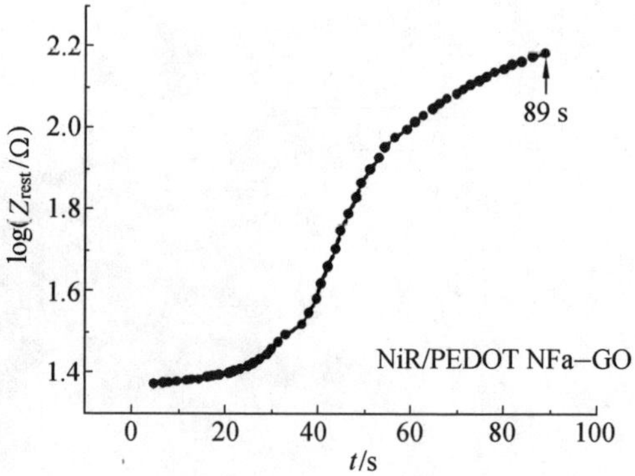

（h）硝酸盐的瞬态响应

续图 6.4

Mutharani 等人利用热敏感聚 N-异丙基丙烯酰胺（PNIPAM）和 PEDOT 组成的导电微凝胶膜，研制了一种温度控制的开关式电化学传感器。该传感器在 5-氟尿嘧啶（5-FU）检测中显示出良好的热敏性和可逆性。当检测 5-FU 实验中，修饰的玻碳电极（PNIPAMPEDOT/GCE）的 CV 曲线在溶液温度高于 PNIPAM 的较低临界溶液温度（LCST）时，表现出较高的氧化峰电流（22.49 μA）。这个氧化峰电流（15.12 μA）在低于 LCST（25 ℃）时明显被抑制，代表“关闭”状态。通过控制溶液温度，实现了 5-FU 在 PNIPAM-PEDOT/GCE 上循环响应的可重复性开关。此外，该可逆开关传感器使测定 5-FU 具有较好的灵敏度，在 40 nm 处的最低检测限为 15 nm。

6.1.4　小结与展望

静电纺丝纳米纤维结构尺寸小、比表面积大等优点对于提高传感器的灵敏度、响应速度和选择性等方面的性能都有着很好的推动与促进作用，也逐渐展示出了其在传感器领域应用的巨大潜力。如本章所述，利用金属、金属氧化物以及导电聚合物制备的纳米纤维能根据电化学原理检测葡萄糖等生物分子以及二氯戊烷、甲醇、甲苯等挥发性有机试剂。但要真正实现纳米纤维在传感器上的大规模应用还需解决以下问题：

（1）掺杂剂较多，掺杂后材料体系复杂，元件寿命短，元件的稳定性差。

（2）纳米纤维烧结致密后，气体渗透性差，导致元件的响应恢复速度变慢，灵敏度下降。

（3）当前静电纺织的工艺仍不成熟，导致静电纺织纳米纤维作为传感器件的稳定性和制备重复性不佳。

6.2　电阻传感器

6.2.1　引言

电阻式传感器基于传感器表面暴露于目标物时电阻的变化的现象，这是由于传感器和目标物之间的灵敏反应。基于纤维的电阻式传感器具有灵敏度高、响应速度快、恢复时间短等优点，其有三种主要策略：无机、有机和复合纳米纤维，本节将依次展开介绍。

6.2.2　无机半导体电阻传感器

基于传统金属氧化物半导体金属传感器制备原理：先将金属氧化物研磨成糊状，然后再将其涂覆到陶瓷管的表面。但是研究表明，这种制备过程会破坏氧化物的形貌和结构，从而降低传感器的性能。而静电纺丝则是一种制备纳米金属氧化物膜的简单实用的方法，制备的金属氧化物具有高孔隙率、高比表面积和小晶粒尺寸等特点，制备出的纳米金属氧化物膜具有工作温度高、响应速度快、恢复时间短、灵敏度高等优点，广泛应用于化学传感器。其中响应特性定义为空气中初始阻力（R_{air}）与目标气敏阻力（R_{gas}）之间的比值。

1. 纯金属氧化物

金属氧化物（如 TiO_2、ZnO、CeO、WO_3、SnO_2、V_2O_3 和 In_2O_3 等）可广泛应用于制造高灵敏度、简易结构、快速响应并恢复的传感器。Li 等人采用静电纺织法制备了多孔 Nb_2O_5–TiO_2 纳米纤维。制备过程中 Nb^{5+}替代 Ti^{4+}、形成 n–n 结并展现出大比表面积等特性。与纯 TiO_2 纳米纤维相比，Nb_2O_5–TiO_2 纳米纤维具有更好的乙醇感测性能、较高的感测响应、较低的检测极限和较低的最佳操作温度。在 250 ℃下展现 500 $mg \cdot L^{-1}$ 乙醇的最高响应，是 300 ℃下纯 TiO_2 纳米纤维的 2.79 倍，这有利于发展小型化和低功耗的商业乙醇传感器。

2. 掺杂的金属氧化物

金属氧化物的表面改性是提高传感器传感性能的有效方法。各种催化剂，如钯、镍、钴、金、铵、镧、钐、氧化锌、氯化镧等已被应用于金属氧化物的表面改性。例如，Noon 等人采用静电纺织法制备掺杂 Pd 的 TiO_2 纳米纤维，掺杂后的 TiO_2 纳米纤维具有气敏特性，例如较低的工作温度（180 ℃）和快速的气体响应（R/R_0=38～2.1 $mg \cdot L^{-1}$ NO_2），如图 6.5 所示。纳米纤维阵列由均匀排列的直纤维组成，平均直径约 200 nm，长度为几毫米（图 6.5（b））。每根纤维由金红石结构的多晶组成，平均晶粒尺寸约 5 nm。制备的传感器在 250 ℃下具有优异的 NO_2 检测性能，包括高响应（20 $mg \cdot L^{-1}$ 下为 90.3）、快速响应及恢复（40 s/18 s），尤其是优异的选择性。Kou 等人在 SnO_2 纳米纤维中掺杂 Rh 元素，有效抑制了 SnO_2 晶粒的生长，对纳米纤维的形貌起到调控作用。同时，Rh 元素的掺入提高了 SnO_2 纳米纤维中的空位氧浓度和化学吸附氧浓度。基于改性后纳米纤维的气体传感器在 200 ℃下对 50 $mg \cdot L^{-1}$ 丙酮气体的响应值是基于单一 SnO_2 纳米纤维气体传感器的 9.6 倍，表现出对丙酮气体优异的选择性和以及良好的响应–恢复特性。以上结果表明，选择适宜的金属离子进行掺杂可以大幅度提升纳米纤维气体传感器中的性能。

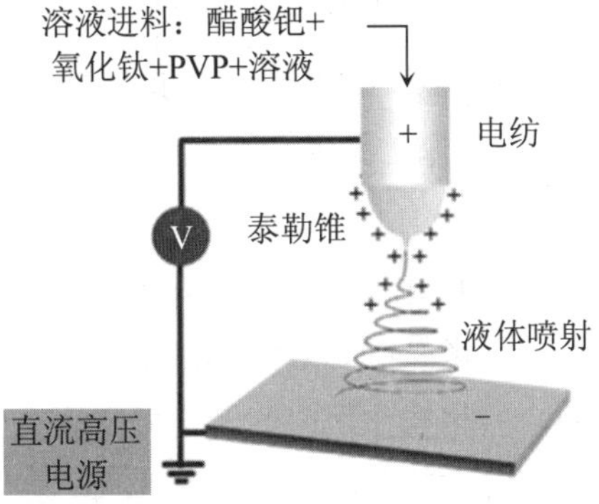

（a）Pd 掺杂 TiO_2 纳米纤维静电纺丝工艺原理图

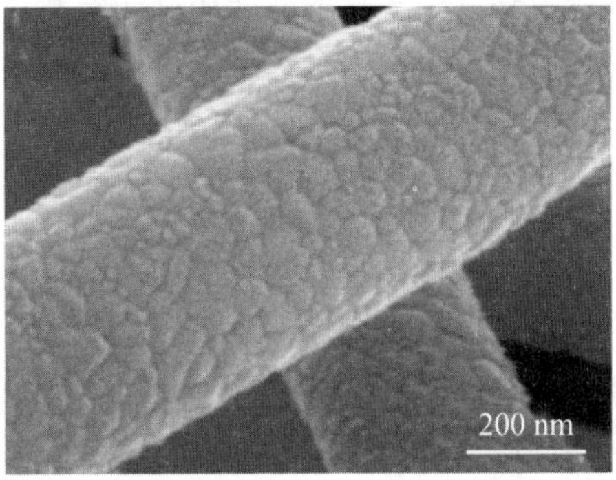

（b）Pd 掺杂 TiO_2 纳米纤维的 FE-SEM 图像

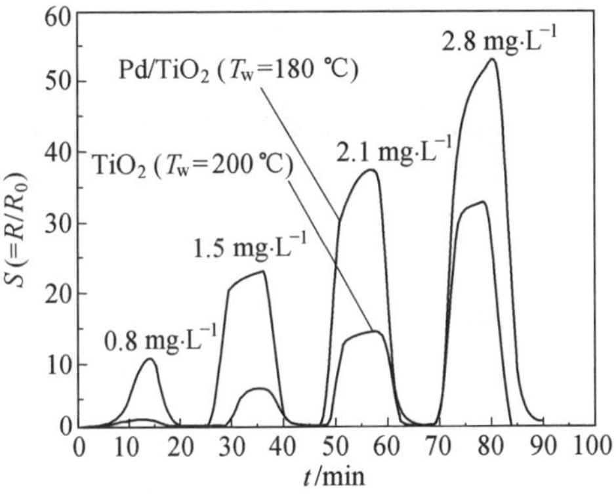

（c）传感器随 NO_2 浓度的响应

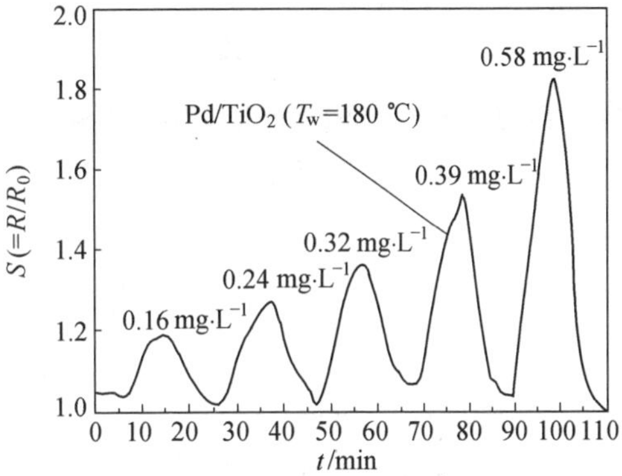

（d）传感器随 NO_2 浓度的响应（$c_{NO2}<0.6\ mg\cdot L^{-1}$ 时）

图 6.5　Pd 掺杂 TiO_2 纳米纤维应用于 NO_2 浓度传感器

6.2.3　有机电阻传感器

基于有机物的纳米纤维具有良好的柔韧性、共轭骨架、易用性和导电性可调等优点，在制造低成本、大尺寸、轻量化、柔性传感器方面具有巨大的潜力，近些年吸引了越来越多的学者研究电阻传感器，以替代金属氧化物纳米纤维。

Aussawasathien 等人采用静电纺丝技术制备了用于湿度传感的高氯酸锂（$LiClO_4$）掺杂的聚氧化乙烯（PEO）静电纺纳米纤维，同时还制备了检测过氧化氢和葡萄糖的甲磺酸（HCSA）掺杂的聚苯胺（PANI）/聚苯乙烯（PS）复合静电纺纳米纤维。制备的聚合物纳米纤维的直径在 400～1 000 nm 之间。由于这些纳米功能聚合物纤维具有很大的比表面积和良好的电学性能，相对于基于这些聚合物的普通薄膜的传感器，用这些聚合物制备的纳米纤维的传感器的灵敏度有了显著的提高。同时，扫描电子显微镜显示，在湿度测量后，$LiClO_4$ 掺杂的纳米纤维有一定的畸变，而甲磺酸掺杂的聚苯胺/聚苯乙烯复合静电纺纳米纤维则没有观察到明显的形态变化。

Ji 等人制备了一种基于锌酞菁（ZnPc）纳米纤维网络的有机室温 NO_2 传感器，器件具有响应快、恢复快的特点。高有序的 ZnPc 纳米纤维和连续的网络大大提高了器件的导电性。超薄敏感层缩短了 ZnPc 与 NO_2 的相互作用时间。另外，研究人员还通过控制沉积 ZnPc 的基片温度，通过调整比表面积，得到了优化的器件，在室温下无须任何处理即可完全恢复。

Jiang 等人制备了一种钯纳米粒子修饰的有机纳米纤维的高灵敏度柔性氢传感器。钯纳米粒子是通过直流磁控溅射法在有机纳米纤维材料上沉积的。制备后又研究了传感器对 5～50 $mg\cdot L^{-1}$ 浓度范围内氢气的检测性能，结果表明该传感器灵敏度高，对低至 5 $mg\cdot L^{-1}$ 的 H_2 的响应高达 6.55%，输出响应随氢气浓度的平方根线性增加。Utarat 等人制备了 3-氨基丙基三乙氧基硅烷、聚丙烯酸和多壁碳纳米管（PAA/MWNTs）复合纳米纤维，并采用静电纺丝法在氟掺杂氧化铟锡涂层玻璃衬底上制备了相应的纳米纤维薄膜。采用层层自组装的葡萄糖氧化酶（GOD）和聚二甲基二甲基氯化铵（PDADMAC）多层膜对电纺 PAA/MWNTs 纳米纤维薄膜进行改性，以用于葡萄糖检测。制备的纳米纤维薄膜在恒电位下具有很好的电化学检测葡萄糖的潜力，这为电化学葡萄糖生物传感器的进一步发展提供了一条有前途的途径。

6.2.4 有机无机复合电阻传感器

为了改善无机纳米纤维工作温度高、电导率低的特点，研究人员又研制了有机无机复合纳米纤维传感器件，它可以在降低工作温度的同时提高灵敏度。当前混合型纳米纤维传感器的研究已取得实质性进展。

Pang 等人采用静电纺丝、原位聚合和煅烧法制备了游离 TiO_2-SiO_2/聚苯胺（TS/PANI）复合纳米纤维。TS/PANI 复合纳米纤维氨传感器能在室温下工作，具有较好的响应值、选择性和重复性。随着 TS 纳米纤维中 TiO_2 含量的增加，TiO_2 与 PANI 之间形成的 P-N 异质结的增加，其氨气传感性能得到改善，与纯 PANI 复合纳米纤维相比，具有更高的响应值和灵敏度。Abdali 等人制备了氨基功能化石墨烯（AMG）

/聚苯胺(PANI）/聚甲基丙烯酸甲酯（PMMA）纳米纤维垫，这种复合材料是一种新型的 CO_2 气体传感材料。制备时首先采用紫外辐射对 PMMA 纳米纤维进行室温下表面处理，然后通过化学氧化聚合在 PMMA 纳米纤维表面沉积 AMG/PANI（图 6.6（a））。结果表明，紫外光辐射通过在 PMMA 表面引入氧化基团，降低了 PMMA 表面的疏水性。在室温下利用不同浓度的 CO_2 气体对基于复合材料的纳米纤维气体响应感应器进行了研究，与 PANI/PMMA 纳米纤维相比，AMG/PANI/PMMA 纳米纤维复合材料在室温下对 CO_2 表现出较好的电阻响应（图 6.6（d）），同时还具有更高的灵敏度和更快的响应时间（图 6.6（e））以及更加优良的 CO_2 选择性（图 6.6（f））。

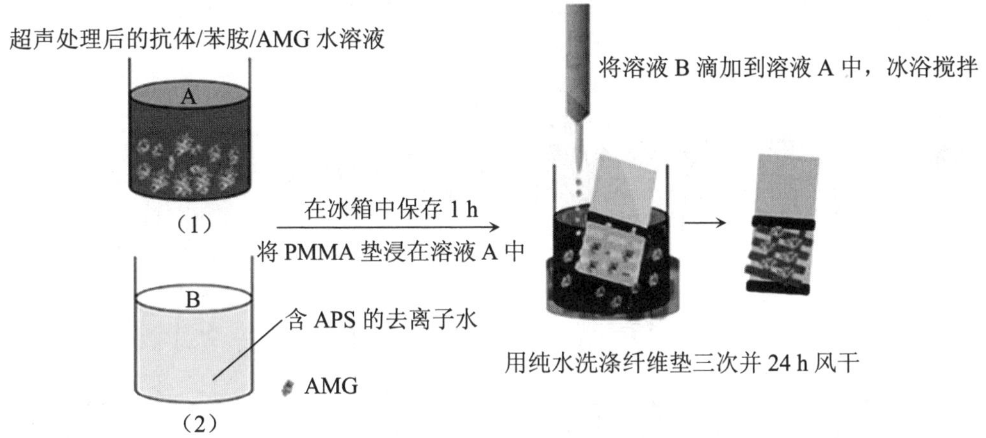

（a）AMG/PANI/PMMA 纳米光纤传感器的制作过程示意图

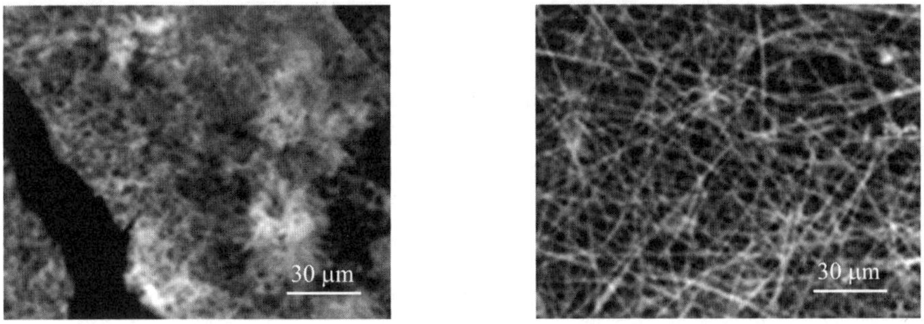

（b）UV 处理前后 AMG/PANI/PMMA 纳米纤维 SEM 图像

图 6.6　CO_2 气体传感器

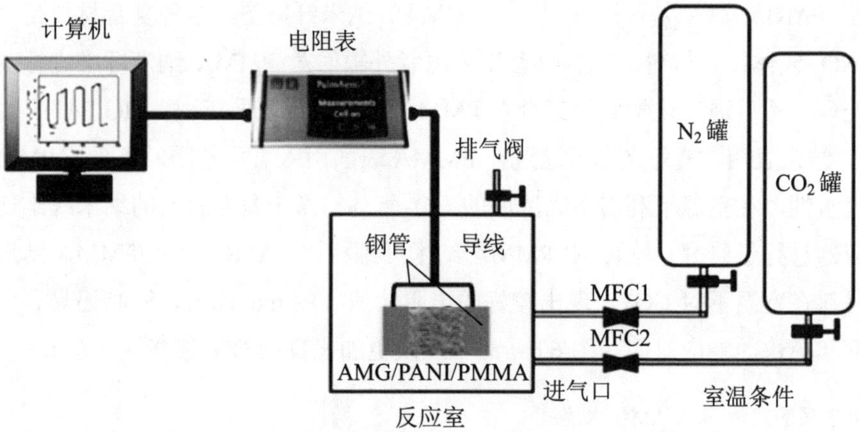

（c）气体传感测量的原理图

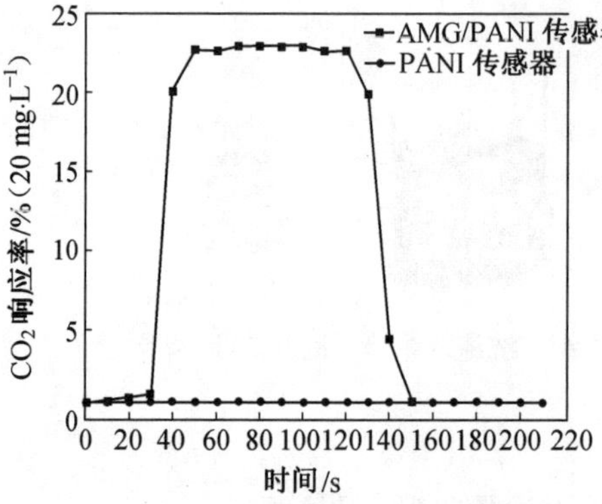

（d）两种传感器对 20 mg·L^{-1} CO_2 气体的动态响应

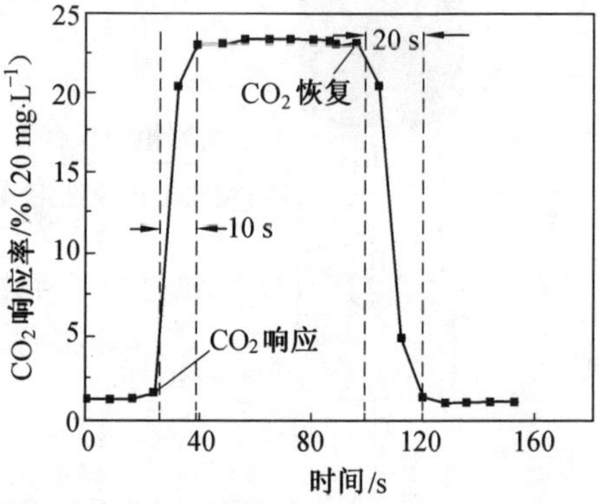

（e）AMG/PANI 纳米纤维气体传感器

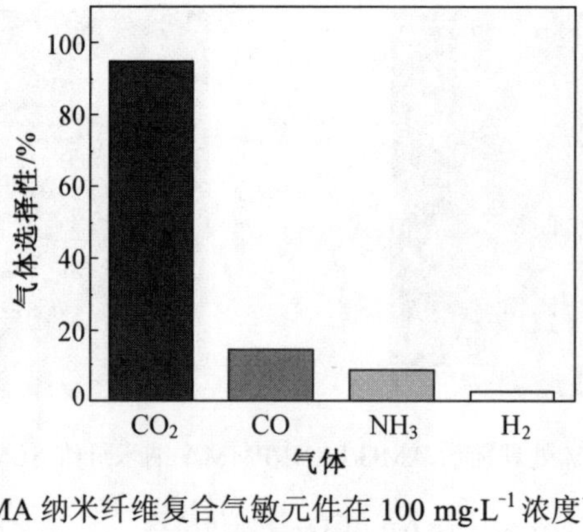

（f）AMG/PANI/PMMA 纳米纤维复合气敏元件在 100 mg·L^{-1} 浓度下对各种气体的选择性

续图 6.6

6.2.5　小结与展望

因为纳米量级功能型聚合物本身具有高表面积和优良的电化学和压电性能，所以纳米纤维传感器的传感性能比起对应的薄膜型传感器有了质的改变，是传感器性能提升的一大途径。正如本章所述，基于无机半导体和有机聚合物两者复合的纳米纤维电阻传感器具有柔韧性好、响应速度快、恢复时间短、灵敏度高等优点，在未来轻量化柔性传感器方面具有巨大的潜力，但当前仍有待解决的问题：

（1）在无机半导体纳米纤维中掺杂的机理仍需进一步深入研究。

（2）有机无机复合纳米纤维传感器在实际使用中的抗拉伸和抗疲劳性能仍需进一步提高。

（3）应用于传感器的纳米纤维制备工艺需进一步改进，进而弥补掺杂和烧结对纳米纤维稳定性和渗透性的削弱，提高由此制备传感器的稳定性和使用寿命。

6.3　生物传感器

6.3.1　引言

静电纺丝提供了适应性强的反应过程，可以用于许多不同的领域中，随着商业上可行产品的开发，很多领域都取得了十分显著的进步。用于生物传感器和用于不同环境应用的生物催化的电纺材料的开发是上述领域之一，在过去的十年中已经取得了重大进展。这些发展导致了复合电纺产品用于大量传感和废物处理系统，传感设备被广泛用于环境中不同物质的分析。数十年来，已经采用了许多先进的传感系统来分析多种介质中的不同分析物。然而，这些系统的缺点是成本高，用于确定化学成分的感测能力有限，用于分析重要的生物化合物的可靠性较低以及无法在线分析。不同生物受体对特定物质的高灵敏度和精确度促使许多研究者开发生物传感器，基于简单蛋白系统开发了生物传感器，该传感器表现出色。

生物传感器的感测能力基于生物敏感剂（例如酶）将分析物转化为可感测的化合物，并产生可测量的信号。由于它们具有加速化学反应的能力，因此基本上可以通过生物敏感剂将分析物转化为新产品。生物传感器仅加速选择性反应的能力使其只能在选择性分析物上工作。因此，它们充当选择性的生物催化剂，并且已经广泛地探索了这种能力以开发出一些用于分析各种分析物的新产品。当生物敏感剂感测到特定分析物时，通常能够产生微弱的信号。通常，需要生物换能器以将由生物感测剂产生的结果转换成由电子显示系统呈现的更可测量的信号。生物传感器与生物敏感剂接触，该生物敏感剂由于分析物对其的作用而发生物理化学变化。这种变化由生物传感器测量，然后将其转换为信号。信号的强度取决于物理化学变化的幅度，而物理化学变化的幅度又取决于分析物的量。因此，该系统能够有选择地确定所研究分析物的浓度。图 6.7 所示为具有基于酶的生物传感器的特定示例的生物传感器的通用工作路径。但是，在此必须提及的是，基于酶的生物传感器属于更广泛的分类系统，其中包括基于免疫球蛋白、核酸、适体等的生物传感器。生物传感器通常可以按其采用的生物传感器的类型来命名。为了使生物传感器系统有效工作，必须将生物传感剂以特定的图案稳定截留或涂覆在支撑介质上，例如不同的多孔物质、水凝胶和纳米纤维。稳定介质的性质对生物传感器的性能有很大影响。以下各节讨论各种携带酶的电纺纳米纤维，其开发方法、酶结合系统以及可能的应用。

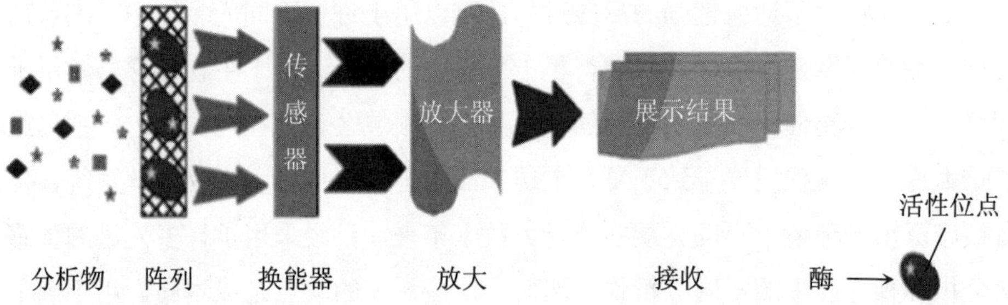

图 6.7 生物传感器的一般工作过程

6.3.2 携带酶的电子纤维

如前所述，酶（或其他生物活性剂）的稳定和固定化是生物传感器发展的前提。可以通过将酶附着或捕获在几种不同类型的微米或纳米尺寸的介质上，来固定和稳定酶。与微米级材料相比，纳米级介质因其高表面积而具有更多优势。目前已经使用了不同类型的纳米材料来固定酶，包括不同类型的纳米颗粒、中孔纳米材料（例如中孔二氧化硅和纳米纤维）。在过去的十年中，已经有关于将酶固定在纳米纤维上的广泛研究。已经发现，纳米纤维可提供更高的承载酶的性能，用于生物传感器和生物催化应用。它们的表面积和孔隙率可以针对特定的酶量身定制，这使它们对酶扩散的限制降低，并提高了分析物的检测效率。此外，可以对其进行物理和化学修饰，以达到所需的活性水平。可以进行不同的化学修饰以适合酶的化学性质和生物传感器必须在其中执行的环境。另外，存在可以电纺成纳米纤维以用于生物传感器开发的多种材料。考虑到这些优点，静电纺丝纳米纤维可用于生产高效、量身定制的纳米传感器。酶已成为生物传感器中最常用的生物活性剂，并且已经研究了许多酶在由电纺纳米纤维开发的生物传感器中的生物活性。通过在静电纺丝过程之后将它们的分子附着在纤维表面上，或者通过将聚合物和酶的溶液以及一些其他添加剂静电纺丝，可以将它们固定在静电纺纳米纤维上。这两种基本技术都已经过研究，以下各节将对其进行简要讨论。

具有酶的静电纺丝纳米纤维可以通过在其表面使用官能团将它们共价键合，将其固定在纳米纤维上。使用这种方法，许多不同的酶已被固定在多种聚合材料上。生物相容性较差的聚合物上的酶固定酶是“生物”物种，与大多数生物合成聚合物等没有生物相容性的物种相容性较差。但是，由于合成聚合物的优异机械性能以及它们上的官能团或它们的修饰形式的存在，人们对它们在支持酶方面的潜在应用进行了广泛的研究。

在聚合物家族的众多成员中，聚丙烯腈（PAN）一直是酶固定领域研究的重点。这是因为其化学结构具有可以与多种物质反应的官能团。Shen 和他的同事采用了一种简单而有效的技术来固定脂肪酶电纺 PAN。纤维和脂肪酶之间的共价键是通过激活 PAN 上的腈基进行酰胺化反应而形成的。该酶通过酰胺键成功地固定在 PAN 纳米纤维上，如图 6.8 所示，并在 PAN 纤维表面形成聚集体。使用该系统可观察到良好的加载效率和较高的活性保留率。

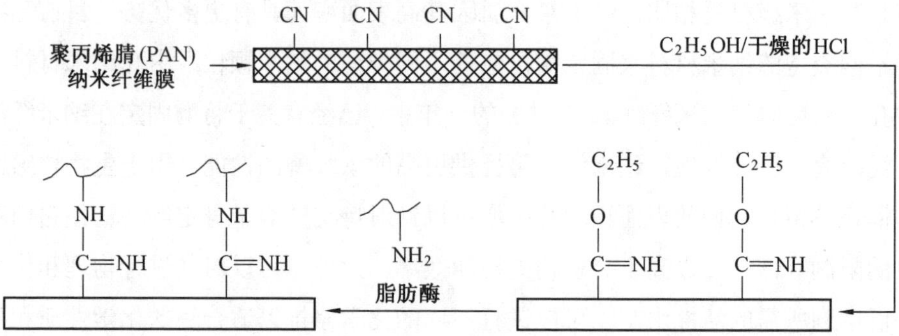

图 6.8　通过酰胺化反应将共价脂肪酶固定在 PAN 上的示意图

同样，聚（丙烯腈-马来酸）（PAN / MA）复合纳米纤维也已通过形成两个大分子的生物相容性层，即壳聚糖（CS）和明胶形成了生物相容性层，用于固定脂肪酶。纳米纤维经过处理，以活化其上的羧基以共价连接大分子，然后使用戊二醛将酶连接到这些修饰的纤维上，戊二醛已被广泛用于向聚合物链中添加醛基以进行共价固定。该过程的示意图如图 6.9 所示。

与未改性的 PAN / MA 纳米复合材料相比，观察到纳米复合材料具有更高的负载量和改善的活性保留。这允许实现所需的生物相容性以及机械强度，而在生物相容性纤维（如 CS 和明胶）中很难达到机械强度。

图 6.9　通过酰胺化反应将共价脂肪酶固定在 PAN/MA 上的示意图

Wang 等人的另一项研究，揭示了固定化的过氧化氢酶在由具有多壁碳纳米管（MWCNT）的聚（丙烯腈-丙烯酸-丙烯酸）复合纳米纤维制成的生物传感器中显示出比无 MWCNT 的生物传感器更好的性能。研究人员使用二甲基氨基丙基-乙基碳二亚胺盐酸盐和羟基琥珀酰亚胺将酶与聚合物链共价键合，从而激活羧基以固定酶。其中主要原因是 MWCNT 的电子转移效率更高。此外，酶分子直接与纳米纤维的共价结合减少了 MWCNT 与酶之间的距离，从而具有更好的传感能力。在一项类似的研究中，卟啉（电子供体）与 PAN 纳米纤维中的碳纳米管（CNT）（电子受体）一起使用，并具有过氧化氢酶共价键合。已证明这种组合可提高酶的活性和稳定性。为了通过改善酶中电子的转移来增强生物传感器系统的灵敏度，Jose 和同事开发了带有 CNT 涂层（用于酶固定）的电纺金纳米纤维。使用聚丙烯腈-金（PAN-Au）盐混合物对纳米纤维进行电纺。随后将它们涂覆有能够共价固定酶葡萄糖氧化酶的羧化 MWCNT。因此，开发了具有非常接近酶的导电纤维的无介体生物传感器，提高电子转移效率，同时消除与氧化还原活性介体有关的缺点。同时该传感器具有非常好的电子传输速率和很高的线性灵敏度。普鲁士蓝（PB）是一种基于六氰合铁酸铁的染料，具有良好的还原能力。Fu 和他的同事们实现了将 CS 和 PB 和 MWCNTs 结合在一起的优势，并使用 3-异氰酸根合丙基三乙氧基硅烷（ICPTES）作为共价偶联剂将葡萄糖氧化酶原位共价固定。在这里，ICPTES 也是溶胶-凝胶的前体。这项研究中使用的 PB 沉积在 MWCNT 的附近。PB/MWCNT、ICPTES、CS 和葡萄糖氧化酶的混合物电纺丝在玻璃碳电极上。该系统用于检测葡萄糖，发现具有良好的灵敏度和低检测限。在 J. Shen 和他的同事进行的一项研究中，还探索了金的高电子转移效率，以开发一种生物传感器，该传感器用于结合过氧化物酶的过氧化氢定量测量，该传感器结合在涂金的电纺二氧化硅纳米纤维上。金涂层被用于增强通过该系统的电子传导性，否则该电子传导性非常差并且导致纳米传感器的性能差。为了开发生物传感器，将分散在 CS 中的金-二氧化硅复合纳米纤维旋涂在氧化铟锡上，然后用酶溶液旋涂在 CS 中。发现开发的纳米传感器对过氧化氢具有良好的传感能力，线性传感范围在 5×10^{-6}～1×10^{-3} mol·L^{-1}之间，检测极限为 2 μmol·L^{-1}。H.Zhu 等人使用类似的方法开发生物传感器，通过在电纺聚乙烯醇（PVA）/聚乙烯亚胺（PEI）/银纳米颗粒复合纳米纤维中添加辣根过氧化物酶，从而检测葡萄糖和谷胱甘肽。纳米纤维是由两种聚合物的混合物制成的。复合纳米纤维被官能化，随后通过还原硝酸银在表面上产生银纳米颗粒，然后通过用酶溶液处理纤维垫来添加酶。发现生产的复合材料系统稳定，可回收且具有快速响应能力。

El-Aassar 和他的同事采用了在电纺聚丙烯腈-甲基丙烯酸甲酯或聚丙烯酰胺-MMA 纳米纤维上使用间隔臂的想法。 在固相支持物之间包含间隔基允许减少固定化介质的空间位阻，从而为结合酶分子提供了一种环境，使酶分子接近自由分子所享受的环境。 研究人员使用聚（AN-co-MMA）的羰基将 PEI 臂共价连接到其上。然后使用戊二醛作为偶联剂，将 PEI 臂与 β-半乳糖苷酶形成共价键。 该技术可以提高 β-半乳糖苷酶的温度稳定性。Wang 和 Hsieh 进行了另一项此类研究。 他们通过添加聚乙二醇（PEG）间隔臂将脂肪酶固定在电纺纤维素纤维中。

聚苯乙烯已基于其机械性能和成型性能而用于酶固定，在有关区域的初步研究中，将改性的聚苯乙烯纳米纤维进行电纺，然后将酶 α-胰凝乳蛋白酶化学连接到其表面。发现酶的负载量约为 10.4%（质量分数），表面覆盖率为 27.4%。评价了开发的系统的酶水解活性，观察到该活性仅降低了 35%，这意味着即使固定化后酶也能发挥良好的性能。在类似的研究中，金（Kim）等人通过共溶在静电纺丝的聚苯乙烯/聚（苯乙烯-马来酸酐共聚物）复合纳米纤维的表面上，改善了溶菌酶的稳定性和活性。该酶通过戊二醛处理交联，不仅通过交联其他酶分子提高了酶的负载量，而且还提高了其稳定性。与未进行任何交联处理相比，发现酶活性增加了 9 倍。而且，它在高温和高 pH 下具有高稳定性。由于其疏水性，酶在聚苯乙烯纳米纤维上的较低负载限制了它们在该领域的应用。但是，可以进行一些修改以克服此问题。Nair 及其同事仅用酒精处理马来酸酐改性的聚苯乙烯纳米纤维即可解决该问题。这使得纤维在负载处理期间保持分散，并且脂肪酶分子易于进入共价位点以与纤维形成共价键。与使用含水酒精处理的酶负载的聚苯乙烯纤维相比，该技术将负载能力提高了 8 倍。

关于将酶固定在聚酰胺纤维上的研究很少，Uzun 和同事共价固定了尼龙 6.6 / MWCNT /（聚 4-（4,7-二（噻吩-2-基）-1H-苯并咪唑-2-基）苯甲醛）（PBIBA）复合物中的葡萄糖氧化酶纳米纤维。导电聚合物 PBIBA 被用作具有良好电活性的共价固定介质。因此，在原本较差的电活性尼龙纳米纤维上获得了具有高电活性的纤维表面。最终的复合系统能够提供良好的灵敏度和高电子转移。

Kumar 和他的同事用苯二胺和戊二醛对静电纺丝的聚甲基丙烯酸甲酯纳米纤维的表面进行功能化，使其能够共价固定化酶，并提高了酶的活性和稳定性。他们将木聚糖酶附着到纤维表面以观察其效率和稳定性。即使在 11 个反应周期后，附着的酶仍保留其 80%的活性。此外，还发现该酶具有良好的热稳定性，因此纳米复合材料可在工业过程中找到应用。

将酶固定在生物相容性/可生物降解的纳米纤维上将酶固定在可生物降解的聚合物上是一个潜在的广泛应用领域。考虑到其形成的纳米复合材料的潜在应用，已将不同的酶固定在可生物降解的纤维上。 Kim 和 Park 将溶菌酶固定在聚（ε-己内酯）/聚（乳酸-乙醇酸共聚物）（PLGA）/聚（乙二醇）-NH_2 复合纳米纤维上。复合物上的胺基用于使用同双功能偶联剂固定模型酶。结果是高酶负载和所得纳米复合材料更高的活性保留率，这使得该产品优于流延膜中含有溶菌酶的产品。同样，Wang 和 Hsieh 在电纺纳米纤维中添加了 PEG 臂以固定脂肪酶。PEG 的添加同时添加了两亲性间隔基和反应性基团，用于与脂肪酶偶联反应。复合结构使脂肪酶在包括酸性/碱性环境和高温在内的不同条件下显示出更高的活性。Basturk 等人使用电纺的聚乙烯醇/聚丙烯酸或 PVA / PAA 纳米纤维固定 α-淀粉酶。通过使用 1,1-羰基二咪唑固定酶，以激活 PVA / PAA 纳米复合材料上的胺基。通过这种发展，α-淀粉酶的温度和 pH 稳定性得到了显著改善。同样，使用这种技术也提高了酶的可重复使用性。由于血红蛋白（Hb）和过氧化物酶之间的结构相似性，Hb 可用于还原过氧化氢，因此在环境控制中会遇到某些介质中的 HbO_2 检测。Li 和同事在含 CNT 的胶原纳米纤维中使用 Hb 代替过氧化物酶。该生物传感器具有良好的生物相容性和稳定性以及对过氧化氢检测的高灵敏度的优势。在另一项研究中，Wu 和 Yin 开发了用于葡萄糖检测的 PB 涂层 CS/PVA 生物传感器。该传感器包含固定的葡萄糖氧化酶，通过定量测量过氧化氢可以很好地检测葡萄糖，过氧化氢是在葡萄糖氧化酶氧化葡萄糖过程中产生的。该传感器是通过将 CS/PVA 醇复合纳米纤维静电纺丝到预先涂有 PB 膜的氧化铟锡涂层的玻璃板上而开发的。再次用 PB 电沉积纳米纤维铟锡氧化物复合材料。将磷酸盐缓冲溶液中的葡萄糖氧化酶添加到纳米纤维垫中，并储存过夜，以稳定两者之间的静电吸引。PB 之所以在这里使用，是因为其出色的电子传输能力和还原过氧化氢的潜力。然而，其在碱性 pH 中的不稳定性一直是当前开发中已经很好解决的问题。发现已开发的复合生物传感器在很窄的测量范围内表现良好。具有仿生复合纳米纤维的电纺纳米纤维较低的机械强度（例如 CS）限制了其以纳米纤维形式应用于生物传感器和生物催化的应用。

综上所述，表 6.1 展示了在生物传感器中可应用的不同种类纳米纤维及其特点。

表 6.1　在生物传感器中可应用的不同种类纳米纤维及其特点

纳米纤维名称	特点
PAN	良好的加载效率和较高的活性保留率
PAN/MA	更高的负载量和机械强度
MWCNT	提高酶的活性和稳定性
聚苯乙烯	酶活性增加了 9 倍
聚甲基丙烯酸甲酯纳米纤维	良好的热稳定性
PEG 臂	解决酶的 pH 稳定性问题

吸附有附着在其上的酶 MeS 的电纺纳米纤维是将酶固定在纳米纤维网上的另一种可能途径。然而，关于通过吸附固定化酶的研究数量有限。对于电纺纳米纤维而言，将良好的机械强度与功能特性相结合通常是一项复杂的任务。Wang 和他的同事通过将脂肪酶固定在具有芯鞘结构的聚（双（甲基苯氧基））磷腈（PMPPH）/ PAN 纳米纤维上来实现这一目标。脂肪酶吸附在 PMPPH 外壳上，如图 6.10 所示。当与单独的 PAN 纤维相比时，观察到更高的吸附能力和活性保留。因此，在研究中获得的复合材料可以实现所需的性能以及易于加工的性能，而单一聚合物是其他复合材料无法实现的。疏水固定剂对脂肪酶特别感兴趣，这是由于改善了脂肪酶疏水环境的活性。为了利用脂肪酶的这种特性，通过将其与聚乙烯吡咯烷酮和 PEG 混合，将其固定在聚砜纳米纤维上。添加这些添加剂以增加纳米纤维与酶的相容性并降低其疏水性。混合的纳米纤维能够从其溶液中吸附酶分子。据观察，将聚砜与生物相容性成分共混可提高酶的活性。Chen 和 Hsieh 完成了另一项通过吸附固定化酶的研究。他们在电纺纤维素纤维上开发了聚丙烯酸接枝物，并将脂肪酶吸附到其上。发现与游离酶相比，吸附的酶具有更好的活性和有机溶剂稳定性。此外，Sakai 和他的同事进行了类似的研究，他们通过吸附将脂肪酶固定在静电纺 PAN 纳米纤维中，发现固

定化的酶显示出改进的催化酯交换反应的活性，这是一种用于将缩水甘油转化为正丁酸乙烯酯的缩水甘油基正丁酸酯的酶。

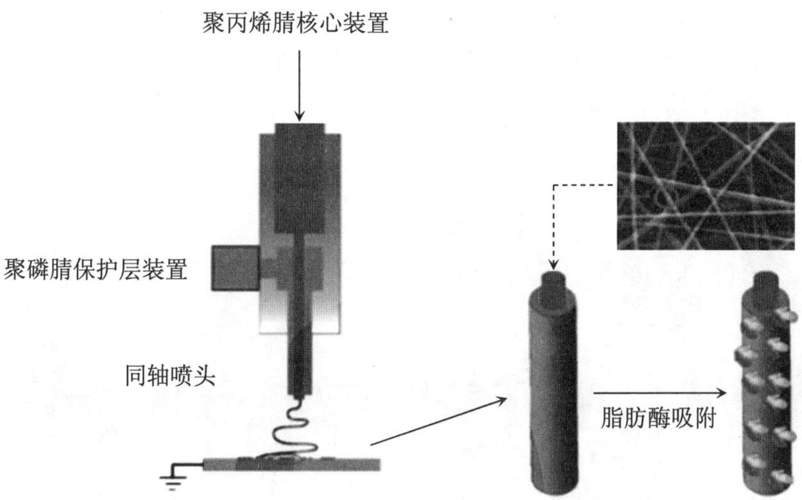

图 6.10　用于将脂肪酶固定在 PMPPH / PAN 复合纳米纤维中的同轴电纺丝装置

目前，关于包裹在纳米纤维中的酶的开发只有很少的研究。这主要是因为酶与静电纺丝溶液的相容性较低。由于大多数酶都可溶于水，因此可以将有限数量的聚合物与它们共电纺丝。然而，已经研究了一些水溶性聚合物的酶固定能力，但是当它们与水接触时，它们也会因溶胀而失去酶分子。解决该问题的一种可能方法是使用交联剂，但它们也具有与之相关的缺点，例如孔隙率和酶活性降低。PVA 因其水溶性而已被广泛用于酶的包封。葡萄糖氧化酶已被固定在 PVA 纳米纤维中，用于开发电流型生物传感器。将该酶掺入静电纺丝溶液中，使其在静电纺丝后成功封装在 PVA 纳米纤维中，如图 6.11 所示。发现该纳米复合材料对葡萄糖感测显示出低响应时间和低检测限，并为开发葡萄糖生物传感器提供了简单的方法。在另一项研究中，采用了类似的方法将纤维素酶掺入 PVA 纳米纤维中，随后将纤维与戊二醛蒸气交联。与流延膜相比，纳米纤维网表现出优异的生物转化效率。通过静电纺丝含有两者的水溶液，脂肪酶也被负载在静电纺丝的 PVA 纤维中。与游离酶相比，纳米复合物中的酶表现出优异的活性，尤其是在较高温度下。为了增强酶的稳定性，将纳米复合材料与戊二醛交联。然而，这导致活性降低。基于通过水溶液的静电纺丝将酶物理包裹在纳米纤维中的缺点，采用了不同的方法，例如添加添加剂，如生物相容性聚合物和表面活性剂，将酶分散在溶剂中并静电纺丝。帕特尔（Patel）等人在电纺介孔二氧化硅纤维中包封的辣根过氧化物酶。纳米纤维由原硅酸盐、酶、葡萄糖和 PVA 的混合物进行电纺丝。发现该复合纳米纤维具有良好的孔隙率，孔径范围为 2～4 nm。此外，发现纤维具有机械柔韧性和可充电性，从而扩大了它们在不同领域的应用范围。

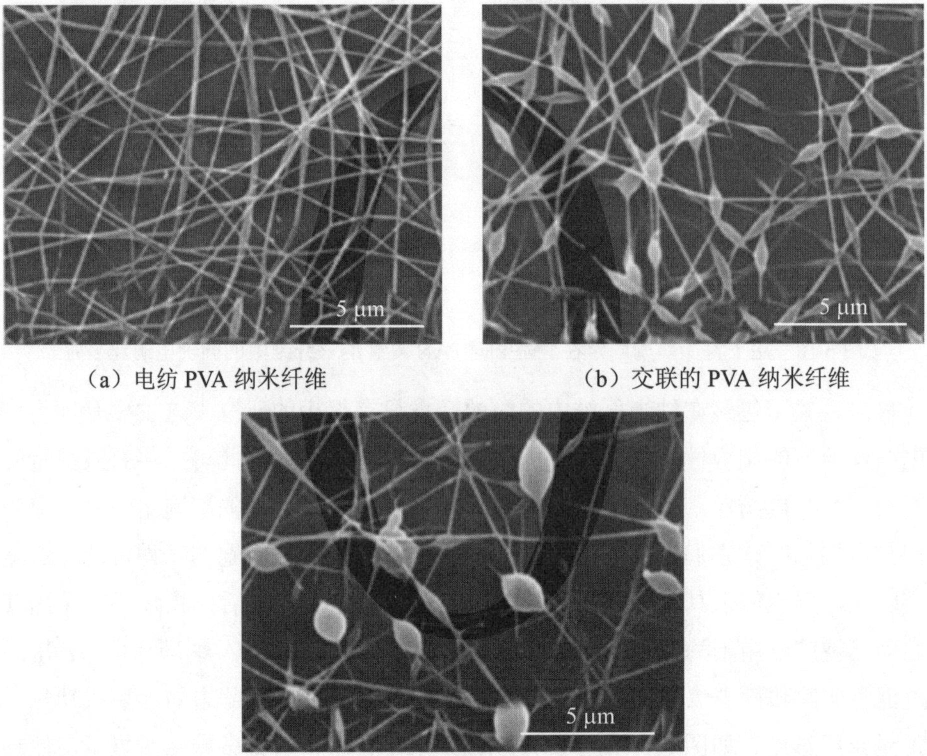

（a）电纺 PVA 纳米纤维　（b）交联的 PVA 纳米纤维

（c）葡萄糖氧化酶纳米纤维

图 6.11　葡萄糖氧化酶已被固定在 PVA 纳米纤维中

6.3.3　生物传感器和生物催化剂

此外，还研究了电纺生物传感器和生物催化剂的潜在环境应用。以下概述了电纺生物传感器和生物催化剂在环境中的应用。过氧化氢是许多工业废水的主要成分。它在废水中的排水必须保持在一定限度以下，并通过不同的环境法规进行规范。传统上已经使用不同类型的传感器来检测废水中的过氧化氢。基于电纺纳米纤维的优势，不同的研究人员已尝试开发用于检测传感器。如前几节所述，已经研究了使用不同类型的电纺生物传感器对其进行定量测量的方法，发现其中大多数提供了有趣的结果，可用于开发商业可行的解决方案来控制工业中过氧化氢的水平。例如，掺入金包被的二氧化硅纳米纤维上的辣根过氧化物酶已用于检测过氧化氢。同样，电纺 PVA / PEI /银纳米颗粒复合纳米纤维也已用于其检测。在另一项研究中，血红蛋白通过静电纺丝直接电纺在玻璃碳电极的表面。所生产的生物传感器不仅对过氧化氢灵敏度高，而且对危害海洋生物并影响氮循环的亚硝酸盐也具有很好的灵敏度。此外，对人的皮肤、眼睛、呼吸道及神经系统有危害的苯酚也展现出很高的灵敏度，必须在制造或使用苯酚的设施中保持安全水平的苯酚。传统上，将不同的传感系统用于其检测和控制。Arecchi 和他的同事们从电纺生物传感器的特性中得到启发，开发了一种新型的生物传感器，用于通过在玻璃碳电极上沉积的聚酰胺 6（尼龙 6）纳米纤维上滴涂酪氨酸酶来检测酚。该生物传感器对酚的检测显示出高灵敏度和低响应时间。在类似的工作中，发现包裹漆酶的 PVA 纳米纤维可生产出高度敏感的生物传感器，用于检测和监测氯酚，氯酚是潜在的致癌物，很可能通过农产品，染料和药物释放到环境中。氯霉素是一种抗生素，因其骨髓毒性而闻名。为了对其进行检测，开发了基于通过电纺丝制备的聚苯乙烯纳米纤维的生物传感器。将包含氯霉素抗体的电纺纳米复合材料沉积在石英晶体电极上。随后，该抗体用巯基丙酸官能化。发现该生物传感器显示出对氯霉素的高灵敏度、良好的响应时间和高捕获选择性。控制各种病原污染物也是静电纺生物传感器的重要环境应用。为了做到这一点，将大肠杆菌抗体固定在负载有金纳米颗粒的电纺硝酸纤维素纳米纤维上。该生物传感器用于检测大肠杆菌细菌，并发现对它具有极好的灵敏度。此外，它的检测时间短，使其可用于许多环境、安全和健康应用的现场检测。与传统的处理不同污染物的系统相比，电纺生物催化剂还具有许多优势，近年来，由于其效率和比较优势，人们对其进行了研究。例如，Niu 及其同事探索了将辣根过氧化物酶封装在静电纺丝的聚丙交酯-乙交酯乙交酯纳米纤维中以降解五氯酚的方法，五氯酚是用于木材防腐的高毒性试剂。酶在纳米纤维中的包封极大地提高了酶的吸附性，从而改善了五氯苯酚的降解。发现所得系统具有良好的五氯苯酚降解能力以及在操作介质中的稳定性。

漆酶是催化氧化许多底物的酶。为了探索漆酶的潜力，不同的研究人员将它们封装在静电纺微纤维和纳米纤维中以提高其功效。Dai 及其同事已经研究了电纺漆酶负载的 PLGA 纳米纤维在其降解土壤中多环芳烃方面的潜在应用。他们开发了核-壳排列，漆酶的核被多孔聚合物壳包裹。与传统系统相比，该复合材料可有效地吸附和降解土壤中的不同多环芳烃，且操作时间短得多。在一项类似的研究中，负载漆酶的 PLGA 纳米纤维用于从水中去除双氯芬酸。已经发现双氯芬酸是一种抗炎药，可以生物放大食物链，因此从制药废液中去除双氯芬酸是至关重要的环境问题。使用戊二醛作为交联剂进行静电纺丝后，将漆酶加载到 PLGA 纳米纤维上。发现所得的系统是一种从废水中去除双氯芬酸的有效、稳定、可重复使用的低成本解决方案。

6.3.4 小结与展望

如前几节所述，纳米纤维已成功用于生物传感器应用中酶的固定化。许多研究已经开发出可能适用于商业纳米传感器产品的解决方案。但是，在某些领域中，将来需要进行更多研究，以使生物传感器和生物催化应用的静电纺丝技术获得最大的输出。电子介体在生物传感器的灵敏度中起着重要作用，因此必须位于设备表面上/附近。因此，有必要研究新颖的技术来定位电子介质，例如 MWCNT 和金属纳米粒子，使其更接近纤维表面。在不影响传感器活动的情况下，避免电子介体是未来需要解决的另一个潜在领域。这是因为预期在生物传感期间介体会从纤维表面浸出，并且可能会产生某些毒性结果，例如在生物医学领域中的应用。如果可以解决这样的问题，那么血液动力植入式设备的开发就可以轻松完成。另一个需要广泛研究的领域是固定化酶的效率降低。酶在纳米纤维上的固定化由于其与成纤维聚合物的相互作用和结合而导致其活性降低。通常较高的负荷可以补偿活动量的减少，但活动量仍然低于预期。因此，在酶合成/修饰和纤维合成/修饰领域都需要进行更多的研究，以实现强结合和高负载以及出色的活性。对于在生物传感器和生物催化中应用酶活化的纳米纤维来说，较低的酶负荷也是一个问题。特别地，酶在疏水性纤维上的负载是非常繁重的任务。一些研究集中在增加酶负载的技术上，例如 Kim 等人的技术。同样，通过电纺马来酸酐改性的聚苯乙烯纳米纤维，通过乙醇处理将其分散在水中，从而增加了酶的负荷。但是，仍然需要简单的适用于多种基材的技术，并且有望在该领域中实现工业上可行的应用。

基于仿生电纺纳米纤维所提供的优势，需要对这一领域进行更多研究，以便更好地理解酶和生物大分子在纳米纤维上的相互作用。 最后，在用于环境应用的电纺生物传感器和生物催化剂的开发中已经取得了长足的进步，但是仍需要进行广泛的研究以探索这些材料的全部潜力。 随着有害和不可控制的污染物数量的增加，电纺生物传感器和生物催化剂的修饰和扩展越来越多，并且需要商业化。

6.4　光学传感器

6.4.1　引言

电纺纳米纤维是目前在许多应用中开发的一类材料，其中一些如今已接近工业生产和商业化。在这种框架下，电纺纳米纤维在光子学中的应用正在成为一个有前途的研究领域，旨在开发新型的微型光源和检测器、激光器和光学传感器。这个领域的发展非常迅速，因为可以通过利用有机发光材料的独特特性（发射的可调性、强吸收性、高量子产率、大斯托克斯位移）的各种方法来生产具有特定的和量身定制的光学特性的光学活性电纺纳米纤维等。产生的荧光纳米结构将卓越的光学性能与高暴露表面（比平薄膜大几个数量级）相结合，这一特性对于传感应用特别有吸引力，因为传感应用中与待检测系统的表面相互作用尤为重要。实际上，纳米纤维中裸露表面的增加可能会大大增强设备的灵敏度。Wang 等人使用电纺纤维的光致发光强度的猝灭来感测重金属离子和硝基芳族化合物的可能性。许多研究旨在改善荧光纳米纤维系统的传感性能，丰富可检测物种的种类，并在基本水平上更深入地研究电纺纳米纤维的光学和光致发光特性。本章提供了背景信息，并概述了发光和荧光电纺纳米纤维领域的最新进展及其在光学传感中的用途。

6.4.2 基于纳米纤维的发光系统

就可加工材料的数量和类型而言，静电纺丝（ES）工艺的高度灵活性得到了近十年来开发的多种发光纳米纤维的证实。所展示的光纤类别的数量正在不断增长，其新颖的系统具有通过光和电激发来发光的特征。电纺荧光纳米纤维可以通过两种方法实现：第一种是使用透明或光学惰性的聚合物，该聚合物可以掺有发光系统（无机量子点、有机发色团和聚合物以及生物发色团）；而第二种是基于在固有发光的共轭聚合物上。这些材料的溶解度、流变学和光学性质具有非常广泛的可变性，这要求开发特定方法以获得具有均匀形态和荧光的电纺纳米纤维。

1. 量子点和染料掺杂的电纺纳米纤维

一种获得发光电纺纳米纤维的非常有效的方法是使用透明的热塑性聚合物或聚合光致抗蚀剂作为基质，并使用无机量子点（QD）或纳米线作为荧光成分（图 6.12（a）、（b）），从为基质聚合物开发的完善程序开始，轻松获得直径为几百纳米的复合纤维，而光致发光性能可以通过具有宽光谱发射特性的无机纳米颗粒来定制范围，从可见光到近红外。这些纳米粒子通常由Ⅱ～Ⅵ族元素（例如 CdS、CdSe、CdTe）组成，直径只有几纳米，更重要的是，电子带隙和光致发光波长可通过改变其大小来调节。实际上，在这些无机系统中，对于接近激子玻尔半径的粒径，量子约束效应是可以理解的，该半径通常为几纳米。通过将颗粒尺寸减小到玻尔半径以下，还可以观察到发射的蓝移。荧光复合纳米纤维可以通过两种主要方法生产：①通过添加胶体量子点和纳米线，通过在聚合物基体中嵌入合适的分子前体，将它们原位合成到 ES 溶液中；②通过在前处理中通过热处理，气相反应或光学和电子束曝光。在前一种方法中，使用胶体 QD 可以精确控制颗粒的直径和尺寸分布，即使这些颗粒的存在可能会稍微改变 ES 溶液的流变性。流变学的变化甚至可能导致 ES 喷丝头的堵塞和工艺失败。这些问题可以通过原位合成来克服，因为聚合物溶液的物理和化学性质因分子前体的存在而被最小地改变。原位合成的另一个优点是可以通过使用基于激光写入或电子束光刻的图案化方法在空间上控制 QD 的形成。QDs 原位合成的主要困难可通过对颗粒直径和尺寸分布的经常受限的控制来表示。表征分子前体分解的效应（包括所产生的原子元素的扩散及其聚集）使这一过程变得复杂。

除了量子点以外，有机发光生色团可以嵌入聚合物基体中以获得发光的电纺纳米纤维（图 6.12（a）、（b））。荧光有机分子具有从紫外线到近红外的发射光，可以轻松地添加到聚合物中，例如聚甲基丙烯酸甲酯（PMMA）、聚苯乙烯（PS）、聚乙烯吡咯烷酮（PVP）和聚环氧乙烷（PEO）。在制备溶液时必须特别注意这些有机分子的发射特性对使用的溶剂及其自身产生的微环境非常敏感。通过这种方法，已经制备出发光的纤维，它们可用于光学传感和其他光子应用，例如光源和微激光器。

图 6.12（e）所示为一个混合发光系统示例，其中有机发光染料被装载在无机沸石 L 晶体中，其取向和组装受到它们包含在直径低至 150 nm 的聚合物电纺纳米纤维中的控制。杂化纳米纤维的特征在于明亮的局部发射，由于 ES 过程引起的沸石晶体取向，该发射被极化了。

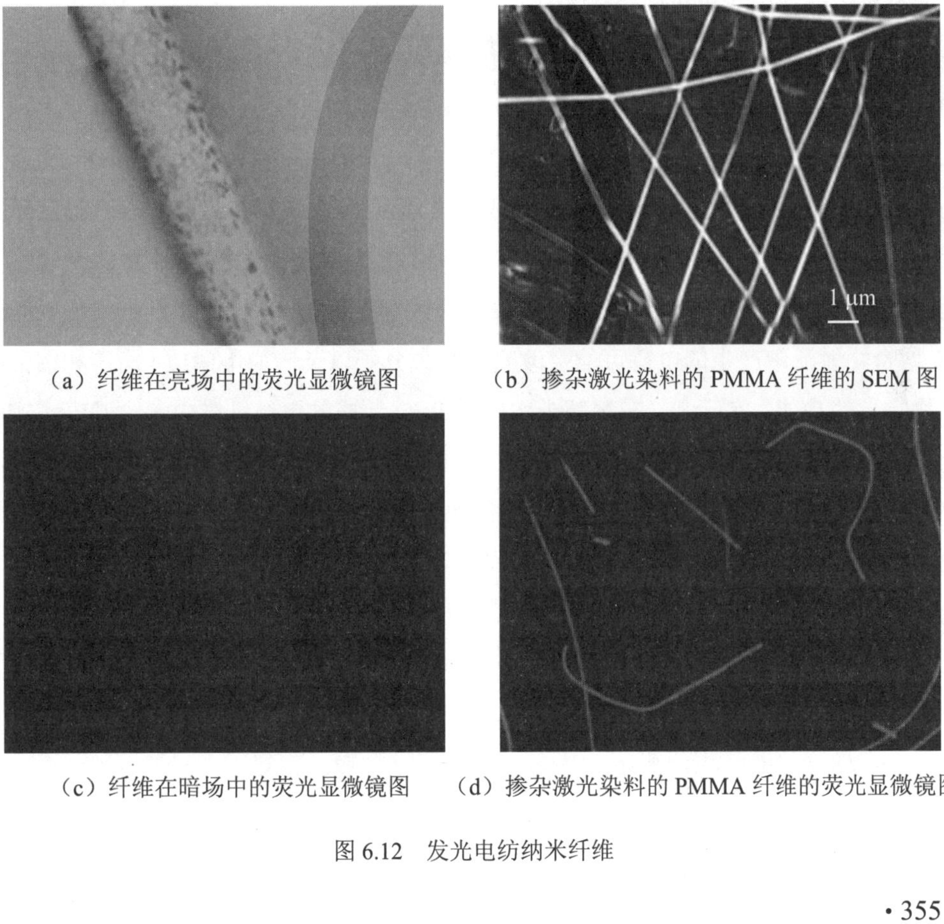

（a）纤维在亮场中的荧光显微镜图　（b）掺杂激光染料的 PMMA 纤维的 SEM 图

（c）纤维在暗场中的荧光显微镜图　（d）掺杂激光染料的 PMMA 纤维的荧光显微镜图

图 6.12　发光电纺纳米纤维

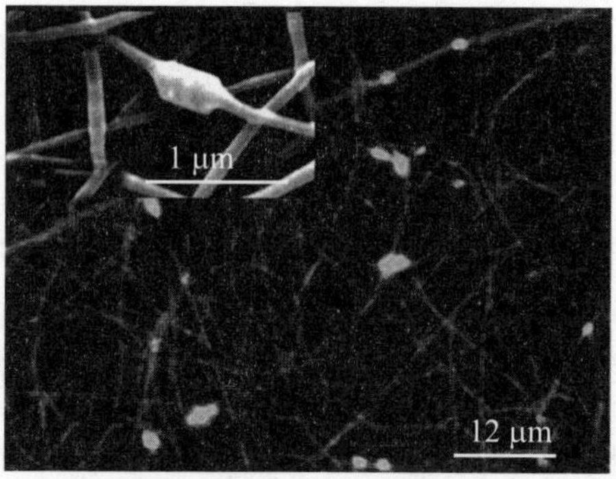

（e）含有负载有红色发光系统的沸石的发光纳米纤维的共聚焦显微镜图像

续图 6.12

2. 嵌入生物发色团的纳米纤维

在可以嵌入电纺纳米纤维的各种系统中，生物发光生色团由于在光学生物传感中的潜在应用而值得特别关注。研究最多的发光生物分子之一是来自水母维多利亚水母的绿色荧光蛋白（GFP）。与大多数包含不同于氨基酸序列的荧光团的荧光蛋白不同，GFP 的发色团由涉及三个氨基酸残基的反应内部产生。此外，GFP 显示出其荧光性质与构象变化之间的相关性，这一性质为在各种传感应用中利用 GFP 铺平了道路。使用生物发色团实现荧光纤维需要特别注意工程化 ES 过程和纤维结构。通常，ES 聚合物射流中产生的电荷可能会改变生物大分子以及诸如蛋白质、病毒和细菌等生物物体的功能。实际上，在溶解于水中的 PEO 制成的电纺纤维中包含 GFP，实际上导致了直径为几百纳米的结构，由于该 GFP 在固体基质中的变性，因此该结构不显示任何荧光。相反，利用共电纺丝技术和将 GFP 蛋白包含在核壳结构中可以有效地保留蛋白的功能性和其荧光性（图 6.13）。在共电纺丝过程中，电荷主要位于外表面，而内部溶液和嵌入式系统不带电，这一条件使人们可以更好地保留嵌入核中的物体的特性。此外，可以在核心中创建更有利的微环境，以保持所用生物发色团的荧光活性。

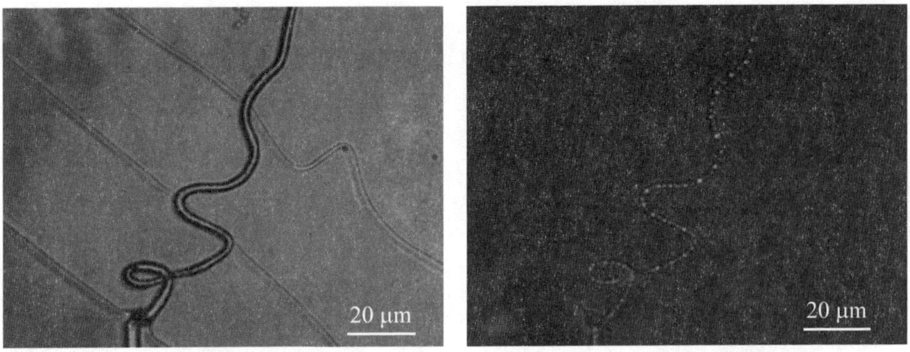

（a）相应的荧光显微照片　　（b）核心中嵌入 GFP 的光纤

图 6.13　嵌入生物发色团的纳米纤维

3. 共轭聚合物制成的纳米纤维

共轭聚合物是一类有机材料，具有许多半导体特有的光电特性。特别是，它们的电子带隙结构可以通过化学合成精细控制，这一特性为实现从紫外线到近红外发射的柔性系统开启了有趣的场景。共轭聚合物的其他性能包括排放效率高、受激发射截面大，并且有可能通过基于解决方案的方法进行处理。文献中为发光共轭聚合物的 ES 开发了许多示例。主要挑战来自这些聚合物的低摩尔质量、显著的链刚性和有限的溶解度，这刺激了特定 ES 方法的发展。类似于 QD 和有机发色团，可以将共轭聚合物与光学透明的基质（例如 PMMA 和 PS）混合，它们可以通过 ES 轻松形成纤维。图 6.14（a）所示为由 PMMA 和聚（9,9-二辛基芴-2,7-（PFO））制备的一种发蓝光的共轭聚合物。纤维沿其长度方向显示明亮且均匀的荧光，并且两种聚合物组分之间具有相分离，这可由高分辨率透射电子显微镜证明。相反，对于 ES 溶液，可以通过使用不同溶剂的混合物获得原始的共轭聚合物纳米纤维。特别地，添加非溶剂例如二甲基亚砜（DMSO），其为相对于通常用于共轭聚合物的溶剂（例如氯仿、甲苯、四氢呋喃）而言，挥发性低并且显示出改善的电性能对于获得原始的共轭聚合物纤维是有效的。

最近，通过 ES 实现了平均直径低至 180 nm 的发蓝光纤维，将共轭聚合物溶液溶解在良好的溶剂（氯仿）中，并添加了有机盐（图 6.14）。所得纤维的光学性质的光谱研究表明，有机盐的存在不会显著改变发射性质（使用的共轭聚合物的发射波长、效率和特征衰减时间）。

（a）PMMA/PFO 电纺纤维 TEM 及荧光显微镜图　（b）MEH-PPV 电纺纳米纤维荧光显微镜图

图 6.14　共轭聚合物的纳米纤维

最后，类似于将 GFP 包含在聚合物纤维中的方法，该方法利用共电纺丝获得由共轭聚合物制成的纤维。在这种情况下，通过使用 PVP 溶液作为壳并使用共轭聚合物溶液作为核，对核壳结构进行电纺。然后，通过乙醇提取溶解 PVP 壳层来实现共轭聚合物的原始纤维。

4. 发光纳米纤维阵列

处理发光纤维时的一个重要问题是产生有序宏观阵列的能力，这正变得越来越重要，因为电纺纳米纤维的大多数新兴应用，包括光学传感，都需要对两种纤维进行精确控制。

形态（特别是直径均匀性图（6.15））和光纤定位是将有源光纤元件连接到激发源和检测器所必需的。为控制光纤定位和实现有序阵列而开发的一种优雅方法，是利用带图案的金属系统作为收集器。最简单的图案化收集器由两个平行的金属条组成，该系统允许收集垂直于金属条单轴排列的纳米纤维。通过对收集器的金属覆盖区域进行适当的设计，还证明了更复杂的几何形状。

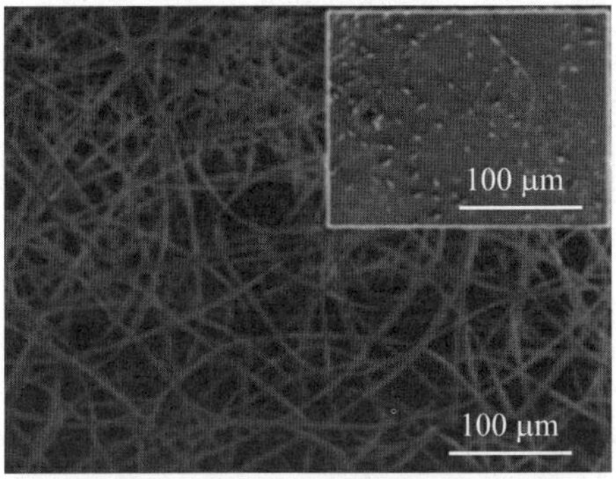

（a）THF、DMSO 电纺纤维 SEM 图

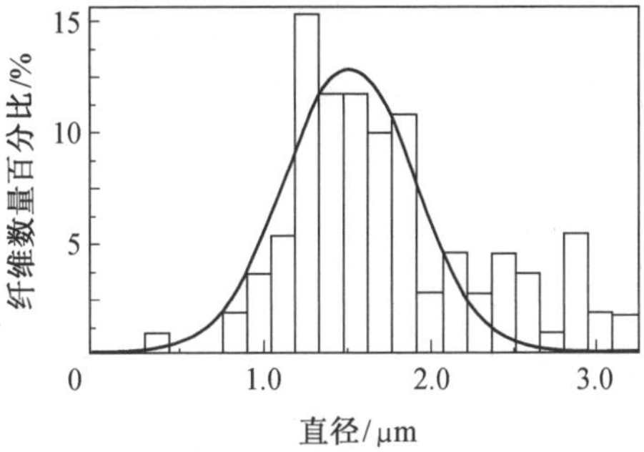

（b）图（a）对应的尺寸分布图

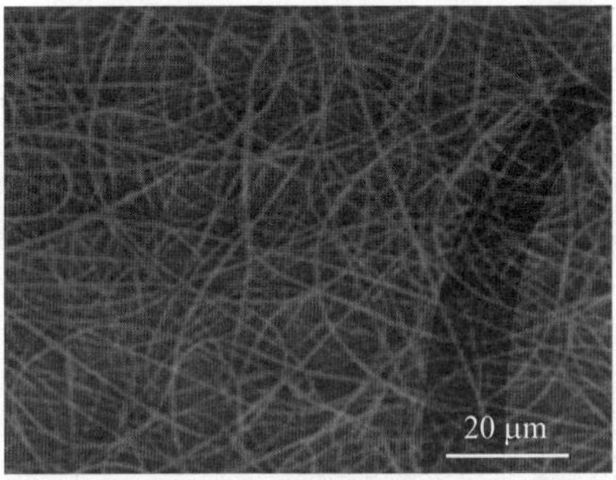

（c）$CHCl_3$、碘化四丁铵电纺纤维 SEM 图

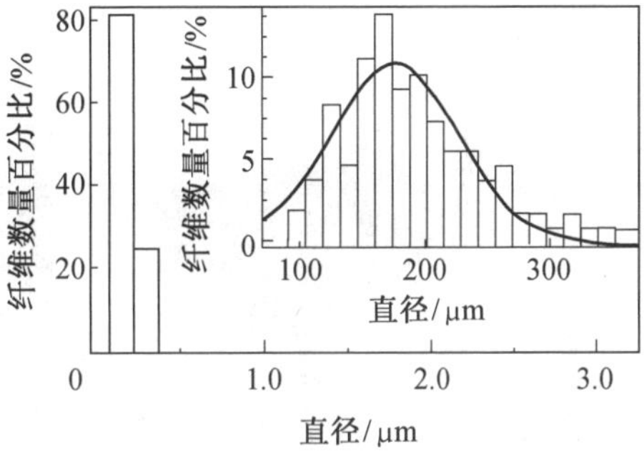

（d）图（c）对应的尺寸分布图

（e）$CHCl_3$、溴化四丁铵电纺纤维 SEM 图

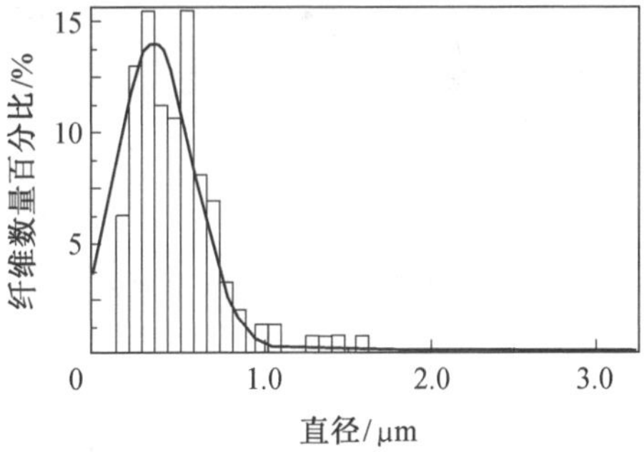

（f）图（e）对应的尺寸分布图

图 6.15　电纺复合聚合物纳米纤维电镜及尺寸分布图

用于获得纳米纤维的有序阵列的 ES 方法的主要缺点是不稳定性，是针头到收集器的射流传播特性导致的。最近，近场静电纺丝（NF–ES）出现了另一种方法，可以精确定位单个纳米纤维并实现具有几乎任意几何形状的纳米纤维阵列。该方法通过在喷射不稳定之前将基底定位在距离针头几毫米或更短的位置，从而利用 ES 喷射的稳定区域。这允许人们通过在垂直于射流传播轴的平面内移动基材来精确沉积单个纳米纤维并实现有序纳米纤维阵列。该技术已应用于不同的材料，例如 PEO，还应用于发光共轭聚合物。由 PEO 和 MEH–PPV 的混合物组成的平行和交叉发光纤维阵列已以微米级空间分辨率沉积在 Si / SiO_2 衬底上（图 6.16）。

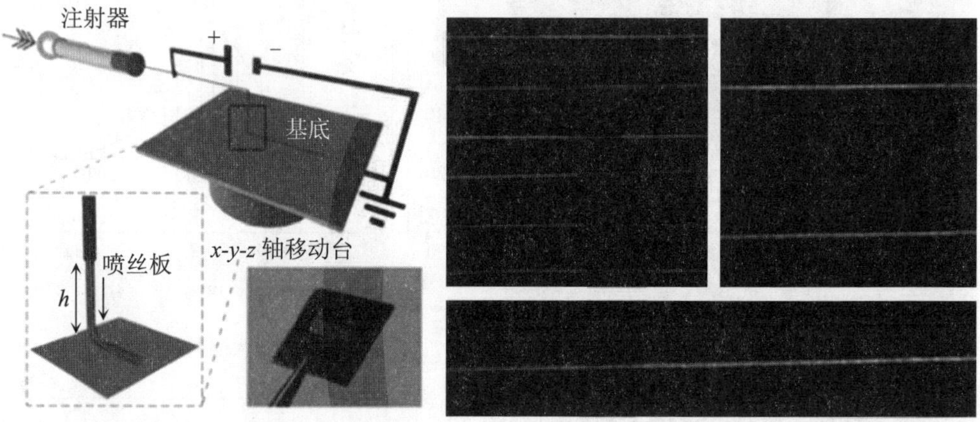

（a）NF–ES 设置方案及纤维阵列光学照片　　（b）平行纳米纤维的共焦荧光显微镜图像

图 6.16　近场静电纺丝（NF–ES）及其制备的纤维阵列

6. 4. 3　电纺纳米纤维的发射特性

研究电纺纳米纤维的发射特性以及与膨体和薄膜性能的比较对于新应用的开发尤为重要。用于制造发光纳米纤维的大多数有机物显示出的发射特性对局部微环境和分子的最终构象非常敏感。实际上，有机分子和聚合物在溶液中的光物理现象中，有许多由特定使用过的溶剂引起的光谱性质变化（峰值波长变化、发射效率和辐射速率的增减等）的例子。当处理固态样品时，这种影响由于加工条件的影响而变得复杂，这可能会导致特定的构象在分子和聚合物中。在 ES 中，这些条件尤其与施加的电场强度、使用的溶剂和环境湿度有关。

1. 电纺纤维的光致发光光谱特性

图 6.17 所示为电纺纳米纤维和由聚[（9,9-二辛基芴基-2,7-二基）-co-（1,4-苯并-{2,1′,3}-噻二唑]（F8BT）制成的旋涂膜的吸收光谱和光致发光光谱之间的比较。F8BT 是一种共轭聚合物，在约 570 nm 处发射。薄膜和纤维之间的主要区别是发射的蓝移，其偏移为 10 nm。对于其他共轭聚合物和发色团，纳米纤维的光致发光峰波长与参考膜相比也有相似的变化。值得一提的是，文献中可用的各种数据也受到散射和自吸收效应的影响，这可能引起吸收和发射光谱的扩大以及光致发光的红移，这与所用材料的固有光学跃迁无关。因此，必须特别注意电纺纳米纤维吸收和发射特性的光学表征。

减少散射和自吸效应的影响方法包括使用可减少散射和波导的积分球和薄样本，以使自吸收贡献尽可能低。通常，可以通过将共轭聚合物分子视为多发色团系统来合理化观察到的发射位移，因为沿着聚合物链的扭曲和扭结的存在会破坏共轭长度并形成一个构象亚基的整体，它们发光，光激发吸收了这种发色团，并且能量转移机制将激发转移到了较低能量的激子（激子迁移），从而决定了最终的光致发光光谱的波长和形状。

聚合物射流的高拉伸速率是典型的 ES 工艺的结果，因此，纤维中大分子的特殊构象可能会影响转移过程。ES 过程可引起聚合物分子沿纤维轴的排列。根据所用聚合物链的结构，这种构型可以增强或减少电子向低能发射亚基的传递。与薄膜相比，影响纤维发射光谱的其他影响通常是链聚集的减少和共轭长度的增加。

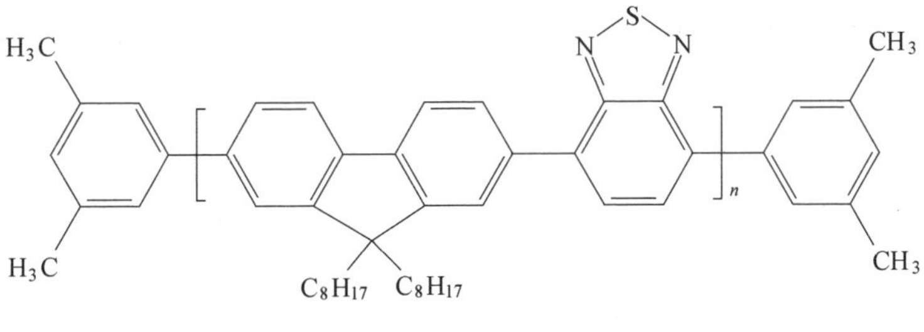

（a）共轭聚合物的化学结构

图 6.17　电纺纳米纤维及旋涂膜

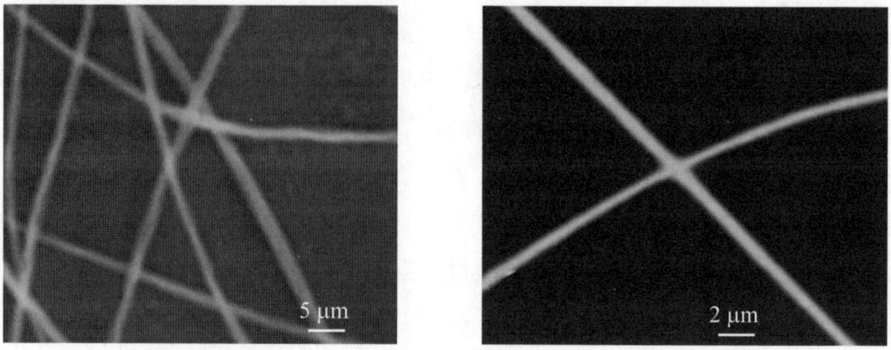

（b）F8BT 电纺纤维的 SEM　　　　（c）共聚焦显微镜图像

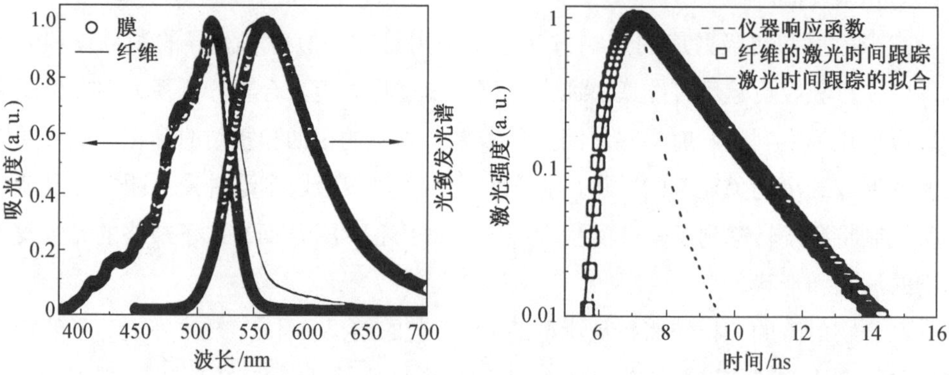

（d）纤维和膜的吸光度与光致发光图谱　　　　（e）纤维的光致发光时间衰减图

续图 6.17

2. 发射量子产率和寿命

对于光子学中的某些应用以及基于荧光系统的光学传感，光致发光量子产率（PLQY）的控制是至关重要的。通常，量子点、低摩尔质量的有机分子和共轭聚合物可能在溶液中显示出很高的 PLQY，而将它们包含在固态系统中会通过激活大量和非常规的非辐射通道而对产率产生不利影响。另外，量子点表面缺陷和陷阱状态也对 PLQY 产生影响。考虑到 ES 过程的特殊处理条件，聚合物电纺纳米纤维 PLQY 的测量与确定允许保留或增加发射光子数量的最佳参数有关。很少有报道对此问题进行过深入研究，这表明与其他固态系统（如薄膜）相比，电纺纳米纤维中的 PLQY

可以得到显著增强。Kuo 和同事测量了由 PMMA 制成的电纺纳米纤维与不同共轭聚合物共混的 PLQY，发现与旋涂膜相比，其 PLQY 增加高达 3 倍。G.Morello 等报道，与细丝相比，原始 F8BT 的电纺纤维的 PLQY 增加了约 50%。PLQY 和光致发光寿命的同时测量也可以确定辐射和非辐射寿命在很大程度上受 ES 过程的影响。衰减通常遵循具有特征寿命 τ_{PL} 的指数趋势，该寿命与固有辐射（τ_r）和非辐射（τ_{nr}）寿命有关：

$$\frac{1}{\tau_{PL}}=\frac{1}{\tau_r}=\frac{1}{\tau_{nr}} \tag{6.1}$$

给定 PLQY，对 Φ 和 τ_{PL} 的测量可以确定辐射寿命和非辐射寿命。该分析已证明，与旋涂膜相比，F8BT 电纺纤维的辐射率提高了 22.5%。因此，ES 提供了一种具有增强发射效率的用于生产固态光致发光结构的方法。

3. 发射的极化

许多报告已经证明，表征 ES 过程的强大拉伸力可能会诱导聚合物链沿纤维轴方向并产生复杂的内部纳米结构。与其他制造方法（旋涂、浇铸）相比，ES 工艺体现出各向异性的光学特性。特别地，当处理包含共轭聚合物作为光活性成分的电纺纳米纤维时，经常会观察到极化发射（图 6.18（a））。在共轭聚合物中，发射偶极跃迁大多沿着聚合物链排列，因此，电纺纤维中的链排列会沿着纤维长度产生整体的光致发光偏振。

在宏观排列的电纺纤维和单个纤维中都观察到偏振发射，其偏振比（平行于并垂直于长纤维轴的偏振的光致发光强度之比）在 2～5 之间。已经证明，机械拉伸后处理可提高极化率，最高可达 25 左右。静电纺丝纤维中聚合物链的微观排列已通过不同方法研究，包括小面积电子衍射、红外光谱和拉曼光谱。聚{[9,9−二（3, 3′−N, N′−三甲基铵）丙基芴基−2,7−二基] −alt−（9, 9 二辛基芴基−2, 7−二基）}/二碘化物盐（PF^+）混合物制成的纳米纤维具有偏振稳态发光，偏振比高达 4（图 6.18（b））。

最近的一项研究利用偏振显微拉曼光谱研究了使用不同溶剂制成的原始 F8BT 电纺纳米纤维的链排列程度（图 6.19）。特别地，已经研究和比较了由四氢呋喃（THF，F8BT 的良好溶剂）溶液或 THF 和阴离子溶剂（DMSO）的混合物制成的纤维。由 THF 制造的纤维具有较高的聚合物链排列度，即使它们通常不连续但显示出线密度约为 10 mm^{-1} 的珠子。与由溶剂混合物制成的纤维相比，由 THF 制成的纤维的极化比增加约两倍。

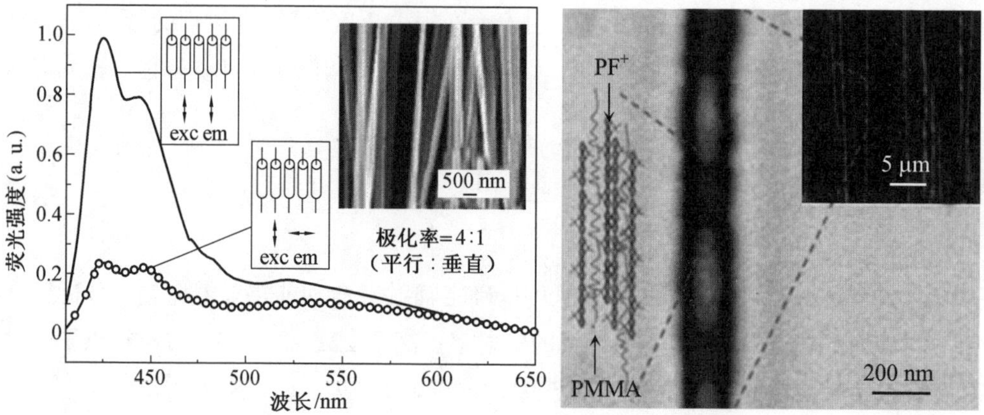

（a）PMMA 纤维的偏振光致发光光谱　（b）PMMA/PF 电纺纤维的 TEM 和荧光显微镜图

图 6.18　PMMA/PF 电纺纳米纤维

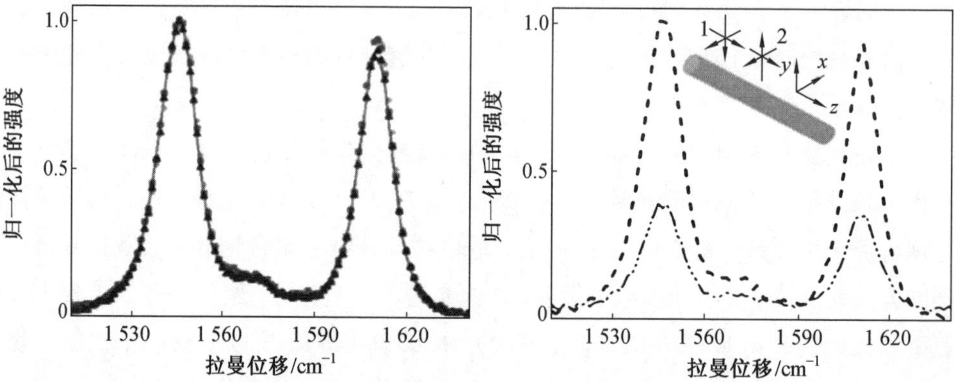

（a）单个 F8BT 纤维的拉曼图　（b）F8BT 薄膜的极化拉曼图

图 6.19　F8BT 电纺纳米纤维

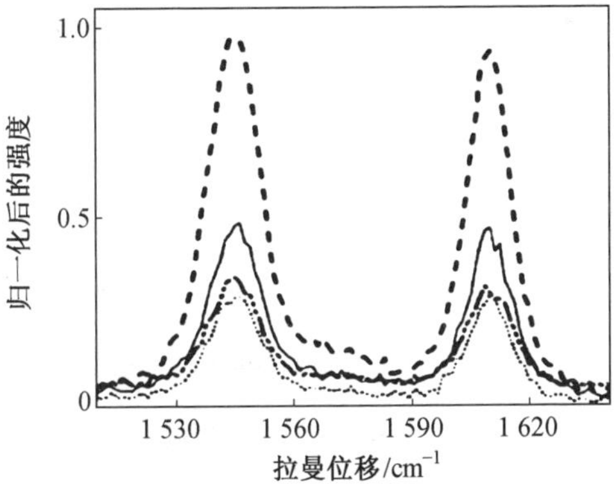

（c）THF 图单个 F8BT 的入射偏振的光谱

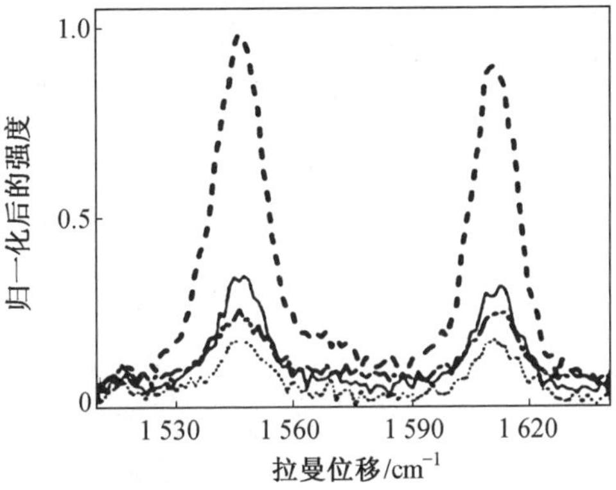

（d）单个 F8BT 纤维使用 THF 的光谱

续图 6.19

6.4.4　通过调制发射强度进行光学传感

上面显示的各种荧光电纺纳米纤维的可用性，在许多情况下与薄膜相比都具有改善的发射特性，并且表面积增加了，这刺激了高性能光学传感器开发的大量研究。光学传感器通常因其高灵敏度、快速响应时间和低功耗而被首选。大多数报告的光学传感系统是基于分析物诱导的电纺纳米纤维的荧光猝灭，该效应可能是由荧光分子与被检测元素之间的能量转移或电子转移引起的。如下所述，该方法主要应用于金属离子、爆炸性分子和生物分子，在检测限和时间响应方面均取得了显著成果。

阳离子（尤其是重金属）的感测和检测对环境和健康相关问题的巨大潜在影响极为重要。这些离子可以强烈地抑制共轭化合物的发射，为光学检测提供了一种有效的机制。为此，掺有荧光化合物和 1,4−二羟基蒽醌的静电纺丝纤维可以有效感测和检测阳离子。最近报道了一种先进的方法，该方法可以将浓度仅为 zmol（1 zmol= 10^{-21} mol）的产物限制在聚合物纳米纤维的连接处。在交叉电纺纤维存在阿托升的容积，可以合成和保留活性发色团（图 6.20（a））。特别是，荧光团−多胺的合成已被用于实现纳米级传感器的阵列，该荧光团的多胺在结合金属离子后被淬灭。对十种金属离子（Al^{3+}、Fe^{3+}、Co^{2+}、Ni^{2+}、Cu^{2+}、Zn^{2+}、Hg^{2+}、Cd^{2+}、Ca^{2+}、Mg^{2+}）有明显检测效果，其中，水中 Co^{2+}（pH=5）的检出限低至 2.0 mg·L^{-1}。由于其可逆性（展示了多达四个循环），它也非常强大，并允许人们在同一垫子中合成具有不同发射波长的荧光团，以并行处理多种分析物。而且，这种垫子表现出良好的机械强度并且适合于沉积将传感器放在丁腈手套上（图 6.20（b）～（e））。J. Orriach 等最近已经开发了一种用于水中 Hg^{2+}的高选择性光学传感器。他们的系统依靠电纺无纺布垫掺有螺环苯基硫代氨基脲若丹明 6G 衍生物（FC1）的化合物，在存在 Hg^{2+}时显示出荧光的开启。与固态膜相比，纳米光纤传感器的响应时间和检测极限都得到了提高。

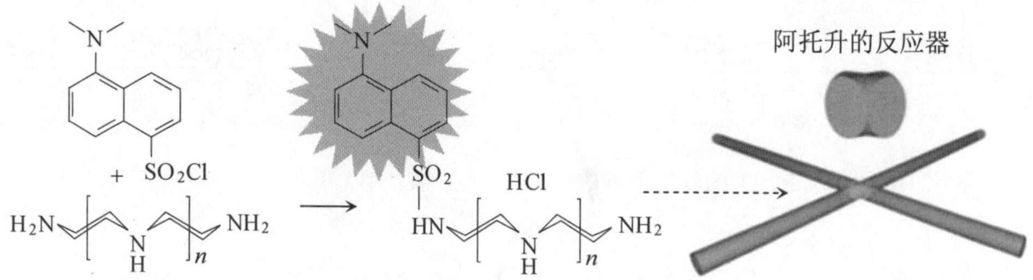

（a）阿托升反应器工作原理的示意图

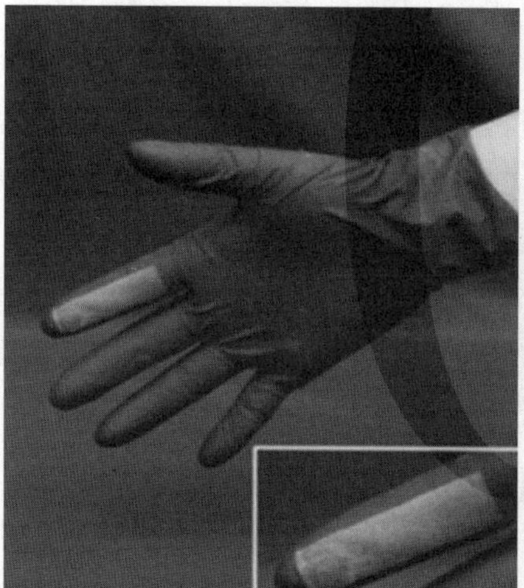

（b）阿托升反应堆垫制成的可穿戴传感器示例

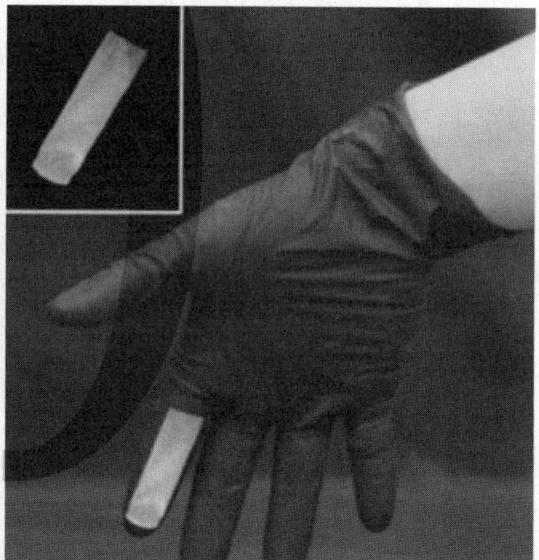

（c）在 365 nm 激发的纤维荧光的图像

图 6.20　电纺纤维的金属离子传感器原理及传感器实物垫子

（d）暴露于 Co^{2+}溶液中的图像

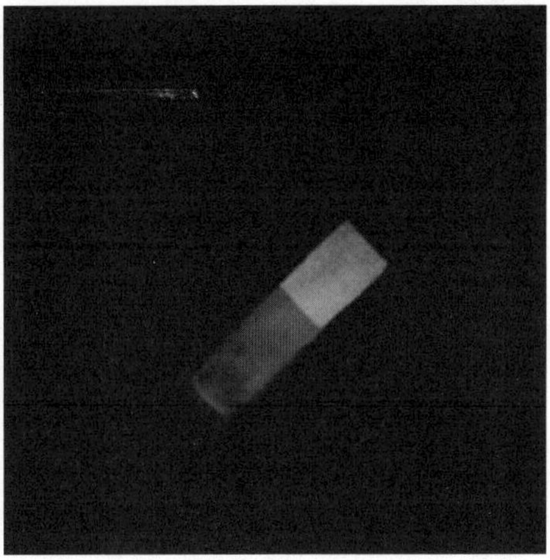
（e）图（d）对应的紫外线激发下的纤维荧光照片

续图 6.20

6.4.5　小结与展望

总之，光学活性电纺纳米纤维构成了一类新颖的、具有吸引人的特性的纳米材料。当前，由电纺纳米纤维制成的荧光垫可以通过许多方法来实现，这些方法基本上包括掺杂荧光 QD、有机 π 共轭分子和生物发色团，或者使用共轭聚合物。通常，这些纳米纤维的特点是与本体系统或薄膜相比具有增强的光学性能，并且还适合开发用于重金属离子、爆炸物和生物分子的光学传感器。迄今为止开发的系统在检测极限、时间响应和相关的低生产成本方面已展示出巨大的潜力。在许多情况下，开发的设备非常简单，它们不需要复杂的分析系统，并且表现出的性能可与其他成熟技术媲美。需要更多的研究工作来提高传感器的特异性并评估其干扰问题。

因此，对纤维成分、形态和光学性质的改进，以及扩展被检测系统的数量和类型，可为开辟新的光学传感方式。静电纺丝的纳米纤维的光物理现象丰富，但未得到很好的利用，在光学传感器领域具有巨大的潜力。

参 考 文 献

[1] FU R, ZHAO D, DING Y, et al. An amperometric l-tryptophan sensor platform based on electrospun tricobalt tetroxide nanoparticles decorated carbon nanofibers[J]. Sensors and Actuators B: Chemical, 2017, 241: 601-606.

[2] LIANG T, QIU Y, GAN Y, et al. Recent developments of high-resolution chemical imaging systems based on light-addressable potentiometric sensors (LAPSs)[J]. Sensors, 2019, 19(19): 4294.

[3] LI W T, ZHANG X D, GUO X. Electrospun Ni-doped SnO_2 nanofiber array for selective sensing of NO_2[J]. Sensors and Actuators B: Chemical, 2017, 244:509-521.

[4] VASSILYEV Y B, KHAZOVA O A, NIKOLAEVA N N. Kinetics and mechanism of glucose electrooxidation on different electrode-catalysts[J]. Journal of Electroanalytical Chemistry, 1985, 196(1): 105-125.

[5] LIU Y, DING Y, ZHANG Y C, et al. Pt/Au nanocorals, Pt nanofibers and Au microparticles prepared by electrospinning and calcination for nonenzymatic glucose sensing in neutral and alkaline environment[J]. Sensors and Actuators B: Chemical, 2012, 171: 954-961.

[6] JIANG L C, ZHANG W D. A highly sensitive nonenzymatic glucose sensor based on CuO nanoparticles modifified carbon nanotube electrode[J]. Biosensors and Bioelectronics, 2010, 25(6): 1402-1407.

[7] CABRITA J F, FERREIRA V C, MONTEIRO O C. Titanate nanofibers sensitized with nanocrystalline Bi_2S_3 as new electrocatalytic materials for ascorbic acid sensor applications[J]. Electrochemical Acta, 2014, 135: 121-127.

[8] ZHANG J, LU H, LU H, et al. Porous bimetallic Mo-W oxide nanofibers fabricated by electrospinning with enhanced acetone sensing performances[J]. Journal of Alloys and Compounds, 2019, 779: 531-542.

[9] MEHRABI P, HUI J, JANFAZA S, et al. Fabrication of SnO_2 Composite nanofiber-based gas sensor using the electrospinning method for tetrahydrocannabinol (THC) detection[J]. Micromachines, 2020, 11(2): 190.

[10] ALI M A, JIANG H, MAHAL N K, et al. Microfluidic impedimetric sensor for soil nitrate detection using graphene oxide and conductive nanofibers enabled sensing interface[J]. Sensors and Actuators B: Chemical, 2017, 239: 1289-1299.

[11] MUTHARANI B, RANGANATHAN P, CHEN S M. Temperature-reversible switched antineoplastic drug 5-fluorouracil electrochemical sensor based on adaptable thermo-sensitive microgel encapsulated PEDOT[J]. Sensors and Actuators B: Chemical, 2019, 304: 127361.

[12] HU L, LI Y. Improved acetone sensing properties of flat sensors based on Co-SnO_2 composite nanofibers[J]. Chinese Science Bulletin, 2011, 56(24):2644-2648.

[13] LI G. Ethanol sensing properties and reduced sensor resistance using porous Nb_2O_5-TiO_2 n-n junction nanofibers [J]. Sensors and Actuators B: Chemical, 2019, 283: 602-612.

[14] MOON J. Pd-doped TiO_2 nanofiber networks for gas sensor applications[J]. Sensors and Actuators B: Chemical, 2010, 149(1): 301-305

[15] KOU X. Superior acetone gas sensor based on electrospun SnO_2 nanofibers by Rh doping[J]. Sensors and Actuators B: Chemical, 2018, 256: 861-869.

[16] JI S. Controllable organic nanofiber network crystal room temperature NO_2 sensor[J]. Organic Electronics, 2013, 14(3): 821-826.

[17] JIANG H. Flexible and highly sensitive hydrogen sensor based on organic nanofibers decorated by Pd nanoparticles[J]. Sensors, 2019,19(6): 1290.

[18] UTARAT P, PHANICHPHANT S, SRIWICHAI S. Preparation of electrospun poly (acrylic acid)/multiwalled carbon nanotubes composite nanofiber for glucose detection[J]. Molecular Crystals and Liquid Crystals, 2019, 688(1): 114-121.

[19] PANG Z, YU J, LI D, et al. Free-standing TiO_2-SiO_2/PANI composite nanofibers for ammonia sensors[J]. Journal of Materials Science: Materials in Electronics, 2017, 29(5): 3576-3583.

[20] ABDALI H, HELI B, AJJI A. Stable and sensitive amino-functionalized graphene/ polyaniline nanofiber composites for room-temperature carbon dioxide sensing[J]. RSC Advances, 2019, 9(70): 41240-41247.

[21] YUN K M, SURYAMAS A B, ISKANDAR F, et al. Morphology optimization of polymer nanofiber for applications in aerosol particle filtration[J]. Separation & Purification Technology, 2010, 75(3): 340-345.

[22] LEE S Y, JANG D H, KANG Y O, et al. Cellular response to poly(vinyl alcohol) nanofibers coated with biocompatible proteins and polysaccharides[J]. Applied Surface Science, 2012, 258(18): 6914-6922.

[23] ALMUHAMED S, KHENOUSSI N, SCHACHER L, et al. Measuring of electrical properties of MWNT-reinforced PAN nanocomposites[J]. Journal of Nanomaterials, 2012, 2012: 1-7.

[24] KHENOUSSI N, ALMUHAMED S, BONNE M, et al. Electrospinning of PAN nanofibers filled with SBA-15 type ordered mesoporous silica[C]. International Symposium on Fibers Interfacing the World, 2013, 54: 71-78.

[25] HEKMATI A H, KHENOUSSI N, NOUALI H, et al. Effect of nanofiber diameter on water absorption properties and pore size of polyamide-6 electrospun nanoweb[J]. Textile Research Journal, 2014, 84(19): 2045-2055.

[26] HEKMATI A H, RASHIDI A, GHAZISAEIDI R, et al. Effect of needle length, electrospinning distance, and solution concentration on morphological properties of polyamide-6 electrospun nanowebs[J]. Textile Research Journal, 2013, 83(14): 1452-1466.

[27] THAVASI V G, SINGH G, RAMAKRISHNA S. Electrospun nanofibers in energy and environmental applications[J]. Energy & Environmental Science, 2008, 1: 205-211.

[28] BANICA F G. Chemical sensors and biosensors: fundamentals and applications[M]. Chichester: Wiley, 2012.

[29] VO-DINH T, CULLUM B. Biosensors and biochips: advances in biological and medical diagnostics[J]. Fresenius Journal of Analytical Chemistry, 2000, 366(6-7): 540-551.

[30] ZHAO M, ZHOU Y, CAI B. The application of porous ZnO 3D framework to assemble enzyme for rapid and ultrahigh sensitive biosensors[J]. Ceramics

International, 2013, 39(8): 9319-9323.

[31] WANG Z G, WAN L S, LIU Z M, et al. Enzyme immobilization on electrospun polymer nanofibers: an overview[J]. Journal of Molecular Catalysis B Enzymatic, 2009, 56(4): 189-195.

[32] BOOTHAPANDI M, RAMANIBAI R. Immunomodulatory effect of natural flavonoid chrysin (5, 7-dihydroxyflavone) on LPS stimulated RAW 264.7 macrophages via inhibition of NF-κB activation[J]. Process Biochemistry, 2019, 84(SEP.): 186-195.

[33] YE P, XU Z K, WU J, et al. Nanofibrous poly(acrylonitrile-co-maleic acid) membranes functionalized with gelatin and chitosan for lipase immobilization[J]. Biomaterials, 2006, 27(22): 4169-4176.

[34] WANG Z G, KE B B, XU Z K. Covalent immobilization of redox enzyme on electrospun nonwoven poly(acrylonitrile-co-acrylic acid) nanofiber mesh filled with carbon nanotubes: a comprehensive study[J]. Biotechnology and Bioengineering, 2007, 97(4): 708-720.

[35] WAN L S, KE B B, WU J, et al. Catalase immobilization on electrospun nanofibers: effects of porphyrin pendants and carbon nanotubes[J]. Journal of Physical Chemistry C, 2007, 111(38): 14091-14097.

[36] JOSE M V, SHARON M, HIRONOBU M, et al. Direct electron transfer in a mediator-free glucose oxidase-based carbon nanotube-coated biosensor[J]. Carbon, 2017, 50(11): 4010-4020.

[37] FU G, YUE X, DAI Z, et al. Glucose biosensor based on covalent immobilization of enzyme in sol-gel composite film combined with Prussian blue/carbon nanotubes hybrid[J]. Biosensors & Bioelectronics, 2011, 26(9): 3973-3976.

[38] SHEN J, YANG X, ZHU Y, et al. Gold-coated silica-fiber hybrid materials for application in a novel hydro-gen peroxide biosensor[J]. Biosensors and Bioelectronics, 2018, 34: 132-136.

[39] ZHU H, DU M L, ZHANG M, et al. Facile fabrication of AgNPs/(PVA/PEI) nanofibers: high electrochemical efficiency and durability for biosensors[J].

Biosensors and Bioelectronics, 2018, 49: 210-215.

[40] EL-AASSAR, Al-DEYAB S S, KENAWY E R, et al. Covalent immobilization of β-galactosidase onto electrospun nanofibers of poly(AN-co-MMA) copolymer[J]. Journal of Applied Polymer Science, 2017, 127: 1873-1884.

[41] WANG Y, YOU L H. Enzyme immobilization to ultra fine cellulose fibers via amphiphilic polyethylene glycol spacers[J]. Journal of Polymer Science Part A: Polymer Chemistry, 2004, 42: 4289-4299.

[42] JIA H, ZHU G, VUGRINOVICH B, et al. Enzyme-carrying polymeric nanofibers prepared via electrospinning for use as unique biocatalysts[J]. Biotechnology Progress, 2002, 18: 1027-1032.

[43] KIM B C. Preparation of biocatalytic nanofibers with high activity and stability via enzyme aggregate coating on polymer nanofibers[J]. Nanotechnology, 2005, 16: S382.

[44] KIM B C, NAIR S, KWAK J H, et al. Improving biocatalytic activity of enzyme-loaded nanofibers by dispersing entangled nanofiber structure[J]. Biomacromolecules, 2007, 8: 1266-1270.

[45] UZUN S D, KAYACI F, UYAR T, et al. Bioactive surface design based on functional composite electrospun nanofibers for biomolecule immobilization and biosensor applications[J]. ACS Applied Materials and Interfaces, 2018, 6: 5235-5243.

[46] KUMAR P, GUPTA A, DHAKATE S R, et al. Covalent immobilization of xylanase produced from Bacillus pumilus SV-85S on electrospun polymethyl methacrylate nanofiber membrane[J]. Biotechnology and Applied Biochemistry, 2013, 60: 162-169.

[47] KIM T G, PARK G T. Surface functionalized electrospun biodegradable nanofibers for immobilization of bioactive molecules[J]. Biotechnology Progress, 2006, 22: 1108-1113.

[48] BAŞTÜRK E, DEMIR S, DAN Z, et al. Covalent immobilization of α-amylase onto thermally crosslinked electrospun PVA/PAA nanofibrous hybrid mem-branes[J]. Journal of Applied Polymer Science, 2013, 127: 349-355.

[49] LI J, MEI H, ZHENG W, et al. A novel hydrogen peroxide biosensor based on hemoglobin-collagen-CNTs composite nanofibers[J]. Colloids and Surfaces B: Bio interfaces, 2014, 118: 77-82.

[50] WU J, FAN Y. Sensitive enzymatic glucose biosensor fabricated by electrospinning composite nanofibers and electrodepositing Prussian blue film[J]. Journal of Electroanalytical Chemistry, 2013, 694: 1-5.

[51] WANG Z G, WANG J Q, XU Z K. Immobilization of lipase from Candida rugosa on electrospun polysulfone nanofibrous membranes by adsorption[J]. Journal of Molecular Catalysis B: Enzymatic, 2006, 42: 45-51.

[52] JIANG X, XIN J, CHENG P, et al. Preparation of coaxial electrospun poly [bis(p-methyl phenoxy)] phosphazene nanofiber membrane for enzyme immobilization[J]. International Journal of Molecular Sciences, 2012, 13: 14136-14148.

[53] CHEN H, SHIEN Y L. Enzyme immobilization on ultrafine cellulose fibers via poly (acrylic acid) electrolyte grafts[J]. Biotechnology and Bioengineering, 2005, 90: 405-413.

[54] SAKAI S, LIU Y P, YAMAGUCHI T, et al. Immobilization of pseudomonas cepacia lipase onto electrospun poly-acrylonitrile fibers through physical adsorption and application to transesterification in nonaqueous solvent[J]. Biotechnology Letters, 2010, 32: 1059-1062.

[55] WU L, YUAN X Y, SHENG J. Immobilization of cellulase in nanofibrous PVA mem-branes by electrospinning[J]. Journal of Membrane Science, 2005, 250: 167-173.

[56] WANG Y, HSIEH Y L. Immobilization of lipase enzyme in polyvinyl alcohol (PVA) nanofibrous membranes[J]. Journal of Membrane Science, 2018, 309: 73-81.

[57] REN G, XU X, LIU Q, et al. Electrospun poly(vinyl alcohol)/glucose oxidase bio composite membranes for biosensor applications[J]. Reactive and Functional Polymers, 2006, 66: 1559-1564.

[58] HERRICKS T E, KIM S H, KIM J, et al. Direct fabrication of enzyme-carrying

polymer nanofibers by electrospinning[J]. Journal of Materials Chemistry, 2005, 15: 3241-3245.

[59] PATEL A C, LI S, YUAN J-M, et al. In situ encapsulation of horseradish per-oxidase in electrospun porous silica fibers for potential biosensor applications[J]. Nano Letters, 2006, 6: 1042-1046.

[60] JIANG H, HU Y, LI YAN, et al. A facile technique to prepare biodegradable coaxial electrospun nano-fibers for controlled release of bioactive agents[J]. Journal of Controlled Release, 2005, 108: 237-243.

[61] DING Y, WANG Y, LI B, et al. Electrospun hemoglobin microbelts based biosensor for sensitive detection of hydrogen peroxide and nitrite[J]. Biosensors and Bioelectronics, 2010, 25: 2009-2015.

[62] MOORCROFT M J, DAVIS J, COMPTON R G. Detection and determination of nitrate and nitrite: a review[J]. Talanta, 2001, 54: 785-803.

[63] ARECCHI A, SCAMPICCHIO M, DRUSCH S. Nanofibrous membrane based tyrosinase-biosensor for the detection of phenolic compounds[J]. Analytica Chimica Acta, 2010, 659: 133-136.

[64] LIN S H, JUANG R S. Adsorption of phenol and its derivatives from water using synthetic resins and low-cost natural adsorbents: a review[J]. Journal of Environmental Management, 2009, 90: 1336-1349.

[65] IGBINOSA E O, ODJADFARE E E, CHIGOR V N, et al. Toxicological profile of chlorophenols and their derivatives in the environment: the public health perspective[J]. The Scientific World Journal, 2013: 1-11.

[66] LIU J, NIU J, YIN L, et al. In situ encapsulation of laccase in nanofibers by electrospinning for development of enzyme biosensors for chlorophenol monitoring[J]. Analyst, 2011, 136: 4802-4808.

[67] SUN M, DING B, LIN J, et al. Three-dimensional sensing membrane functionalized quartz crystal microbalance biosensor for chloramphenicol detection in real time[J]. Sensors and Actuators B: Chemical, 2011, 160: 428-434.

[68] YUNIS A. Chloramphenicol toxicity: 25 years of research[J]. The American Journal

of Medicine, 1989, 87: 44-48.

[69] LUO Y, NARTKER S, WIEDERODER M, et al. Novel biosensor based on electrospun nanofiber and magnetic nanoparticles for the detection of E. coli O157: H7[J]. IEEE Transactions on Nanotechnology, 2012, 11: 676-681.

[70] NIU J, XU J, DAI Y, et al. Immobilization of horseradish peroxidase by electrospun fibrous mem-branes for adsorption and degradation of pentachlorophenol in water[J]. Journal of Hazardous Materials, 2019, 246: 119-125.

[71] DAI Y, YIN L, NIU J. Laccase-carrying electrospun fibrous membranes for adsorption and degradation of PAHs in shoal soils[J]. Environmental Science and Technology, 2011, 45:10611-10618.

[72] SATHISHKUMAR P, CHAE J C, UNNITHAN A R, et al. Laccase-poly (lactic-co-glycolic acid) (PLGA) nanofiber: Highly stable, reusable, and efficacious for the transformation of diclofenac[J]. Enzyme and Microbial Technology, 2012, 51: 113-118.

[73] ILIC S, DRMIC D, FRANJIC S, et al. Pentadecapeptide BPC 157 and its effects on a NSAID toxicity model: diclofenac-induced gastrointestinal, liver, and encephalopathy lesions[J]. Life Sciences, 2018, 88: 535-542.

[74] LUANA P, CAMPOSEO A, TEKMEN C, et al. Industrial upscaling of electrospinning and applications of polymer nanofibers: a review[J]. Macromolecular Materials and Engineering, 2013, 298: 504-520.

[75] WANG X, CHRISTOPHER D, LEE S H, et al. Electrospun nanofibrous membranes for highly sensitive optical sensors[J]. Nano Letters, 2002, 2(11): 1273-1275.

[76] CAMPOSEO A, PERSANO L, PISIGNANO, et al. Light-emitting electrospun nanofibers for nanophotonics and optoelectronics[J]. Macromolecular Materials & Engineering, 2013, 298(5): 487-503.

[77] LIU H, EDEL J B, BELLAN L M, et al. Electrospun polymer nanofibers as subwavelength optical waveguides incorporating quantum dots[J]. Small, 2010, 2(4): 495-499.

[78] LU X, WANG C, WEI Y. One-dimensional composite nanomaterials: synthesis by

electrospinning and their applications[J]. Small, 2009(5): 2349-2370.
[79] WEI S, SAMPATJI J, GUO Z, et al. Nanoporous poly(methyl methacrylate)-quantum dots nanocomposite fibers toward biomedical applications[J]. Polymer, 2011, 52(25): 5817-5829.
[80] SHIRASAKI Y, SUPRAN G J, BAWENDI M G, et al. Emergence of colloidal quantum-dot light-emitting technologies[J]. Nature Photonics, 2013, 7(1): 13-23.
[81] VOSSMEYER T, KATSIKAS L, GIERSIG M, et al. CdS nanoclusters: synthesis, characterization, size dependent oscillator strength, temperature shift of the excitonic transition energy, and reversible absorbance shift[J]. Chemical Physics, 1994, 98: 7665-7673.
[82] BENEDETTO F D, CAMPOSEO A, PERSANO L, et al. Light-emitting nanocomposite CdS-polymer electrospun fibres via in situ nanoparticle generation[J]. Nanoscale, 2011, 3: 4234-4239.
[83] LU X, ZHAO Y, WANG C. Fabrication of PbS nanoparticles in polymer-fiber matrices by electrospinning[J]. Advanced functional materials, 2005, 17:2485-2488.
[84] PERSANO L, CAMPOSEO A, BENEDETTO F D, et al. CdS-polymer nanocomposites and light-emitting fibers by in situ electron-beam synthesis and lithography[J]. Advanced Materials, 2012, 24: 5320-5326.
[85] PERSANO L, MOLLE S, GIRARDO S, et al. Soft nanopatterning on light-emitting inorganic-organic composites[J]. Advanced Functional Materials, 2008, 18: 2692-2698.
[86] CAMPOSEO A, POLO M, NEVES A A R, et al. Multi-photonin situ synthesis and patterning of polymer-embedded nanocrystals[J]. Journal of Materials Chemistry, 2012, 22: 9787-9793.
[87] NER Y, GROTE J G, STUART J A, et al. Enhanced fluorescence in electrospun dye-doped DNA nanofibers[J]. Soft Matter, 2008, 4: 1448-1453.
[88] PAGLIARA S, CAMPOSEO A, POLINI A, et al. Electrospun light-emitting nanofibers as excitation source in microfluidic devices[J]. Lab on a Chip, 2009, 9: 2851-2856.

[89] CAMPOSEO A, BENEDETTO F D, STABILE R, et al. Laser emission from electrospun polymer nanofibers[J]. Small, 2009, 5: 562-566.

[90] DAS A J, LAFARGUE C, LEBENTAL M, et al. Three-dimensional microlasers based on polymer fibers fabricated by electrospinning[J]. Applied Physics Letters, 2011, 99: 263303.

[91] CUCCHI I, SPANO F, GIOVANELLA U, et al. Fluorescent electrospun nanofibers embedding dye-loaded zeolite crystals[J]. Small, 2007, 3: 305-309.

[92] SPARKS J S, SCHELLY R C, SMITH W L, et al. The covert world of fish biofluorescence: a phylogenetically widespread and phenotypically variable phenomenon[J]. PLoS One, 2017, 9: e83259.

[93] TSIEN R Y. The green fluorescent protein[J]. Annu Review of Biochemistry, 1998, 67: 509-544.

[94] BIZZARRI R, SERRESI M, LUIN S, et al. Green fluorescent protein based pH indicators for in vivo use: a review. [J]. Analytical and Bioanalytical Chemistry, 2009, 393: 1107-1122.

[95] YARIN A L, ZUSSMAN E, WENDORFF J H, et al. Material encapsulation and transport in core-shell micro/nanofibers, polymer and carbon nanotubes and micro/nanochannels[J]. Journal of Materials Chemistry, 2012, 17: 2585-2599.

[96] DROR Y, SALALHA W, AVRAHAMI R, et al. One-step production of polymeric microtubes by co-electrospinning[J]. Small, 2007, 3: 1064-1073.

[97] MCGEHEE M D, HEEGER A J. Semiconducting (conjugated) polymers as materials for solid state lasers[J]. Advanced Materials, 2000, 12: 1655-1668.

[98] KUO C C, LIN C H, CHEN W C, et al. Morphology and photophysical properties of light-emitting electrospun nanofibers prepared from poly(fluorene) derivative/PMMA blends[J]. Macromolecules, 2007, 40: 6959-6966.

[99] ZHONG W, LI F, CHEN L, et al. A novel approach to electrospinning of pristine and aligned MEH-PPV using binary solvents[J]. Journal of Materials Chemistry, 2016, 22: 5523-5530.

[100] CAMPOSEO A, GREENFELD I, TANTUSSI F, et al. Local mechanical properties

of electrospun fibers correlate to their internal nanostructure[J]. Nano Letters, 2013, 13: 5056-5062.

[101] FASANO V, POLINI A, MORELLO G, et al. Bright light emission and waveguiding in conjugated polymer nanofibers electrospun from organic salt added solutions[J]. Macromolecules, 2013, 46: 5935-5942.

[102] LI D, BABEL A, JENEKHE S A, et al. Nanofibers of conjugated polymers prepared by electrospinning with a two-capillary spinneret[J]. Advanced Materials, 2004, 16:2062-2066.

[103] LI D, XIA Y N. Electrospinning of nanofibers: reinventing the wheel[J]. Advanced Materials, 2004, 16: 1151-1170.

[104] DAN L, GONG O, MCCANN J T, et al. Collecting electrospun nanofibers with patterned electrodes[J]. Nano Letters, 2005, 5: 913-916.

[105] SUN D, CHANG C, LI S, et al. Near-field electrospinning[J]. Nano Letters, 2006, 6: 839-842.

[106] BISHT G S, CANTON G, MIRSEPASSI A, et al. Controlled continuous patterning of polymeric nanofibers on three-dimensional substrates using low-voltage near-field electrospinning[J]. Nano Letters, 2011, 11: 1831-1837.

[107] CHANG C, LIMKRAILASSIRI K, LIN L. Continuous near-field electrospinning for large area deposition of orderly nanofiber patterns[J]. Applied Physics Letters, 2008, 93: 123111.

[108] CHANG C, TRAN V H, WANG J, et al. Direct-write piezoelectric polymeric nanogenerator with high energy conversion efficiency[J]. Nano Letters, 2010, 10: 726-731.

[109] CAMILLO D, FASANO V, RUGGIERI F, et al. Near-field electrospinning of light-emitting conjugated polymer nanofibers[J]. Nanoscale, 2017, 5: 11637-11642.

[110] JOSHUA C B, MATTHEW C, BRAZARD J, et al. Conformation and energy transfer in single conjugated polymers[J]. Accounts of Chemical Research, 2012, 45: 1992-2001.

[111] NGUYEN T Q, DOAN V, SCHWARTZ B J, et al. Conjugated polymer aggregates

in solution: control of interchain interactions[J]. Chemical Physics, 1999, 110: 4068-4078.

[112] RICHARD-LACROIX M, PELLERIN C. Molecular orientation in electrospun fibers: from mats to single fibers[J]. Macromolecules, 2013, 46: 9473-9493.

[113] MORELLO G, POLINI A, GIRARDO S, et al. Enhanced emission efficiency in electrospun polyfluorene copolymer fibers[J]. Applied Physics Letters, 2013, 102: 211911.

[114] PAGLIARA S, CAMPOSEO A, CINGOLANI R, et al. Hierarchical assembly of light-emitting polymer nanofibers in helical morphologies[J]. Applied Physics Letters, 2009, 95: 263301.

[115] CAMPOY-QUILES M, ISHII Y, SAKAI H, et al. Highly polarized luminescence from aligned conjugated polymer electrospun nanofibers[J]. Applied Physics Letters, 2008, 92: 213305.

[116] ISHII Y, MURATA H. True photoluminescence spectra revealed in electrospun light-emitting single nanofibers[J]. Chemistry of Materials, 2012, 22: 4695-4703.

[117] GREENHAM N C, SAMUEL I D W, HAYES G R, et al. Measurement of absolute photoluminescence quantum efficiencies in conjugated polymers[J]. Chemical Physics Letters, 1995, 241: 89-96.

[118] HWANG I, SCHOLES G D. Electronic energy transfer and quantum-coherence in π-conjugated polymers[J]. Chemistry of Materials, 2011, 23: 610-620.

[119] PISIGNANO D. Polymer nanofibers[J]. Royal Society of Chemistry, 2019: 50-119.

[120] NGUYEN T Q, WU J, DOAN V, et al. Control of energy transfer in oriented conjugated polymer-mesoporous silica composites[J]. Science, 2000, 288:652-656.

[121] CHEN C A, WALLACE J U, WEI K H, et al. Light-emitting organic materials with variable charge injection and transport properties[J]. Chemistry of Materials, 2006, 18: 204-213.

[122] LEE C C, LAI S Y, SU W B, et al. Relationship between the microstructure development and the photoluminescence efficiency of electrospun poly (9,9-dioctylfluorene-2,7-diyl) fibers[J]. Journal of Physical Chemistry C, 2017, 117:

20387-20396.

[123] BELLAN L M, CRAIGHEAD H G. Molecular orientation in individual electrospun nanofibers measured via polarized Raman spectroscopy[J]. Polymer, 2008, 49: 3125-3129.

[124] STACHEWICZ U, BAILEY R J, WANG W, et al. Size dependent mechanical properties of electrospun polymer fibers from a composite structure[J]. Polymer, 2012, 53: 5132-5137.

[125] KUO C C, WANG C T, CHEN W C. Highly-aligned electrospun luminescent nanofibers prepared from polyfluorene/PMMA blends: fabrication, morphology, photophysical properties and sensory applications[J]. Macromolecular Materials and Engineering, 2008, 293: 999-1008.

[126] PAGLIARA S, VITIELLO M S, CAMPOSEO A, et al. Optical anisotropy in single light-emitting polymer nanofibers[J]. Journal of Physical Chemistry C, 2011, 115: 20399-20405.

[127] DING B, WANG M, WANG X, et al. Electrospun nanomaterials for ultrasensitive sensors[J]. Materials Today, 2010, 13: 16-27.

[128] GUO X, YING Y, TONG L. Photonic nanowires: from subwavelength waveguides to optical sensors[J]. Accounts of Chemical Research, 2014, 47: 656-666.

[129] WANG M, MENG G, HUANG Q, et al. Electrospun 1,4-DHAQ-doped cellulose nanofiber films for reusable fluorescence detection of trace Cu^{2+} and further for Cr^{3+}[J]. Journal of Environmental Science and Technology, 2012, 46: 367-373.

[130] ANZENBACHER P, LI F, PALACIOS M A, et al. Toward wearable sensors: fluorescent attoreactor mats as optically encoded cross-reactive sensor arrays[J]. Angewandte Chemie, 2012, 124: 2395-2398.

[131] ORRIACH-FERNA' NDEZ F J, MEDINA-CASTILLO A L, DIAZ-GOMEZ J E, et al. A sensing microfibre mat produced by electrospinning for the turn-on luminescence determination of Hg^{2+} in water samples[J]. Sensors and Actuators B: Chemical, 2018, 195: 8-14.

[132] HENG L, WANG B, ZHANG Y, et al. Sensing mechanism of nanofibrous membranes for fluorescent detection of metal ion[J]. Journal of Nanoscience and Nanotechnology, 2012, 12: 8443-8447.

[133] ANZENBACHER P, PALACIOS M A. Polymer nanofibre junctions of attolitre volume serve as zeptomole-scale chemical reactors[J]. Nature Chemistry, 2009, 1: 80-86.

第 7 章　纳米纤维在其他领域的应用

经过几十年的发展，静电纺丝技术越发成熟，制造成本低廉、合成工艺简单、合成纤维种类繁多，被认为是最具大规模生产纳米纤维的重要方法之一。通过静电纺丝技术可以获得几纳米到几微米的超细纤维，电纺纤维除了具有微观一维尺寸和高比表面积等特征外，还具有稳定的力学性能、孔径小、均一性好和非常高的空隙率等特点，使其在个体防护和电磁屏蔽等领域显示出巨大的应用潜力。本章重点介绍了静电纺丝技术在个体防护和电磁屏蔽领域的应用及应该关注的发展方向，并指出这项技术在这些领域中未来的转化方向。

7.1　个人防护装备

7.1.1　个人防护装备的种类及其特点

防护装备可以定义为保护人体免受外部环境威胁的一种专业装备，例如子弹化学和生物制剂、火、冷和热等；防护装备广泛应用在各专业领域，诸如军事、体育、医院和工业领域等。纳米技术或纳米纤维复合材料可赋予防护装备高强度和更好的耐用性，并可开发诸多功能性纺织品，与传统防护装备相比，纳米纤维防护装备不仅具有高强度、舒适、透气、隔离效率高、质量轻等特点，还具有分解化学物质、自愈合等功能，目前正广泛应用于各种防护设备（图 7.1）。

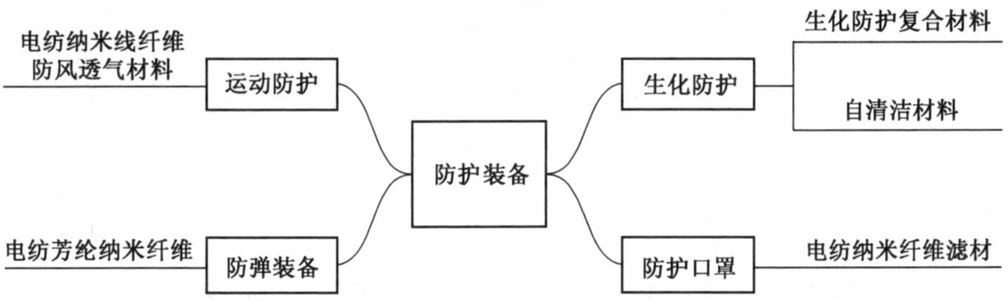

图 7.1　纳米纤维在个人防护装备中的应用

7.1.2　运动防护装备

对于运动防护装备而言，最主要的性能为透气性、舒适性和防护功能，纳米纤维织物恰能满足上述需求。如图 7.2 所示，纳米纤维膜对水蒸气扩散阻力最小，与微孔聚四氟乙烯（PTFE）膜的流动阻力相当。

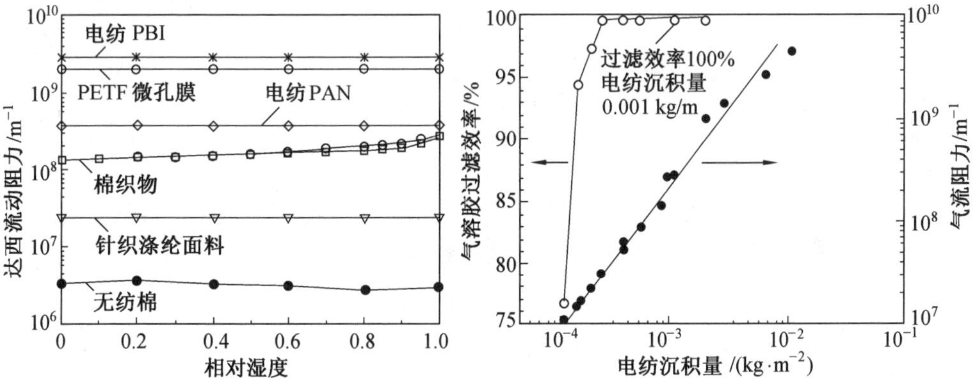

（a）不同纤维织物的达西流动阻力　　（b）电纺纤维的过滤效率和气流阻力

图 7.2　纤维的气流阻力

电纺纳米纤维网由于其高比表面积和相互连接的孔隙率，在纳米技术应用中非常重要。在这项研究中，Anzenbacher 研究了静电纺丝持续时间对聚氨酯（PU）电纺网某些物理和机械性能的影响，以用于潜在的应用，例如防护服和膜。Lee 和 Obendorf 研究了静电纺丝纳米纤维层状织物的传输特性。他们将聚氨酯纳米纤维纺在无纺布上（图 7.3（a）～（c）），结果表明，与无纺布相比，这种复合织物的透气性能和防护性能都得到了有效提高。Yuan 研究了在各种测试条件下纤维直径和纤维膜孔隙率对纳米纤维透气性的影响。通过研究发现，增加电纺纤维的直径和膜的孔径可以提高透气性能。此外，进一步研究发现样品的透气性能随着湿度的降低或空气流速的增大显著提高。Hong 通过控制纳米纤维膜的孔径，提高了用于户外运动服面料的聚氨酯纳米纤维膜的透气性。Gorji 等发现聚氨酯纳米纤维膜的透气性遵循菲克第一扩散定律（图 7.3（d））。

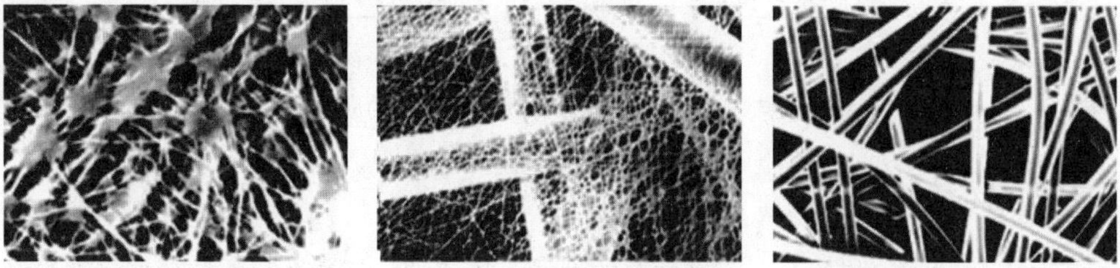

（a）微孔膜　　（b）层状织物体系　　（c）纺黏非织造布

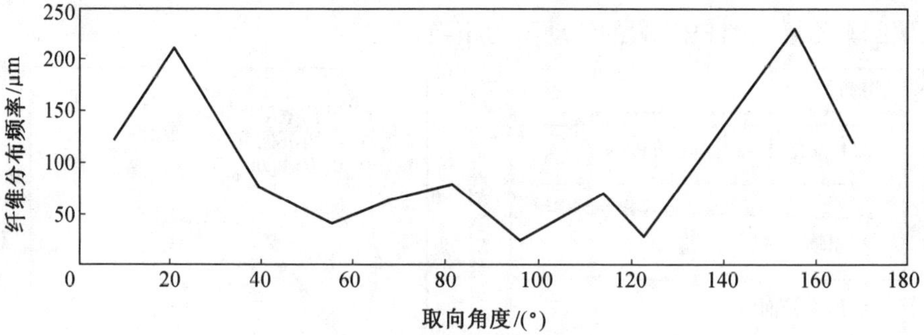

（d）典型的纤维取向分布

图 7.3　膜和纳米纤维网的 SEM 图及取向分布

通过调节纺丝的工艺能够有效赋予材料一些新的性能，静水压力是评估防水透气膜性能的一个重要指标，随着静电纺丝时间的增加，聚氨酯纳米纤维膜厚度增加，膜的静水压力增加但透气性基本不变（图 7.4（a））。Bagherzadeh 等将其制备的多层纳米纤维织物与 Gortex™商业保护性织物比较发现，采用纳米纤维膜取代 PTFE 后不仅具有增强的防风性能和防水性能，而且还提高了膜的透气性，还可以通过改变纳米纤维密度，开发不同防护等级的膜材（图 7.4（b））。不仅如此，研究人员发明的形状记忆聚氨酯电纺纳米纤维已成功应用于透气防水膜，这种膜具有形状记忆功能，其热量和水分的传输效率随人体温度变化而变化（图 7.4（c））。

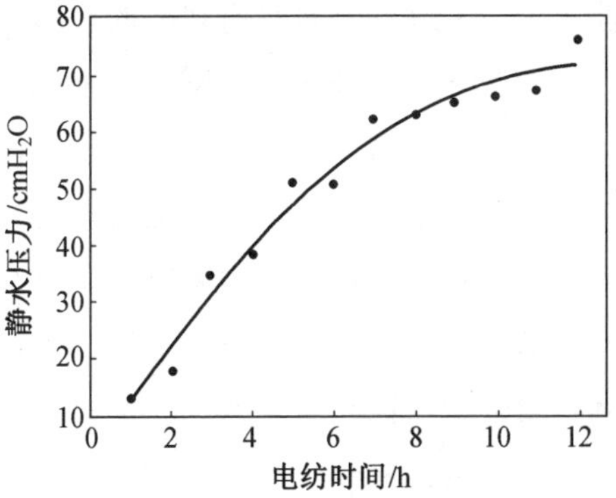

（a）电纺时间对静水压力的影响

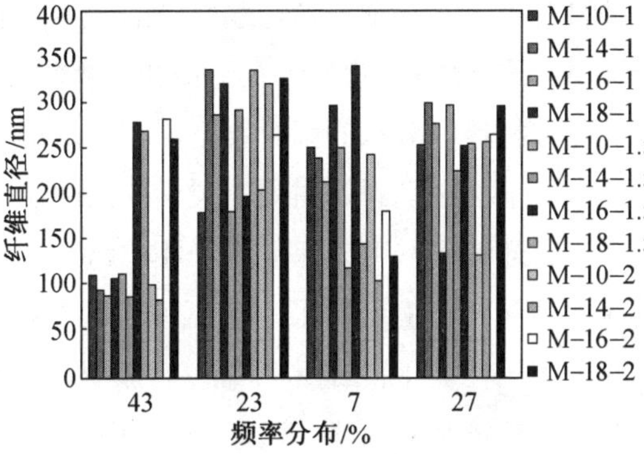

（b）不同样品的纳米纤维直径直方图

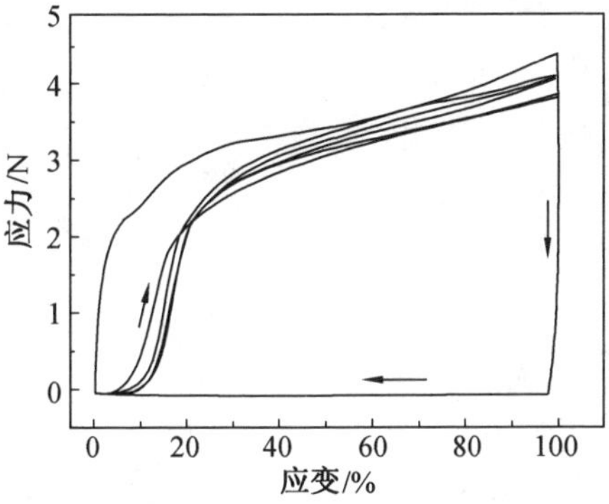

（c）聚氨酯电纺纳米纤维应力-应变曲线

图 7.4　纤维膜的静水压力、纤维直径及力学性能

（1 cmH_2O=0.980 665 Pa）

7.1.3　化学和生物防护装备

在军事行动、恐怖袭击或事故中，人们受到有毒物质污染的威胁。因此，为了保护在危险环境条件下士兵和专业工作人员的健康，需要多功能的防护材料来防护和净化有毒的生物和化学物质。纳米纤维织物孔径小能有效防护气溶胶形式的生物化学毒物，良好的透气性又能保证人体的舒适。Gibson 等直接将纳米纤维涂料涂布在含有活性炭成分聚氨酯泡沫上，研究发现样品的气流阻力和过滤性能与纳米线纤维涂层的质量有关，当纳米纤维涂层的质量达到 1 $g \cdot m^{-2}$ 时可以完全防止气溶胶颗粒的渗透。Chen 等开发了一种聚合物-聚合物纳米复合膜，与目前化学防护服使用的

标准材料相比，该复合膜在水/试剂渗透性方面高出 10 倍，并且具有较高的水蒸气透过率（透气性），此外，在干燥和水合状态下均具有与基质膜相似的高机械硬度。这种复合材料由疏水的聚合物基材和亲水的聚电解质纳米离子凝胶组成，疏水主体基质提供了机械坚固、耐用、灵活的阻隔层，而聚电解质提供了高度透水（透气）的膜，其连续形态可确保在离子相中跨膜运输。这种复合材料中的纳米凝胶相可以响应电刺激而收缩可扩张，从而可通电来调控防护等级（图 7.5（a）、（b））。Gorji 等开发了一种多孔膜/纳米纤维复合膜材，这种致密的膜可防止农药渗透，同时可通过溶液扩散机制使水蒸气通过，从而提供良好的透气性和舒适度（图 7.5（c）～（e））。

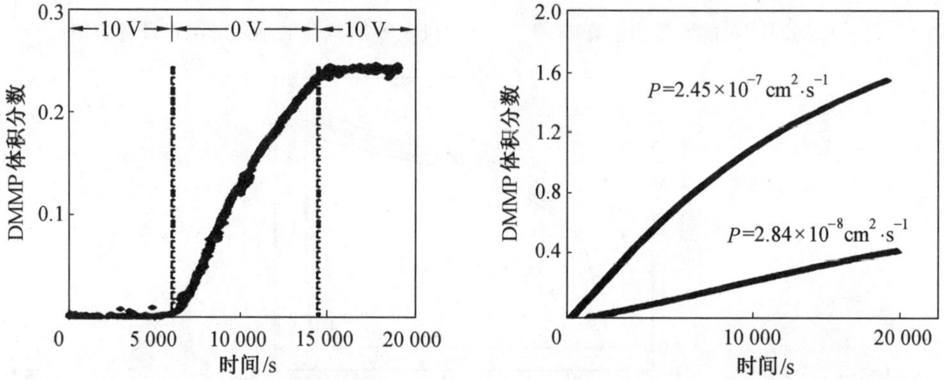

（a）复合材料的电敏运输有损伤的 PETE 膜　　（b）无损伤及穿透原始的 PETE 膜

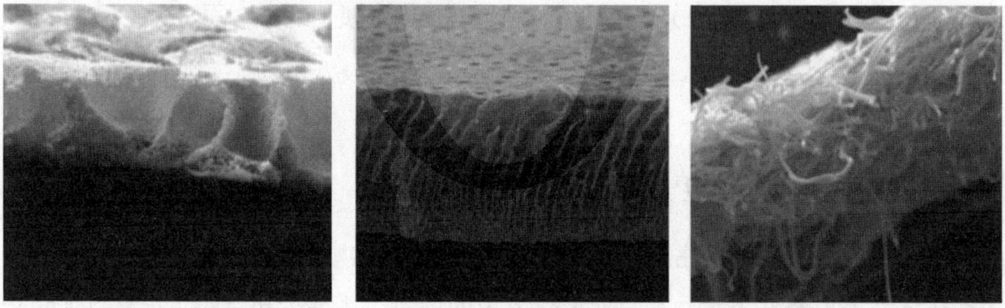

（c）PU 多孔膜　　（d）聚砜多孔膜　　（e）填充有离子水凝胶

图 7.5　防护化学毒物的电纺纤维膜（DMMP 为甲基膦酸二甲酯）

在纳米纤维中嵌入纳米颗粒（Ag、MgO、Ni、Ti 等）可使纳米纤维具备催化分解有毒有机物的自清洁功能，这种具有自清洁功能的纳米纤维织物可用于制作生化防护服。Liu 等以水和 HCOOH 为溶剂采用静电纺丝制备了四种尼龙-6/多金属盐（$H_{3+n}PMo_{12-n}V_nO_{40}\cdot xH_2O$ 和 $H_3PW_{12}O_{40}\cdot xH_2O$）复合纳米纤维膜，研究了复合纤维膜对芥子毒气（HD）的降解能力。研究发现，这种膜对芥子毒气有良好的净化功能，降解率高达 41.55%。纳米纤维膜同时表现出对气溶胶的良好过滤性能。制备的复合纳米纤维膜在 HD 的原位降解中显示出潜在的应用前景，并且可以用作保护布（图 7.6）。

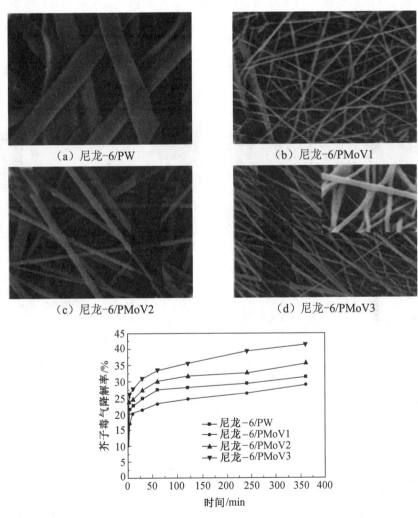

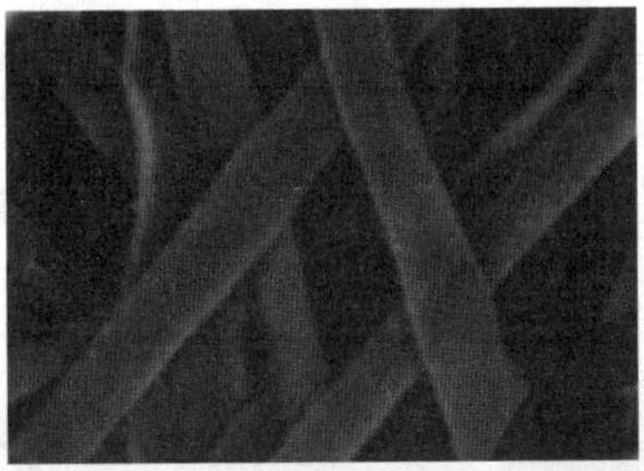

（a）尼龙-6/PW

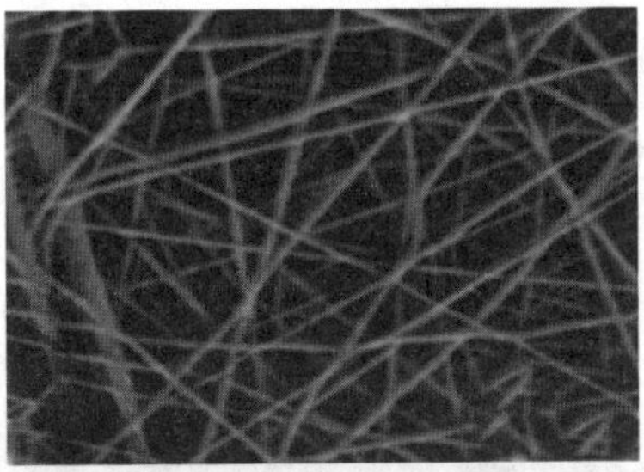

（b）尼龙-6/PMoV1

（c）尼龙-6/PMoV2

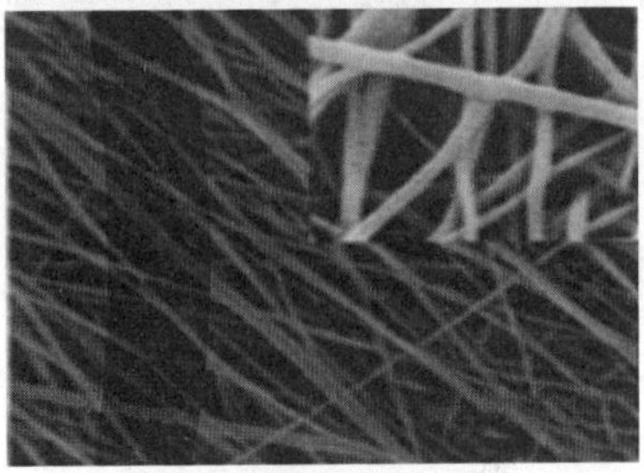

（d）尼龙-6/PMoV3

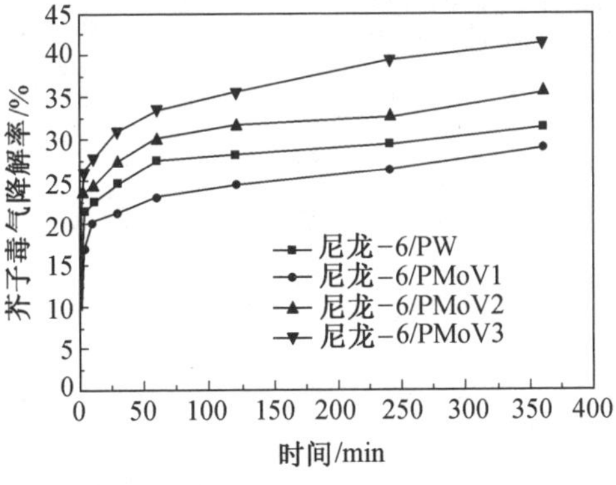

（e）芥子毒气在尼龙-6/POMs 纳米纤维膜上的降解率

图 7.6　纤维膜的 SEM 图像和芥子毒气在尼龙-6/POMs 纳米纤维膜上的降解率

根据前人报道，通过环糊精和邻碘代苯甲酸对纳米纤维膜进行功能化，可使纳米纤维膜具备分解剧毒物质对氧磷的功能，这种材料可用于农业防护服。Sundarrajan等用含 MgO 的不同聚合物制备了纳米线、纤维膜并研究了其对神经毒剂对氧磷的分解效果（图 7.7）。研究表明，不同膜的净化效果按照从小到大的顺序：PVC-MgO＜PVDF＜PSU＜PVDF-MgO＜PSU-MgO。他们还比较了掺入 PSU 膜中的不同纳米颗粒对净化效果从小到大的顺序的影响：木炭（1）＜PSU-Al_2O_3（1.5）＜PSU-MgO（2.1）＜Al_2O_3 纳米颗粒（2.8）＜MgO 纳米颗粒（5.4）。结论是，含 5%MgO 的复合膜的活性是目前使用的木炭的两倍。

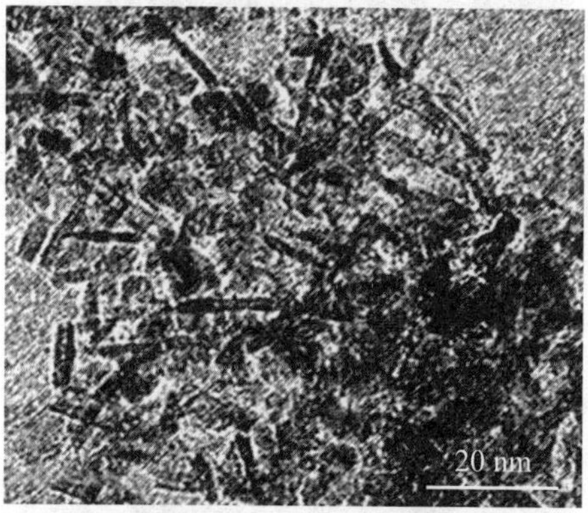

（a）MgO 纳米颗粒的 TEM 图像

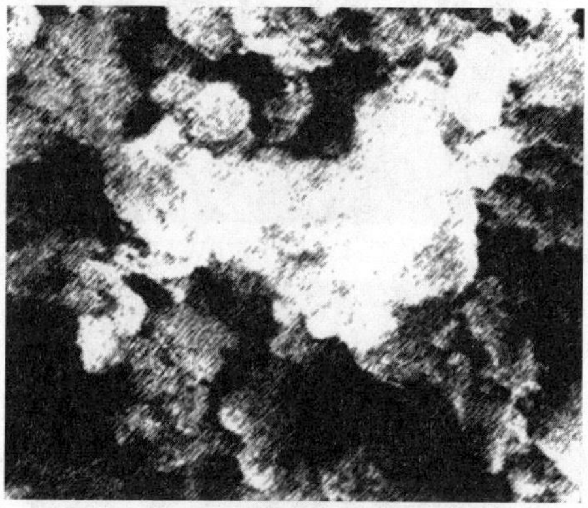

（b）MgO 纳米颗粒的 SEM 图像

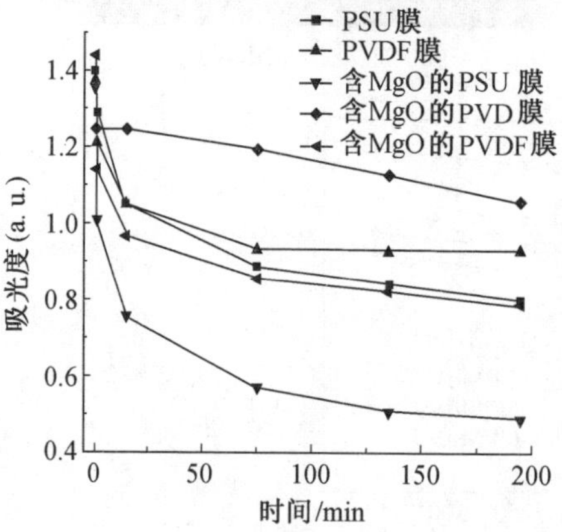

（c）不同聚合物膜的吸光度变化图

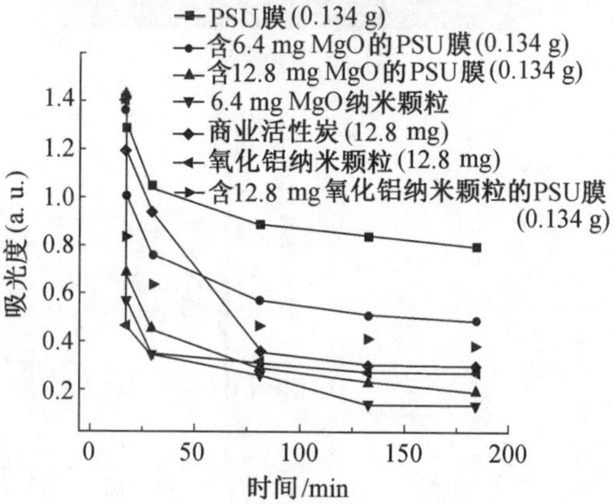

（d）不同材料对氧磷的吸光度变化图

图 7.7　具备分解剧毒物质对氧磷功能的纳米纤维膜

7.1.4 防弹装备

防弹材料的要求是：具有良好的强度，具有良好的变形和冲击波传递能力，模量适中，断裂伸长率较低，保证子弹在接触到人体皮肤前能量被耗散。芳纶的强度和模量较高，且具有优异的耐高温和耐摩擦性能，在防弹衣、头盔和盾牌中得到了广泛的应用，但其性能在潮湿和光照条件下逐步下降。根据前人报道，直径在 100～500 nm 之间的间位芳族聚酰胺纳米纤维具有良好的热性能和机械性能。芳纶纳米纤维增强的聚丙烯酸酯薄膜具有很高的透明度，机械性能良好，弹性模量优于单壁碳纳米管复合材料。由凯夫拉尔衍生的芳族聚酰胺纳米纤维与聚环氧乙烷分层组装而成的复合材料表现出高模量、柔韧性、低密度等特点，可用于制作防弹衣。

Rein 等首次采用静电纺丝方法，将具有不同介电常数和电导率的溶剂混合物制成超大分子量聚乙烯的超细纳米纤维。他们研究了由超大分子量聚合物生产高取向纳米纤维的可能性，并提出了制备具有改善性能的超强、多孔和单组分纳米复合纤维的新方法。他们的另一项工作更进一步研究发现了高温下 UHMWPE 和 CNT 的亚稳态互溶条件，通过静电纺丝工艺制造了用 CNT 增强的 UHMWPE 微米级纤维。通过精准的控温控湿条件下由静电纺丝制备的微米级纤维的强度为 6.6 GPa 和 6% 的断裂伸长率。Baniasadi 等报告了一种电纺 PVDF-TrFE 纳米纤维的高拉伸压电结构（图 7.8（a））。研究结果表明，扭转过程不仅增加了破坏应变，而且还增加了整体强度和韧性。这种扭曲结构提高了材料的强度和韧性，机械性能的这种提高很可能是纳米纤维之间的相互作用增加，摩擦和范德瓦尔斯相互作用以及压电效应导致的表面电荷相互作用所致。这种材料的能量吸收值达到了 98 $J\cdot g^{-1}$，而常用于制造防弹背心的 Kevlar®最多可吸收 80 $J\cdot g^{-1}$，研究人员希望这种材料结构有高应力点自我增强的功能，将来可用于军用飞机或其他防护装备中（图 7.8（b））。

此外，Fan 等将氢氧化钾溶解在二甲亚砜（DMSO）溶液中用来分解芳族聚酰胺纤维，制备出芳族聚酰胺纳米纤维（ANF），使用 ANF 合成了芳族聚酰胺纳米纤维功能化的石墨烯片（ANFGS），ANFGS 可作为增强聚合物的新型纳米填料。在聚甲基丙烯酸甲酯（PMMA）添加质量分数为 0.7% 的 ANFGS 制备得到的 ANFGS/PMMA 复合膜的拉伸强度和弹性模量分别达到 63.2 MPa 和 3.42 GPa，分别增加了 84.5% 和 70.6%。他们还发现 ANFGS/PMMA 复合材料的热稳定性会随着 ANFGS 量的增加而提高，且 ANF 对紫外线有一定的吸收作用，ANFGS/PMMA 膜具有一定程度的紫外线屏蔽功能，可用于紫外防护。

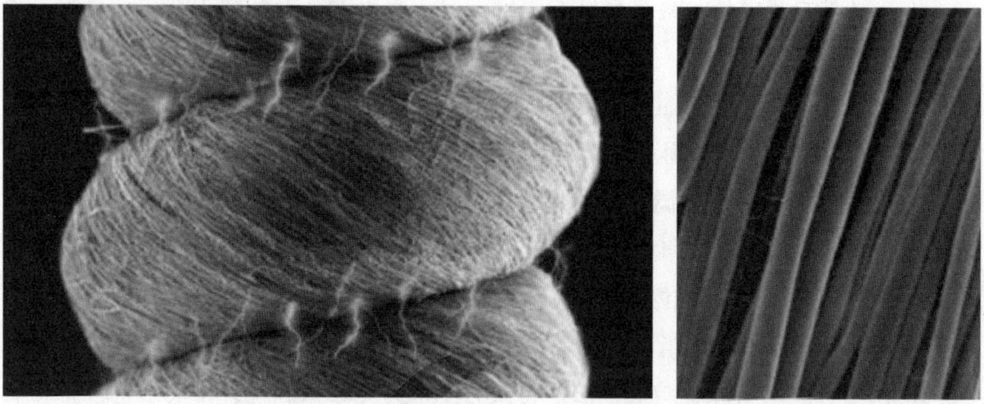

（a）通过静电纺丝制造的高强度纤维

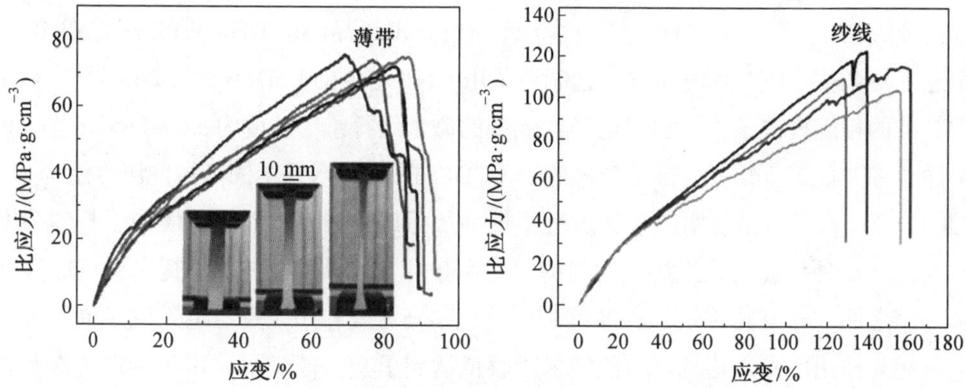

（b）关于薄带和纱线的比应力与应变关系的拉伸实验结果

图 7.8　高强度纤维

7.1.5　防护面罩

自 2020 年以来，人们对医用防护口罩的需求越来越高，同时医护人员、化工从业者等特殊环境工作下的人员更是对防护面罩提出了更高的要求。目前，市场上出现诸多的防护口罩生产品牌，例如 3M、绿盾口罩、阳普等，为消费者提供了多种选择。然而，大多数防护面罩的质量和厚度一般不均匀，这影响了它们的收集效率和压降，因此在开发过程中应考虑渗透性、过滤性能和结构的均匀性。现有的高效空气颗粒过滤面罩可以过滤 0.3 μm 以上的颗粒，效率达到了 99.97%，但它们不足以过滤较小的病原体，如病毒等。当纤维尺寸减小时，能穿透的最大颗粒尺寸（MPPS）减小，意味着总过滤效率变得更高。纤维用于颗粒捕获（特别是亚微米级颗粒）的有效比表面积是决定过滤效率的主要因素，具有以上优点的网状纳米纤维被用于各种空气过滤器件，也被用于提高传统聚合物膜料的过滤性能。在纳米级过滤膜的研究中发现，使用电纺纳米的薄膜结构具有比常规膜更高的渗透通量。电纺纳米纤维膜，由于高比表面积、孔隙的良好互连性，以及在纳米尺度上结合活性化学或功能的潜力，在过滤空气中微粒方面是非常高效的。

虽然随着纤维尺寸的减小，压降（纳米纤维膜对气体的阻力）增大，但截留率和惯性冲击效率的提高弥补了压降的上升。 在对亚微米级尺寸颗粒的过滤中，在较低的压降和较小的纤维尺寸下可以实现相同的过滤效率，或者在相同的压降下更小的纤维尺寸可以实现更好的过滤效率。对于纳米尺度的纤维，必须考虑纤维表面气体流动的影响。对于直径小于 500 nm 的纤维，空气在纤维表面滑动，则纤维上的阻力小于表面没有气体流动的微米尺寸纤维的情况，纤维表面的气体流动使压降降低。这是因为更多的粒子在纤维附近移动会导致更高的扩散、拦截和惯性撞击的效率。根据滑移流模型，直径为 200～300 nm 的纤维是过滤器件的理想材料。

纳米纤维介质非稳态过滤模拟结果表明，通过减小纤维直径，具有相同压降的过滤介质的最低过滤效率提高。Yun 等通过静电纺丝聚丙烯腈（PAN）纤维制备了一种过滤介质，这种纤维的直径比商用过滤器更均匀。他们发现，质量小得多的电纺滤光片可以获得与商用滤光片相当的纳米粒子穿透值。另外，随着过滤层厚度的增加，纳米粒子的穿透力可能会降低。结果表明，用直径 80～200 nm 的 nylon-6 纤维（基重 10.75 $g \cdot m^{-2}$）制成的纳滤器的过滤效率优于商品化的高效率空气颗粒过滤器（图 7.9（a）、(b)）。当前研究表明可以通过纳米纤维的表面修饰和功能化制备高选择性过滤的亲和膜。Lala 课题组发现纳米纤维具有抗菌功能，可用作抗菌的过滤器。实验使用醋酸纤维素、聚丙烯腈和聚氯乙烯与不同数量的硝酸银进行紫外线照射处理，使银纳米颗粒得到增强。实验使用 DMF 作为常规溶剂，这有助于在室温下进行自发的银纳米粒子还原，然后再进行紫外辐射（图 7.9（c）、(e)）。Zu 等制备了表面改性聚砜（PSU）无纺布，来开发一种新型的亲和膜，实验中采用空气等离子体处理聚砜纤维网，利用硝酸铈铵引发甲基丙烯酸（MAA）接枝共聚，从而在纤维表面引入羧基。他们使用甲苯胺蓝 O 染料，与羧基形成稳定的络合物，作为聚甲基丙烯酸接枝聚砜纤维网捕获的模型靶分子。实验中他们还研究了甲苯胺蓝 O 染料的吸附等温线和速率。与传统的微滤膜相比，这种新型的亲和膜具有更小的压降（Δp=4.83～10.34 kPa）和更高的通量。

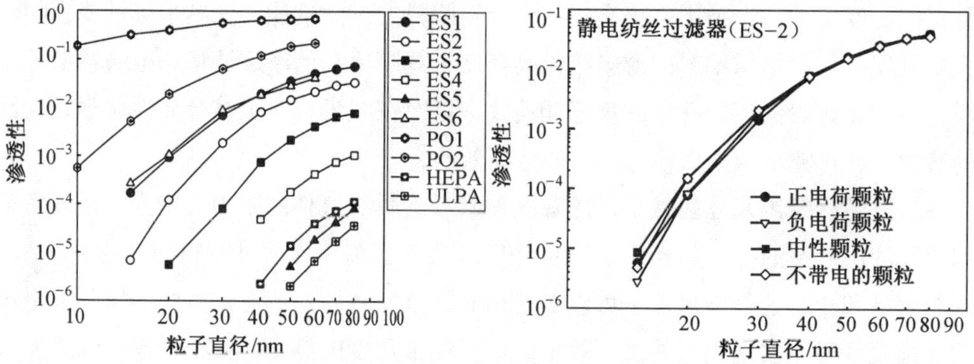

（a）纳米颗粒在不同过滤介质的渗透性　　（b）通过静电纺丝过滤器纳米粒子的渗透性

（c）纤维素（对照组）　　（d）未处理　　（e）处理后

图 7.9　纤维膜的渗透性及抗菌性能的 SEM 图

为了获得抗菌的纳米纤维，研究者采用等离子体预处理、紫外诱导 4-乙烯基吡啶接枝共聚和接枝吡啶基团与已溴化合物的四元化等方法。由于具有高效的抗菌活性，这些新型纳米纤维在高性能滤光片、防护纺织品和生物医疗器械等方面具有潜在的应用前景。Vrieze 等人测试了负载杀菌剂的静电纺丝聚酰胺纳米纤维的过滤性能，结果表明用杀菌剂对纳米纤维进行功能化处理后，其去除效率高于处理前的过滤效率。Tan 课题组研究了 N-卤胺添加剂对尼龙纳米纤维抗菌性能的影响，结果表明，使用 N-卤胺添加剂和活性氯成分对尼龙纳米纤维处理后，在 5～40 min 的短暂接触时间后，观察到大肠杆菌和金黄色葡萄球菌的总数减少。Son 开发了一种用银纳米粒子直接静电纺丝醋酸纤维素溶液制备抗菌超细纤维的可行方法：用少量硝酸银直接静电纺丝醋酸纤维素（CA）溶液，然后光还原，用这种方法得到了超细抗菌 CA 纤维，添加剂与材料性能的关系如图 7.10（a）～（c）所示。Yao 课题组研究了表面改性电纺聚偏氟乙烯-六氟丙烯（PVDF-HFP）纤维膜的抗菌活性。他们评估了改良 PVDF-HFP 纤维膜对革兰氏阳性金黄色葡萄球菌和革兰氏阴性大肠杆菌的抗菌活性，如图 7.10（d）～（f）所示。结果表明，用四元化吡啶基团修饰的纳米纤维对两种细菌均具有良好的抗菌性能，杀灭率高达 99.999 9%。这些都证实了电纺纤维在防护面罩领域的巨大应用前景。

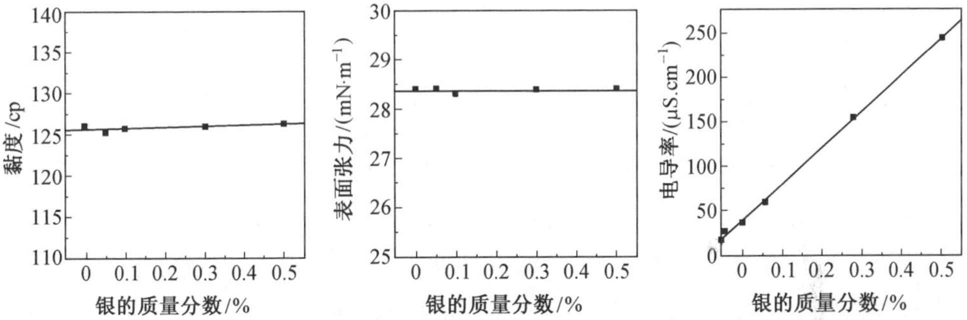

（a）质量分数为 10%CA 溶液的黏度　（b）表面张力的变化　（c）电导率的变化

（d）未经处理　（e）60 ℃下季铵化 24 h　（f）60 ℃下季铵化 96 h

图 7.10　表面改性电纺聚偏氟乙烯-六氟丙烯（PVDF-HFP）纤维膜

7.1.6 小结和展望

电纺纤维在个人防护未来的潜在应用包括将弹性体膜直接应用于服装系统，从而消除了诸如层压和固化之类的昂贵制造步骤。可以将纤维直接静电纺丝到通过 3D 人体扫描获得的 3D 屏幕形式上。科学家目前正在使用基于激光的光学数字化系统来记录测试对象的体表坐标。该信息可以与计算机辅助设计和制造（CAD/CAM）流程集成在一起，以允许将电纺服装喷涂到数字化表格上，从而实现定制合身的无缝化防护装。此外，局部结构与性能的改变导致了纤维整体性能的改变。多孔超细纤维可以改变单根纤维的结构与性能，多孔结构所带来的新性质赋予了静电纺纤维新的应用价值。多孔纤维也是今后个体防护领域值得关注方向之一。

7.2 纳米纤维电磁屏蔽

7.2.1 引言

随着当代信息技术的迅猛发展，频率为吉赫兹（GHz）的电磁波被广泛应用于军用及民用领域等。在军事上，雷达利用吉赫兹电磁波对目标进行定位，当其所发出的电磁波遇到金属物体时，一部分电磁波会产生反射现象，根据这部分反射电磁波可以得知目标方位及距离。而为了保护军事目标安全，提升武器装备战场生存能力，要求入射电磁波在达到物体表面时尽可能降低电磁波的反射率，以防止雷达探测定位。因此，隐身技术的研究开发得到了各军事强国的关注。隐身技术是指通过武器装备自身结构设计或将吸波材料涂覆于装备表面实现入射电磁波低反射率的技术。在生活中，各类电子设备与通信设施，如手机、电脑、电磁炉、通信基站等已经成为人们日常生活中必不可少的组成部分，其在服务于人类的同时所产生的兆赫兹电磁波辐射也对人体健康构成了严重威胁。据研究表明：人体细胞分子在电磁场作用下会受到被迫振动摩擦，所产生的热能能够对于眼睛、皮肤等造成损伤，而长期处于电磁波辐射的环境中，会造成体内基因的突变以及日积月累形成不可恢复的伤害。为了达到军事隐身目的及解决日常生活中的电磁波辐射问题，电磁吸波材料作为有效的防护手段，其通过介电损耗和磁损耗将电磁波转换为热能或其他形式的能量消耗掉。

在第二次世界大战期间，德国等国家开始研究电磁波吸波材料，首先采用了活性炭粉与橡胶混合制成涂料应用于潜艇和战机表面，进行了简单而有效的防护。美国人后续也尝试使用铁、铝、铜等金属粉与黏合剂混合制备了涂层隐身材料。20 世纪 50 年代初期，对隐身技术有了的初步探索。60 年代开始，美国和苏联等均展开了有计划的隐身技术研发并取得了长足的进步，D-21、SR-71、A-12 等隐身战机相继面世。发展到 90 年代，隐身技术开始真正应用在了现代战争中。现如今，随着电子信息技术的高速发展，军事装备也越来越趋于电子信息化，只有拥有电磁控制权，才能在危机中占据主动。因此，隐身技术这项军事竞赛也一直没有停下脚步。美国为了进一步提升战机隐身性能，又相继研发了 F-22（猛禽）和 F35（闪电）隐身战机；我国为了提升军事竞争力也陆续研发出了 J-12/J-13/J-14/J-20 等歼击系列高性能隐身战机；其他国家的新型隐身战机还有俄罗斯的苏霍伊 PAK FA（T-50）战机、日本 X-2（心神）战机、韩国 KFX C100 战机、瑞典 SAAB FS2020 战机等。

7.2.2　电磁波的概念

电磁波是以波动的形式传播的电磁场，其传播方向、磁场方向和电场方向三者相互垂直，因此，电磁波也属于横波。通过麦克斯韦方程组可以揭示其电场与磁场、场源与场及媒质与场之间的变化规律和相互关系。假设用 $\boldsymbol{r}$ 表示三维空间位置矢量，t 表示时间变量，则麦克斯韦方程组的微分形式可以表示为

$$\nabla\cdot\boldsymbol{H}(\boldsymbol{r},t)=\boldsymbol{J}(\boldsymbol{r},t)+\frac{\partial\boldsymbol{D}(\boldsymbol{r},t)}{\partial t} \tag{7.1}$$

$$\nabla\cdot\boldsymbol{E}(\boldsymbol{r},t)=-\frac{\partial\boldsymbol{B}(\boldsymbol{r},t)}{\partial t} \tag{7.2}$$

$$\nabla\cdot\boldsymbol{B}(\boldsymbol{r},t)=0 \tag{7.3}$$

$$\nabla\cdot\boldsymbol{D}(\boldsymbol{r},t)=\rho(\boldsymbol{r},t) \tag{7.4}$$

式中 $\boldsymbol{H}(\boldsymbol{r}, t)$——电场强度矢量，A/m；

$\boldsymbol{J}(\boldsymbol{r}, t)$——电流密度矢量，$A/m^2$；

$\boldsymbol{D}(\boldsymbol{r}, t)$——电位移矢量，$C/m^2$；

$\boldsymbol{E}(\boldsymbol{r}, t)$——电场强度矢量，V/m；

$\boldsymbol{B}(\boldsymbol{r}, t)$——磁感应强度矢量，T；

$\boldsymbol{\rho}(\boldsymbol{r}, t)$——电荷密度矢量，$C/m^3$。

电磁波由高频率到低频率，主要分为：伽马射线、X 射线、紫外线、可见光、红外线、微波和无线电波，其频率和波长分布如图 7.11 所示。军事上，雷达波的频率一般处于 2～18 GHz 范围内，所对应波长为 15～1.67 cm，对其进一步细分又可分为 S（2～4 GHz，15～7.5 cm）、C（4～8 GHz，7.5～3.75 cm）、X（8～12 GHz，3.75～2.5 cm）、Ku（12～18 GHz，2.5～1.67 cm）波段。日常生活中，手机、微波炉、电磁炉等电器所发出的电磁波频率一般处于 1～5 GHz 之间。因此，对应电磁吸波材料的研发也主要针对上述的 1～18 GHz 频段。目前，所报道的电磁吸波材料在 X 波段以上已能达到高效的吸收效果，而在 X 波段以下却依旧不尽如人意。其主要原因就在于高频段的电磁波波长较短，而低频段的电磁波波长相对较长，要求所对应的电磁吸波材料厚度、结构和组分材料的选择上应有所差异。

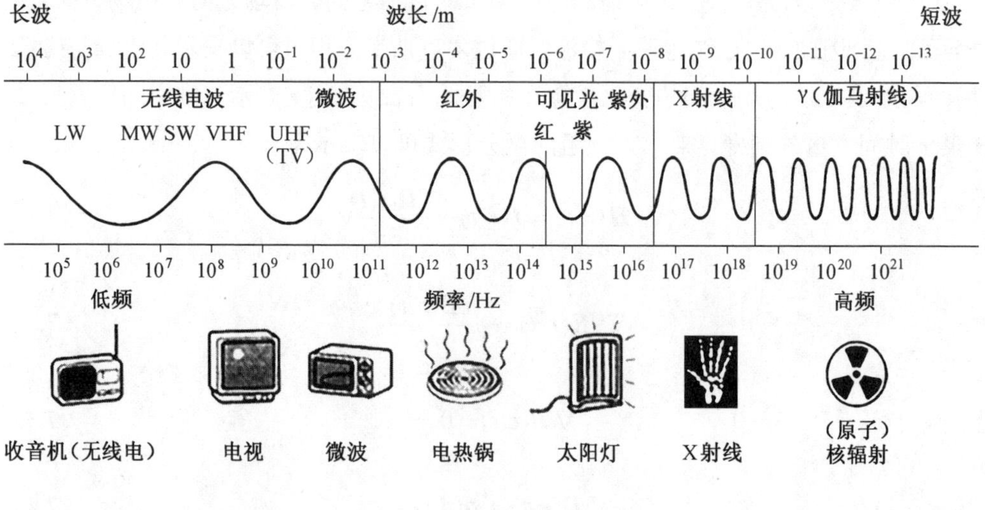

图 7.11　电磁波分布图

7.2.3　吸波性能评价

对材料吸波能力的评判主要基于材料的复介电常数（ε_r）和复磁导率（μ_r）这两个最重要的电磁参数进行考量，其表示方式如下：

$$\varepsilon_r = \varepsilon' - j\varepsilon'' \qquad \mu_r = \mu' - j\mu'' \tag{7.5}$$

式中，复介电常数和复磁导率的实部（ε'，μ'）和虚部（ε''，μ''）分别代表对于电磁波的储存和损耗能力。后续对材料电磁吸波性能的评价均围绕上述 ε_r 和 μ_r 展开。

1. 材料损耗能力

根据 ε_r 和 μ_r 的实部和虚部值，可分别计算得出材料的介电损耗能力和磁损耗能力，且其均以正切损耗角来表示，正切损耗值越大，相应的损耗能力也就越强：

$$\tan\delta_\varepsilon = \frac{\varepsilon''}{\varepsilon'} \qquad \tan\delta_\mu = \frac{\mu''}{\mu'} \tag{7.6}$$

此外，综合考虑材料的介电损耗和磁损耗能力，整体的电磁波损耗值又可表示为以下形式：

$$\alpha = \frac{\sqrt{2}\pi f}{c} \times \sqrt{(\mu''\varepsilon'' - \mu'\varepsilon') + \sqrt{(\mu''\varepsilon'' - \mu'\varepsilon')^2 + (\mu'\varepsilon'' - \mu''\varepsilon')^2}} \tag{7.7}$$

2. 反射损耗值

材料反射损耗值（reflection loss）是衡量其吸波能力强弱的关键指标。其以一般以英文字母简写 RL 表示，单位为 dB，且认为 RL≤−10 dB 的频宽为材料的有效吸波频段。根据传输线理论，电磁波由阻抗为 Z_0 的自由空间垂直入射到阻抗为 Z 的介质材料表面时，反射率 RL 计算公式如下：

$$\mathrm{RL} = 20\log\left|\frac{Z - Z_0}{Z + Z_0}\right| \tag{7.8}$$

对于有限厚度的单层吸波介质，采用以下公式进行波阻抗 Z 的计算：

$$Z = Z_0\sqrt{\frac{\mu_r}{\varepsilon_r}}\tanh\left(\mathrm{i}\frac{2\pi fd}{c}\sqrt{\mu_r\varepsilon_r}\right) \tag{7.9}$$

式中 Z_0——空气阻抗，Z_0=377 Ω；

f——电磁波频率；

d——吸波层厚度；

c——光速。

因此，在已知材料 ε_r 和 μ_r 值的情况下，通过上述公式，可模拟计算得出不同厚度下材料的反射损耗值，且其值越低，所代表材料的吸收性能越佳。

3. 阻抗匹配

根据式（7.8）和式（7.9），存在三类阻抗极端情况值得注意。第一类为材料阻抗 Z=0 的情况，此时会出现入射电磁波全反射现象，对应材料无任何吸收性能。第二类为 $Z=Z_0$，这一状态为材料追求的零反射理想状态，而此条件下要求材料 $\varepsilon'=\mu'$ 且 $\varepsilon'=\mu'$。然而，实际中吸波材料复介电常数的实部和虚部值一般远大于复磁导率所对应的值。因此，对于目前众多吸波材料而言，调节材料复介电常数值处于合理范围是实现良好阻抗匹配的关键点。第三类为 $Z\to\infty$，材料表现为全透射状态，此时入射电磁波可全部穿透材料，然而在材料中不产生任何损耗，基于这一特性，此类材料一般被用作吸波材料基材。由上述可知，对于电磁吸波材料的设计来说，$Z=Z_0$ 即满足阻抗匹配是最佳的情况，此时，入射电磁波绝大部分或全部透入吸波体，从而可以被电磁吸波剂最大限度地加以吸收。然而，零反射并不代表吸波性能强，绝缘性能良好的材料即为一类透波材料，虽然其具有优良的阻抗匹配特性，但却几乎没有电磁吸波能力。因此，设计性能优良的电磁吸波材料，阻抗匹配是需要考量的重要因素之一。而另一需要考量的重要因素是需使吸波剂沿着电磁波的传播路径有效分布，使其包围电磁波的所有通道，又能容许电磁波通过，在通过的途径中逐步被衰减吸收。

4. 四分之一波长理论

以吸波体的为例，当垂直入射的电磁波进入厚度为 t_m 的吸波体后，除被吸收剂吸收之外，部分透进的电磁波遇到金属板后反射回入射表面，从入射面观察到的入射波与反射波的波程差$\Delta\geqslant 2t_m$，当$\Delta\geqslant 2t_m$ 时，则$\Delta=2t_m=(2n+1)\lambda/2$ 时有电磁能量相消现象，即

$$2t_m=(2n+1)\frac{\lambda}{2} \tag{7.10}$$

$$t_{\mathrm{m}} = (2n+1)\frac{\lambda}{4} \tag{7.11}$$

当电磁波在吸波体中传播时，引入表征介质属性的 ε_{r} 和 μ_{r} 后，有

$$\lambda = \frac{c}{f\sqrt{\varepsilon_{\mathrm{r}}\mu_{\mathrm{r}}}} \tag{7.12}$$

所以

$$t_{\mathrm{m}} = (2n+1)\times\frac{c}{4f\sqrt{\varepsilon_{\mathrm{r}}\mu_{\mathrm{r}}}} \tag{7.13}$$

取 n=0 时，即

$$t_{\mathrm{m}} = \frac{c}{4f\sqrt{\varepsilon\mu}} \tag{7.14}$$

因此，四分之一波长理论用以计算材料最低反射损耗频率(f)与厚度(t_{m})的关系，当频率与材料厚度满足上式关系式时会产生电磁波相消现象，所对应的即为最低反射损耗值，且由上式可知随着厚度的增加，最低反射损耗值逐渐移向低频。这一理论对于强电磁吸波材料的设计也具有重要的指导意义。

7.2.4　吸波材料

静电纺丝纳米纤维符合吸波材料“薄、轻、宽、强”的发展目标，通过纺丝液的合理配制，可以将电损耗材料、磁损耗材料和电阻材料有效结合，从而制备得到多重损耗机制的电磁吸波纳米纤维。常用的纺丝液主要有聚己内酯、聚氯乙烯、聚乙烯吡咯烷酮等高分子材料，将其溶于有机溶剂中并与电磁吸波粒子均匀混合进行纺丝制备得到纳米吸波纤维。例如，将四氧化三铁纳米粒子加入聚己内酯或聚氯乙烯纺丝液中混合均匀并通过静电纺丝技术可纺出直径在 1 000 nm 以下的纳米纤维且具有良好的电磁吸波效果。此外，高分子纺丝液还可作为碳源使用，经后续高温碳化可以得到性能较好的碳膜基体，进一步将其与高性能吸波剂复合可以得到电磁吸波性能优良的吸波膜材，因此具有重大的研究价值。Mordina 等通过共纺聚丙烯腈/$FeCl_3$ 和聚甲基丙烯酸甲酯的溶液，然后在高温下稳定和碳化制备出了 Fe_3O_4 纳米颗粒嵌入介孔碳的复合纳米纤维材料，结果表明具有质量分数为 25%的碳纳米纤维（由质量分数为 5%的 Fe_3O_4 组成）的 7.5 mm 吸收体厚度下，发现吸收带宽最大为 4.33 GHz，反射损耗为−25 dB。Xu 等制备了一种轻质多孔纤维素纳米纤维（CNF）/碳纳米管

（CNT）泡沫，如图 7.12 所示，该泡沫由孔直径在 30～90 μm 之间的蜂窝状垂直多孔结构和 1.7～50 nm 的纳米孔结构组成，得益于其独特的结构，该材料的有效吸收带宽达到 29.7 GHz，其特定的微波吸收性能超过 80 000 $dB·cm^{-2}·g^{-1}$，远远超过了先前报道的 CNT 的复合吸波材料。由于该材料具有良好连通的多孔结构以及 CNF-CNF 和 CNF-CNT 分子链之间的强氢键，其具有超低密度（9.2 $mg·cm^{-3}$）和良好的抗疲劳性能。

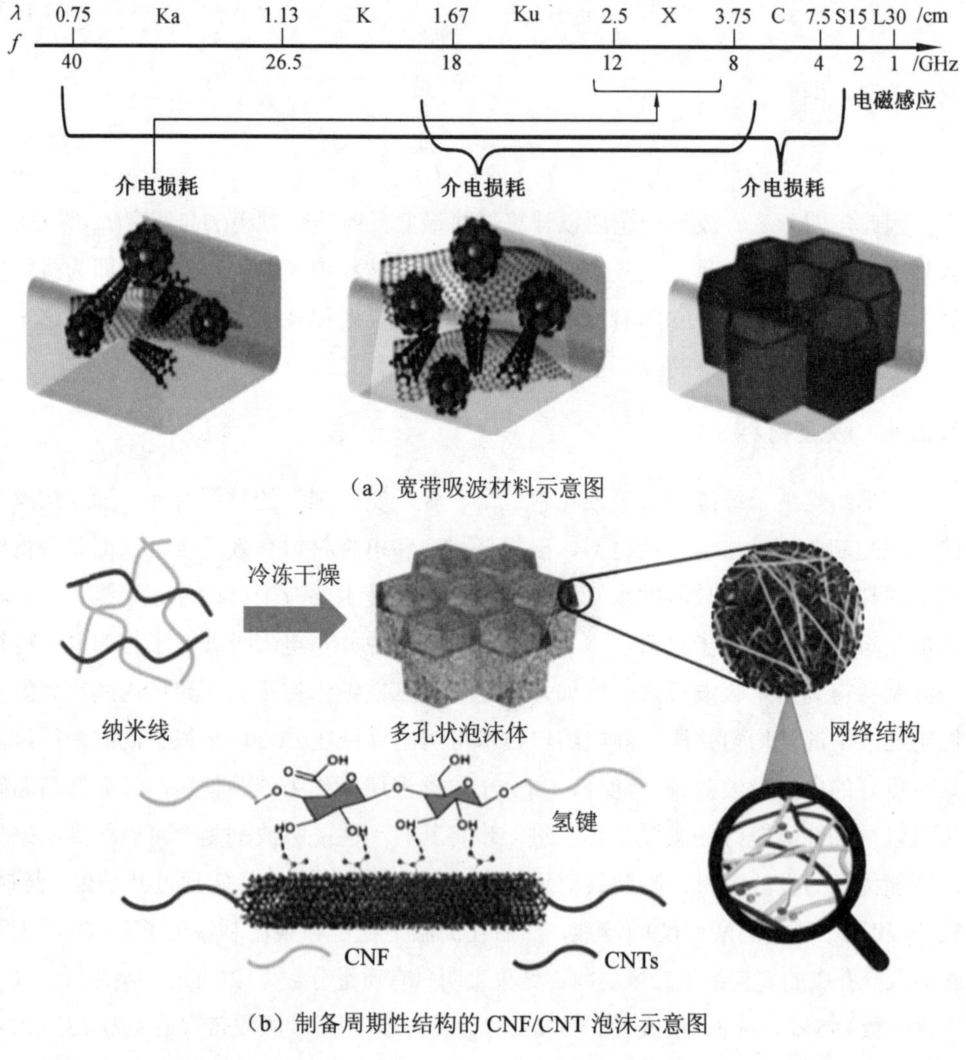

（a）宽带吸波材料示意图

（b）制备周期性结构的 CNF/CNT 泡沫示意图

图 7.12　轻质多孔纤维素纳米纤维（CNF）/碳纳米管（CNT）泡沫

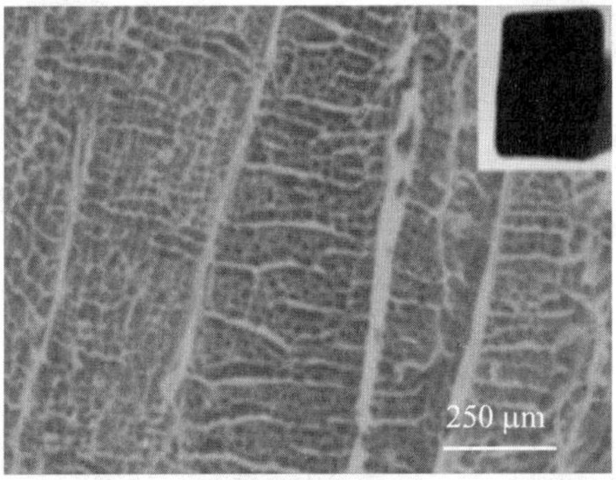

（c）CNF/CNT 泡沫的 SEM 图

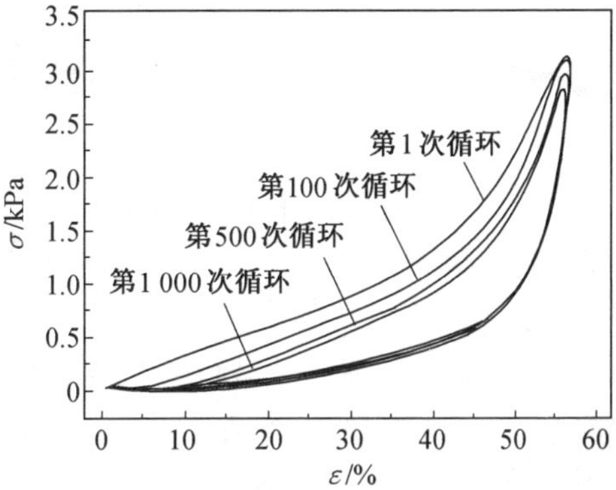

（d）CNF/CNT 泡沫的抗疲劳测试

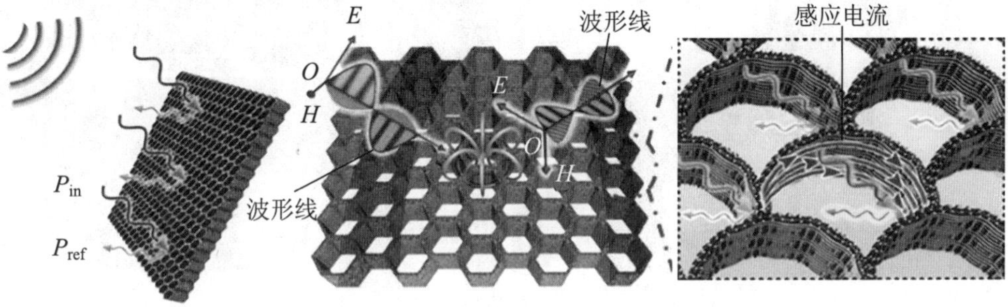

（e）CNF/CNT 泡沫的微波吸收机理

续图 7.12

为了提高 SiC 纳米纤维的吸波性能，Xu 等在 SiC 纳米纤维的表面设计了双界面以形成 SiC@C@PPy 异质结构。如图 7.13 所示，内层碳和外层聚吡咯（PPy）分别通过气体蚀刻和氧化聚合合成。受益于双界面的极化增强和复合材料中完善的导电网络，SiC@C@PPy 纳米纤维负载复合材料显示出理想的电磁衰减和阻抗匹配特性。SiC@C@PPy 纳米纤维的最佳有效吸收带宽（EAB）在 2.78 mm 的厚度下可达到 8.4 GHz，填充剂质量分数为 15%。此外，通过调制样本的厚度，其有效吸收带宽可以覆盖整个 X（8～12 GHz）和 Ku（12～18 GHz）频段。这些研究结果证实了通过静电纺丝技术合成的纤维是一种良好的吸波材料，也说明了静电纺丝技术在这些领域中具有广阔的应用前景。

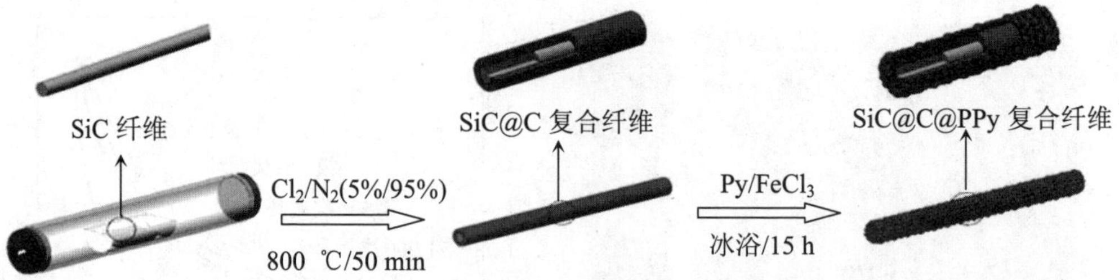

（a）SiC@C@PPy 的合成过程示意图

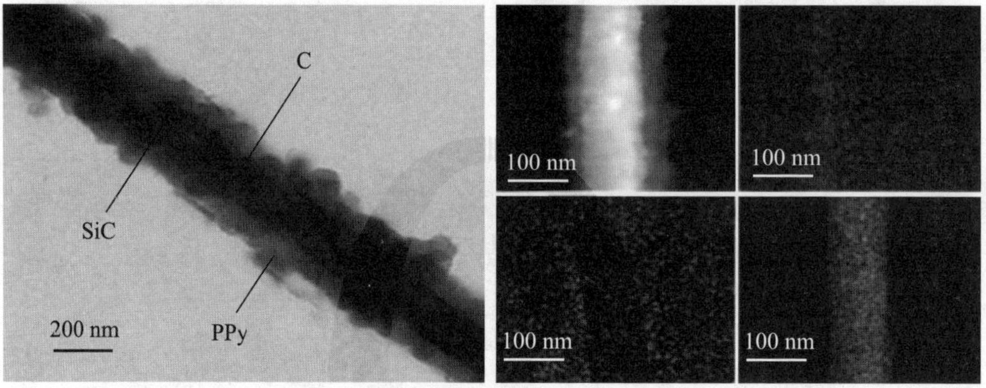

（b）SiC@C@PPy 透射电镜图及 EDS 面扫描图

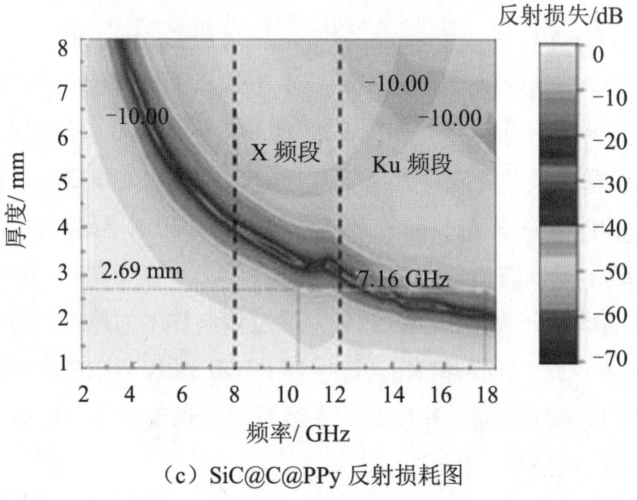

（c）SiC@C@PPy 反射损耗图

图 7.13 SiC@C@PPy 纳米纤维复合材料

7.2.5 小结和展望

静电纺丝技术能够实现纤维的形貌调控和阵列制备,并且通过聚合物模板实现了对纤维波长、振幅与阵列图案的可控制备，电纺纤维及其复合材料的制备具有重要理论研究和实际应用的价值。通过静电纺丝技术探究导电高聚物主链结构、室温电导率、掺杂剂性质等因素是电磁屏蔽材料的重要方向。将导电高聚物与无机磁损耗物质或超微粒子复合可望发展成为一种新型的轻质宽频吸波材料。

参考文献

[1] LEE S, OBENDORF S K. Transport properties of layered fabric systems based on electrospun nanofibers[J]. Fibers & Polymers, 2007, 8(5): 501-506.

[2] YUAN W. Effect of fiber diameter and web porosity on breathability of nanofiber mats at various test conditions[D]. Austin: The University of Texas, 2014.

[3] HONG S K, LIM G, CHO S J. Breathability enhancement of electrospun microfibrous polyurethane membranes through pore size control for outdoor sportswear fabric[J]. Sensors and Materials, 2015, 27(1): 77-85.

[4] GORJI M, JEDDI A A A, GHAREHAGHAJI A A. Fabrication and characterization of polyurethane electrospun nanofiber membranes for protective clothing applications[J]. Journal of Applied Polymer Science, 2012, 125(5): 4135-4141.

[5] BAGHERZADEH R, LATIFI M, NAJAR S S, et al. Transport properties of multi-layer fabric based on electrospun nanofiber mats as a breathable barrier textile material[J]. Textile Research Journal, 2012, 82(1): 70-76.

[6] ZHUO H, HU J, CHEN S. Electrospun polyurethane nanofibres having shape memory effect[J]. Materials Letters, 2008, 62(14): 2078-2080.

[7] ZHOU H. Coaxial electrospun polyurethane core-shell nanofibers for shape memory and antibacterial nanomaterials[J]. Express Polymer Letters, 2011, 5(2): 182-187.

[8] PHILLIP G, HEIDI S, DONALD R. Transport properties of porous membranes based on electrospun nanofibers[J]. Colloids & Surfaces A Physicochemical & Engineering

Aspects, 2001, 187: 469-481.

[9] CHEN H, RAHMATHULLAH A M, PALMESE G R, et al. Polymer-polymer nanocomposite membranes as breathable barriers with electro-sensitive permeability[J]. ACS Symposium Series, 2009, 1016: 307-322.

[10] GORJI M, BAGHERZADEH R, FASHANDI H. Electrospun nanofibers in protective clothing[M]. Electrospun Nanofibers, 2016, 243: 571-598.

[11] LIU F, LU Q, JIAO X, et al. Fabrication of nylon-6/POMs nanofibrous membranes and the degradation of mustard stimulant research[J]. RSC Advances, 2014, 4(78): 41271-41276.

[12] VORONTSOV A V, LION C, SAVINOV E N, et al. Pathways photocatalyticgas phase destruction of HD stimulant 2-chloroethyl sulfide[J]. Journal of Catalysis, 2003, 220(2): 414-223.

[13] SUNDARRAJAN S, RAMAKRISHNA S. Fabrication of nanocomposite membranes from nanofibers and nanoparticles for protection against chemical warfare stimulants[J]. Journal of Materials Science, 2007, 42(20): 8400-8407.

[14] YAO L, LEE C, KIM J. Fabrication of electrospun meta-aramid nanofibers in different solvent systems[J]. Fibers & Polymers, 2010, 11(7): 1032-1040.

[15] YANG M, CAO K, YEOM B, et al. Aramid nanofiber-reinforced transparent nanocomposites[J]. Journal of Composite Materials, 2015, 49(15): 1873-1879.

[16] REIN D M, SHAVIT H L, KHALFIN R L, et al. Electrospinning of ultrahigh-molecular-weight polyethylene nanofibers[J]. Journal of Polymer Science Part B: Polymer Physics, 2007, 45(7): 766-773.

[17] REIN D M, COHEN Y, LIPP J, et al. Elaboration of ultra-high molecular weight polyethylene/carbon nanotubes electrospun composite fibers[J]. Macromolecular Materials and Engineering, 2010, 295(11): 1003-1008.

[18] BANIASADI M, HUANG J, XU Z, et al. High-performance coils and yarns of polymeric piezoelectric nanofibers[J]. ACS Applied Materials & Interfaces, 2015, 7(9): 5358-5366.

[19] FAN J, SHI Z, ZHANG L, et al. Aramid nanofiber-functionalized graphene

nanosheets for polymer reinforcement[J]. Nanoscale, 2012, 4(22): 7046-7055.

[20] BARHATE R S, LOONG C K, RAMAKRISHNA S. Preparation and characterization of nanofibrous filtering media[J]. Journal of Membrane Science, 2006, 283(1-2): 209-218.

[21] YANG C. Aerosol filtration application using fibrous media—an industrial perspective[J]. Chinese Journal of Chemical Engineering, 2012, 20(1): 1-9.

[22] KOSMIDER K, SCOTT J. Polymeric nanofibres exhibit an enhanced air filtration performance[J]. Filtration & Separation, 2002, 39(6): 20-22.

[23] YUN K M, HOGAN C J, MATSUBAYASHI Y, et al. Nanoparticle filtration by electrospun polymer fibers[J]. Chemical Engineering Science, 2007, 62(17): 4751-4759.

[24] LALA N L, RAMASESHAN R, BOJUN L, et al. Fabrication of nanofibers with antimicrobial functionality used as filters: protection against bacterial contaminants[J]. Biotechnology & Bioengineering, 2007, 97(6): 1357-1365.

[25] MA Z, KOTAKI M, RAMAKRISHNA S. Surface modified nonwoven polysulphone (PSU) fiber mesh by electrospinning: a novel affinity membrane[J]. Journal of Membrane Science, 2006, 272(1/2): 179-187.

[26] SANDER D V, NELE D, KAREL L, et al. Filtration performance of electrospun polyamide nanofibres loaded with bactericides[J]. Textile Research Journal, 2012, 82(1): 37-44.

[27] TAN K, OBENDORF S K. Fabrication and evaluation of electrospun nanofibrous antimicrobial nylon 6 membranes[J]. Journal of Membrane Science, 2007, 305(1-2): 287-298.

[28] SON W K, YOUK J H, LEE T S, et al. Preparation of antimicrobial ultrafine cellulose acetate fibers with silver nanoparticles[J]. Macromolecular Rapid Communications, 2004, 25(18): 1632-1637.

[29] YAO L, LEE C, KIM J. Fabrication of electrospun meta-aramid nanofibers in different solvent systems[J]. Fibers & Polymers, 2010, 11(7): 1032-1040.

[30] PATEL C R P, TRIPATHI P, SINGH S, et al. New emerging radially aligned carbon

nano tubes comprised carbon hollow cylinder as an excellent absorber for electromagnetic environmental pollution[J]. Journal of Materials Chemistry C, 2016, 4 (23):5483-5490.

[31] AKMAN O, KAVAS H, BAYKAL A, et al. Magnetic metal nanoparticles coated polyacrylonitrile textiles as microwave absorber[J]. Journal of Magnetism and Magnetic Materials, 2013, 327: 151-158.

[32] 黎南. 国外舰船隐身技术的发展动向与分析[J]. 舰船电子工程, 2013, 33 (4): 15-17.

[33] 李新, 陆萍, 张德磊. 雷达波吸收涂料的研究进展[J]. 广东化工, 2018, 45 (5): 141-148.

[34] ZHANG Y, HUANG Y, CHEN H, et al. Composition and structure control of ultralight graphene foam for high-performance microwave absorption[J]. Carbon, 2016, 105: 438-447.

[35] LIU W, LI H, ZENG Q, et al. Fabrication of ultralight three-dimensional graphene networks with strong electromagnetic wave absorption properties[J]. Journal of Materials Chemistry C, 2015, 3 (7):3739-3747.

[36] 杨新兴, 李世莲, 尉鹏, 等. 环境中的电磁波污染及其危害[J]. 前沿科学, 2014, 8 (1): 13-26.

[37] JIANG Y, CHEN Y, LIU Y, et al. Lightweight spongy bone-like graphene@SiC aerogel composites for high-performance microwave absorption[J]. Chemical Engineering Journal, 2018, 337 (1): 522-531.

[38] YAN L, HONG C, SUN B, et al. In situ growth of core-sheath heterostructural SiC nanowire arrays on carbon fibers and enhanced electromagnetic wave absorption performance[J]. ACS Applied Materials & Interfaces, 2017, 9 (7): 6320-6331.

[39] 徐剑盛, 周万城, 罗发, 等. 雷达波隐身技术及雷达吸波材料研究进展[J]. 材料导报, 2014, 28 (9): 46-49.

[40] 许学春. 隐身导弹与隐身技术的应用[J].飞航导弹, 2013, 000(5): 87-90.

[41] 郭晓琴. 磁性纳米粒子负载石墨烯的结构调控及吸波机理研究[D]. 郑州: 郑州大学, 2016.

[42] LI N, HUANG G W, LI Y Q, et al. Enhanced microwave absorption performance of coated carbon nanotubes by optimizing the Fe_3O_4 nanocoating structure[J]. ACS Applied Materials & Interfaces, 2017, 9 (3): 2973-2983.

[43] JI R, CAO C, CHEN Z, et al. Solvothermal synthesis of $Co_xFe_{3-x}O_4$ spheres and their microwave absorption properties[J]. Journal of Materials Chemistry C, 2014, 2(29): 5944-5953.

[44] QING Y, MIN D, ZHOU Y, et al. Graphene nanosheet- and flake carbonyl iron particle-filled epoxy-silicone composites as thin-thickness and wide-bandwidth microwave absorber[J]. Carbon, 2015, 86: 98-107.

[45] CHEN Y H, HUANG Z H, LU M M, et al. 3D Fe_3O_4 nanocrystals decorating carbon nanotubes to tune electromagnetic properties and enhance microwave absorption capacity[J]. Journal of Materials Chemistry A, 2015, 3 (24): 12621-12625.

[46] LIU T, XIE X, PANG Y, et al. Co/C nanoparticles with low graphitization degree: a high performance microwave-absorbing material[J]. Journal of Materials Chemistry C, 2016, 4 (8): 1727-1735.

[47] WU T, LIU Y, ZENG X, et al. Facile hydrothermal synthesis of Fe_3O_4/C core-shell nanorings for efficient low-frequency microwave absorption[J]. ACS Applied Materials & Interfaces, 2016, 8 (11): 7370-7380.

[48] LI N, HUANG G W, LI Y Q, et al. Enhanced microwave absorption performance of coated carbon nanotubes by optimizing the Fe_3O_4 nanocoating structure[J]. ACS Applied Materials & Interfaces, 2017, 9 (3): 2973-2983.

[49] NAJIM M, MODI G, MISHRA Y K, et al. Ultra-wide bandwidth with enhanced microwave absorption of electroless Ni-P coated tetrapod-shaped ZnO nano- and microstructures[J]. Physical Chemistry Chemical Physics, 2015, 17(35): 22923-22933.

[50] YANG Z, LI Z, YU L, et al. Achieving high performance electromagnetic wave attenuation: a rational design of silica coated mesoporous iron microcubes[J]. Journal of Materials Chemistry C, 2014, 2 (36): 7583-7588.

[51] LV H, JI G, ZHANG H, et al. Facile synthesis of a CNT@Fe@SiO_2 ternary

composite with enhanced microwave absorption performance[J]. RSC Advances, 2015, 5 (94): 76836-76843.

[52] LV H, LIANG X, CHENG Y, et al. Coin-like alpha-Fe_2O_3@$CoFe_2O_4$ core-shell composites with excellent electromagnetic absorption performance[J]. ACS Applied Materials & Interfaces, 2015, 7 (8): 4744-4750.

[53] LIU W, LIU J, YANG Z, et al. Extended working frequency of ferrites by synergistic attenuation through a controllable carbothermal route based on prussian blue shell[J]. ACS Applied Materials & Interfaces, 2018, 10 (34): 28887-28897.

[54] PAN J J, SUN X, WANG T, et al. Porous coin-like Fe@MoS_2 composite with optimized impedance matching for effcient microwave absorption[J]. Applied Surface Science, 2018, 457: 271-279.

[55] HOU Y, CHENG L, ZHANG Y, et al. Electrospinning of Fe/SiC hybrid fibers for highly efficient microwave absorption[J]. ACS Applied Materials & Interfaces, 2017, 9 (8): 7265-7271.

[56] CHISCAN O, DUMITRU I, POSTOLACHE P, et al. Electrospun PVC/Fe_3O_4 composite nanofibers for microwave absorption applications[J]. Materials Letters, 2012, 68: 251-254.

[57] MORDINA B, KUMA R, TIWARI R K, et al. Fe_3O_4 Nanoparticles embedded hollow mesoporous carbon nanofibers and polydimethylsiloxane-based nanocomposites as efficient microwave absorber[J]. The Journal of Physical Chemistry C, 2017, 121(14): 7810-7820.

[58] XU H, YIN X, LI M, et al. Ultralight cellular like foam from cellulose nanofiber/carbon nanotubes self-assemblies for ultra-broadband microwave absorption[J]. ACS Applied Materials & Interfaces, 2019, 11(25): 22628-22636.

[59] XU C. Dual-interfacial polarization enhancement to design tunable microwave absorption nanofibers of SiC@C@PPy[J]. ACS applied electronic materials, 2020, 2(6): 1505-1513.

名词索引

附录　部分彩图

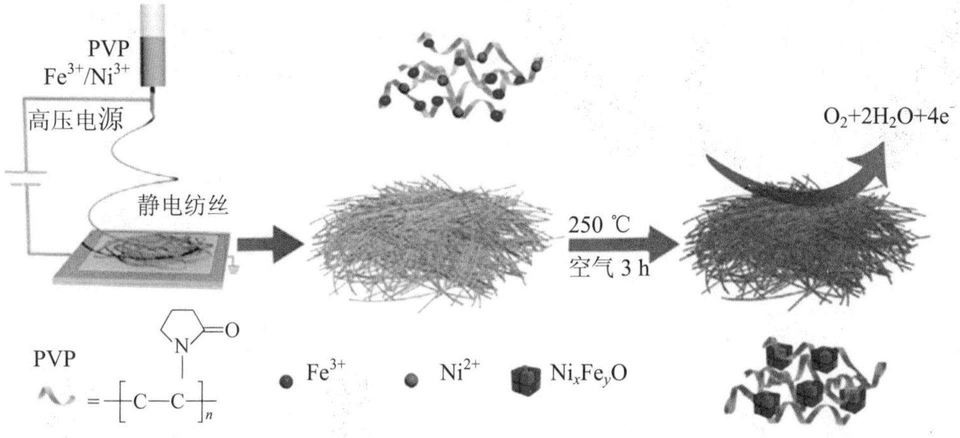

（a）

图 3.11

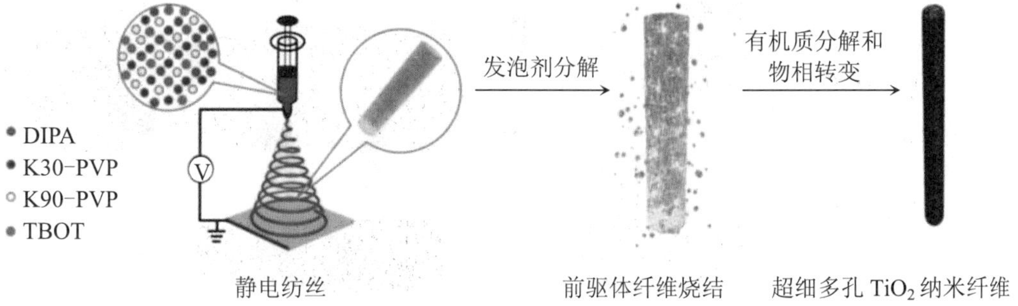

（a）

图 3.27

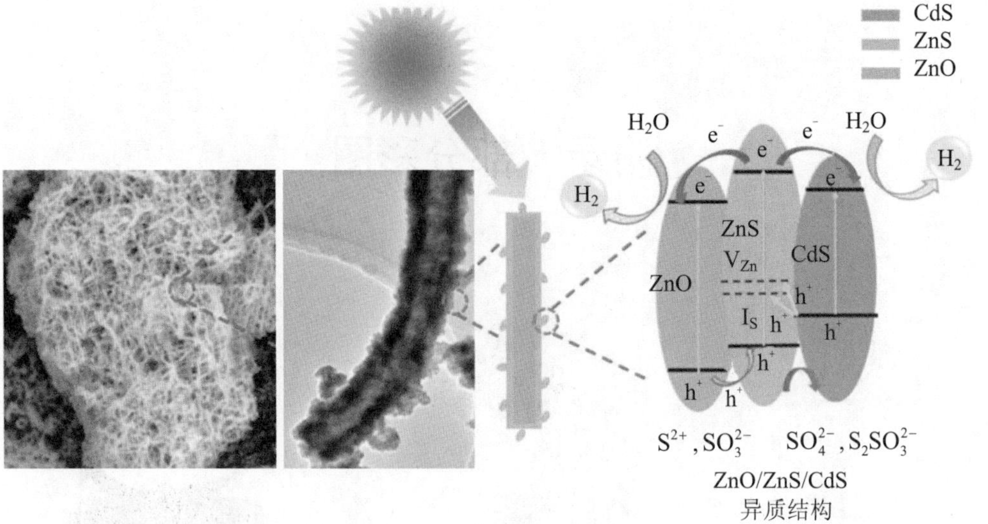

（a）

图 3.30

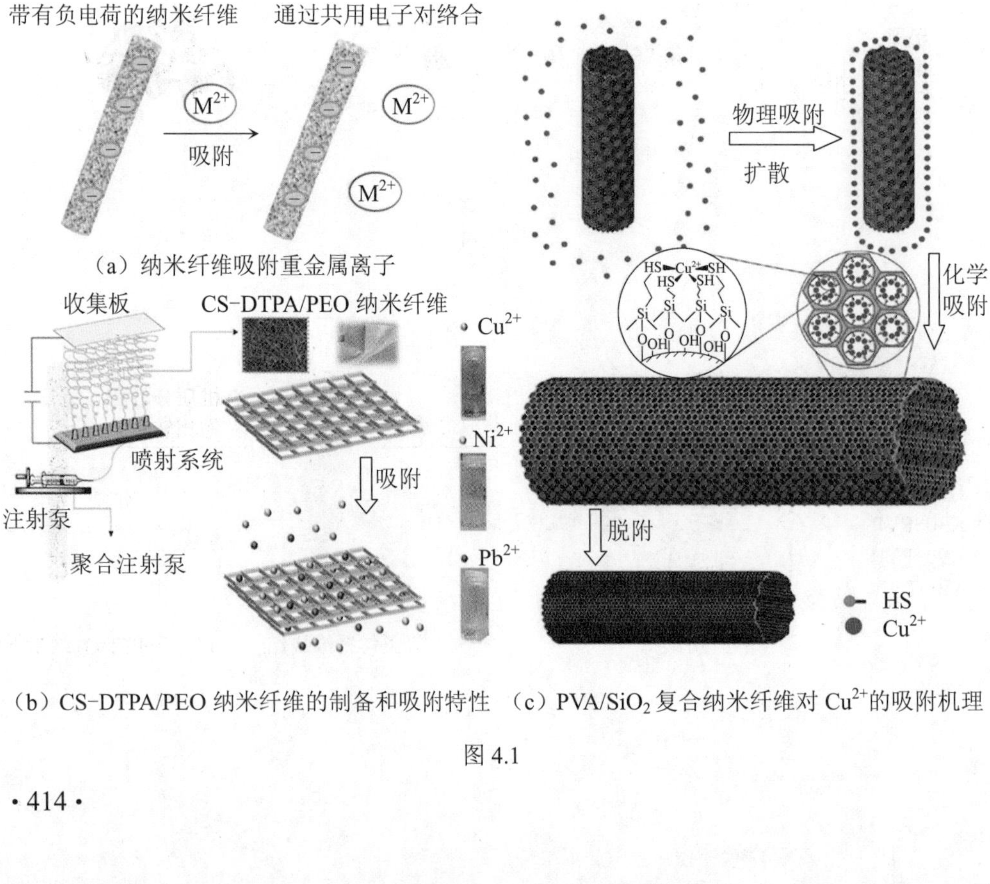

（a）纳米纤维吸附重金属离子

（b）CS-DTPA/PEO 纳米纤维的制备和吸附特性 （c）PVA/SiO_2 复合纳米纤维对 Cu^{2+} 的吸附机理

图 4.1

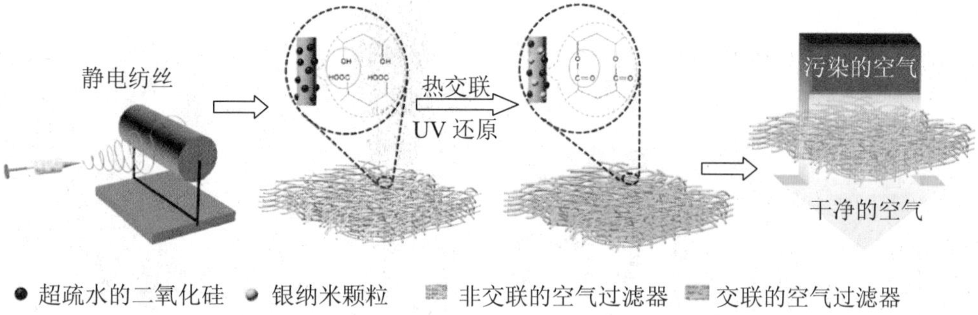

（a）电纺、UV 还原和热交联组合工艺制备 PVA/PAA/SiO_2-Ag 纳米纤维膜示意图

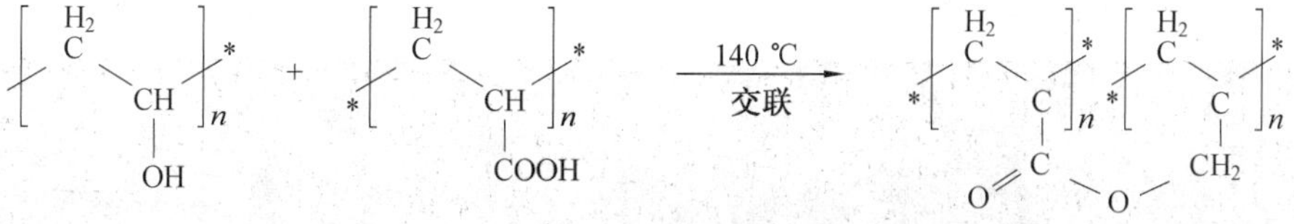

（b）PVA/PAA 间通过酯化反应的热交联机理

图 4.10

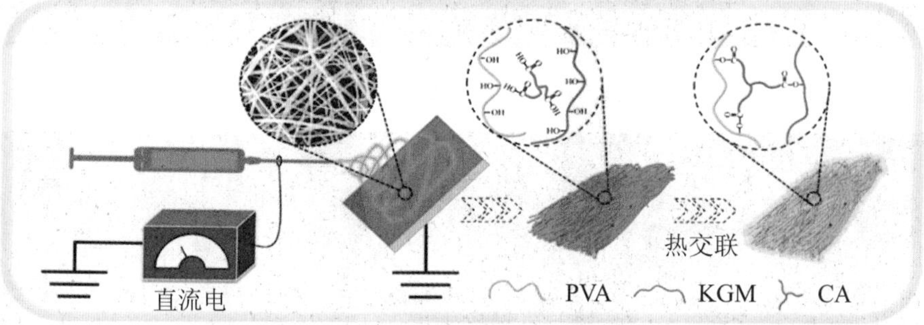

（a）ZnO @ PVA/KGM 膜的制备

（b）空气过滤功能

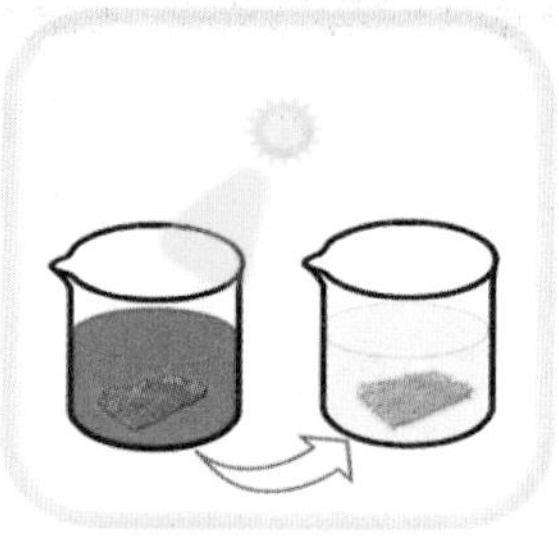

（c）光催化降解功能

（d）抑制细菌功能

图 4.11

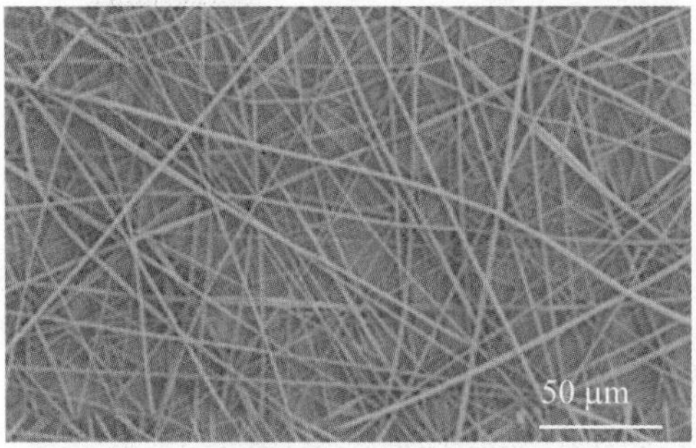

（a）纤维随机排列的 SEM 图谱

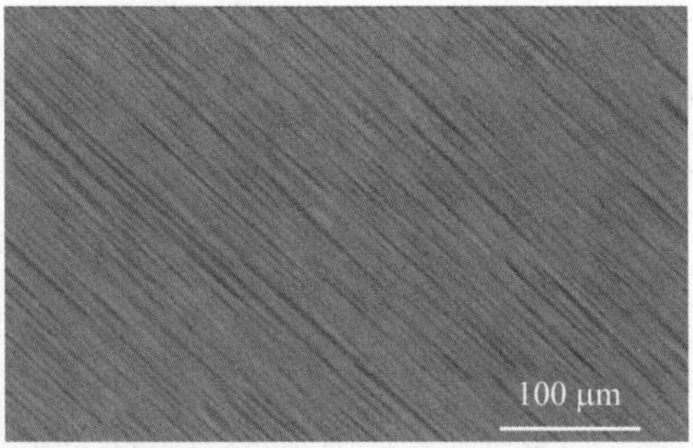

（b）单轴排列的 SEM 图谱

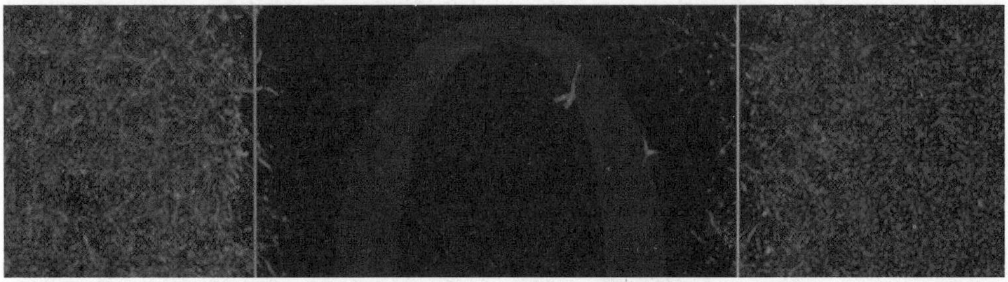

（c）随机排列的 PLA 纤维上的细胞迁移

（d）单轴排列的 PLA 纤维上细胞迁移

图 5.1

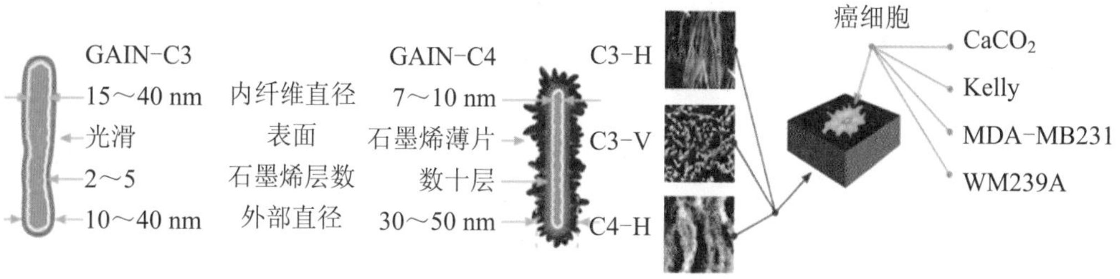

（a）C3 和 C4 类型的支架及其差异示意图以及对应的 SEM 图谱

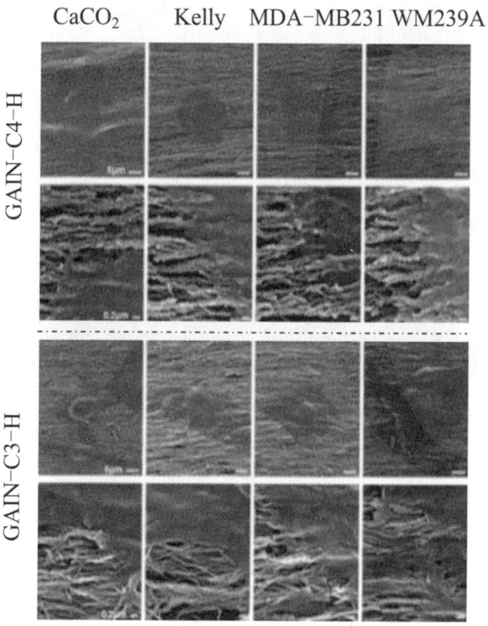

（b）C3-H 和 C4-H 支架上的 SEM 图谱

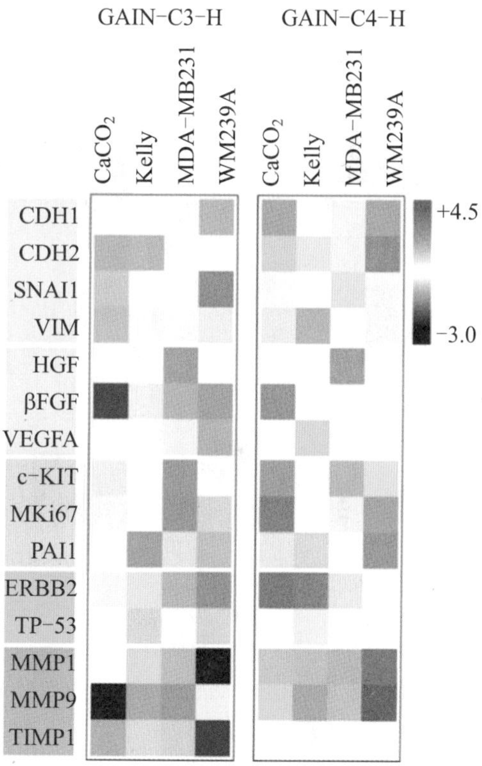

（c）癌细胞类型的分组基因表达的热图

图 5.16